Applied Stochastic Hydrogeology

Applied Stochastic Hydrogeology

Editor:

Dr. Tanjina Nur, Ph.D.

ArclerPress

www.arclerpress.com

Applied Stochastic Hydrogeology

Editor: Dr. Tanjina Nur, Ph.D.

© 2017 Arcler Press LLC
708 3rd Avenue, 6th Floor
New York
NY 10017
United States of America
www.arclerpress.com

ISBN: 978-1-68094-370-2

Library of Congress Control Number: 2016950950

About the Editor

Dr. Tanjina Nur, Ph.D.

Tanjina finished her PhD in Civil and Environmental Engineering in 2014 from University of Technology Sydney (UTS). Now she is working as Post-Doctoral Researcher in the Centre for Technology in Water and Wastewater (CTWW) and published about eight International journal papers with 80 citations. Her research interest is wastewater treatment technology using adsorption process.

Contents

List of Contributors .... *xi*

Preface .... *xvii*

Introduction .... *xix*

Chapter 1 Fully Stochastic Distributed Methodology for Multivariate Flood Frequency Analysis 1

1. Introduction 1
2. Methodology 2
3. Case Study and Dataset 7
4. Results and Discussion 8

Conclusions 12

Acknowledgments 13

Author Contributions 13

Conflicts of Interest 13

References 13

Chapter 2 An Analysis of the Occurrence Probabilities of Wet and Dry Periods Through A Stochastic Monthly Rainfall Model 17

1. Introduction 17
2. Case Study 20
3. Results and Discussion 23

Conclusions 27

References 27

Chapter 3 An Indirect Simulation-Optimization Model For Determining Optimal Tmdl Allocation Under Uncertainty 33

1. Introduction 33
2. Materials and Methods 34
3. Results 36
4. Discussion 37

Conclusions 39

References 40

Chapter 4 Stochastic Urban Pluvial Flood Hazard Maps Based Upon a Spatial-Temporal Rainfall Generator 43

1. Introduction 43
2. Pilot Location: The Cranbrook Catchment 44
3. Spatial-Temporal Stochastic Rainfall Generator 44
4. Hydraulic Model 45
5. Stochastic Flood Hazard Mapping 45
6. Conclusions and Outlook 46

Acknowledgments 47

Author Contributions 47

Conflicts of Interest 47

References 47

Chapter 5 Stochastic Flocculation Model for Cohesive Sediment Suspended in Water 49

1. Introduction 49
2. Materials and Methods 50

3. Results and Discussion ... 51
Conclusions ... 54
Acknowledgments ... 55
References ... 55

Chapter 6 Seismic-Reliability-Based Optimal Layout of a Water Distribution Network ... 57
1. Introduction ... 57
2. Methodology ... 58
3. Study Network ... 61
4. Application Results ... 62
5. Summary and Conclusions ... 64
Acknowledgments ... 64
Author Contributions ... 64
References ... 65

Chapter 7 Analysis of Urban Drainage Networks Using Gibbs' Model ... 67
1. Introduction ... 67
2. Methods ... 68
3. Results and Discussion ... 71
Conclusions ... 74
Acknowledgments ... 75
References ... 75

Chapter 8 Behavior of a Fully-Looped Drainage Network and the Corresponding Dendritic Networks ... 77
1. Introduction ... 77
2. Methodology ... 78
3. Results and Discussion ... 80
Conclusions ... 84
Acknowledgments ... 84
References ... 84

Chapter 9 Quantitative Estimation of the Impact of Precipitation and Land Surface Change on Hydrological Processes Through Statistical Modeling ... 87
1. Introduction ... 87
2. Materials and Methods ... 88
3. Results ... 90
4. Discussions ... 96
Conclusions ... 96
Acknowledgments ... 96
References ... 97

Chapter 10 The Influence of Statistical Uncertainty in the Hydraulic Boundary Conditions on the Probabilistically Computed High Water Level Frequency Curve in the Rhine Delta ... 99
1. Introduction ... 99
2. Methods ... 100
3. Results ... 102
4. Conclusions and Recommendations ... 106
Acknowledgments ... 107
Appendix A ... 107
Appendix B ... 110
References ... 110

Chapter 11 **Fuzzy-Boundary Intervals Combined With Multi-Stage Stochastic Programming Model 113**
1. Introduction 113
2. Mathematical Formulation of the FBISP Method 115
3. Formulation of Optimization Problem for the Alfeios River Basin 119
4. Wadi Water and Agriculture Future Scenarios 125
6. Results 126
7. Discussion and Conclusions 133
Acknowledgments 135
References 135

Chapter 12 **Assessment of Groundwater Chemistry and Status in a Heavily Used Semi-Arid Region With Multivariate Statistical Analysis 139**
1. Introduction 139
2. Study Area 140
3. Materials and Methods 142
5. Results and Discussion 143
Conclusions 148
Acknowledgments 149
Appendix 149
References 150

Chapter 13 **Uncertainty Analysis of Multiple Hydrologic Models Using the Bayesian Model Averaging Method 153**
1. Introduction 153
2. Methods 154
3. Materials 155
4. Results and Discussion 157
Conclusions 160
Acknowledgments 161
References 161

Chapter 14 **Modeling the Projected Changes of River Flow in Central Vietnam Under Different Climate Change Scenarios 163**
1. Introduction 163
2. Data and Methods 164
3. Results and Discussion 166
4. Summary and Conclusions 172
Acknowledgments 173
References 173

Chapter 15 **Modelling Hydrology of a Single Bioretention System With Hydrus-1D 175**
1. Introduction 175
2. Methods 176
3. Results and Discussion 178
Conclusions 183
Acknowledgments 184
References 184

Chapter 16 **Saltwater Intrusion in Coastal Aquifers: A Primary Case Study Along The Adriatic Coast Investigated Within a Probabilistic Framework 185**
1. Introduction 185
2. Characterization of the Area of Interest 186

3. Materials and Methods ... 187
4. Results and Discussion ... 191
Conclusions ... 194
Acknowledgments ... 194
References ... 194

Chapter 17 Forecasting Daily Precipitation Using Hybrid Model of Wavelet-Artificial Neural Network and Comparison With Adaptive Neurofuzzy Inference System ... 197
1. Introduction ... 197
2. Materials and Methods ... 198
Conclusions ... 208
Conflict of Interests ... 208
References ... 208

Chapter 18 The Effect of Training Image and Secondary Data Integration With Multiple-Point Geostatistics in Groundwater Modelling ... 211
1. Introduction ... 211
2. Methodology ... 213
3. Result and Discussion ... 218
4. Capture Zones ... 220
Conclusions ... 223
References ... 223

Index ... 227

List of Contributors

Isabel Flores-Montoya
Department of Civil Engineering: Hydraulic, Energy and Environment, Technical University of Madrid, Madrid 28040, Spain

Álvaro Sordo-Ward
Department of Civil Engineering: Hydraulic, Energy and Environment, Technical University of Madrid, Madrid 28040, Spain

Luis Mediero
Department of Civil Engineering: Hydraulic, Energy and Environment, Technical University of Madrid, Madrid 28040, Spain

Luis Garrote
Department of Civil Engineering: Hydraulic, Energy and Environment, Technical University of Madrid, Madrid 28040, Spain

Tommaso Caloiero
National Research Council—Institute for Agricultural and Forest Systems in Mediterranean (CNR-ISAFOM), Via Cavour 4/6, 87036 Rende (Cs), Italy

Beniamino Sirangelo
Department of Environmental and Chemical Engineering (DIATIC), University of Calabria, Via P. Bucci 41C, 87036 Rende (CS), Italy

Roberto Coscarelli
National Research Council—Research Institute for Geo-Hydrological Protection (CNR-IRPI), Via Cavour 4/6, 87036 Rende (Cs), Italy

Ennio Ferrari
Department of Computer Engineering, Modeling, Electronics, and Systems Science (DIMES), University of Calabria, Via P. Bucci 41C, 87036 Rende (CS), Italy

Feng Zhou
Sino-France Institute of Earth Systems Science, Laboratory for Earth Surface Processes, College of Urban and Environmental Sciences, Peking University, Beijing 100871, China

Yanjun Dong
Laboratory of the Pear River Estuarine Dynamics and Associated Process Regulation, Pearl River Hydraulic Research Institute, Guangzhou 510611, China

Jing Wu
Center for Environmental Science, University of Maryland, Cambridge, MD 21613, USA

Jiangli Zheng
Laboratory of the Pear River Estuarine Dynamics and Associated Process Regulation, Pearl River Hydraulic Research Institute, Guangzhou 510611, China

Yue Zhao
Department of Water Environmental Planning, Chinese Academy for Environmental Planning, Beijing 100012, China

Nuno Eduardo Simões
MARE, Department of Civil Engineering, University of Coimbra, Coimbra 3030-788, Portugal

Susana Ochoa-Rodríguez
Department of Civil and Environmental Engineering, Imperial College London, London SW7 2AZ, UK

Li-Pen Wang
Department of Civil Engineering, KU Leuven, Heverlee (Leuven) 3001, Belgium

Rui Daniel Pina
MARE, Department of Civil Engineering, University of Coimbra, Coimbra 3030-788, Portugal
Department of Civil and Environmental Engineering, Imperial College London, London SW7 2AZ, UK

Alfeu Sá Marques
MARE, Department of Civil Engineering, University of Coimbra, Coimbra 3030-788, Portugal

Christian Onof
Department of Civil and Environmental Engineering, Imperial College London, London SW7 2AZ, UK

João P. Leitão
Eawag, Swiss Federal Institute of Aquatic Science and Technology, Dübendorf 8600, Switzerland

Hyun Jung Shin
Department of Oceanography, Inha University, Incheon 402-751, Korea

Minwoo Son
Department of Civil Engineering, Chungnam National University, Daejeon 305-764, Korea

Guan-hong Lee
Department of Oceanography, Inha University, Incheon 402-751, Korea

Do Guen Yoo
Research Center for Disaster Prevention Science and Technology, Korea University, Seoul 136-713, Korea

Donghwi Jung
Research Center for Disaster Prevention Science and Technology, Korea University, Seoul 136-713, Korea

Doosun Kang
Department of Civil Engineering, Kyung Hee University, Yongin-si, Gyeonggi-do 446-701, Korea

Joong Hoon Kim
School of Civil, Environmental and Architectural Engineering, Korea University, Anam-ro 145, Seongbuk-gu, Seoul 136-713, Korea

Yongwon Seo
Department of Civil Engineering, Yeungnam University, 280 Daehak-ro, Gyeongsan 712-749, Korea

Junshik Hwang
Department of Civil Engineering, Yeungnam University, 280 Daehak-ro, Gyeongsan 712-749, Korea

Seong Jin Noh
Korea Institute of Civil Engineering and Building Technology, 283 Goyangdae-ro, Goyang 411-712, Korea

Young-Ho Seo
Department of Civil and Environmental Engineering, Seoul National University, 1 Gwanak-ro, Gwanak-gu, Seoul 151-742, Korea

Young-Oh Kim
Department of Civil and Environmental Engineering, Seoul National University, 1 Gwanak-ro, Gwanak-gu, Seoul 151-742, Korea

Muhammad Saifullah
College of Hydrology and Water Resources, Hohai University, Nanjing 210098, China
National Cooperative Innovation Center for Water Safety & Hydro-Science, Hohai University, Nanjing 210098, Chin

Zhijia Li
College of Hydrology and Water Resources, Hohai University, Nanjing 210098, China
National Cooperative Innovation Center for Water Safety & Hydro-Science, Hohai University, Nanjing 210098, Chin

Qiaoling Li
College of Hydrology and Water Resources, Hohai University, Nanjing 210098, China
National Cooperative Innovation Center for Water Safety & Hydro-Science, Hohai University, Nanjing 210098, Chin

Muhammad Zaman
College of Water Conservancy and Hydropower Engineering, Hohai University, Nanjing 210098, China

Sarfraz Hashim
College of Hydrology and Water Resources, Hohai University, Nanjing 210098, China

Hua Zhong
Nanjing Hydraulic Research Institute, Nanjing 210029, China
State Key Laboratory of Hydrology–Water Resources and Hydraulic Engineering, Nanjing 210098, China

Pieter van Gelder
Safety and Security Science Group, Faculty of Technology, Policy and Management, Delft University of Technology, Delft 2628 BX, The Netherlands

Wen Wang
State Key Laboratory of Hydrology–Water Resources and Hydraulic Engineering, Nanjing 210098, China
College of Hydrology and Water Resources, Hohai University, Nanjing 210098, China

Gaoxu Wang
Nanjing Hydraulic Research Institute, Nanjing 210029, China
State Key Laboratory of Hydrology–Water Resources and Hydraulic Engineering, Nanjing 210098, China

Yongzhi Liu
Nanjing Hydraulic Research Institute, Nanjing 210029, China
State Key Laboratory of Hydrology–Water Resources and Hydraulic Engineering, Nanjing 210098, China

Shuai Niu
Nanjing Hydraulic Research Institute, Nanjing 210029, China
State Key Laboratory of Hydrology–Water Resources and Hydraulic Engineering, Nanjing 210098, China

Eleni Bekri
Chair of Hydrology and River Basin Management, Technische Universität München, Arcisstrasse 21, München 80333, Germany
Environmental Engineering Laboratory, Department of Civil Engineering, University of Patras, Patras 26504, Greece

Markus Disse
Chair of Hydrology and River Basin Management, Technische Universität München, Arcisstrasse 21, München 80333, Germany

Panayotis Yannopoulos
Environmental Engineering Laboratory, Department of Civil Engineering, University of Patras, Patras 26504, Greece

Xuedi Zhang
School of Environmental Science and Engineering, Chang'an University, No. 126, Yanta Road, Xi'an 710054, Shaanxi, China
Key Laboratory of Subsurface Hydrology and Ecology in Arid Areas, Ministry of Education, No. 126 Yanta Road, Xi'an 710054, Shaanxi, China

Hui Qian
School of Environmental Science and Engineering, Chang'an University, No. 126, Yanta Road, Xi'an 710054, Shaanxi, China
Key Laboratory of Subsurface Hydrology and Ecology in Arid Areas, Ministry of Education, No. 126 Yanta Road, Xi'an 710054, Shaanxi, China

Jie Chen
School of Environmental Science and Engineering, Chang'an University, No. 126, Yanta Road, Xi'an 710054, Shaanxi, China
Key Laboratory of Subsurface Hydrology and Ecology in Arid Areas, Ministry of Education, No. 126 Yanta Road, Xi'an 710054, Shaanxi, China

Liang Qiao
School of Environmental Science and Engineering, Chang'an University, No. 126, Yanta Road, Xi'an 710054, Shaanxi, China
Key Laboratory of Subsurface Hydrology and Ecology in Arid Areas, Ministry of Education, No. 126 Yanta Road, Xi'an 710054, Shaanxi, China

Leihua Dong
State Key Laboratory of Water Resources and Hydropower Engineering Science, Wuhan University, Wuhan 430072, China
National Research Center for Sustainable Hydropower Development, China Institute of Water Resources and Hydropower Research, Beijing 100038, China

Lihua Xiong
State Key Laboratory of Water Resources and Hydropower Engineering Science, Wuhan University, Wuhan 430072, China

Kun-xia Yu
State Key Laboratory of Water Resources and Hydropower Engineering Science, Wuhan University, Wuhan 430072, China

Tuan B. Le
Department of Civil and Environmental Engineering, University of Texas at San Antonio, One UTSA Circle, San Antonio, TX 78249, USA

Hatim O. Sharif
Department of Civil and Environmental Engineering, University of Texas at San Antonio, One UTSA Circle, San Antonio, TX 78249, USA

Yingying Meng
College of Water Sciences, Beijing Normal University, Beijing 100875, China
Beijing Water Science and Technology Institute, Beijing 100048, China

Huixiao Wang
Beijing Water Science and Technology Institute, Beijing 100048, China

Jiangang Chen
Beijing Water Science and Technology Institute, Beijing 100048, China

Shuhan Zhang
Beijing Water Science and Technology Institute, Beijing 100048, China

Giada Felisa
Dipartimento di Ingegneria Civile, Chimica, Ambientale e dei Materiali (DICAM), Università di Bologna, Viale Risorgimento 2, Bologna 40136, Italy

Valentina Ciriello
Dipartimento di Ingegneria Civile, Chimica, Ambientale e dei Materiali (DICAM), Università di Bologna, Viale Risorgimento 2, Bologna 40136, Italy

Vittorio Di Federico
Dipartimento di Ingegneria Civile, Chimica, Ambientale e dei Materiali (DICAM), Università di Bologna, Viale Risorgimento 2, Bologna 40136, Italy

Abazar Solgi
Water Resources Engineering, Faculty of Water Science, Shahid Chamran University of Ahvaz, Ahvaz 65981-34199, Iran

Vahid Nourani
Water Resources Engineering, Faculty of Civil Engineering, University of Tabriz, Tabriz, Iran

Amir Pourhaghi
Water Resources Engineering, Faculty of Water Science, Shahid Chamran University of Ahvaz, Ahvaz 65981-34199, Iran

X. L. He
Department of Geosciences and Natural Resource Management, Copenhagen University, Copenhagen, Denmark

T. O. Sonnenborg
Geological Survey of Denmark and Greenland (GEUS), Copenhagen, Denmark
F. Jørgensen
Geological Survey of Denmark and Greenland (GEUS), Copenhagen, Denmark

K. H. Jensen
Department of Geosciences and Natural Resource Management, Copenhagen University, Copenhagen, Denmark

Preface

Stochastic hydrogeology has matured into a distinct branch or discipline within hydrogeology. Stochastic hydrogeology is an area of research that developed considerably over the last 20 years. Applied Stochastic Hydrogeology is a book that allows newcomers to the field an easy introduction. The book covers the fundamentals and practical aspects of geostatistics and stochastic hydrogeology, coupling theoretical and practical aspects, with examples, case studies and guidelines for applications, and provides a summary and review of the major developments in these areas. The first part of the book discusses fully stochastic distributed methodology for flood frequency analysis and an analysis of the occurrence probabilities of wet and dry periods using stochastic monthly rainfall model. Then we present an indirect simulation-optimization model for determining optimal TMDL allocation under uncertainty and stochastic urban pluvial flood hazard maps based upon a spatial-temporal rainfall generator. Stochastic flocculation model for cohesive sediment suspended in water and seismic-reliability-based optimal layout of a water distribution network are also discussed. In the middle part of the book, we focus on analysis of urban drainage networks using Gibbs' model in South Korea and quantitative estimation of the impact of precipitation and behavior of a fully-looped drainage network and the corresponding dendritic networks. Land surface change on hydrological processes through statistical modeling and the influence of statistical uncertainty in the hydraulic boundary conditions are also discussed. We also try to find out optimizing water allocation under uncertain system conditions for water and agriculture future scenarios in Alfeios River Basin (Greece) and assessment of groundwater chemistry and status in a heavily used semi-arid region with multivariate statistical analysis. The last chapters of the book describe uncertainty analysis of multiple hydrologic models using the bayesian model and modeling the projected changes of river flow in central Vietnam under different climate change scenarios. We also present modelling hydrology of a single bioretention system with HYDRUS-1D and saltwater intrusion in coastal aquifers within a Probabilistic Framework. At the end of the book, we also discuss forecasting daily precipitation using hybrid model and the effect of training image and secondary data integration with multiple-point geostatistics in groundwater modelling

Editor

Dr. Tanjina Nur, Ph.D.

INTRODUCTION

The term "stochastics" derives from the Greek word "Stochasticos" which in turn is derived from "Stochazesthai", which is derived from Stochos. The word Stochos means "target", while the word Stochazesthai has the following meanings: (a) to shoot (an arrow) at a target, (b) to guess or conjecture (the target), (c) to imagine, think deeply, bethink, contemplate, cogitate, meditate. In the modern sense "stochastic" in stochastic methods refers to the random element incorporated in these methods.

Stochastic methods thus aim at predicting the value of some variable at non-observed times or at non-observed locations, while also stating how uncertain we are when making these predictions

But why should we care so much about the uncertainty associated with our predictions? The following example (Figure 1) shows a time series of observed water table elevations in a piezometer and the outcome of a groundwater model at this location. Also plotted are the differences between the data and the model results. We can observe two features. First, the model time series seems to vary more smoothly then the observations. Secondly, there are noisy differences between model results and observations. These differences, which are called residuals, have among others the following causes:

- *Observation errors*: Is it rarely possible to observe a hydrological variable without error. Often, external factors influence an observation, such as temperature and air pressure variations during observation of water levels;
- *Errors in boundary conditions, initial conditions and input*: Hydrological models only describe part of reality, for example groundwater flows in a limited region. At the boundaries of the model values of the hydrological variables (such groundwater heads or fluxes) have to be prescribed. These boundary values cannot be observed everywhere, so there is likely to be some error involved. Also, if a model describes the variation of a hydrological system in time, then the hydrological variables at time step zero must be known as it determines how the system will be evolve in later time steps. Again, the initial values of all the hydrological variables at all locations are not exactly known and are estimated with error. Finally, hydrological models are driven by inputs such as rainfall and evaporation. Observing rainfall and evaporation for larger areas is very cumbersome and will usually be done with considerable error;
- *Unknown heterogeneity and parameters*: Properties of the land surface and subsurface are highly heterogeneous. Parameters of hydrological systems such as surface roughness, hydraulic conductivity and vegetation properties are therefore highly variable in space and often also in time. Even if we were able to observe these parameters without error, we cannot possibly measure them everywhere. In many hydrological models parameters are assumed homogeneous, i.e. represented by a single value for the entire (or part of the) model region. Even if models take account of the heterogeneity of parameters, this heterogeneity is usually represented by some interpolated map from a few locations where the parameters have been observed. Obviously, these imperfect representations of parameters lead to errors in model results;

- *Scale discrepancy*: Many hydrological models consist of numerical approximations of solutions to partial differential equations using either finite element or finite difference methods. Output of these models can at best be interpreted as average values for elements or model blocks. The outputs thus ignore the within element or within block variation of hydrological variables. So, when compared to observations that represent averages for much smaller volumes (virtually points), there is discrepancy in scale that will yield differences between observations and model outcomes.
- *Model or system errors*: All models are simplified versions of reality. They cannot contain all the intricate mechanisms and interactions that operate in natural systems. For instance, saturated groundwater flow is described by Darcy's Law, while in reality it is not valid in case of strongly varying velocities, in areas of partly nonlaminar flow (e.g. faults) or in areas of very low permeability and high concentrations of solvents. Another example is when a surface water model uses a kinematic wave approximation of surface water flow, while in reality subtle slope gradients in surface water levels dominate the flow. In such cases, the physics of reality differ from that of the model. This will cause an additional error in model results.

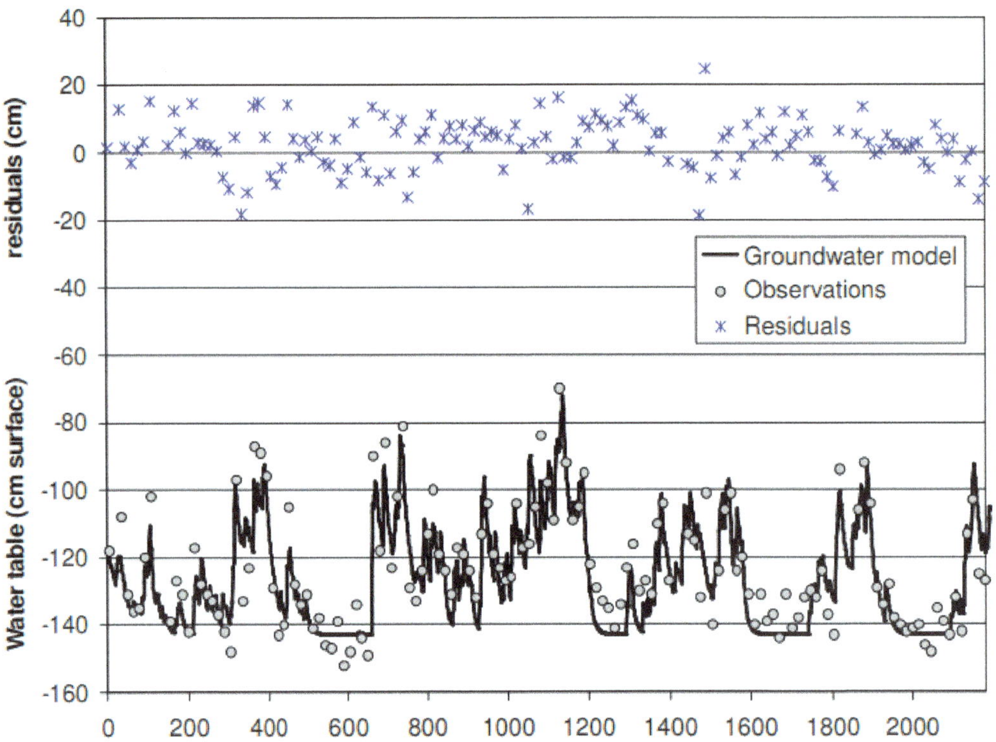

Figure 1: Observed water table depths and water table depths predicted with a groundwater model.

There are two distinct ways of dealing with errors in hydrological model outcomes:

- *Deterministic hydrology*: In deterministic hydrology one is usually aware of these errors. They are taken into account, often in a primitive way, during calibration of models. During this phase of the modelling process one tries to find the parameter values of the model (e.g. surface roughness or hydraulic conductivity) such that the magnitude of the residuals is minimized. After calibration of the model, the errors are not explicitly taken into account while performing further calculations with the model. Errors in model outcomes are thus ignored.
- *Stochastic Hydrology*: Stochastic hydrology not only tries to use models for predicting hydrological variables, but also tries to quantify the errors in model outcomes. Of course, in practice we do not know the exact values of the errors of our model predictions; if we knew them, we could correct our model outcomes for them and be totally accurate. What we often do know, usually from the few measurements that we did take, is some probability distribution of the errors. We will define the probability distribution more precisely in the next chapters. Here it suffices to know that a probability distribution tells one how likely it is that an error has a certain value.

To make this difference more clear, Figure 2 is shown. Consider some hydrological variable z, say soil moisture content, whose value is calculated (at some location and at some time) by a unsaturated zone model. The model output is denoted as $\breve{z}$. We then consider the error $e = \breve{z} - z$. Because we do not know it exactly we consider it as a so called random variable E whose exact value we do not know but of which we do know the probability distribution. So in case of deterministic hydrology modelling efforts would only yield $\breve{z}$ (Figure 2a), while stochastic hydrology would yield both $\breve{z}$ (and the probability distribution of the (random) error E (Figure 2a).

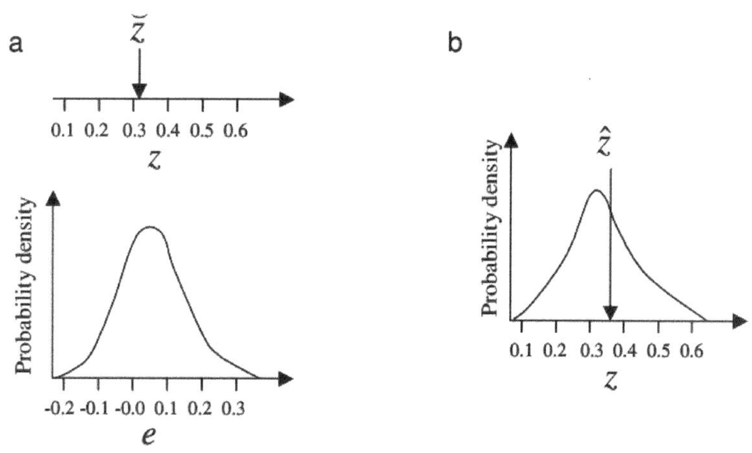

Figure 2: Stochastic Hydrology.

Most of the methods used in stochastic hydrology do not consider errors in model outcomes explicitly. Instead it is assumed that the hydrological variable z itself is a random variable Z. This means that we consider the hydrological variable (e.g. soil moisture) as one for which we cannot know the exact value, but for which we can calculate the probability distribution (Figure 2b). The probability distribution of Figure 2b thus tells us that although we do not know the soil moisture content exactly, we do know that it is more likely to be around 0.3 then around 0.2 or 0.5. Models that provide probability distributions of target variables instead of single values are called stochastic models. Based on the probability distribution it is usually possible to obtain a so called best prediction $\hat{z}$, which is the one for which the errors are smallest on average. Incidentally, the value of the best prediction does not have to be the same as the deterministic model outcome $\breve{z}$.

There are a number of clear advantages in taking the uncertainty in model results into account, i.e. using stochastic instead of deterministic models.

- That model outcomes often give a much smoother picture of reality. This is because models are often based on an idealized representation of reality with simple processes and homogenous parameters. However, reality is usually messy and rugged. This may be a problem when interest is focused on extreme values: deterministic models typically underestimate the probability of occurrence of extremes, which is rather unfortunate when predicting for instance river stages for dam building. Stochastic models can be used with a technique called "stochastic simulation" which is able to produce images of reality that are rugged enough to get the extreme statistics right.
- The value of the best prediction $\hat{z}$ does not have to be the same as the deterministic model outcome $\breve{z}$. This is particularly the case when the relation between model input (e.g. rainfall, evaporation) or model parameters (e.g. hydraulic conductivity, manning coefficient) and model output is non-linear (this is the case in almost all hydrological models) and our deterministic assessment of model inputs and parameters is not error free.
- By explicitly accounting for the uncertainty in our prediction we may in fact be able to make better decisions. A classical example is remediation of polluted soil, where stochastic methods can be used to estimate the probability distribution of pollutant concentration at some non-visited location. Given a critical threshold above which regulation states that remediation is necessary, it is possible to calculate the probability of a false positive decision (we decide to remediate, while in reality the concentration is below the threshold) and that of a false negative. Given these probabilities and the associated costs (of remediation and health risk) it is then possible for each location to decide whether to remediate such that the total costs and health risk are minimized.
- Stochastic methods where a relation is established between the uncertainty in model outcomes and the number of observations in time and space used to either parameterize or calibrate the model. If such a relation exists, it can be used for monitoring network design. For example, in groundwater exploration wells are drilled to perform pumping tests for the estimation of transmissivities and to observe hydraulic heads. The transmissivity observations can be used to make an initial map of transmissivity used in the groundwater model. This initial map can subsequently be updated by calibrating the groundwater model to head observations in the wells.
- Certain stochastic methods are able to quantify the uncertainty in groundwater head predicted by the model in relation to the number of wells drilled, their location and how often they have been observed. These stochastic methods can therefore be used to perform monitoring network optimization: finding the optimal well locations and observation times to minimize uncertainty in model predictions.
- Stochastic methods are advantageous over deterministic methods is related to the previous one. Stochastic methods enable us to relate the uncertainty in model outcomes to different sources of uncertainty (errors) in input variables, parameters and boundary conditions. Therefore, using stochastic analysis we also know which (error) source contributes the most to the uncertainty in model outcomes, which source comes second etc. If our resources are limited, stochastic

hydrology thus can guide us where to spend our money (how many observations for which variable or parameter) to achieve maximum uncertainty reduction at minimum cost.

Hydrogeology

Hydrogeology is a branch of geology, the study of rocks and the structures that are formed over past periods of time. The way that sediments have been deposited to create different layers of rock beneath the surface, or the way that rocks have been heated and folded over millions of years to create complex structures are the subjects of geology. Hydrogeology looks at how water interacts with geological systems. But there is more to hydrogeology than wet rocks. Water is a vital natural resource for people all around the world - whether it is piped to homes or drawn out of wells.

Importance of Hydrogeology

Though many parts of the globe have abundant sources of water and adequate rainfall, there are still places where the only source is underground water. In such places, hydrogeologists are bound to trace appropriate reserves and evaluate the quantity of water that can be hauled out without causing any damage to the sub-terranean aquifers or the ecosystems. By knowing the movement of underground water, hydrogeologists can also predict the hazards of pollution and ways to protect these resources. Without the know-how of hydrogeology, there is no way we can forecast the condition of the water sources. Thus, hydrogeology is a very important aspect in the protection of the most necessary natural resource.

Groundwater - A Hidden Resource

When we think of fresh water we probably think of rivers, lakes and perhaps reservoirs. Yet even when we consider the huge lakes of North America or mighty rivers like the Amazon and the Ganges, over 97% of the planet's fresh water is to be found under the surface of the earth in the form of groundwater.

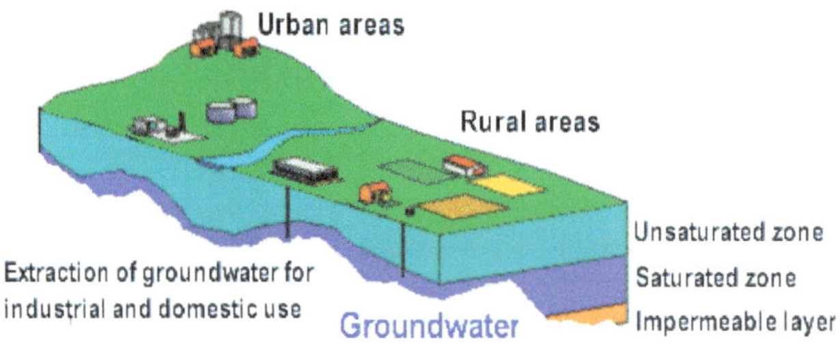

In many parts of the world groundwater is the main source of water for day to day use. Boreholes driven down into the saturated layer under the ground yield water for irrigation and urban use. At least 1,500 million urban dwellers are supplied with water from groundwater reserves. It is also extensively used as a low cost rural water supply. Groundwater has many benefits. It is cheap to develop because of its naturally good quality and widespread occurrence. It is reliable even in times of drought due to the large amounts stored under the surface. It is also protected against catastrophic events - if natural disasters or war disrupt surface water distribution, then groundwater reserves can easily be developed.

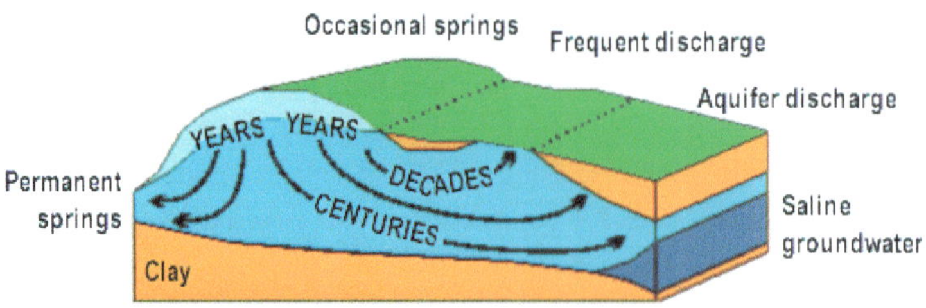

Some areas near the surface may not be saturated with water and only have an intermittent supply of water. But, many areas deeper under the surface are saturated with water accumulated over hundreds or even thousands of years. These areas often supply permanent springs and can be tapped with boreholes to provide water. When we draw groundwater from deep aquifer supplies, we are tapping into water locked away under the surface and filtered through layer upon layer of rock. This is one of the reasons why groundwater is so reliable. The water in these areas may even be protected against recent surface pollution, providing a high quality supply where none was previously available.

Hydrogeology in Relation to Other Fields

Hydrogeology, as stated above, is a branch of the earth sciences dealing with the flow of water through aquifers and other shallow porous media (typically less than 450 m or 1,500 ft below the land surface.) The very shallow flow of water in the subsurface (the upper 3 m or 10 ft) is pertinent to the fields of soil science, agriculture and civil engineering, as well as to hydrogeology. The general flow of fluids (water, hydrocarbons, geothermal fluids, etc.) in deeper formations is also a concern of geologists, geophysicists and petroleum geologists. Groundwater is a slow-moving, viscous fluid (with a Reynolds number less than unity); many of the empirically derived laws of groundwater flow can be alternately derived in fluid mechanics from the special case of Stokes flow (viscosity and pressure terms, but no inertial term).

The mathematical relationships used to describe the flow of water through porous media are the diffusion and Laplace equations, which have applications in many diverse fields. Steady groundwater flow (Laplace equation) has been simulated using electrical, elastic and heat conduction analogies. Transient groundwater flow is analogous to the diffusion of heat in a solid, therefore some solutions to hydrological problems have been adapted from heat transfer literature.

CHAPTER 1

FULLY STOCHASTIC DISTRIBUTED METHODOLOGY FOR MULTIVARIATE FLOOD FREQUENCY ANALYSIS

Isabel Flores-Montoya, Álvaro Sordo-Ward, Luis Mediero and Luis Garrote

Keywords: Derived flood frequency curve, stochastic rainfall model, distributed, event-based, rainfall–runoff model, probabilistic, initial soil moisture

INTRODUCTION

Characterizing the extreme behavior of basin response is necessary for hydraulic infrastructure design, territorial planning, flood management and risk analysis. The magnitude of a flood and its consequences may depend not only on the peak flow but also on the volume, duration and temporal distribution of the hydrograph. Therefore, a multivariate analysis may be needed in some cases. Statistical analyses are traditionally applied in engineering practice to obtain a unique design flood.

The methodologies traditionally applied in engineering practice are statistical analyses with the aim of obtaining a unique design flood. Statistical analyses consist in fitting the most appropriate extreme-value distribution to the observed series of maximum annual discharges, in order to estimate the peak flow associated to a given return period (usually high, e.g., 500 to 10,000 years, in the case of dams). These studies need long and robust observed series to obtain accurate quantile estimates. However, they are usually either not available or not long enough to ensure extrapolations with acceptable uncertainty.

Fortunately, long rainfall series are more often available than discharge ones. This makes it possible to estimate the areal rainfall depth associated to a given return period and to transform it deterministically into a design flood event. However, this approach presents some drawbacks. On the one hand, it erroneously assumes that the return periods associated with the rainfall depth and the derived flood event are the same [1,2,3]. On the other hand, some variables such as the shape and duration of the hyetograph, the initial soil moisture content in the basin or the parameters that characterize the runoff generation processes are arbitrarily defined by the designer. These deterministic assumptions lead to different values of the design flood and hamper the estimation of the real probability of exceedance of the event, as well as its uncertainty.

Some of these drawbacks have been overcome by the derived flood frequency approach proposed by Eagleson [4] that combines a storm event model with a rainfall–runoff (RR) model to obtain the peak flow frequency curve.

Recently, new rainfall and hydrological models have been developed allowing new modifications to the derived flood frequency approach. Some authors have maintained the event-based approach combining a simple stochastic storm generator [5,6] or a more complex model such as the Stochastic Storm Transposition (SST) method [7,8] with semi-distributed [9] or distributed [10] RR models. However, the problem of the characterization of the basin initial state remains unsolved.

This problem can be solved by a continuous simulation approach that combines a stochastic rainfall model such us the Generalized Neyman–Scott Rectangular Pulses (GNSRP) [11,12,13,14,15,16,17,18,19,20] with a continuous RR model. Continuous simulation needs more input data and computation requirements than event-based simulation. In order to simulate long discharge series without an excessive computational cost, most authors have chosen semi-distributed models (ARNO [21]; TOPMODEL [22,23,24,25,26,27]; and HEC-HMS [28]) or lumped models (MISDc [29,30,31,32,33]) or simple

Citation: Isabel Flores-Montoya, Álvaro Sordo-Ward, Luis Mediero and Luis Garrote, Fully Stochastic Distributed Methodology for Multivariate Flood Frequency Analysis, doi:10.3390/w8060225.

and conceptual distributed models (FEST [34] and TETIS [35,36]). A semi-continuous approach has also been developed by replacing observed rainfall events with stochastic ones in a continuous simulation of observed data (SCHADEX [37,38,39]), a Simplified Continuous Rainfall–Runoff Model (SCRRM) which used soil moisture data provided by ground, satellite and reanalysis data [40] and a Hybrid-CE approach which combines a long continuous simulation of the rainfall and a short continuous simulation run as inputs of an event-based rainfall–runoff model [41]. These approaches try to overcome the limitations of event-based models without increasing significantly the computational requirements.

In this context, a new methodology called MODEX is proposed in this study. It combines a spatial-temporal stochastic rainfall generator based on the GNSRP, the RainSim V3, with a spatially distributed, physically-based and event-based RR model, (RIBS) [42,43] through the identification and selection of the annual maximal storm events by the Exponential method and the maximum accumulated rainfall criteria.

The RainSim V3 rainfall model simulates continuous rainfall series that represent the spatial-temporal variability of the observed rainfall keeping its monthly statistics and extreme behavior. The RIBS model is event-based so independent rainfall events have to be identified. The Exponential method is used for this purpose [44,45]. Two consecutive events are independent when they are caused by different weather systems, so the occurrence of the second wet period is not conditional on the first occurrence. This is important from the point of view of flood risk because the consequences of a wet period after another wet period are much worse than after a long dry period. The storm selection reduces the number of simulations and ensures that only the most significant storm events are considered.

Event-based models have some advantages over the continuous models. They could represent better the extreme behavior of peak flows requiring less input data (neither temperature nor evapotranspiration) and the calibration process is simpler due to the smaller number of parameters to calibrate. Furthermore, event-based models require less computation effort so a more complex model, such as a distributed and physically-based one with a higher temporal resolution (hourly or less), can be used. That way the basin spatial variability can be represented; real processes such as infiltration or percolation can be simulated; and complete hydrographs with different shapes and durations can be obtained. These are the main reasons for the selection of RIBS event-based RR model.

The main drawback of event-based models is the initial state characterization, which has a significant influence on the results [46,47,48,49,50,51,52,53]. This paper tries to overcome this problem. Mediero et al. [54] proposed a calibration process for the RIBS model (called here semi-probabilistic), which considers the model parameters as random variables characterized by a probability distribution though a Monte Carlo simulation but using a unique initial state. Based on this process, this paper proposes two other calibration–simulation approaches: the deterministic and the probabilistic (Table 1). The deterministic approach is a simplification of the semi-probabilistic one, which uses a unique value of the model parameters. On the other hand, the probabilistic approach characterizes the initial state and the model parameters through a probability distribution and uses the values obtained through a Monte Carlo simulation. That way the probabilistic approach tries to overcome the initial state characterization problem.

Table 1. Summary of the three calibration–simulation approaches.

Approach	Model Parameters		Initial State	
	Calibration	Simulation	Calibration	Simulation
Deterministic	M Monte Carlo cases	Associated with a 50% probability of exceedance	Random	Related to a intermediate recharge
Semi-probabilistic		Randomized probabilistic values		
Probabilistic			Each MC case with all Initial State	Randomized probabilistic values

This methodology has been applied to the Corbès and Générargues basins, in the Southeast of France. The results obtained for each calibration–simulation approaches have been compared to show which approach is better. The probabilistic approach offer a really good fit and better than the others, especially when the peak flow is considered. This means the proposed methodology can be used to characterize the extreme behavior of the basin overcoming the main drawbacks of event-based models and also take into account the basin variability.

METHODOLOGY

The methodology involves four steps (see Figure 1): (i) stochastic generation of continuous series of rainfall through the RainSim V3 model; (ii) identification and selection of independent storm events from the rainfall series by applying the Exponential method; (iii) rainfall–runoff transformation through the distributed event-based RIBS model, obtaining series of flood hydrographs; and (iv) validation of the methodology and analysis of the results, obtaining a multivariate frequency analysis of the hydrograph characteristics and an uncertainty estimation of the results. These four steps are detailed in the following sections.

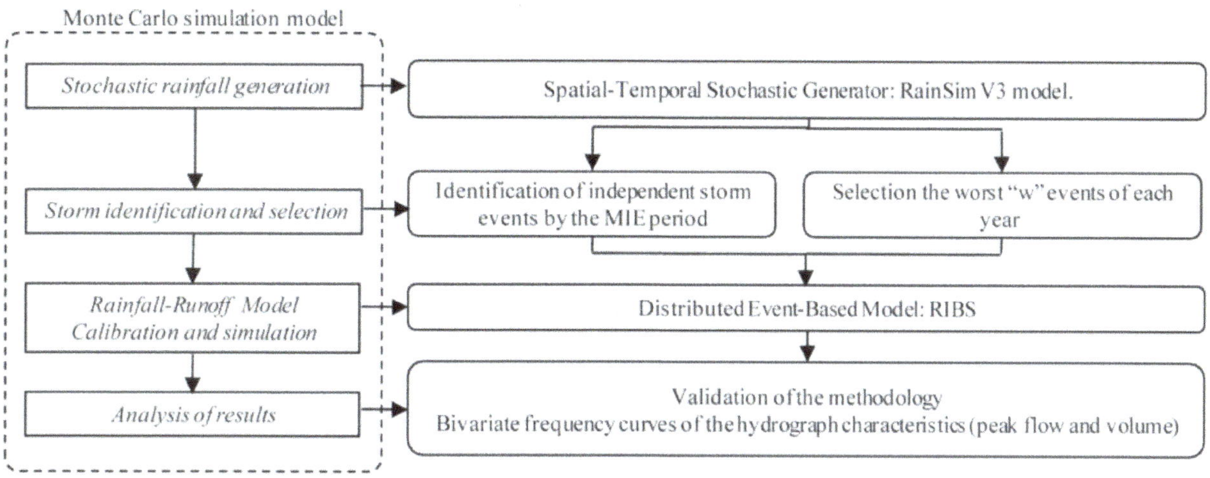

Figure 1. General conceptual framework of the MODEX methodology.

Stochastic Rainfall Generation

Stochastic rainfall models generate arbitrarily long rainfall series that allow characterizing the extreme behavior of the basin and may be used as input in RR models.

The RainSim V3 model simulates the origins of storms based on a uniform Poisson distribution (parameter λ). Each event is composed of a random number of C rainfall cells. Each cell is delayed from the origin of the event according to an exponential distribution with parameter ν and distributed in the space following a Poisson distribution with parameter ρ. Each cell has an independent duration, intensity and radius, which are exponentially distributed (parameters γ, η and ξ, respectively,). The RainSim parameters are summarized in Table 2. The intensity of rainfall at each point and time step is given by the sum of the intensities of the active cells at such instant.

Table 2. RainSim V3 calibration parameters.

Parameters	Description	Units
λ^{-1}	Mean waiting time between adjacent storm origins	(h)
β^{-1}	Mean waiting time for rain cell origins after storm origin	(h)
η^{-1}	Mean duration of rain cell	(h)
ξ^{-1}	Mean intensity of a rain cell	(mm/h)
γ^{-1}	Mean radius of rain cells	(km)
ρ	Spatial density of rain cell centers	(km^{-2})
Φ	Vector of scale factors, Φ_m, one for each rain gauge, m	(-)

The RainSim V3 model consists of three modules (analysis, fitting and simulation) and runs in four steps. First, the analysis module calculates the monthly values of a set of statistics which characterize the observed rainfall. Second, the fitting module obtains the value of the model parameters that generate simulated rainfall series whose statistics are the most similar to the calculated from the observed series. The third module simulates synthetic rainfall series based on the adjusted parameters. Lastly, the analysis module is applied again to the synthetic series to compare the statistics from simulated and observed series and check the consistency of simulated ones.

The RainSim V3 model allows the simulation of continuous series of rainfall of N number of years for a set of rain gauges with an arbitrary time step. A detailed description of the model described above may be found in Burton et al. [19].

Storm Events Identification and Selection

As mentioned above, the selection of an event-based RR model requires the identification of independent storm events by fixing a minimum inter-event period (MIP) between two events (Figure 2). Several methods (such as arbitrary separation, auto-correlation, the exponential method and the range of correlation) have been used to identify individual storms [55,56,57,58].

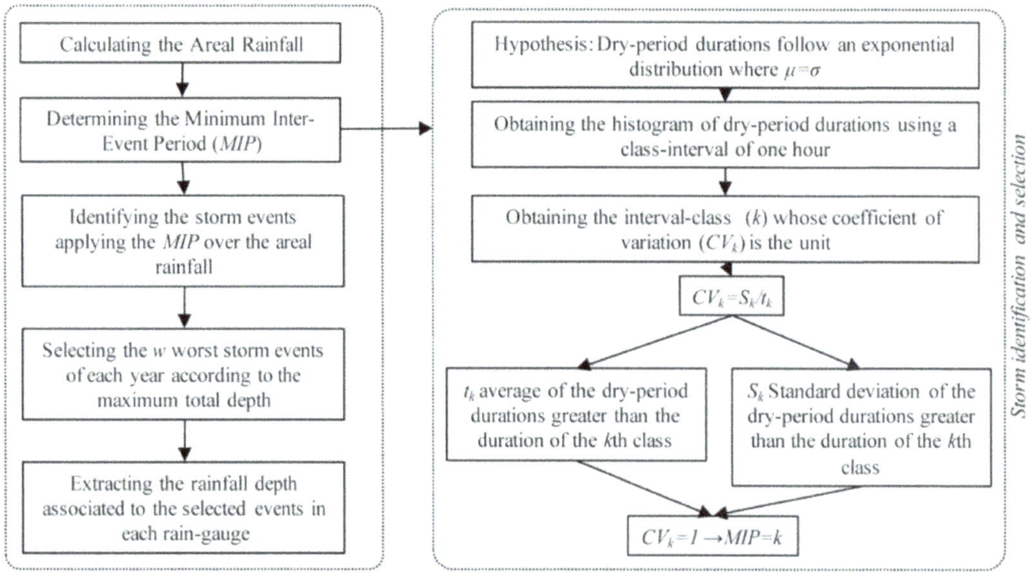

Figure 2. Flow chart that illustrates the steps carried out in storm identification and selection.

The Exponential method (Figure 3) is used here because it better ensures the independency between events [45]. This method is derived from the Poisson process, which describes the arrival of the storm origins. Consequently, the duration of dry periods can be approximated by an exponential distribution with a mean equal to the standard deviation, $\mu=\sigma$, and therefore with a coefficient of variation, $CV=\sigma/\mu$, equal to the unity. A class-interval width is defined and the histogram of dry period durations is obtained. The CV associated to each class-interval is determined from:

$$CV_k = \frac{s_k}{\overline{t_k}} \tag{1}$$

where k is the number of the histogram class-interval, sk is the standard deviation of dry period durations greater than the durations of the kth class and tk is its mean. The MIP is the dry period duration at which CV=1. A detailed description of the Exponential method may be found in Restrepo-Posada and Eagleson [44].

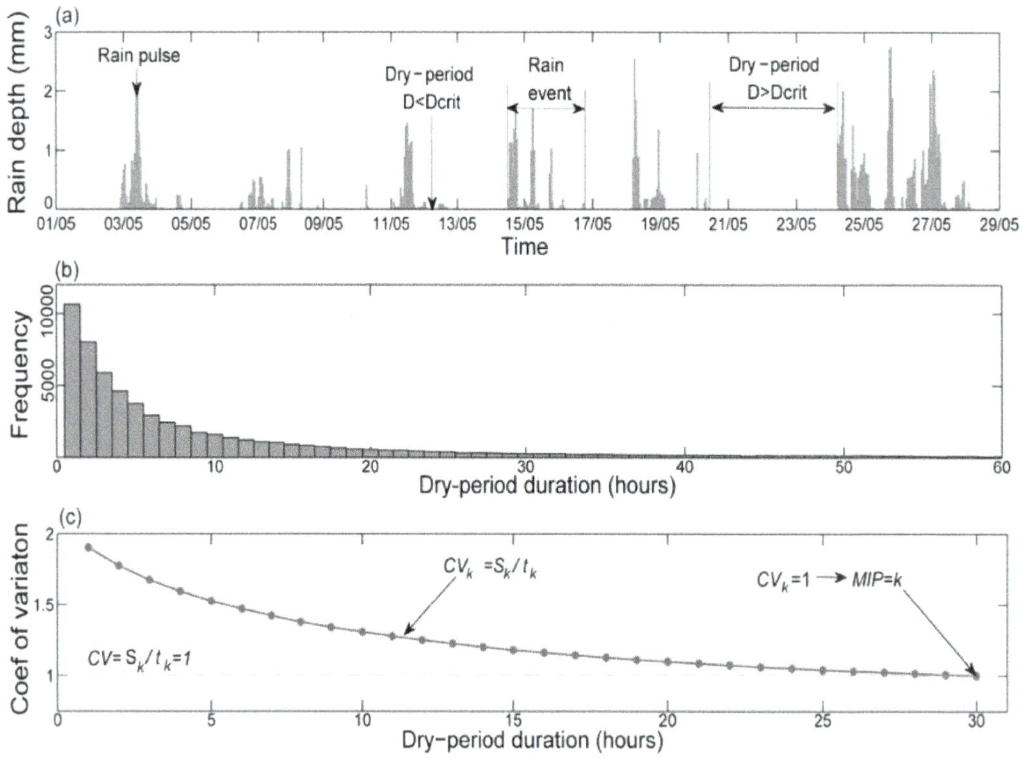

Figure 3. (**a**) Rainfall intensity series where dry periods are identified; (**b**) dry-period duration histogram; and (**c**) intersection of the curve of CV_k with the line CV=1that indicates the minimum inter-event period that ensures storm event independence.

The exponential method is applied to the basin areal rainfall considering a minimum rainfall threshold of 0.5 mm/h. The storm events that are separated by dry periods equal or longer than the MIP are identified. In order to reduce the number of simulations, a sub-set of w events is selected from each year. As the objective is to characterize the basin extreme behavior, the selection criteria chosen is the maximum accumulated rainfall.

Rainfall–Runoff Model

The RIBS model simulates the response of the basin to spatially distributed rainfall. It is based on a digital terrain-based model and information provided in raster-style grids that are combined to determine the operating parameters. It consists of two modules: one that involves runoff generation and another of runoff propagation. The runoff generation module represents the soil characteristics by the Brooks-Corey parameterization where the saturated hydraulic conductivity, Ks, decreases exponentially in depth, y, in directions parallel, p, and normal, n, to the land surface accordingly to the following expression.

$$Ks_n(y) = K0_n e^{-fy}\left(\frac{\theta-\theta_r}{\theta_s-\theta_r}\right)^{\varepsilon} \tag{2}$$

$$Ks_p(y) = K0_p e^{-fy}\left(\frac{\theta-\theta_r}{\theta_s-\theta_r}\right)^{\varepsilon} \tag{3}$$

where K0n and K0p (mm/h) represent the saturated hydraulic conductivity in the directions normal and parallel to the surface, respectively, and have to be defined for each soil class; f (mm^{-1}) controls the reduction of the saturated hydraulic conductivity with depth; and θ, θ_r and θ_s represent, the soil moisture, the soil residual moisture and the saturated soil moisture for each soil class, respectively. In addition, ε is the soil porosity index [59]. Surface runoff occurs when either the rainfall intensity exceeds the infiltration capacity or the soil is completely saturated.

The runoff propagation module simulates the runoff routing through hill slopes and river channels. The water velocities in riverbed and hill-slope, v_s and vh are considered to be uniform throughout the basin for each time t, and are related to each other through the dimensionless parameter Kv:

$$K_v = v_s(t)/v_h(t) \tag{4}$$

The riverbed velocity is given by the following equation:

$$v_s(t) = C_v\left[Q(t)/Q_{ref}\right]^r \tag{5}$$

where Q(t) is the flow at the basin outlet at time t, Q_{ref} is a reference flow, r is a parameter that accounts for the degree of nonlinearity in the basin, and C_v is a calibration coefficient.

The initial state of the basin is characterized by the water table position in each cell, $N_{wt}(x, y)$. Initial states are obtained through a groundwater model [60], which considers only water movement in the saturated zone below the water level. This model leads to the balance of the drainage base flows when a long-term recharge, Ri, is applied uniformly to the basin. Larger recharge rates generate higher water table positions and an initial state closer to the soil saturation, θ≈θs, while smaller recharge rates produce lower water table positions and consequently an initial state closer to the dry soil condition, θ≈θr, (see Figure 4). Thus the variable that characterizes the initial state is Ri. A representative set of S initial states are considered, which means an enough large number of Ri values. The recharge values used to generate them cover the whole feasible range. The set of initial states generates a set of hydrographs uniformly distributed between the one obtained with dry soil and that obtained with saturated soil. A detailed description of the RIBS model may be found in Garrote and Bras [42,43].

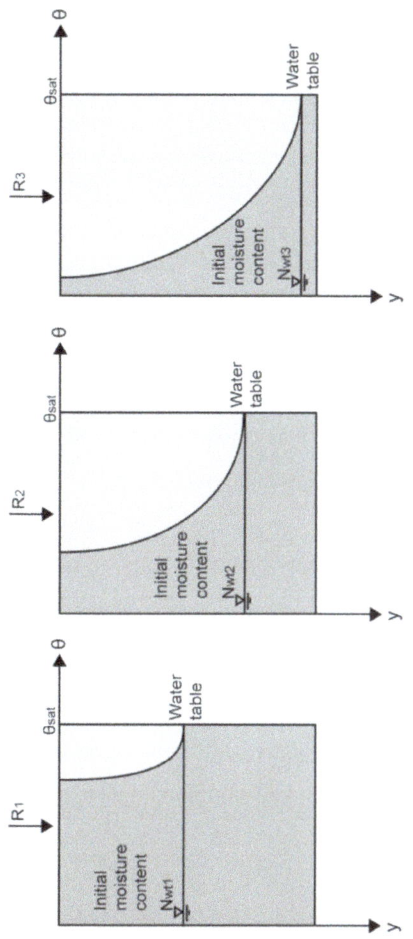

Figure 4. Initial state and water table position in a cell for different recharges $R_1 > R_2 > R_3$: (**a**) R_1 generates an initial state with a high water table and a soil moisture close to the saturation soil moisture; (**b**) R_2 generates an initial state with an intermediate water table and a soil moisture; and (**c**) R_3 generates an initial state with a low water table and a soil moisture close to the residual soil moisture.

Calibration and Synthetic Hydrograph Simulation

The sensitivity analysis carried out by Garrote and Bras [42,43] and Mediero et al. [46] concluded that the parameters f, C_v and K_v, have the most influence on the results of the RIBS model. Also the initial state, or Ri that generates it, is determinant for the results.

The calibration process consists on obtaining the parameter values that allow the model representing the best real system behavior by minimizing an objective function that quantifies the error between simulations and observations. However, a given set of parameters that produces good results for one event may generate worse results for another event, because of differences in the catchment response among flood events. A probabilistic approach accounts for this fact and characterizes each model parameter by a probability density function. The model is run with randomized combinations of model parameter values obtained through the probability distributions functions obtained as a result of the calibration process, to account the uncertainties inherent to the model.

This study seeks to analyze and compare three calibration–simulation processes: deterministic, semi-probabilistic and probabilistic. The process proposed by Mediero [46], called here semi-probabilistic, consists on the simultaneous minimization of a set of objective functions, obtaining a set of optimal parameter value combinations based on the Pareto dominance. Each parameter is characterized by the probability distribution that best represents the set of values obtained as a result.

The choice of the objective functions depends on what aspects of the hydrograph have to be optimized. Given that the proposed methodology is applied to a flood frequency analysis, the objective functions selected represent the peak flow and hydrograph volume:

1. Relative error of peak flow (REP), that takes into account the relative difference between the observed (Q^{obs}_{max}) and simulated peak flows (Q^{obs}_{max}).

$$REP = \frac{\left|Q^{sim}_{max} - Q^{obs}_{max}\right|}{Q^{obs}_{max}} \qquad (6)$$

2. Volume relative square error (VRSE), that quantifies the relative squared difference between volumes of observed (V^{obs}) and simulated hydrographs (V^{sim}):

$$VRSE = \frac{\left(V^{sim} - V^{obs}\right)^2}{V^{obs^2}} \qquad (7)$$

The calibration process is applied to K representative observed events. The number of extreme events that have enough rainfall and flow data to be used in the calibration process is usually reduced and cannot be considered representative from a point of view. However, models are usually calibrated with a reduce sample and then validate if the calibrated model reproduce the basin behavior for other events although the reduce number of events used in the calibration process.

A Monte Carlo simulation of M cases is carried out for each observed event, using a uniform distribution of the parameter values (f, C_v and K_v). The range of each parameter values is determined through a previous sensitivity analysis. A random initial state among the S considered is used as input. As a result, the optimal values of each parameter that minimize the two objective functions are obtained (Figure 5).

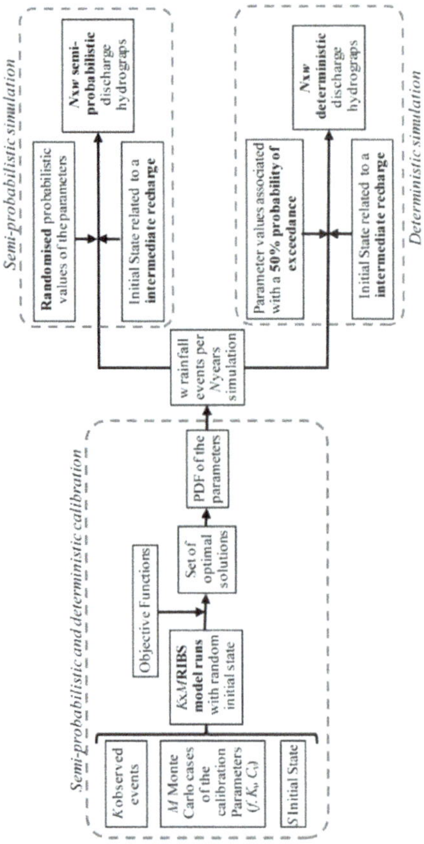

Figure 5. Semi-probabilistic and deterministic calibration–simulation process.

The optimal values of each parameter are adjusted to several probability distribution functions. The chi-square and Kolmogorov–Smirnov tests are used to select the function that best represents the parameter variability. Finally, a set of synthetic events are simulated by considering randomized probabilistic values of the parameters according to the distributions adjusted and an initial moisture content corresponding to an intermediate recharge (Figure 5).

The deterministic calibration–simulation approach is a simplification of the semi-probabilistic approach. In this case, the calibration process is the same but the synthetic events are simulated with a deterministic value of the parameters which corresponds to an exceedance probability of 50% and the initial state of moisture corresponding to an intermediate recharge (Figure 5).

Finally, the probabilistic calibration–simulation approach modifies both the calibration and the simulation processes to characterize the basin initial state by the probability density distribution of the recharge, Ri. Each Monte Carlo case is simulated using all the initial states, selecting the one that leads to a simulated hydrograph with the volume closest to the observed, as the optimal initial state. The set of recharge values associated with the optimal initial states for all Monte Carlo cases and events are adjusted to a nonparametric probability distribution. Then, the multi-objective optimization is applied to all of the Monte Carlo cases simulated with the optimal initial state. The set of Pareto optimal values of the parameters and their probability distributions are obtained. Each synthetic event is simulated with a randomized probabilistic value of the calibration parameters and the initial state associated to a randomized probabilistic value of the recharge (Figure 6).

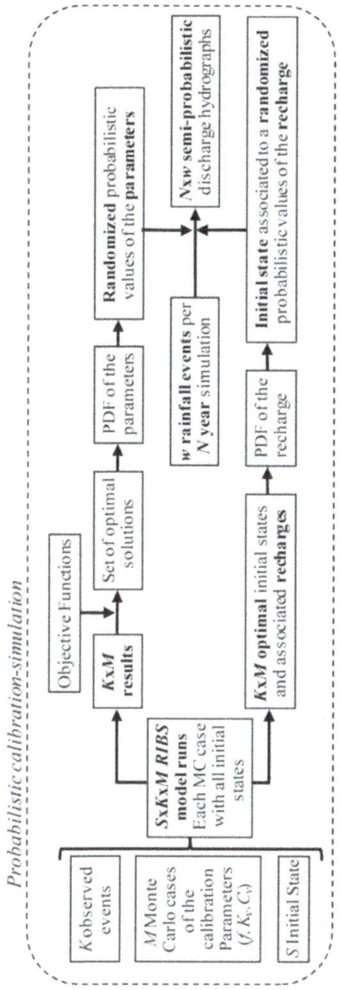

Figure 6. Probabilistic calibration–simulation process.

Analysis of the Results: Validation

Both, rainfall and RR model results are validated through the comparison between the observed and simulated nonparametric frequency curves. The observed data may also be adjusted to a parametric extreme probability distribution. The Gumbel and Generalised Extreme Value (GEV) distributions are adjusted to the observed data by the L-moments approach, choosing the one which fits best based on the chi-squared test. Due to the length of the observed series, the extrapolation of this function will be representative, approximately, up to a return period of 100 years (Klemeš 2000). Its 95% confidence intervals are also obtained. Therefore, in order to compare the observed series (of only some decades of years) with the simulated series (of thousands of years), P short-series of around 100 years are obtained by random sampling. The median and 95% uncertainty bounds derived from the P short-series are calculated.

The validation consists of comparing the observed, adjusted and median simulated frequency curves, checking if the observed and adjusted ones are contained between the uncertainty bounds. In addition, a quantitative analysis of the goodness-of-fit is performed by means of an objective functions that quantify the error, the Nash–Sutcliffe model efficiency coefficient (NSE). The NSE calculates the relative value of the variance of the residuals in relation to the variance of the observed values. Its optimal value is one.

$$NSE\,(\theta) = 1 - \frac{\sum_{Tr=2}^{Tr=100}\left[x_{Tr} - x'_{Tr}\,(\theta)\right]^2}{\sum_{Tr=2}^{Tr=100}\left[x_{Tr} - \overline{x}\right]^2} \tag{8}$$

where x_{Tr} represents the value of the adjusted distribution associated with the return period T_r, x'_{Tr} (θ) represents the value of the median simulated distribution associated with the return period T_r and $\overline{x}$ represents the average of the observed values. The interval of return periods considered is between two and 100 years, where the adjusted distribution may be considered representative (Klemeš 2000 [61]).

CASE STUDY AND DATASET

The methodology has been applied to two sub-basins of the Gardons River, Southeastern France: Saint-Jean-du-Gard and Mialet. The Gardons River rises in the Cevennes Mountains and runs through Languedoc-Roussillon, flowing into the Rhône near Beaucaire located between Avignon and Arles. The two sub-basins are located at the headwater of the Gardons River.

The Saint-Jean-du-Gard sub-basin is monitored at the Corbès gauging station and the Mialet sub-basin at the Générargues gauging station. Both gauging stations are located close to the confluence of the two rivers (see Figure 7). The drainage area at Corbès station is 262.3 km², whereas the one at Générargues site is 245.3 km². Both the sub-basins have a typical Mediterranean hydrological regime with major level rises in autumn (in October) and winter (in January), and low levels in summer (in July). A daily rainfall data series of 62 years (1948–2009) is available from eight gauges located in and around the basins (Figure 7) and also from hourly rainfall series (1988–2008) that are available only for two of them (PH_0194 and PH_0206). The flow data are obtained from the two discharge stations that provide hourly and daily data for the period 1971–2010.

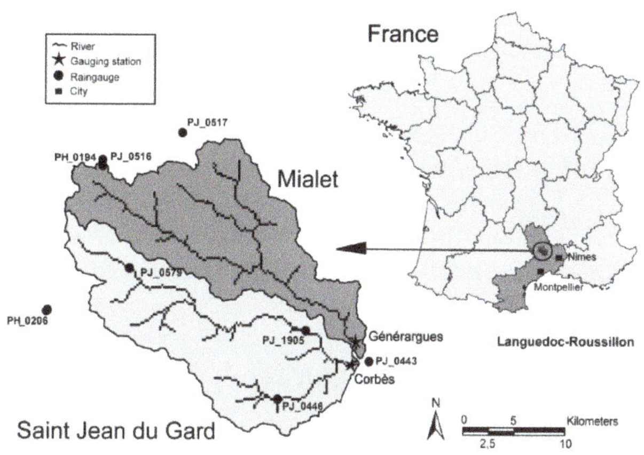

Figure 7. Location of the Saint-Jean-du-Gard and the Mialet basins.

From the data series, only 10 representative flood events have enough rainfall and flow data to calibrate the RIBS model (see Table 3). Although 10 events cannot be considered representative from this point of view, the model validation results show that the model reproduces quite well the basin behavior.

Event		Maximum Intensity (mm/h)	Peak Flow (m^3/s)		Hydrograph Volume (hm^3)	
			Générargues	Corbès	Générargues	Corbès
1	3 February 1994	15.80	149.00	140.00	13.31	13.75
2	22 September 1994	30.40	438.00	375.00	20.53	19.93
3	2 November 1994	44.00	385.00	372.00	63.94	69.11
4	17 September 1995	19.60	427.00	479.00	16.45	19.97
5	24 November 1995	15.15	262.00	189.00	23.25	25.19
6	11 January 1996	17.82	316.00	359.00	44.24	55.26
7	25 February 1996	8.87	105.00	117.00	14.24	18.10
8	11 January 1998	36.45	317.00	287.00	21.95	27.08
9	19 October 1999	19.24	324.00	290.00	15.18	15.27
10	23 November 2002	26.14	294.00	277.00	20.22	22.46

Table 3. Flood events used for calibration of the RIBS model.

Drainage areas and slope directions maps, which are necessary for the RIBS model, are obtained from a digital elevation model with 200 × 200 m^2 cells. Twenty-three soil types are considered based on the Coordination of Information on the Environment Land Cover database [62], obtaining the soil parameters used in the RIBS model according to the Brooks-Corey parameterization.

RESULTS AND DISCUSSION

Rainfall Model Calibration

The RainSim V3 model is calibrated from the available observed series for the eight rain gauges. Selected monthly statistics are: the mean 24 h rainfall accumulation (mean); the probability that an 24 h accumulation is less than 1 mm (pdry1.0); the variance of the 24 h accumulation (var24); the correlation of 24 h accumulation of two time series of the same site offset by a lag of 24 h (corr24); the skewness coefficient of 24 h accumulations (skew24); the dry–dry transition probability of 24 h accumulations (pdd0) and the variance; and the skewness coefficient of 48 h accumulation (var48 and skew48). From these statistics, the optimal values of the calibration parameters (λ, β, η, ξ, γ, ρ, and Φ) are obtained.

Based on the optimal parameter values, a continuous series of 1000 years of hourly rainfall is generated. The monthly statistics from the synthetic series are calculated and compared with those of the observed series. The results show a good fit for most of the rain gauges. As an example, the results for the rain gauge PH_0194 are provided in Figure 8.

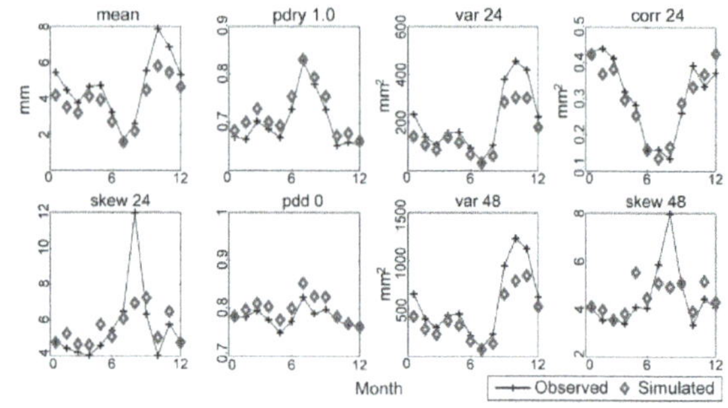

Figure 8. Calibration of RainSim V3 in the Gardons basin. Statistical values (rain gauge PH_0194).

Rainfall Model Validation

Once the rainfall model is calibrated, a series of hourly rainfall for 6000 years for the eight rain gauges are generated and the series of annual maximum daily rainfall (AMR) are obtained from the observed and simulated rainfall series. Validation of the model is carried out with the series of 62 years of observed AMR and 300 short-series of 125 years sampled from the 6000-year series of simulated AMR by random sampling. The results of the validation are shown in Figure 9 and Table 4.

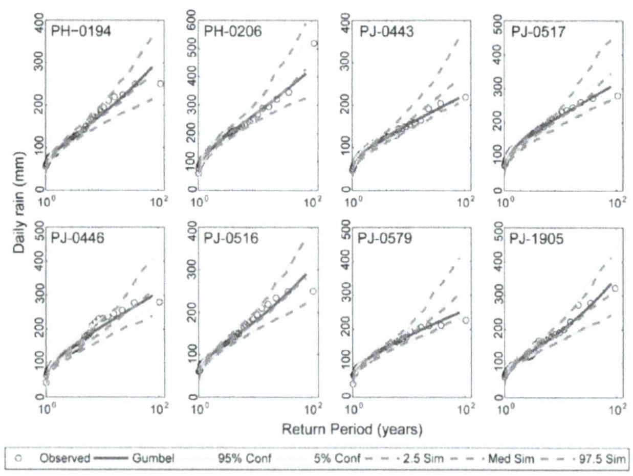

Figure 9. Frequency curves of the AMR for the eight rain gauges selected: observed data (circles); fitted distribution to observed data (solid line) with its 95% confidence interval (dotted line); and median of the 300 62-year series obtained from the simulation series with its 2.5% and 97.5% percentiles (dotted-dashed lines).

Table 4. NSE values calculated for the AMR for the eight rain gauges.

Station	PH_0194	PH_0206	PJ_0443	PJ_0517	PJ_0446	PJ_0516	PJ_0579	PJ_1905
NSE	0.99518	0.99364	0.94334	0.93327	0.9422	0.99498	0.95852	0.99248

As it can be seen, the results are very good for all of the rain gauges (see Table 4) with NSE coefficients greater than 0.93.

RIBS Calibration

The RIBS model is calibrated using the three calibration–simulation processes, considering 10 representative observed events, 1000 Monte Carlo combinations of the calibration parameters (f, Cv and Kv) and 25 initial states. The following results are obtained: (a) a probability distribution for each parameter that is common for the deterministic and semi-probabilistic approaches; and (b) a probability distribution for each parameter and for the recharge (initial condition) in the case of the probabilistic approach.

Synthetic Events Simulation

After performing the three calibration–simulation processes, the three series of 6000 years with five events per year were simulated (6000 years × 5 events = 30,000 hydrographs in total), obtaining three series of 30,000 hydrographs: deterministic, semi-probabilistic and probabilistic. These series enable the calculation of the nonparametric frequency curves of annual maximum peak flows and their associated hydrograph volumes (see Figure 10). This calculation is based on the Gringorten plotting position formula [63].

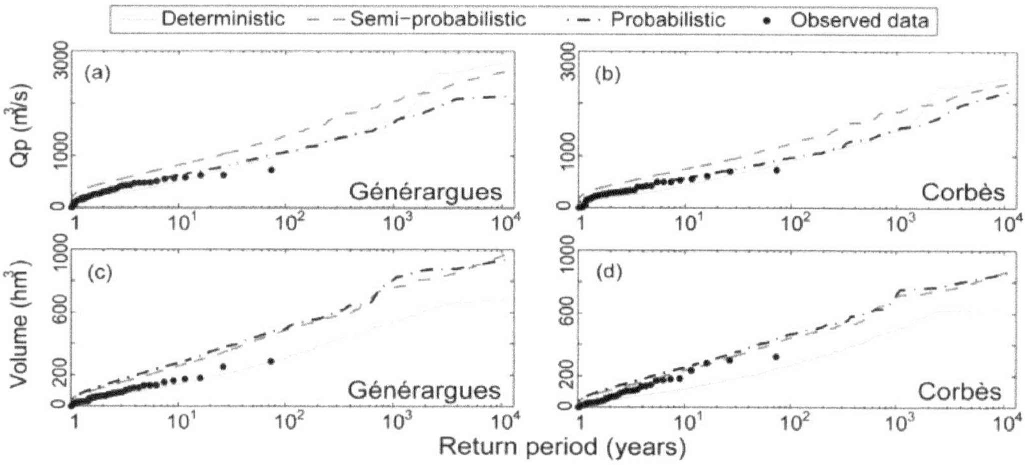

Figure 10. Frequency curves of peak flows (**a**, **b**) and the associated hydrograph volume (**c**, **d**). Deterministic (solid line), semi-probabilistic (dashed line) and probabilistic (dash-dot line) and observed data (circles). Results for the Générargues (**a**,**c**) and Corbès (**b**, **d**) basins.

Although the three curves fit the observations sufficiently well for low and medium return periods, the uncertainty increases for the high ones and the simulated curves spread out, with increasing the differences between themselves. In order to determine which type of calibration–simulation process would best represent the behavior of the basin, the validation is carried out.

Methodology Validation

The validation is applied to the 62-year series of observed annual maximum peak flow and hydrograph volume and 300 short-series of 125 years sampled from the three series of 6000 years of maximum annual peak flow and hydrograph volume. The validation results are compared to determine which calibration–simulation approach represents better the basin behavior (Figure 11 and Table 5).

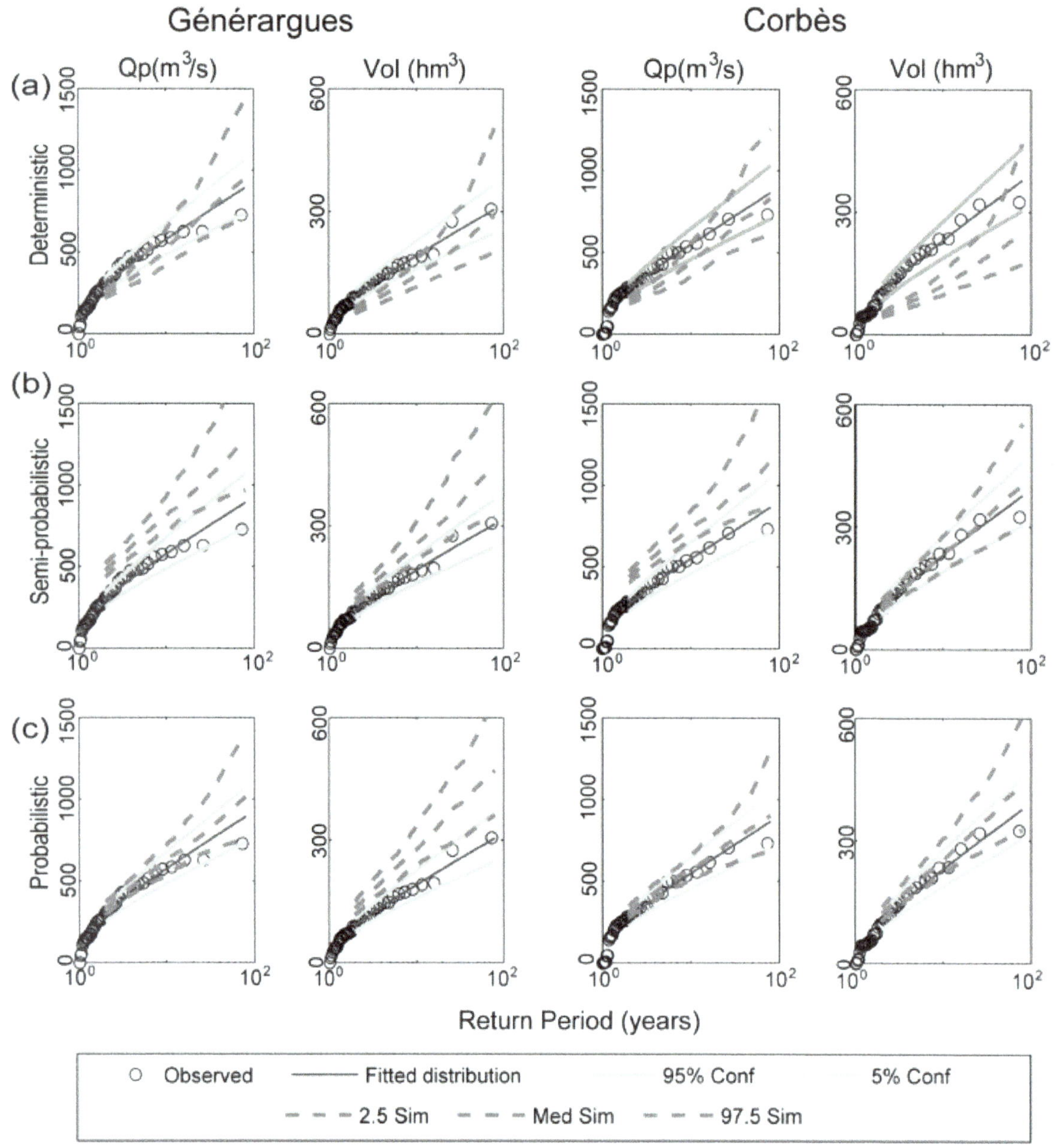

Figure 11. Frequency curves of the peak flow and the associated hydrograph volume for the Génèrargues and the Corbès basins: (**a**) Deterministic; (**b**) semi-probabilistic; and (**c**) probabilistic. Observed data (circles); Distribution fitted to the observed series (solid line) with its 95% confidence limits (dotted line); median and 2.5% and 97.5% percentiles of model simulations (dashed lines).

Variable	Objective Function	Station	Simulation		
			Deterministic	Semi-Probabilistic	Probabilistic
Qmax	NSE	Géneérargues	0.458	*−1.764*	**0.91**
		Corbès	0.07	*−0.989*	**0.968**
VolQmax	NSE	Géneérargues	**−0.195**	−0.231	*−1.255*
		Corbès	*−1.598*	**0.972**	0.908

Table 5. NSE values calculated for the simulated peak flows and hydrograph volumes by the deterministic, semi-probabilistic and probabilistic approaches. The bold text shows the approach that fits best and the italics text shows the worst fit.

The results show that, in the case of peak flows, the probabilistic calibration offers the best fit with NSE coefficients larger than 0.9. The deterministic approach presents an acceptable fit with NSE coefficients smaller than 0.5 while the semi-probabilistic calibration offers a poor fit with NSE coefficients smaller than zero. However, in the case of volumes, the results vary from one basin to another. At Génèrargues a truly good fit is not achieved by any of the approaches. The cause may be that the considered impervious area by the RR model may be greater than the actual because the RIBS model assumes that the cells through which the river flows are impervious. However, for Corbès, the semi-probabilistic approach leads to the best results (NSE coefficient equal to 0.97), although the probabilistic calibration achieves also a good fit (NSE coefficient equal to 0.91). Therefore, it may be concluded that the best calibration–simulation approach for the cases analyzed is the probabilistic one, which characterizes the initial state and considers the variability of the basin behavior and the uncertainty inherent to the model. In contrast, the other two calibrations do not represent this variability either entirely or partially, which leads to an overestimation or underestimation of the variables that characterize the hydrographs.

Bivariate and Uncertainty Analysis

Once the proposed methodology is validated and the probabilistic approach is chosen, a bivariate and uncertainty analysis is carried out. The model results are the series of 6000 years with five complete hydrographs per year that allow characterizing the extreme behavior of the basin. To conduct the bivariate analysis, the peak flows and volumes of the simulated hydrographs are obtained and represented in the flow-volume space (Figure 12). These graphs show that the range of peak flows associated with a hydrograph volume increase with the volume value although the larger the hydrograph volume is, the smaller the number of events is, appearing them more scattered.

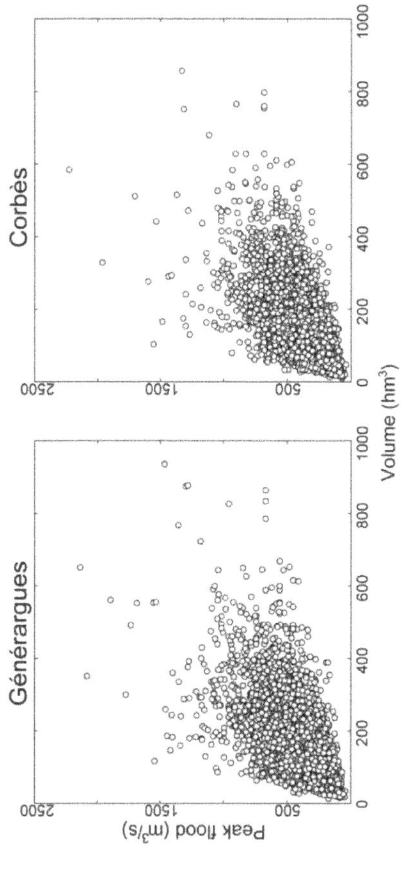

Figure 12. Peak flow and hydrograph volumes of the simulated hydrograph for the Générargues and the Corbès basins.

The uncertainty analysis consists on determining the variability in the estimates of the variables associated with a given return period. Hydrograph series of different length (2000, 1500, 1200, 1000, 750, 500, 250 and 125) are obtained from the 6000 years series by random sampling. The nonparametric frequency curves are obtained from each series based on the Gringorten formula and the relationship between the exceedance probability and the return period. Only the uncertainty of the peak flow estimations for the Corbès basin is presented here (see Figure 13).

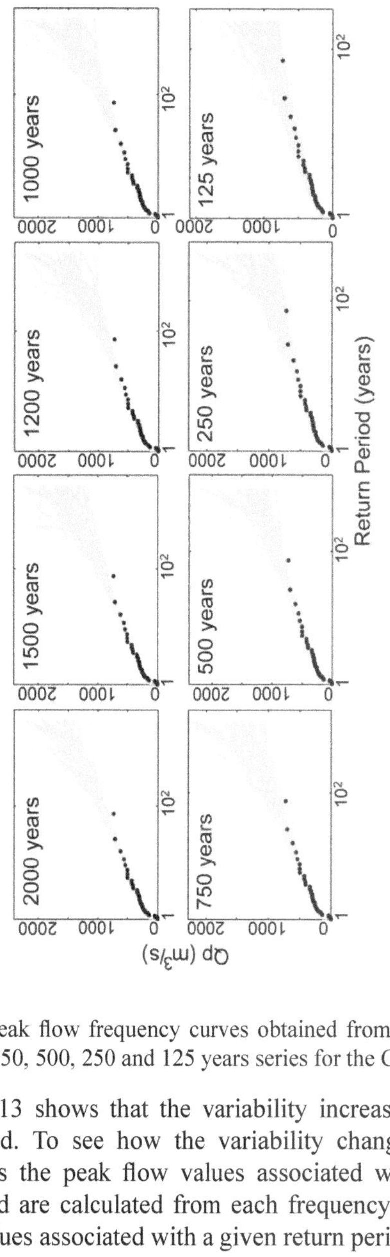

Figure 13. Peak flow frequency curves obtained from 2000, 1500, 1200, 1000, 750, 500, 250 and 125 years series for the Corbès basin.

Figure 13 shows that the variability increases with the return period. To see how the variability change with the length series the peak flow values associated with a given return period are calculated from each frequency curve. The different values associated with a given return period and with a series length are statistically analyzed obtaining the 25% percentile (Q1), the 50% percentile (Q2 or median) and the 75% percentile (Q3) quartiles. The results are represented in a box-plot presented in Figure 14. This figure shows that the variability decreases with the length series. Moreover, the median values are bracketed between the confidence intervals, so the results are good enough.

In order to quantify, the uncertainty of the estimations, the interquartile range (IQR) is calculated. The IQR is the difference between the upper and lower quartiles and it is a measure of the statistical dispersion (Table 6).

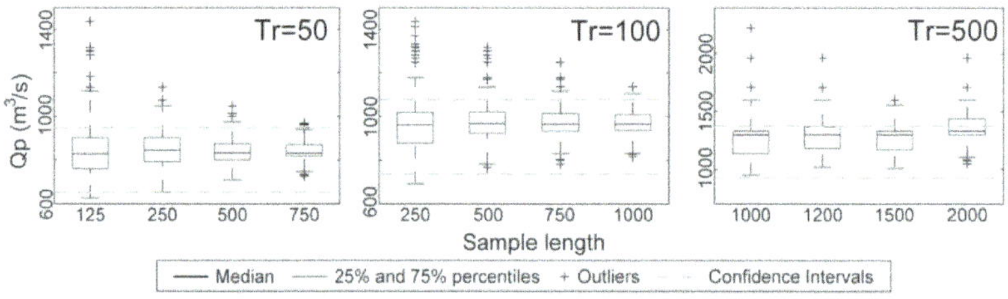

Figure 14. Median, 25% and 75% percentiles and outlier values of the peak flows associated with return periods of 50, 100 and 500 years calculated from different series length for the Corbès basin.

Table 6. Values of the first (Q_1), second (Q_2), third (Q_3) quartiles, the interquartile range IQR, and the relative value of the IQR against Q_2 and against the mean of Q_1 and Q_3.

Tr = 500 Years						
Data length	Q1	Q2	Q3	IQR	IQR/Q2	IQR/((Q1 + Q3)/2)
1000	1135.0	1300.0	1331.0	196.0	15.1%	15.9%
1250	1183.0	1301.0	1372.0	189.0	14.5%	14.8%
1500	1171.0	1300.0	1331.0	160.0	12.3%	12.8%
2000	1300.0	1331.0	1438.0	138.0	10.4%	10.1%
Tr = 100 years						
Data length	Q1	Q2	Q3	IQR	IQR/Q2	IQR/((Q1 + Q3)/2)
250	878.3	961.7	1018.0	139.7	14.5%	14.7%
500	923.0	968.6	1022.0	99.0	10.2%	10.2%
750	932.5	965.0	1013.0	80.5	8.3%	8.3%
1000	936.7	965.4	1008.0	71.3	7.4%	7.3%
Tr = 50 years						
Data length	Q1	Q2	Q3	IQR	IQR/Q2	IQR/((Q1 + Q3)/2)
125	761.3	828.8	903.4	142.1	17.1%	17.1%
250	794.1	846.4	903.4	109.4	12.9%	12.9%
500	801.0	833.7	876.2	75.2	9.0%	9.0%
750	819.3	831.8	870.6	51.2	6.2%	6.1%

Figure 14 and Table 5 show that the uncertainty decreases with the length of the series and increases with the return period. This means that the longer the simulated series are, the more accurate the estimations associated to high return period are. This estimated uncertainty may be taken into account in the infrastructure design.

CONCLUSIONS

This study presents a stochastic methodology for the generation of flood event series of arbitrary length by the combination of a stochastic rainfall generator and a distributed rainfall–runoff event-based model through the identification of independent storm events and the selection of a sub-set of events for each year with the criterion of the higher accumulated rainfall.

The combination of the continuous stochastic rainfall generator with the identification of independent events through the exponential method allows an accurate simulation of the rainfall over the basin and the characterization of storm events considering spatial and temporal variability.

The probabilistic calibration–simulation approach proposed here overcomes the main drawback of event-based models, the basin initial state characterization. This approach considers the recharge (initial state of the basin) as a random variable and obtains its probability density function. The results are compared with those obtained for the other approaches (deterministic and semi-probabilistic). The comparison shows that the probabilistic approach better represents the basin behavior than the other two because of the characterization of the initial state, the representation of the runoff generation and propagation processes variability and the uncertainty inherent to the model through a probabilistic calibration and simulation.

The application of this methodology results in a set of complete hydrographs that allows the calculation of the frequency curves of all the variables that characterize the hydrographs and the development of a multivariate analysis.

The main pros and cons of the proposed methodology are:

Pros:

- The application of the RainSim model allows the generation of arbitrarily long hourly rainfall series, even from daily-observed rainfall series.
- The use of a distributed physically-based rainfall–runoff event-based model allows simulating extreme hydrograph series of thousands of years with an hourly resolution.
- The probabilistic approach characterizes the basin initial state and solves the main problem of event-base model.
- This approach also takes into account the basin variability through the runoff generation and propagation.
- The extreme hydrograph series obtained as results can be used in hydraulic infrastructure design, territorial planning, flood management and risk analysis.

Cons:

- The RIBS model assumes that the river cells are impervious, which means that all rain that falls on these cells transforms into runoff. Because of this, the runoff volume produced in the river cells could be greater than the real one and prevent obtaining a good fit.
- To calibrate the RIBS model, it is necessary to dispose of extreme rainfall events and the associated flood events with an hourly resolution. The number of these events is usually reduced, which may difficult the calibration process or make that the results are not quite representative from a statistical point of view
- It should be noted that the study is based on limited case studies and needs more verification in order to validate the proposed methodology

Considering these pros and cons, the methodology is appropriate if there are enough rainfall and flow data to calibrate the models, as well as, if there is a good characterization of the morphological and hydraulic characterization of the catchment available.

Therefore, the methodology presented here may be a useful tool to characterize the hydrological extreme behavior of basins, which, in turn, is essential for the design of hydraulic infrastructures, territorial planning, flood management and risk analysis.

ACKNOWLEDGMENTS

The authors wish to acknowledge support from project MODEX Physically-based modeling of extreme hydrologic response under a probabilistic approach. Application to Dam Safety Analysis (CGL2011-22868), funded by the Spanish Ministry of Science and Innovation (now the Ministry of Economy and Competitiveness) and from the project DRIHM-Distributed Research Infrastructure for Hydro-Meteorology, funded by the European Commission through Grant Agreement No. 283568. The authors are also grateful for the financial contribution made by the COST Office grant ES0901 "European procedures for flood frequency estimation (FloodFreq)" and for the support and advice provided by Emmanuel Paquet during the stay of the first author at the Division Technique Générale (DTG) of Électricité de France in Grenoble in October 2013.

AUTHOR CONTRIBUTIONS

Isabel Flores-Montoya, Álvaro Sordo-Ward, Luis Mediero and Luis Garrote conceived and designed the experiments; and Isabel Flores-Montoya performed the experiments and analyzed the data; Isabel Flores-Montoya wrote the paper with the contribution of Álvaro Sordo-Ward, Luis Mediero and Luis Garrote.

CONFLICTS OF INTEREST

The authors declare no conflict of interest.

REFERENCES

1. Adams, B.; Howard, C.D. Design storm pathology. Can. Water Resour. J. **1986**, 11, 49–55.
2. Alfieri, L.; Laio, F.; Claps, P. A simulation experiment for optimal design hyetograph selection. Hydrol. Process. **2008**,22, 813–820.
3. Viglione, A.; Blöschl, G. On the role of storm duration in the mapping of rainfall to flood return periods. Hydrol. Earth Syst. Sci. **2009**, 13, 205–216.
4. Eagleson, P.S. Dynamics of flood frequency. Water Resour. Res. **1972**, 8, 878–898.
5. Sordo-Ward, A.; Garrote, L.; Martín-Carrasco, F.; Dolores Bejarano, M. Extreme flood abatement in large dams with fixed-crest spillways. J. Hydrol. **2012**, 466, 60–72.
6. Sordo-Ward, A.; Garrote, L.; Bejarano, M.D.; Castillo, L.G. Extreme flood abatement in large dams with gate-controlled spillways. J. Hydrol. **2013**, 498, 113–123.
7. Foufoula-Georgiou, E. A probabilistic storm transposition approach for estimating exceedance probabilities of extreme precipitation depths. Water Resour. Res. **1989**, 25, 799–815.
8. Wilson, L.L.; Foufoula-Georgiou, E. Regional rainfall frequency analysis via stochastic storm transposition. J. Hydraul. Eng. **1990**, 116, 859–880.
9. Franchini, M.; Helmlinger, K.R.; Foufoula-Georgiou, E.; Todini, E. Stochastic storm transposition coupled with rainfall-runoff modeling for estimation of exceedance probabilities of design floods. J. Hydrol. **1996**, 175, 511–532.
10. England, J.F., Jr. Frequency Analysis and Two-Dimensional Simulations. Ph.D. Thesis, Colorado State University, Fort Collins, CO, USA, 2006.
11. Rodriguez-Iturbe, I.; Cox, D.; Isham, V. Some models for rainfall based on stochastic point processes. Proc. R. Soc. Lond. Ser. A **1987**, 410, 269–288.
12. Rodríguez-Iturbe, I.; Febres de Power, B.; Valdés, J. Rectangular pulses point process models for rainfall: Analysis of empirical data. J. Geophys. Res. **1987**, 92, 9645–9656.
13. Cowpertwait, P.S. A generalized point process model for rainfall. Proc. R. Soc. Lond. Ser. A Math. Phys. Sci. **1994**, 447, 23–37.
14. Cowpertwait, P.S. A generalized spatial-temporal model of rainfall based on a clustered point process. Proc. R. Soc. Lond. Ser. A Math. Phys. Sci. **1995**, 450, 163–175.
15. Cowpertwait, P.S.P.; O'Connell, P.E.; Metcalfeb, A.V.; Mawdsley, J.A. Stochastic point process modelling of rainfall. I. Single-site fitting and validation. J. Hydrol. **1996**, 175, 17–46.
16. Cowpertwait, P.S. A spatial-temporal point process model with a continuous distribution of storm types. Water Resour. Res. **2010**, 46.
17. Kilsby, C.G.; Jones, P.D.; Burton, A.; Ford, A.C.; Fowler, H.J.; Harpham, C.; James, P.; Smith, A.; Wilby, R.L. A daily weather generator for use in climate change studies. Environ. Model. Softw. **2007**, 22, 1705–1719.
18. Leonard, M.; Lambert, M.F.; Metcalfe, A.V.; Cowpertwait, P.S.P. A space-time Neyman–Scott rainfall model with defined storm extent. Water Resour. Res. **2008**, 44.

19. Burton, A.; Kilsby, C.G.; Fowler, H.J.; Cowpertwait, P.S.P.; O'Connell, P.E. RainSim: A spatial–temporal stochastic rainfall modelling system. Environ. Model. Softw. **2008**, 23, 1356–1369.
20. Burton, A.; Fowler, H.J.; Kilsby, C.G.; O'Connell, P.E. A stochastic model for the spatial-temporal simulation of nonhomogeneous rainfall occurrence and amounts. Water Resour. Res. **2010**, 46, W11501.
21. Hashemi, A.M.; Franchini, M.; O'Connell, P.E. Climatic and basin factors affecting the flood frequency curve: PART I? A simple sensitivity analysis based on the continuous simulation approach. Hydrol. Earth Syst. Sci. Discuss. **2000**, 4, 463–482.
22. Cameron, D.S.; Beven, K.J.; Tawn, J.; Blazkova, S.; Naden, P. Flood frequency estimation by continuous simulation for a gauged upland catchment (with uncertainty). J. Hydrol. **1999**, 219, 169–187.
23. Cameron, D.; Beven, K.J.; Tawn, J.; Naden, P. Flood frequency estimation by continuous simulation (with likelihood based uncertainty estimation). Hydrol. Earth Syst. Sci. **2000**, 4, 23–34.
24. Cameron, D.; Beven, K.; Naden, P. Flood frequency estimation under climate change (with uncertainty). Hydrol. Earth Syst. Sci. **2001**, 4, 393–405.
25. Blazkova, S.; Beven, K.J. Frequency version of TOPMODEL as a tool for assessing the impact of climate variability on flow sources and flood peaks. J. Hydrol. Hydromech. **1995**, 43, 392–411.
26. Blažkova, S.; Beven, K. Flood frequency estimation by continuous simulation for a catchment treated as ungauged (with uncertainty). Water Resour. **2002**, 38, 1–14.
27. Blazkova, S.; Beven, K. Flood frequency estimation by continuous simulation of subcatchment rainfalls and discharges with the aim of improving dam safety assessment in a large basin in the Czech Republic. J. Hydrol. **2004**, 292, 153–172.
28. Haberlandt, U.; Ebner von Eschenbach, A.D.; Buchwald, I. A space-time hybrid hourly rainfall model for derived flood frequency analysis. Hydrol. Earth Syst. Sci. **2008**, 12, 1353–1367.
29. Brocca, L.; Melone, F.; Moramarco, T. Distributed rainfall-runoff modelling for flood frequency estimation and flood forecasting. Hydrol. Process. **2011**, 25, 2801–2813.
30. Brocca, L.; Liersch, S.; Melone, F.; Moramarco, T.; Volk, M. Application of a model-based rainfall-runoff database as efficient tool for flood risk management. Hydrol. Earth Syst. Sci. **2013**, 17, 3159–3169.
31. Camici, S.; Tarpanelli, A.; Brocca, L.; Melone, F.; Moramarco, T. Design soil moisture estimation by comparing continuous and storm-based rainfall-runoff modeling. Water Resour. Res. **2011**, 47.
32. Ciabatta, L.; Brocca, L.; Massari, C.; Moramarco, T.; Gabellani, S.; Puca, S.; Wagner, W. Rainfall-runoff modelling by using SM2RAIN-derived and state-of-the-art satellite rainfall products over Italy. Int. J. Appl. Earth Obs. Geoinform.**2016**, 48, 163–173.
33. Loizu, J.; Álvarez-Mozos, J.; Casalí, J.; Chahor, Y.; Brocca, L. Catorce años de simulación en continuo de caudales horarios en las cuencas navarras de los ríos Cidacos y Arga. Estudio comparativo de dos modelos hidrológicos. Available online: http://www.uco.es/jia2015/ponencias/b/b011.pdf (accessed on 26 February 2016).
34. Rulli, M.C.; Rosso, R. An integrated simulation method for flash-flood risk assessment: 1. Frequency predictions in the Bisagno River by combining stochastic and deterministic methods. Hydrol. Earth Syst. Sci. Discuss. **2002**, 6, 267–284.
35. Francés, F.; García-Bartual, R.; Bussi, G. High return period annual maximum reservoir water level quantiles estimation using synthetic generated flood events. In Risk Analysis, Dam Safety, Dam Security and Critical Infrastructure Management; CRC Press: Boca Raton, FL, USA, 2011; pp. 185–190.
36. Ocio, D.; Stocker, C.; Eraso, Á.; Cowpertwait, P. Regionalized extreme flows by means of stochastic storm generation coupled with a distributed hydrological model. The case of the basque country. In E-Proceedings of the 36th IAHR World Congress, The Hague, The Netherlands, 28 June–3 July 2015.
37. Paquet, E.; Garavaglia, F.; Garçon, R.; Gailhard, J. The SCHADEX method: A semi-continuous rainfall–runoff simulation for extreme flood estimation. J. Hydrol. **2013**, 495, 23–37.
38. Brigode, P.; Bernardara, P.; Paquet, E.; Gailhard, J.; Garavaglia, F.; Merz, R.; Micovic, Z.; Lawrence, D.; Ribstein, P. Sensitivity analysis of SCHADEX extreme flood estimations to observed hydrometeorological variability. Water Resour. Res. **2014**, 50, 353–370.
39. Lawrence, D.; Paquet, E.; Gailhard, J.; Fleig, A.K. Stochastic semi-continuous simulation for extreme flood estimation in catchments with combined rainfall–snowmelt flood regimes. Nat. Hazards Earth Syst. Sci. **2014**, 14, 1283–1298.
40. Massari, C.; Brocca, L.; Barbetta, S.; Papathanasiou, C.; Mimikou, M.; Moramarco, T. Using globally available soil moisture indicators for flood modelling in Mediterranean catchments. Hydrol. Earth Syst. Sci. **2014**, 18, 839–853.
41. Li, J.; Thyer, M.; Lambert, M.; Kuczera, G.; Metcalfe, A. An efficient causative event-based approach for deriving the annual flood frequency distribution. J. Hydrol. **2014**, 510, 412–423.
42. Garrote, L.; Brass, R.L. A distributed model for real-time flood forecasting using digital elevation models. J. Hydrol.**1995**, 167, 279–306.
43. Garrote, L.; Brass, R.L. An integrated software environment for real-time use of a distributed hydrologic model. J. Hydrol. **1995**, 167, 307–326.
44. Restrepo-Posada, P.J.; Eagleson, P.S. Identification of independent rainstorms. J. Hydrol. **1982**, 55, 303–319.
45. Bonta, J.V.; Rao, A.R. Factors affecting the identification of independent storm events. J. Hydrol. **1988**, 98, 275–293.

46. Cosby, B.J.; Hornberger, G.M.; Clapp, R.B.; Ginn, T. A statistical exploration of the relationships of soil moisture characteristics to the physical properties of soils. Water Resour. Res. **1984**, 20, 682–690.

47. Refsgaard, J.C.; Thorsen, M.; Jensen, J.B.; Kleeschulte, S.; Hansen, S. Large scale modelling of groundwater contamination from nitrate leaching. J. Hydrol. **1999**, 221, 117–140.

48. Castillo, V.M.; Gomez-Plaza, A.; Martinez-Mena, M. The role of antecedent soil water content in the runoff response of semiarid catchments: A simulation approach. J. Hydrol. **2003**, 284, 114–130.

49. Zehe, E.; Becker, R.; Bárdossy, A.; Plate, E. Uncertainty of simulated catchment runoff response in the presence of threshold processes: Role of initial soil moisture and precipitation. J. Hydrol. **2005**, 315, 183–202.

50. Noto, L.V.; Ivanov, V.Y.; Bras, R.L.; Vivoni, E.R. Effects of initialization on response of a fully-distributed hydrologic model. J. Hydrol. **2008**, 352, 107–125.

51. Moore, R.J.; Cole, S.J.; Bell, V.A.; Jones, D.A. Issues in Flood Forecasting: Ungauged Basins, Extreme Floods and Uncertainty; IAHS-AISH Publication: Wallingford, UK, 2006; pp. 103–122.

52. Vivoni, E.R.; Entekhabi, D.; Bras, R.L.; Ivanov, V.Y. Controls on runoff generation and scale-dependence in a distributed hydrologic model. Hydrol. Earth Syst. Sci. Discuss. **2007**, 4, 983–1029.

53. Berthet, L.; Andréassian, V.; Perrin, C.; Javelle, P. How crucial is it to account for the antecedent moisture conditions in flood forecasting? Comparison of event-based and continuous approaches on 178 catchments. Hydrol. Earth Syst. Sci. **2009**, 13, 819–831.

54. Mediero, L.; Garrote, L.; Martín-Carrasco, F.J. Probabilistic calibration of a distributed hydrological model for flood forecasting. Hydrol. Sci. J. **2011**, 56, 1129–1149.

55. Sifalda, V. Entwicklung eines Berechnungsregens für die Bemessung von Kanalnetzen. Wasser Abwasser **1963**, 114, 435–440.

56. Yen, B.C.; Chow, V.T. Design hyetographs for small drainage structures. J. Hydraul. Div. **1980**, 106, 1055–1076.

57. Wenzel, H.G.; Voorhees, M.L. An Evaluation of the Urban Design Storm Concept; UILU-WRC-81–0164; University of Illinois at Urbanna-Champaign, Water Resour. Center: Champaign, IL, USA, 1981.

58. Grace, R.A.; Eagleson, P.S. The Synthesis of Short-Time Increment Rainfall Sequences; Hydrodynamics Laboratory, Dept. of Civil Engineering, Massachusetts Institute of Technology: Cambridge, UK, 1966; p. 105.

59. Cabral, M.C.; Garrote, L.; Bras, R.L.; Entekhabi, D. A kinematic model of infiltration and runoff generation in layered and sloped soils. Adv. Water Resour. **1992**, 15, 311–324.

60. Cabral, M.C.; Bras, R.L.; Tarboton, D.; Entekhabi, D. A distributed, physically-based, rainfall-runoff model incorporating topography for real-time flood forecasting (No. 332). Available online: www.researchgate.net (accessed on 24 January 2013).

61. Klemeš, V. Tall tales about tails of hydrological distributions: Parts I and II. J. Hydrol. Eng. **2000**, 5, 227–239.

62. CORINE Land Cover. Coordination of Information on the Environment (2006); European Environment Agency: Copenhagen, Denmark, 2006.

63. Gringorten, I.I. A plotting rule for extreme probability paper. J. Geophys. Res. **1963**, 68, 813–814.

CHAPTER 2

AN ANALYSIS OF THE OCCURRENCE PROBABILITIES OF WET AND DRY PERIODS THROUGH A STOCHASTIC MONTHLY RAINFALL MODEL

Tommaso Caloiero , Beniamino Sirangelo, Roberto Coscarelli and Ennio Ferrari

Keywords: Monthly rainfall, stochastic model, dry and wet periods, Calabria

INTRODUCTION

A natural temporary imbalance of water availability, consisting of persistent lower-than-average or higher-than-average precipitation, can cause extreme dry and wet conditions that adversely impact agricultural yields, water resources, infrastructure and human systems. For example, dry conditions have severe consequences because they can give rise to agricultural yield losses and water shortages [1,2]. They also damage natural ecosystems [3,4] and forestry [5,6] and lead to soil degradation and desertification [7,8], social alarm [9], famine and impoverishment [10,11]. Conversely, wet conditions can cause flood, damage crops, reduce yields, and contribute to groundwater contamination [12,13,14,15,16,17]. For these reasons, the analysis of the distribution of dry and wet periods has always been a concern for researchers [18,19,20,21,22,23,24,25,26]. In general, climate anomalies are quantitatively assessed through indices that allow scientists to characterize them in terms of intensity, duration, frequency, recurrence probability and spatial extent [27,28]. Among the several indices, two of the most commonly used are the standardized precipitation index (SPI), which transforms monthly precipitation time series into a standardized normal distribution, and the drought severity index (DSI), which uses accumulated monthly precipitation anomalies. In particular, the SPI has found widespread application in different countries of the world [29,30,31,32,33,34,35], in the Mediterranean basin [36,37,38] and also in Central [39] and Southern Italy [40,41,42,43,44,45,46]. The SPI is easier to calculate than more complex indices because it is based on precipitation alone for estimating wet or dry conditions [38,47]. Moreover, these conditions can be monitored by the SPI on a variety of time scales from sub-seasonal to inter-annual scales [43].

Generally, the probabilistic structure of dry and wet periods, especially with a long duration, cannot be properly investigated due to the limited number of events in the historical series [48]. Moreover, complete data are strictly required to perform the analysis of wet and dry periods, because missing values may significantly influence estimates of event duration and the character of their alternation [49]. In order to overcome such a difficulty, the probabilistic behavior of dry and wet periods characteristics can be derived analytically, assuming a given stochastic structure of the underlying hydrological and meteorological series [24,50,51,52,53,54,55,56,57,58,59]. This has led to the development of stochastic models frequently used to produce long rainfall series that are statistically similar to historical records (e.g., [60]). In particular, mathematical modeling of hydrological data as a stochastic process is of interest to a variety of hydrological areas such as flood forecasting, reservoir operations and agricultural planning [61,62,63]. Current state-of-the-art stochastic precipitation models at a single site or for multiple sites can reproduce a wide range of statistics from hourly scale extremes to larger

Citation: Tommaso Caloiero, Beniamino Sirangelo, Roberto Coscarelli and Ennio Ferrari An Analysis of the Occurrence Probabilities of Wet and Dry Periods through a Stochastic Monthly Rainfall Model, doi:10.3390/w8020039.

aggregation periods [64,65,66,67,68,69,70,71,72,73,74,75]. In particular, numerous approaches for the stochastic modeling of daily rainfall data are available in the hydrological and climatological literature [76,77,78,79,80,81,82,83,84]. These models are widely used because they are easy to formulate and fast to implement [26]. Stochastic modeling also allows for accounting of potentially wider range of conditions that may not be necessarily represented in poorly sampled data. Moreover, by means of stochastic simulators, effectively intrinsic variability, not associated with forced signals (e.g., by global warming), can be generated; this variability is important because it places uncertainties on projections. However, very little work has been done on stochastic generation of monthly rainfall data [85] because in the past low attention was paid to totals at this time aggregation.

The aims of this study are:

(a) to propose a stochastic procedure for modeling precipitation at monthly scale;

(b) to analyze dry and wet periods through the application of the standardized precipitation index (SPI) to a set of monthly rainfall series generated by the proposed model through a Monte Carlo procedure. These results have been compared with the ones obtained through the application of the drought severity index (DSI).

In particular, the model has been applied to a set of monthly rainfall series observed in Southern Italy (the region of Calabria).

METHODOLOGY

Standardized Precipitation Index

In this study, dry and wet periods were expressed using the SPI [86] on different time scales. Indeed, it is generally agreed that the SPI on short-term scales (e.g., 3 or 6 months) describes drought affecting vegetation and agricultural practices, while on long-term scales (e.g., 12 or 24 months) it is a broad proxy for water resource management [40,41,42]. A detailed description of the SPI has been proposed previously [87]. Although a classification restricted only to drought periods has been originally proposed [86], it has become customary to use the index to classify wet periods as well. Table 1 reports the climatic classification according to the SPI, provided by the National Drought Mitigation Center (NDMC).

Table 1. Climate classification according to the SPI values [86].

SPI Value	Class	Probability (%)
SPI ⩾ 2.0	Extremely wet	2.3
1.5 ⩽ SPI < 2.0	Severely wet	4.4
1.0 SPI < 1.5	Moderately wet	9.2
0.0 SPI < 1.0	Mildly wet	34.1
−1.0 ⩽ SPI < 0.0	Mild drought	34.1
−1.5 ⩽ SPI < −1.0	Moderate drought	9.2
−2.0 ⩽ SPI < −1.5	Severe drought	4.4
SPI < −2.0	Extreme drought	2.3

Despite its widespread application, some limitations of the SPI calculation must be addressed. In fact, as evidenced by a previous study [88], the SPI is an index with several univariate probability distribution recommendations that typically rely on the Kolmogorov–Smirnov test. However, this test has been shown to be relatively insensitive in previous statistical analyses [89] and requires time-consuming Monte Carlo simulation of critical values when applied to distributions derived from the data [90,91]. Moreover, probability distribution fitting for the SPI is also complicated by the presence of periods with zero precipitation, as has been previously described [92], and current procedures do not adequately capture the likelihood of zero precipitation events.

Drough Severity Index (DSI)

Based on the accumulated monthly rainfall deficit concept of a previous report [93], the drought severity index (DSI) is evaluated through the precipitation anomalies (mm) defined in this work with reference to the 1961–1990 mean [25,94,95]. Two indices, DSI3 and DSI6, can be used by means of different termination rules. Considering the DSI3, if the precipitation anomaly in month t is denoted as X_t and is negative and the precipitation in the preceding 3-month period (*i.e.*, $t-1$, $t-2$, $t-3$) is also lower than its mean, then a drought sequence can start. In this case, the positive value assumed for DSI3 is proportional to the deficit in month t. Considering the next month $(t+1)$, if its precipitation deficit is $-Y$ mm, and the mean monthly precipitation total for the preceding 3 months has not been exceeded, then DSI3 for the month $(t+1)$ is $X+Y$. On the contrary, if the precipitation anomaly is positive, the drought can continue provided that the 3-monthly mean total has not been exceeded. The termination of a drought event occurs when the 3-monthly mean total is exceeded, thus assigning a value of zero to DSI3. DSI6 is calculated identically, using the 6-month mean to identify drought termination. In order to allow comparisons between different rain gauges, the DSI values can be standardized by dividing the absolute deficit by the site mean annual precipitation, and then multiplying by 100, thus expressing the accumulated precipitation deficit as a percentage of the annual mean total precipitation.

Stochastic Modelling of Monthly Rainfall

Let $H_{i,j}$ be the sequence of random variables describing the monthly total rainfall, where j represents the month and $i$represents the year, counted from a non-specified origin $i = 0$. Defined N_j as the number of days of the j-month and I_0 as a generic reference value of the daily rainfall, the dimensionless random variables,

$$X_{i,j} = \frac{H_{i,j}}{N_j I_0} \tag{1}$$

indicate a sequence that can be described as a discrete cyclostationary parameter stochastic process with period P equal to 12 (months).

The model proposed in this paper is useful to describe pluviometric regimes in which there is a scarce probability that the$H_{i,j}$ values should be equal to zero. With this condition, it can be assumed that, for $X_{i,j} > 0$ and, by adopting the transformation $Y_{i,j} = X_{i,j}^{\lambda}$ for $\lambda > 0$, the sequence of random variables Z_k:

$$Z_k = \frac{Y_{i,j} - \mu_{Y,j}}{\sigma_{Y,j}} \text{ with } k = 12i + j \tag{2}$$

is a standardized stationary Gaussian stochastic process.

The two functions $\mu_{Y,j} = E_i(Y_{i,j})$ and $\sigma_{Y,j} = \sqrt{E_i\left(Y_{i,j}^2\right) - \mu_{Y,j}^2}$ are appointed to the deseasonalization of the monthly rainfall process, while the transformation $Y_{i,j} = X_{i,j}^{\lambda}$ is finalized to the Gaussianization of the same process. The mean μY,jand the variance σ²Y,j functions can be described by means of the truncated Fourier series:

$$\mu_{Y,j} \cong \tilde{\mu}_{Y,j} = \frac{1}{2}a_0^{(\mu)} + \sum_{m=1}^{N_h^{(\mu)}} \left[a_m^{(\mu)}\cos\left(\frac{\pi}{6}jm\right) + b_m^{(\mu)}\sin\left(\frac{\pi}{6}jm\right)\right] \tag{3}$$

$$\sigma_{Y,j}^2 \cong \tilde{\sigma}_{Y,j}^2 = \frac{1}{2}a_0{}^{(\sigma^2)} + \sum_{m=1}^{N_h^{(\sigma^2)}} \left[a_m^{(\sigma^2)}\cos\left(\frac{\pi}{6}jm\right) + b_m^{(\sigma^2)}\sin\left(\frac{\pi}{6}jm\right)\right] \tag{4}$$

Where $N_h^{(\mu)}$ and $N_h^{(\sigma^2)}$ are the number of harmonics used for the mean and variance functions, respectively, and $a_0^{(\cdot)}, a_m^{(\cdot)}, b_m^{(\cdot)}$ are the Fourier coefficients for each of the mean and variance functions. Given a fixed value of the parameter λ and considering a period of *N* years, on the basis of a sample of observed monthly rainfall, $h_{i,j}$, it is possible to evaluate a transformed sample $y_{i,j} = x_{i,j}^{\lambda}$. The parameters $a_0^{(\mu)}, a_m^{(\mu)}, b_m^{(\mu)}$ of the function $N_h^{(\mu)}$, for a fixed number of harmonics N(μ)h, can then be estimated with the use of the least squares method by minimizing the function expressed by the following equation:

$$S_\mu^2\left(a_0^{(\mu)}, a_1^{(\mu)}, b_1^{(\mu)}, \ldots\right) = \sum_{i=0}^{N-1}\sum_{j=1}^{12}\left[y_{i,j} - \frac{1}{2}a_0^{(\mu)} - \sum_{m=1}^{N_h^{(\mu)}} f_\mu\left(j, m; a_m^{(\mu)}, b_m^{(\mu)}\right)\right]^2 \tag{5}$$

where $f_\mu\left(j, m; a_m^{(\mu)}, b_m^{(\mu)}\right) = a_m^{(\mu)}\cos(jm\pi/6) + b_m^{(\mu)}\sin(jm\pi/6)$.

The estimation of the coefficients for the truncated Fourier series expansion of the mean function can be performed by solving a system of $1 + 2N_h^{(\mu)}$ linear algebraic equations in $1 + 2N_h^{(\mu)}$ unknowns, obtained by equaling to zero the partial derivatives of $S_\mu^2\left(a_0^{(\mu)}, a_1^{(\mu)}, b_1^{(\mu)}, \ldots\right)$, with respect to $a_0^{(\mu)}, a_m^{(\mu)}, b_m^{(\mu)}$. Rarely, *i.e.* when the temporal span of the sample is a multiple of the period P and when the series do not have any missing monthly data, the trigonometric interpolation theory provides the estimations of the coefficients $a_0^{(\mu)}, a_m^{(\mu)}, b_m^{(\mu)}$ in explicit form.

An analogous procedure can be performed for the estimation of the function $\tilde{\sigma}_{Y,j}^2$, for a fixed number of harmonics $N_h^{(\sigma^2)}$.

The estimation of the number of harmonics to be used for the mean and for the variance should be done by testing both the hypotheses $H_0^{(\mu)}$ and $H_0^{(\sigma^2)}$ that the sample mean and variance $m_{Y,j}$ and $s_{Y,j}^2$, derive from statistical Gaussian universes with mean and variance respectively equal to $\tilde{\mu}_{Y,j}$ and $\tilde{\sigma}_{Y,j}^2$. Thus, by assuring a parameter parsimony criterion, and given a fixed value of the parameter λ, the number of harmonics can be detected as the smallest numbers that give values of $\tilde{\mu}_{Y,j}$ and $\tilde{\sigma}_{Y,j}^2$, for which $H_0^{(\mu)}$ and $H_0^{(\sigma^2)}$ cannot be rejected with a specified significance level α.

The extremes of the non-rejection intervals of the tests can be easily evaluated [96]. In fact, given that $Y_{i,j}$ are Gaussian variables, the distributions of the sample means follow a Student's law and the sample variances are distributed following a gamma distribution, with scale and shape parameters equals to 2 and $(N_j - 1)/2$, respectively.

The procedure for identifying the number of harmonics, which must be used for the mean and the variance, has to be repeatedly applied in order to evaluate the values λ^ of the parameter λ, which makes Gaussian the random variable Z_k and, more specifically, minimizes the function expressed by the following equation:

$$S_G^2(\lambda) = \frac{g_{1,Z}^2(\lambda)}{\sigma_{g_{1,z}}^2} + \frac{[g_{2,Z}(\lambda) - 3]^2}{\sigma_{g_{2,z}}^2} \tag{6}$$

where $g_{1,Z}$ and $g_{2,Z}$ are the sample skewness and kurtosis coefficients, evaluated on the basis of the N_K observed values of Z_k, while $\sigma_{g_{1,z}}^2$ and $\sigma_{g_{2,z}}^2$ indicate their sample variances:

$$\sigma_{g_{1,z}}^2 = 6N_K(N_K - 1)/[(N_K - 2)(N_K + 1)(N_K + 3)] \tag{7}$$

$$\sigma_{g_{2,z}}^2 = 24N_K(N_K - 1)^2/\left[(N_K - 2)\left(N_K^2 - 9\right)(N_K + 5)\right] \tag{8}$$

The values of skewness, $g_{1,Z}(\hat{\lambda})$, and kurtosis, $g_{2,Z}(\hat{\lambda})$ can then be used to test the Gaussianity of the process of theZ_t. The Gaussianity hypothesis cannot be rejected if $g_{1,Z}(\hat{\lambda})$ and $g_{2,Z}(\hat{\lambda})$ fall within the intervals:

$$\left(-g_{1,Z;1-\alpha/2}^{(N_K)}; g_{1,Z;1-\alpha/2}^{(N_K)}\right) \quad (9)$$

$$\left(g_{2,Z;\alpha/2}^{(N_K)}; g_{2,Z;1-\alpha/2}^{(N_K)}\right) \quad (10)$$

Where $g_{1,Z;1-\alpha/2}^{(N_K)}$ is the percentile 1-α/2 of the sample skewness distribution of the standardized Gaussian law with dimension N_K, while $g_{2,Z;\alpha/2}^{(N_K)}$ and $g_{2,Z;1-\alpha/2}^{(N_K)}$ are the percentiles $\alpha/2$ and $1-\alpha/2$ of the sample kurtosis distribution of the standardized Gaussian law with dimension N_K.

The required percentiles can be evaluated through Monte Carlo simulation techniques after having analyzed the correlative features of the variable Z_k.

Correlative Structure

Usually, the sequence of random variables Z_k with k = 1, 2, ..., shows a weak correlative structure which, if significant, can be modeled as an autoregressive process of order p. Being Z_k standardized Gaussian variables, and given a white noise standardized Gaussian process W_k, then:

$$Z_k = \psi_0 W_k + \sum_{l=1}^{p} \phi_l Z_{k-l} \quad (11)$$

Using the sample values $r_{Z,l}$ of the autocorrelation coefficients of lag l = 1, ..., p of the sequence Z_k, it is possible to obtain the estimations ф^l of the parameters φ_l by solving the Yule-Walker system, and to estimate $\hat{\psi}_0 = \sqrt{1-\sum_{l=1}^{p}\hat{\phi}_l r_{Z,l}}$ [97].

The p-order of the autoregressive process can be fixed as the minimum value for which cannot be rejected the hypothesis $H_{0,p}^{(\rho_\nu)}$ that the sample biases $w_{p,k} = \left(z_k - \sum_{l=1}^{p}\hat{\phi}_l z_{k-l}\right)/\hat{\psi}_0$, with $k = p + 1$, $p + 2$, are uncorrelated for lag ν = 1, 2, ... The hypothesis $H_{0,p}^{(\rho_\nu)}$ can be tested, at a significance level α, by using the Anderson test [98,99]. The test must be applied also in the case p = 0, for which $W_{0,k} = Z_k$ and $w_{0,k} = z_k$, in order to verify the hypothesis that the process Z_k can be considered as white noise.

CASE STUDY

The study area (the region of Calabria) is a peninsula with an oblong shape occupying the southern part of Italy. It has a surface of 15,080 km^2 and a coastline of 738 km on the Ionian and Tyrrhenian Seas. Calabria has a length of about 250 km and a width ranging between 31 and 111 km. The maximum elevation is 2267 m a.s.l., while the average elevation is 597 m a.s.l. [100].

Because of its geographic position and mountainous nature, Calabria has a high climatic variability with a typically dry summer subtropical climate, also known as the Mediterranean climate [101]. Its coastal zones are characterized by mild winters and hot summers with little precipitation. In particular, the Ionian side, which is influenced by currents coming from Africa, has high temperatures with short and heavy precipitation, while the Tyrrhenian side is influenced by western air currents and presents milder temperatures and considerable orographic precipitation. In the inland zones there are colder winters (with snow) and fresher summers (with some precipitation) than those marking the coastal zones [102].

Monthly precipitation data have been collected and published in the region of Calabria by the former Italian Hydrographic Service. In this work, particular attention has been given to the problems arising from the low quality and inhomogeneities of the data series. Thus, the monthly database used in the further analysis was a part of the high-quality one presented in a previous study [103], in which a multiple application of the Craddock test [104] to detect the inhomogeneities was performed. In particular, in the present work, a set of 12 monthly total precipitation series, which were found to be homogeneous for the period 1916–2011, were selected. In Figure 1, the study area and the characterization of the rainfall series through box plots are presented. The main features of the rain gauges are shown in Table 2, together with the percentages of missing data. The percentage of gaps in the series ranges between 1.5 and 12.4%, and is mostly detected during the Second World War period.

Table 2. Main features of the selected rain gauges.

Code	Rain Gauge	Longitude (Degrees)	Latitude (Degrees)	Elevation (m a.s.l.)	First Year	Last Year	Missing Data (%)
970	Cassano allo Ionio	16.319	39.783	250	1921	2011	10.4
1010	Cosenza	16.265	39.287	242	1916	2011	3.5
1030	San Pietro in Guarano	16.314	39.346	660	1922	2011	2.5
1100	Cecita ex Acquacalda	16.538	39.400	1180	1923	2011	7.1
1700	Isola di Capo Rizzuto	17.094	38.961	90	1922	2011	10.6
2310	Capo Spartivento	16.056	37.927	48	1921	2011	12.4
2510	Scilla	15.720	38.252	73	1939	2011	2.3
2600	Cittanova	16.078	38.352	407	1916	2011	11.5
2760	Joppolo	15.905	38.592	185	1929	2011	2.3
2830	Filadelfia	16.293	38.787	550	1920	2011	3.3
2890	Tiriolo	16.510	38.940	690	1941	2011	2.9
3160	Campotenese C.C.	16.068	39.873	965	1922	2011	1.5

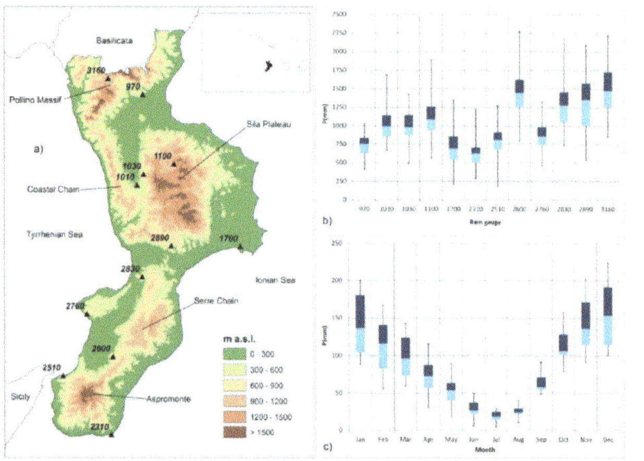

Figure 1. Localization of the selected rain gauges on a DEM of the region of Calabria (**a**) and characterization of the rainfall series through box plots: (**b**) annual rainfall for each rain gauge; (**c**) monthly rainfall distribution for the whole set of rain gauges. The bottom and top of the box are the first and third quartiles, the band inside the box is the median, the ends of the whiskers represent the minimum and maximum of all of the data.

Estimation of the Model Parameters

All the monthly total rainfall series present some months with no data. For this reason, the estimation of the truncated

Fourier coefficients for mean and variance functions have been made, as mentioned in Section 2.2, with the use of the least squares method by solving systems of linear algebraic equations, assuming I_0 = 1 mm. Instead, in the procedure for the evaluation of the parameter λ, the minimum of the function in Equation (6) has been evaluated by applying the Brent algorithm, which was preceded by the search of a bracketing interval for the minimum [105].

For the 12 rain gauges selected in this work, Table 3 and Table 4 report the estimated values of λ^, of the number of harmonics N(μ)h and N(σ2)h, and of the Fourier coefficients. As to what concerns the mean function (Table 3), 2 harmonics have been evaluated for all the rain gauges with the exception of the Capo Spartivento rain gauge, for which 3 harmonics are needed.

Table 3. Coefficient λ^ and Fourier coefficients for the mean function $\tilde{\mu}_{Y,j}$.

Rain Gauge	$\hat{\lambda}$	$N_h^{(\mu)}$	$\hat{a}_0^{(\mu)}/2$	$\hat{a}_1^{(\mu)}$	$\hat{b}_1^{(\mu)}$	$\hat{a}_2^{(\mu)}$	$\hat{b}_2^{(\mu)}$	$\hat{a}_3^{(\mu)}$	$\hat{b}_3^{(\mu)}$
Campotenese	0.466	2	1.730	0.649	0.265	0.009	−0.203	—	—
Capo Spartivento	0.338	3	0.962	0.504	0.089	−0.080	−0.151	−0.021	0.050
Cassano allo Jonio	0.473	2	1.256	0.473	0.158	−0.022	−0.121	—	—
Cecita	0.442	2	1.448	0.543	0.251	0.011	−0.129	—	—
Cittanova	0.401	2	1.551	0.587	0.246	−0.026	−0.199	—	—
Cosenza	0.477	2	1.429	0.637	0.248	0.005	−0.183	—	—
Filadelfia	0.477	2	1.626	0.675	0.274	0.027	−0.184	—	—
Isola Capo Rizzuto	0.315	2	1.000	0.450	0.135	−0.020	−0.167	—	—
Joppolo	0.498	2	1.361	0.620	0.169	−0.006	−0.196	—	—
S.Pietro in Guarano	0.513	2	1.480	0.692	0.278	0.011	−0.214	—	—
Scilla	0.484	2	1.295	0.553	0.090	−0.080	−0.175	—	—
Tiriolo	0.405	2	1.451	0.591	0.210	−0.019	−0.138	—	—

Table 4. Fourier coefficients for the variance function.

Rain Gauge	$N_h^{(\sigma^2)}$	$\hat{a}_0^{(\sigma^2)}/2$	$\hat{a}_1^{(\sigma^2)}$	$\hat{b}_1^{(\sigma^2)}$	$\hat{a}_2^{(\sigma^2)}$	$\hat{b}_2^{(\sigma^2)}$	$\hat{a}_2^{(\sigma^2)}$	$\hat{b}_2^{(\sigma^2)}$
Campotenese	2	0.345	0.112	0.020	0.010	0.048	—	—
Capo Spartivento	2	0.140	0.024	−0.036	−0.029	−0.017	—	—
Cassano allo Jonio	2	0.205	0.039	−0.010	0.003	0.041	—	—
Cecita	2	0.232	0.051	−0.002	0.019	0.015	—	—
Cittanova	1	0.224	0.005	−0.026	—	—	—	—
Cosenza	2	0.259	0.082	0.024	−0.007	0.051	—	—
Filadelfia	3	0.353	0.024	0.008	0.037	0.022	−0.058	0.005
Isola Capo Rizzuto	3	0.157	0.014	−0.062	−0.035	0.006	−0.017	−0.001
Joppolo	1	0.282	0.033	−0.004	—	—	—	—
S.Pietro in Guarano	2	0.310	0.094	0.032	−0.006	0.031	—	—
Scilla	1	0.241	0.015	−0.017	—	—	—	—
Tiriolo	2	0.236	0.045	0.021	−0.010	0.029	—	—

Different results have been obtained for the variance function (Table 4). In fact, 1 harmonic has been evaluated for 3 out of 12 rain gauges, 2 harmonics for 7 rain gauges, and 3 harmonics for the other 2 rain gauges. Moreover, the comparisons among the observed values, $m_{Y,j}$ and $s^2_{Y,j}$, the modeled values, $\tilde{\mu}_{Y,j}$ and $\tilde{\sigma}^2_{Y,j}$, and the dimension of the non-rejection intervals of the hypotheses H(μ)0 and H(σ2)0 are shown in Figure 2 for the Cosenza and Campotenese C.C. rain gauges.

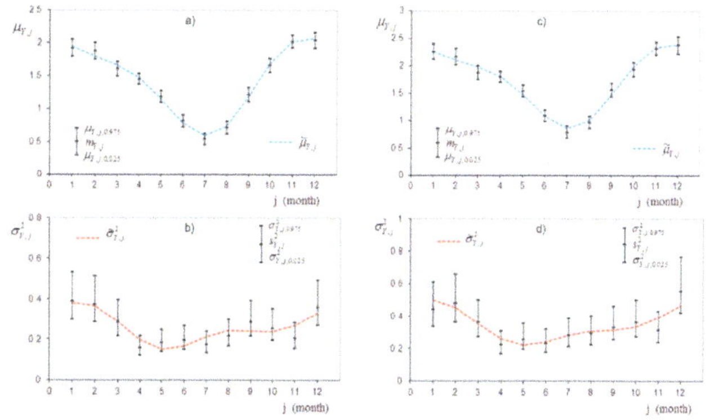

Figure 2. Comparison among the observed values, the modeled values, and the ranges (identified by the whiskers of the error bar) of the non-rejection regions of the hypotheses H(μ)0 and H(σ2)0, for the mean (blue) and the variance (red) functions, evaluated for the Cosenza (**a**,**b**) and the Campotenese C.C. (**c**,**d**) rain gauges.

By using the estimated values of the parameters, and by means of Equation (2), it was possible to build the observed sample z_k, with zero mean and unit variance, and the sample coefficients of skewness, $g_{1,Z}$, and kurtosis, $g_{2,Z}$. These values have been used to test the Gaussianity of the process Z_t, by verifying the non-rejection of the hypotheses $H_0^{(g_1)}$ and $H_0^{(g_2)}$ (Table 5). Moreover, the comparisons between the cumulative frequency of the observed values and the cumulative distribution function (cdf) of the standardized Gaussian law have been reported in Figure 3, for the Cecita ex Acquacalda and Scilla rain gauges. The non-rejection of the hypotheses $H_0^{(g_1)}$ and $H_0^{(g_2)}$ and the good fitting with the observed values (Figure 3) confirm the goodness of the Gaussianization of the Z_k process.

Table 5. Gaussianity test based on the coefficients of skewness and kurtosis.

Rain Gauge	N_K	$g_{1,Z}$	$\pm g_{1,Z,0.975}^{(N_K)}$	$g_{2,Z,0.025}^{(N_K)}$	$g_{2,Z}$	$g_{2,Z,0.975}^{(N_K)}$
Campotenese	1055	−0.006	±0.148	2.731	2.813	3.315
Capo Spartivento	957	−0.005	±0.155	2.718	2.857	3.333
Cassano allo Jonio	969	−0.005	±0.154	2.720	2.746	3.331
Cecita	985	0.012	±0.153	2.722	3.114	3.328
Cittanova	1020	−0.001	±0.151	2.726	3.164	3.321
Cosenza	1112	−0.028	±0.144	2.738	2.814	3.304
Filadelfia	1067	−0.001	±0.147	2.732	2.994	3.313
Isola Capo Rizzuto	962	−0.013	±0.155	2.719	2.828	3.332
Joppolo	968	0.008	±0.154	2.720	3.184	3.331
S.Pietro in Guarano	1047	0.005	±0.149	2.730	3.046	3.316
Scilla	857	0.000	±0.162	2.706	2.951	3.352
Tiriolo	827	0.001	±0.165	2.702	3.021	3.357

For all the rain gauges, the sequences of observed values z_k show low linear correlation coefficients, but they are not low enough to consider the process Z_k uncorrelated. In fact, the application of the Anderson test, with a lag $\nu_{max} = 24$, to the $z_k$series, evidenced that the hypothesis of uncorrelated process is non-rejectable only for the Cassano allo Jonio, Capo Spartivento and Joppolo rain gauges.

For the other 9 rainfall series, in order to describe the correlative structure, it was sufficient to adopt an autoregressive model of order $p = 1$. In fact, after the estimation of the parameters ϕ^1=r1,Z and $\hat{\psi}_0 = \sqrt{1 - r_{1,Z}^2}$, the application of the Anderson test, always till a lag $\nu_{max} = 24$, to the sequence of the sample bias w1,k, showed the non-rejectability of the hypothesis of uncorrelated process. Table 6 shows a synthesis of the application of the Anderson test to the selected 12 rain gauges, limited only to the lag $\nu = 1$.

Table 6. Anderson test applied to the autocorrelation coefficient of lag $\nu = 1$.

Rain Gauge	$r_{Z,1,0.025}^{(N_{K,0})}$	$r_{Z,1}$	$r_{Z,1,0.975}^{(N_{K,0})}$	$H_{0,0}^{(P_1)}$	$r_{W_1,1,0.025}^{(N_{K,1})}$	$r_{W_1,1}$	$r_{W_1,1,0.975}^{(N_{K,1})}$	$H_{0,1}^{(P_1)}$
Campotenese	−0.062	0.070	0.060	rejected	−0.062	−0.004	0.060	not rejected
Capo Spartivento	−0.065	0.005	0.063	not rejected	—	—	—	—
Cassano allo Jonio	−0.065	0.042	0.063	not rejected	—	—	—	—
Cecita	−0.064	0.067	0.062	rejected	−0.064	0.003	0.062	not rejected
Cittanova	−0.063	0.113	0.061	rejected	−0.063	−0.004	0.061	not rejected
Cosenza	−0.060	0.065	0.058	rejected	−0.060	0.004	0.058	not rejected
Filadelfia	−0.062	0.117	0.060	rejected	−0.062	−0.008	0.060	not rejected
Isola Capo Rizzuto	−0.065	0.073	0.063	rejected	−0.065	0.007	0.063	not rejected
Joppolo	−0.064	0.061	0.062	not rejected	—	—	—	—
S.Pietro in Guarano	−0.062	0.097	0.060	rejected	−0.062	−0.011	0.060	not rejected
Scilla	−0.069	0.154	0.066	rejected	−0.069	−0.017	0.066	not rejected
Tiriolo	−0.070	0.103	0.068	rejected	−0.070	−0.013	0.068	not rejected

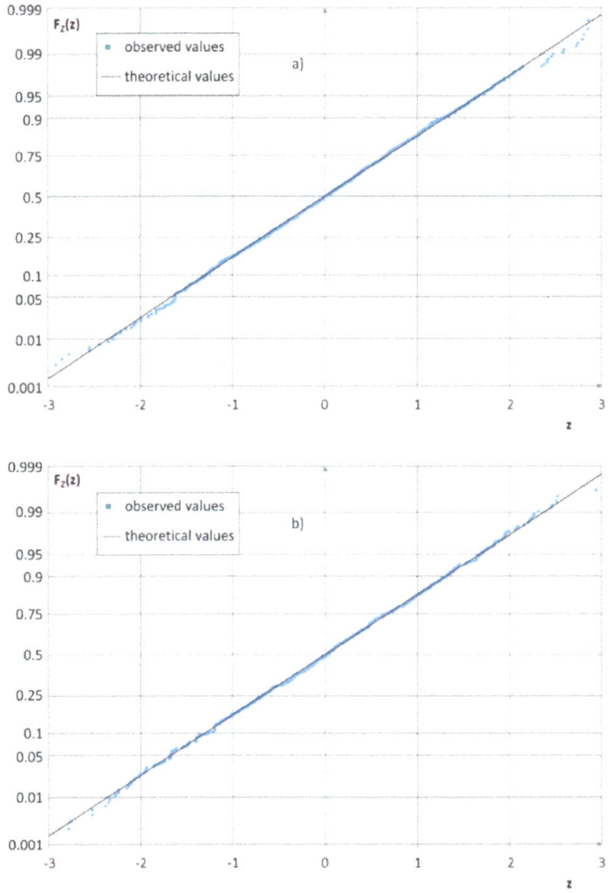

Figure 3. Comparison between the cumulative frequency of the observed values and the cdf of the standardized Gaussian law for the Cecita ex Acquacalda (**a**) and Scilla (**b**) rain gauges.

RESULTS AND DISCUSSION

After the parameter estimation by means of the selected rainfall database, 10^4 years long synthetic series have been generated for each rain gauge through a Monte Carlo procedure, reproducing the proposed model.

Then, the SPI was evaluated at two different time scales: 3-month SPI, which reflects short- and medium-term moisture conditions, and 12-month SPI, which is linked to long-term precipitation patterns and can impact on streamflows, reservoir levels, and even groundwater levels. The results obtained in terms of the occurrence probabilities of wet and dry conditions were compared with the class probability shown in Table 1.

In Figure 4, the results of the 12-month SPI for two rain gauges are shown (Isola di Capo Rizzuto and Capo Spartivento). In this figure, by means of the simulated data, the SPI values corresponding to the probability of extreme (2.3%), severe (4.4%) and moderate (9.2%) conditions have been detected for each month. As a result, the probabilities proposed in another publication [86] in the evaluation of the SPI were quite different from the ones obtained by the application of the SPI to the 10^4 simulated years. For example, for the Isola di Capo Rizzuto rain gauge (Figure 4a), the occurrence probabilities of the extreme values confirmed the correspondent SPI values previously proposed [86], while the severe and moderate probabilities values showed a different behavior. Similarly, for the Capo Spartivento rain gauge (Figure 4b), the extreme and moderate probabilities differed from the correspondent SPI values proposed in Table 1, while only the severe probabilities agree with them. This is very important, since for most practical applications the SPI is essential in terms of classification (Table 1) and not as arithmetic values [87].

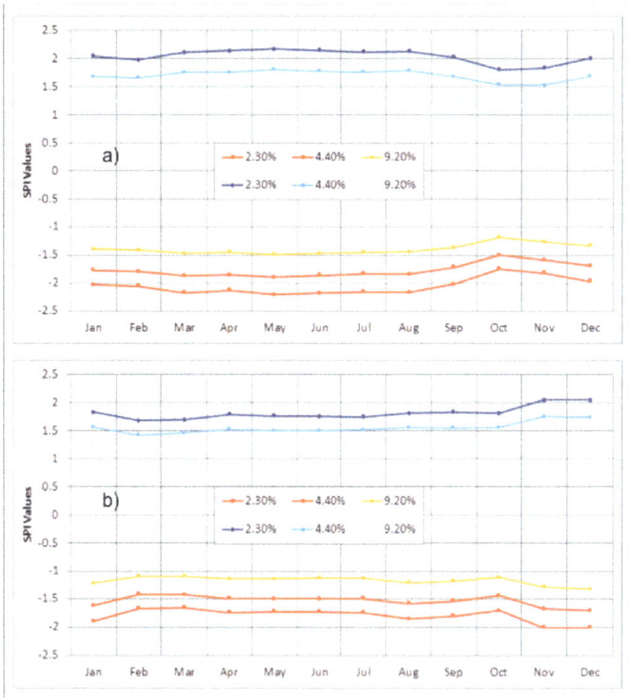

Figure 4. Values of the 12-month SPI for fixed monthly occurrence probabilities of wet and dry conditions, evaluated for the simulated data of the Isola di Capo Rizzuto (**a**) and the Capo Spartivento (**b**) rain gauges. According to the theoretical values proposed in another publication [86], each color identifies the prefixed value of the occurrence probability of wet (blue scale) and dry (red scale) conditions.

As regards to the 3-month-SPI, Figure 5 presents some box plots that summarized the regional monthly occurrence probabilities of the different wet and dry classes. In particular, the box plots corresponding to extreme, severe and moderate dry and wet conditions have been shown together with the theoretical values proposed by [86]. From these analyses it has emerged that in Calabria, as in other Italian regions [43], the highest probability to detect extreme drought (Figure 5a) or severe drought (Figure 5b) conditions, were mainly observed in the wet seasons and, in particular, in the autumn period. In fact, the autumn and the spring months showed the highest values of the mean occurrence probabilities of drought, while the minimum values have been detected in August.

Moreover, in the summer period, the occurrence probabilities of drought conditions showed a high regional variability, with large difference values between the several rain gauges. Conversely, the highest probability to detect extremely wet (Figure 5d) or severely wet (Figure 5e) conditions, were mainly observed in winter. No remarkable differences emerged for the moderate dry (Figure 5c) and wet (Figure 5f) conditions.

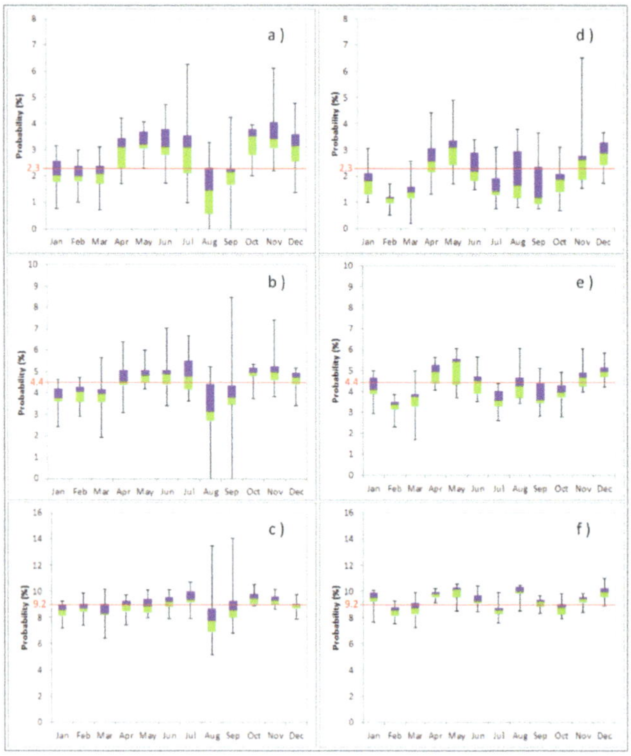

Figure 5. Box-plots of the regional monthly occurrence probabilities of extreme drought (a); severe drought (b); moderate drought (c); extremely wet (d), severely wet (e) and moderately wet (f) conditions for the 3-month SPI. The red lines indicate the theoretical values proposed by [86]. The bottom and top of the box are the first and third quartiles, the band inside the box is the median, the ends of the whiskers represent the minimum and maximum of all of the data.

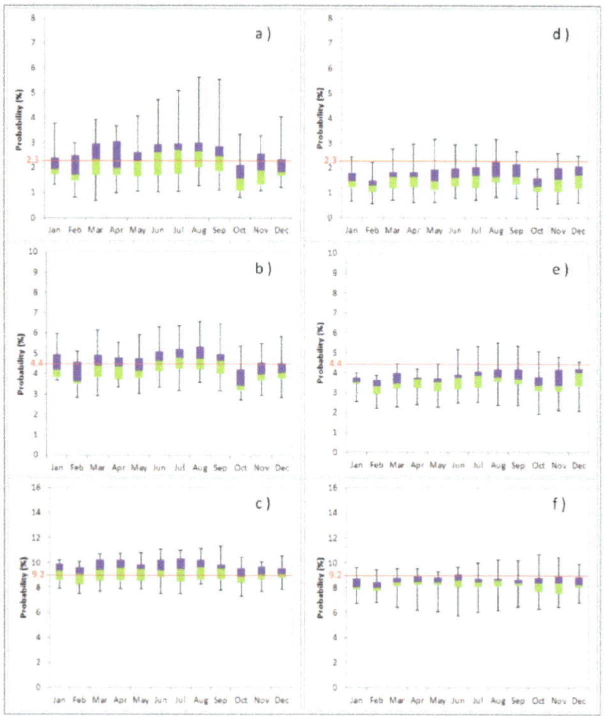

Figure 6. Box-plots of the regional monthly occurrence probabilities of extreme drought (a); severe drought (b); moderate drought (c), extremely wet (d); severely wet (e) and moderately wet (f) conditions for the 12-month SPI. The red lines indicate the theoretical values proposed by [86]. The bottom and top of the box are the first and third quartiles, the band inside the box is the median, the ends of the whiskers represent the minimum and maximum of all of the data.

With regards to the comparison between the occurrence probabilities of wet and dry conditions evaluated from the simulated data and the class probability shown in Table 1, the different dry and wet mean occurrence values generally fluctuate around the values proposed by [86], but higher probability of extreme drought (Figure 5a) can be observed.

Similarly to Figure 5, also for the 12-month SPI, the box plots with the regional monthly occurrence probabilities of the different wet and dry classes have been evaluated (Figure 6). As opposed to Figure 5, in these box plots there are few differences between the occurrence probabilities evaluated in the different months, in particular for the dry conditions. In fact, highest probabilities to detect extreme drought (Figure 6a) or severe drought (Figure 6b) conditions can be observed in the summer period. Moreover, only for the extreme drought (Figure 6a) and in the summer period, the occurrence probabilities of drought conditions showed a high regional variability. The most important result of the 12-month SPI analysis emerged from the comparison between the occurrence probabilities of wet and dry conditions evaluated from the simulated data and the class probability shown in Table 1. In fact, the extreme (Figure 6a) and the severe drought (Figure 6b) mean occurrences values generally fluctuate around the values proposed by [86], and the moderate drought (Figure 6c) presented slightly higher mean occurrences values than these. By contrast, the wet conditions (Figure 6d–f) always showed lower mean occurrences values than those presented in Table 1.

These results indicate greater probability of dry conditions than wet conditions when long-term precipitation patterns are considered, with consequences on streamflows, reservoir levels, and groundwater levels.

Generally, from the box plots of Figure 5 and Figure 6, it can be observed that the 3-month SPI does not greatly vary in comparison to the 12-month SPI. The only exception are the box plots which refer to the extreme wet conditions (Figure 5d and Figure 6d) where, in 10 out of 12 months, the probability of the 3-month SPI evidenced a greater spread than the 12-month SPI. These results agree with the ones shown in Figure 7 and with past studies in the region of Calabria [41,42], which evidenced that there is a great spatial heterogeneity of the SPI12 while the SPI3 shows a spatial homogeneity. Moreover, as a further result, the model spreads higher in summer than in the other seasons. This result can be due to the Mediterranean climate of the region, which shows high climatic variability with a typically dry summer subtropical climate. The rain gauges with the higher spread in summer are located in the Ionian side of the region, which is influenced by currents coming from Africa and is characterized by short and heavy precipitation in particular in the summer period.

Figure 7 shows the box plots with the occurrence probabilities of the different wet and dry classes evaluated for the whole data series.

In particular, for the 3-month SPI (Figure 7a), there are no marked differences between the evaluated mean occurrences probabilities and the ones previously proposed [86], and the same happens between the occurrence probability of the corresponding dry/wet classes. Instead, for the 12-month SPI

(Figure 7b) strong differences between the evaluated mean occurrences probabilities and the ones shown in Table 1 have been detected, with the drought and wet conditions which present respectively higher and lower values than the ones previously proposed [86].

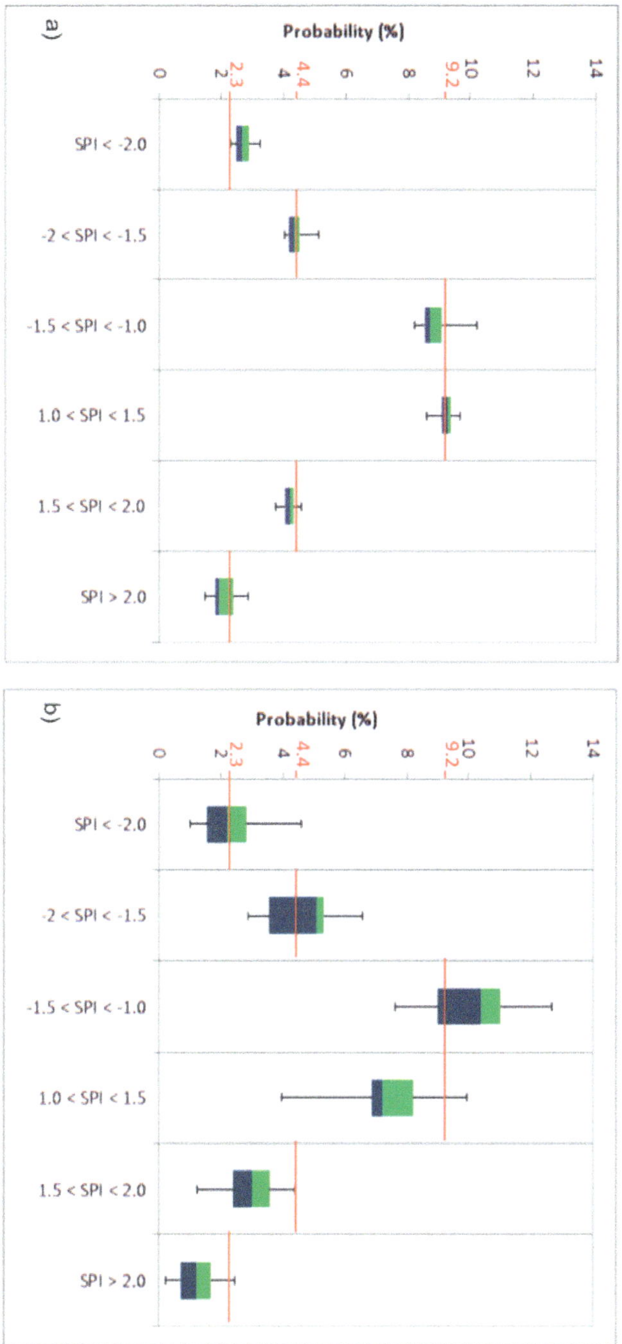

Figure 7. Box-plots of the regional occurrence probabilities of extreme, severe and moderate drought and wet conditions for the 3-month SPI (**left**) and the 12-month SPI (**right**). The lines indicate the theoretical values proposed by [86]. The bottom and top of the box are the first and third quartiles, the band inside the box is the median, the ends of the whiskers represent the minimum and maximum of all of the data.

Detailed results for each rain gauge have been shown in Table 7. The results confirm more chance for dry conditions than wet conditions when long-term precipitation patterns are considered. In fact, for almost all the rain gauges, the lowest probability values have been detected for the wet conditions, while the highest probability values have been detected for the dry conditions. For example, for the Cosenza rain gauge (code 1010), the probabilities that the 12-month SPI belong to the extreme classes (SPI < −2 and SPI > 2) are 4.2% and 2.0%, respectively. The same happens for the severe SPI classes (−2 < SPI < −1.5 and 1.5 < SPI < 2) where dry conditions show higher probabilities (5.9%) than wet conditions (4.0%). A similar behavior is shown by the 3-month SPI, even though with lower marked differences between dry and wet conditions (Table 7). These results, which evidence the highest probability values evaluated for the dry conditions, are visually described inFigure 8. The different colors of the figure represent several probability classes. In fact, the colors associated to the ranges with higher probabilities can be observed only for dry conditions. On the contrary, the lower probability classes can be identified mainly for the wet conditions.

Table 7. Probability of the simulated data to fall within each SPI class evaluated for each rain gauge.

Months	SPI Class	970	1010	1030	1100	1700	2310	2510	2600	2760	2830	2890	3160
3	SPI < −2	2.6	3.0	3.3	2.9	2.3	2.7	2.5	2.7	2.3	2.3	2.9	2.6
	−2<SPI< −1.5	4.2	4.3	5.1	4.5	4.1	4.6	4.0	4.2	4.0	4.9	4.5	4.4
	−1.5 < SPI < −1	8.4	8.7	9.1	9.1	10.2	9.2	8.5	8.6	8.2	8.8	9.1	8.5
	1 < SPI < 1.5	9.4	9.3	9.5	9.0	8.6	9.1	9.2	9.4	9.3	9.7	9.0	9.3
	1.5 < SPI < 2	4.2	4.3	4.6	4.0	4.3	4.3	4.0	4.3	3.8	4.6	4.0	4.1
	SPI > 2	1.7	1.9	2.4	1.9	2.7	2.9	1.5	2.4	1.5	2.4	2.0	1.8
12	SPI < −2	1.7	4.2	3.1	2.4	2.7	1.7	2.2	1.3	2.4	2.2	1.9	1.3
	−2 < SPI < −1.5	3.6	5.9	5.1	4.7	4.9	4.1	4.2	3.5	4.3	4.0	4.4	3.5
	−1.5 < SPI < −1	8.5	10.6	9.8	9.9	10.1	9.3	9.2	8.0	9.1	8.0	10.1	8.5
	1 < SPI < 1.5	8.8	8.3	8.5	8.0	8.0	8.2	8.2	8.9	9.0	9.9	6.4	7.7
	1.5 < SPI < 2	3.5	4.0	3.9	3.3	3.8	3.8	3.3	3.5	3.8	4.7	2.3	2.7
	SPI > 2	1.2	2.0	1.6	1.4	2.5	1.9	1.2	1.5	1.7	2.1	0.7	0.9

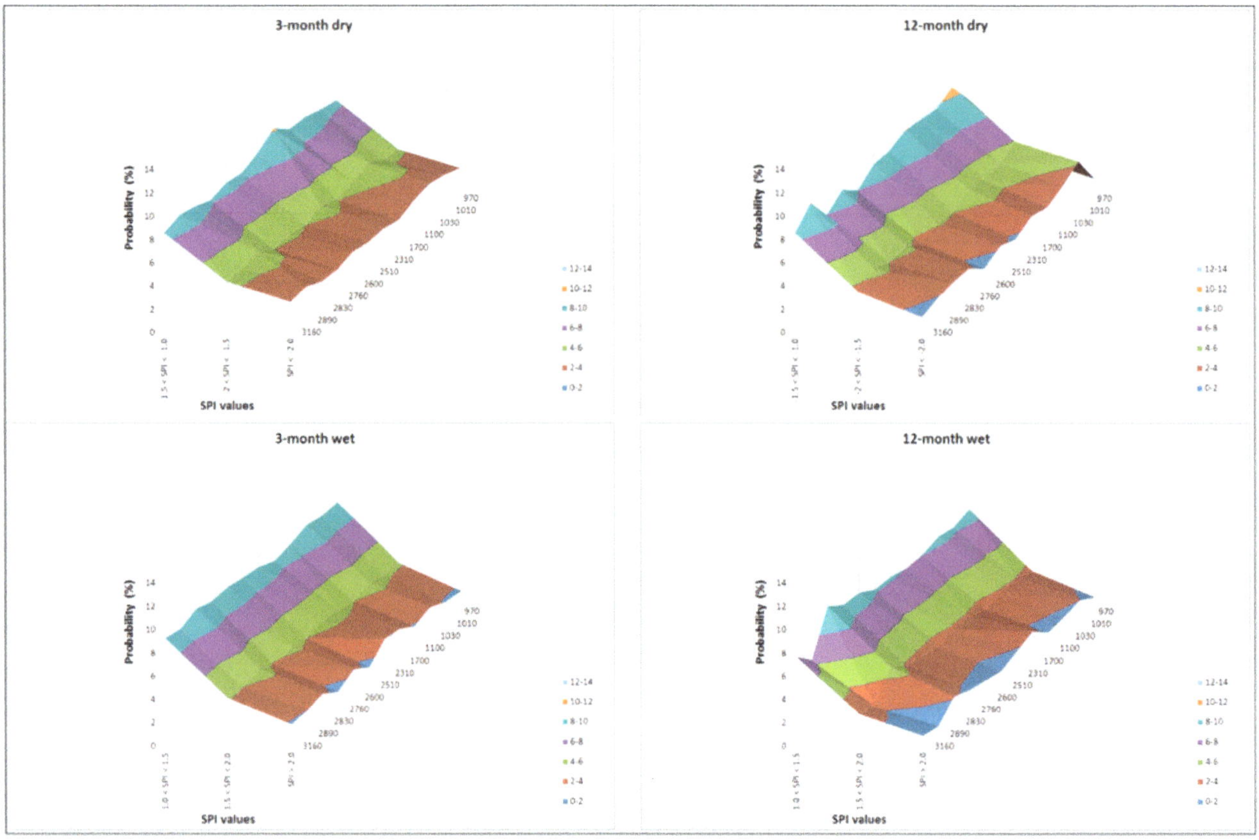

Figure 8. Occurrence probabilities of extreme, severe and moderate drought and wet conditions for the 3-month SPI and the 12-month SPI for each rain gauge.

Table 8. Percentages of the simulated monthly data to fall within each DSI class evaluated for each rain gauge.

DSI	DSI Class	970	1010	1030	1100	1700	2310	2510	2600	2760	2830	2890	3160
	Negative	0.5	0.4	0.4	0.5	0.5	0.5	0.3	0.4	0.4	0.4	0.3	0.3
DSI3	Positive	56.1	53.5	57.0	50.5	55.9	59.5	58.9	63.1	56.0	55.6	62.5	65.2
	Zero	43.4	46.0	42.6	49.0	43.6	40.0	40.7	36.6	43.6	44.0	37.1	34.5
	Negative	0.9	1.0	0.9	0.9	0.9	0.8	0.8	0.8	0.9	0.9	0.8	0.7
DSI6	Positive	52.0	48.3	52.7	43.5	46.9	54.5	55.0	59.3	50.5	49.3	59.7	64.0
	Zero	47.1	50.7	46.4	55.6	52.3	44.7	44.2	39.9	48.6	49.8	39.5	35.3

Some results obtained from the application of the SPI are confirmed by the DSI, evaluated for each rain gauge on two time scales (3 and 6 months) using the synthetic monthly precipitation series. Table 8 compares the two indices (DSI3 and DSI6) at each location, in terms of percentages of months for which the drought severity index was negative, positive or equal to zero (no rainfall deficit). As was the case in a previous publication [25], for all the rain gauges a six-monthly rule (DSI6) hinders the development of a drought sequence. In fact, the deficit emerges considering rainfall behavior over three, rather than six months with a maximum percentage decrease of 9% (ID 1700) passing from DSI3 to DSI6. The different results obtained with DSI6 also appear when DSI values equal to zero are considered. For DSI3 the percentages of months with positive values are always greater than those equal to zero. On the contrary, for DSI6 there are rain gauges for which months with DSI equal to zero are greater than those with positive values. Some of the DSI features confirm the results of SPI. For example, the probabilities of extreme drought for SPI on large scale (in this study, 12 month-SPI) are often lower than those relative to 3-month SPI. Moreover, some of the rain gauges with the highest probabilities of extreme drought are the same, showing high percentages with positive values of DSI3 (*i.e.*, ID 1030, 2310, 2600, 2890, 3160).

The results of this paper confirm that, due to the long-term trend of global warming, there is a higher chance of dry conditions than of wet conditions [106,107]. This is a critical issue for an agricultural region, such as Calabria, that suffers climate change [108,109,110], which is a major driver of agricultural and meteorological drought. In fact, precipitation and temperature anomalies caused by climate change will induce agro-meteorological drought [111]. In addition, water is usually significantly contaminated by organic matter derived from agricultural production and sewage, which may further aggravate the agricultural drought risk [112].

CONCLUSIONS

Unlike most natural hazards such as earthquakes and cyclones, both of which can strike quickly, drought does not usually have a sudden beginning or end. It is an insidious hazard caused by a period of abnormally dry weather, persisting long enough to produce a serious hydrologic imbalance. While a drought is unlikely to cause human deaths in most developed countries, a drought in a developing country without adequate access to aid can be devastating. In fact, drought can often be a natural hazard with the biggest economic impact, resulting in very costly and dramatic impacts on the environment such as stock losses, vulnerability to fires (especially in forested areas), crop damage, soil erosion, power blackouts if your community is reliant on electricity from hydro dams, and water supply shortages. The severity of a drought depends upon the degree of moisture deficiency, duration, and size of the affected area.

Monitoring dry and wet periods using meteorological indices, such as rainfall, is an essential component for drought preparedness. The variability of the rainfall is intrinsically present in its process, and it places uncertainties on projections. The use of the stochastic approach and the simulation procedure can effectively generate the variability of the rainfall process, and addresses the problem of quantity and reliability of the data used to fit the rainfall distribution. In the present paper, a model of monthly precipitation has been proposed. It adopts variable transformations, finalized to the deseasonalization and to Gaussianization of the monthly rainfall process, and includes a procedure for testing the autocorrelation. The model provided a good representation of the monthly rainfall for the selected 12 rain gauges of Calabria (Southern Italy). For this reason, 10^4-year-long synthetic series have been generated for each rain gauge through a Monte Carlo procedure, and dry and wet periods in the region of Calabria were analyzed using the SPI applied to the simulated series. The index was calculated at two different time scales: 3-month SPI, which reflects short and medium-term moisture conditions, and 12-month SPI, which is linked to long-term precipitation patterns and can have an impact on streamflows, reservoir levels, and even groundwater levels. The occurrence probabilities of extreme, severe and moderate wet and dry conditions have been evaluated for each month, and compared with the corresponding probability classes commonly adopted in the literature. The comparison evidenced some differences, in particular for the 12-month SPI, for which higher probability values for dry conditions and lower probability values for wet conditions have been detected. Analogous results, as further confirmation of the good representation provided by the proposed model, have been obtained by means of the application of the DSI, based on the accumulated monthly rainfall deficits evaluated through the precipitation anomalies.

The advantage of the proposed model is that the stochastic approach overcomes the problems of quantity and reliability of the data, and can well reproduce the statistical characteristics of observed data, thus allowing a better prediction of the occurrence probabilities of extreme dry/wet conditions. Moreover, the stochastic model has the advantage of being applicable everywhere and for any gauge, because it does not depend on station altitude and climatic zone. For its characteristics, the model can also be applied to precipitation monthly grids, but this application depends on the grid size. In fact, the whole procedure is highly time-consuming, and its application is not recommended for too detailed grid sizes. Anyway, the proposed model can be considered an attractive tool for management decision-making, allowing the identification of a drought risk. Finally, properly adapted, the model can also be applied together with the climate projection obtained from global circulation models as a reliable tool for drought estimation in a changing climate.

Author Contributions

Beniamino Sirangelo conceived and designed the stochastic modeling; Tommaso Caloiero analyzed the data and performed the applications; Roberto Coscarelli and Ennio Ferrari contributed to the results and discussion; all the Authors wrote the paper.

Conflicts of Interest

The authors declare no conflict of interest.

REFERENCES

1. Austin, R.B.; Cantero-Martínez, C.; Arrúe, J.L.; Playán, E.; Cano-Marcellán, P. Yield rainfall relationships in cereal cropping systems in the Ebro river valley of Spain. *Eur. J. Agron.* **1998**, *8*, 239–248.
2. Quiring, S.M.; Papakryiakou, T.N. An evaluation of agricultural drought indices for the Canadian prairies. *Agric. Forest Meteorol.* **2003**, *118*, 49–62.
3. Kogan, F. Droughts of the late 1980s in the United States as derived from NOAA Polar-Orbiting Satellite data. *Bull. Am. Meteorol. Soc.* **1995**, *76*, 655–668.
4. Kogan, F. Global drought watch from space. *Bull. Am. Meteorol. Soc.* **1997**, *78*, 621–636.
5. Abrams, M.D.; Ruffuer, M.C.; Morgan, T.A. Tree-ring responses to drought across species and contrasting sites in the ridge and valley of central Pennsylvania. *Forest Sci.* **1998**, *44*, 550–558.
6. Orwing, D.A.; Abrams, M.D. Variation in radial growth responses to drought among species site and canopy strata.*Trees* **1997**, *11*, 474–484.
7. Coscarelli, R.; Caloiero, T.; Minervino, I.; Sorriso-Valvo, M. Map of sensitivity to desertification of an high productivity area in Southern Italy. *J. Maps* **2015**.
8. Nicholson, S.E.; Tucker, C.J.; Ba, M.B. Desertification drought and surface vegetation: An example from the west African Sahel. *Bull. Am. Meteorol. Soc.* **1998**, *79*, 815–829.
9. Morales, A.; Olcina, J.; Rico, A.M. Diferentes persepciones de la sequía en España: Adaptación catastrofismo e intentos de corrección. *Investig. Geogr.* **2000**, *23*, 5–46.
10. García, R.V. *Nature Pleads Not Guilty*; Pergamon Press: Oxford, UK, 1984.
11. Kanti, B. Coping mechanisms practised by drought victims (1994/1995) in north Bengal, Bangladesh. *Appl. Geogr.* **1998**,*18*, 355–373.
12. Cooper, G.; McGechan, M.; Vinten, A. The influence

of a changed climate on soil workability and available workdays in Scotland. *J. Agric. Eng. Res.* **1997**, *68*, 253–269.

13. Earl, R. Prediction of trafficability and workability from soil moisture deficit. *Soil Tillage Res.* **1997**, *40*, 155–168.

14. Falloon, P.D.; Betts, R.A. The impact of climate change on global river flow in HadGEM1 simulations. *Atmos. Sci. Lett.***2006**, *7*, 62–68.

15. Finlayson, J.; Betteridge, K.; MacKay, A.; Thorrald, B.; Singleton, P.; Costall, D. A simulation model of the effects of cattle treading on pasture production on North Island, New Zealand, Hill Land. *N. Z. J. Agric. Res.* **2002**, *45*, 255–272.

16. Montanarella, L. Trends in land degradation in Europe. In *Climate and Land Degradation*; Sivakumar, M.V.K., Ndegwa, N., Eds.; Springer: Heidelberg, Germany, 2007; pp. 83–104.

17. Webb, J.; Anthony, S.; Brown, L.; Lyons-Visser, H.; Ross, C.; Cottril, B.; Johnson, D.; Scholefield, D. The impact of increasing the length of the cattle grazing season on emissions of ammonia and nitrous oxide and on nitrate leaching in England and Wales. *Agric. Ecosyst. Environ.* **2005**, *105*, 307–321.

18. Anagnostopoulou, C.H.R.; Maheras, P.; Karacostas, T.; Vafiadis, M. Spatial and temporal analysis of dry spells in Greece. *Theor. Appl. Climatol.* **2003**, *74*, 77–91.

19. Caloiero, T.; Coscarelli, R.; Ferrari, E.; Sirangelo, B. Analysis of dry spells in Southern Italy (Calabria). *Water* **2015**, *7*, 3009–3023.

20. Chapman, T. Stochastic models for daily rainfall in the Western Pacific. *Math. Comput. Simul.* **1997**, *43*, 351–358.

21. Deni, S.M.; Jemain, A.A.; Ibrahim, K. The best probability models for dry and wet spells in Peninsular Malaysia during monsoon seasons. *Int. J. Climatol.* **2010**, *30*, 1194–1205.

22. Di Giuseppe, E.; Vento, D.; Epifani, C.; Esposito, S. Analysis of dry and wet spells from 1870 to 2000 in four Italian sites. *Geophys. Res. Abstr.* **2005**, *7*, 07712.

23. Dobi-Wantuch, I.; Mika, J.; Szeidl, L. Modeling wet and dry spells with mixture distributions. *Meteorol. Atmos. Phys.***2000**, *73*, 245–256.

24. Sirangelo, B.; Caloiero, T.; Coscarelli, R.; Ferrari, E. A stochastic model for the analysis of the temporal change of dry spells. *Stoch. Environ. Res. Risk. Assess.* **2015**, *29*, 143–155.

25. Phillips, I.D.; McGregor, G.R. The utility of a drought index for assessing the drought hazard in Devon and Cornwall, South West England. *Meteorol. Appl.* **1998**, *5*, 359–372.

26. Wilks, D.S. Interannual variability and extreme-value characteristics of several stochastic daily precipitation models.*Agric. Forest Meteorol.* **1999**, *93*, 153–169.

27. Tsakiris, G.; Pangalou, D.; Vangelis, H. Regional drought assessment based on the Reconnaissance Drought Index (RDI). *Water Resour. Manag.* **2007**, *21*, 821–833.

28. Wilhite, D.A.; Hayes, M.J.; Svodoba, M.D. Drought monitoring and assessment in the US. In *Drought and Drought Mitigation in Europe*; Voght, J.V., Somma, F., Eds.; Kluwers: Dordrecht, The Netherlands, 2000; pp. 149–160.

29. Khan, S.; Gabriel, H.F.; Rana, T. Standard precipitation index to track drought and assess impact of rainfall on watertables in irrigation areas. *Irrig. Drain. Syst.* **2008**, *22*, 159–177.

30. Logan, K.E.; Brunsell, N.A.; Jones, A.R.; Feddema, J.J. Assessing spatiotemporal variability of drought in the US central plains. *J. Arid Environ.* **2010**, *74*, 247–255.

31. Manatsa, D.; Mukwada, G.; Siziba, E.; Chinyanganya, T. Analysis of multidimensional aspects of agricultural droughts in Zimbabwe using the Standardized Precipitation Index (SPI). *Theor. Appl. Climatol.* **2010**, *102*, 287–305.

32. Patel, N.R.; Yadav, K. Monitoring spatio-temporal pattern of drought stress using integrated drought index over Bundelkhand region; India. *Nat. Hazards* **2015**, *77*, 663–677.

33. Raziei, T.; Saghafian, B.; Paulo, A.A.; Pereira, L.S.; Bordi, I. Spatial patterns and temporal variability of drought in western Iran. *Water Resour. Manag.* **2009**, *23*, 439–455.

34. Xingcai, L.; Zongxue, X.U.; Bo, L. Spatio-temporal characteristics of Standardized Precipitation Index in the Taihu Basin during 1951–2000. *Wuhan Univ. J. Nat. Sci.* **2009**, *14*, 518–524.

35. Zhai, L.; Feng, Q. Spatial and temporal pattern of precipitation and drought in Gansu province northwest China. *Nat. Hazards* **2009**, *49*, 1–24.

36. Livada, I.; Assimakopoulos, V.D. Spatial and temporal analysis of drought in Greece using the Standardized Precipitation Index (SPI). *Theor. Appl. Climatol.* **2007**, *89*, 143–153.

37. Sönmez, F.K.; Kömüscü, A.Ü.; Erkan, A.; Turgu, E. An analysis of spatial and temporal dimension of drought vulnerability in Turkey using the Standardized Precipitation Index. *Nat. Hazards* **2005**, *35*, 243–264.

38. Vicente-Serrano, S.M. Differences in spatial patterns of drought on different time sales an analysis of the Iberian Peninsula. *Water Resour. Manag.* **2006**, *20*, 37–60.

39. Vergni, L.; Todisco, F. Spatio-temporal variability of precipitation temperature and agricultural drought indices in Central Italy. *Agric. Forest Meteorol.* **2011**, *151*, 301–313.

40. Bonaccorso, B.; Bordi, I.; Cancelliere, A.; Rossi, G.; Sutera, A. Spatial variability of drought: An analysis of SPI in Sicily. *Water Resour. Manag.* **2003**, *17*, 273–296.

41. Buttafuoco, G.; Caloiero, T. Drought events at different timescales in Southern Italy (Calabria). *J. Maps* **2014**, *10*, 529–537.

42. Buttafuoco, G.; Caloiero, T.; Coscarelli, R. Analyses of drought events in Calabria (Southern Italy) using standardized precipitation index. *Water Resour. Manag.* **2015**, *29*, 557–573.

43. Capra, A.; Scicolone, B. Spatiotemporal variability of drought on a short–medium time scale in the Calabria Region (Southern Italy). *Theor. Appl. Climatol.* **2012**, *3*,

471–488.

44. Capra, A.; Consoli, S.; Scicolone, B. Long-term climatic variability in Calabria and effects on drought and agrometeorological parameters. *Water Resour. Manag.* **2013**, *27*, 601–617.
45. Mendicino, G.; Senatore, A.; Versace, P. A Groundwater Resource Index (GRI) for drought monitoring and forecasting in a Mediterranean climate. *J. Hydrol.* **2008**, *357*, 282–302.
46. Rossi, G.; Cancelliere, A. Early warning of drought: Development of a drought bulletin for Sicily. In Proceedings of the 2nd International Conference: New Trends in Water and Environmental Engineering for Safety and Life: Ecocompatible Solutions for Aquatic Environments, Capri, Italy, 24–28 June 2002.
47. Wu, H.; Hayes, M.J.; Wilhite, D.A.; Svoboda, M.D. The effect of the length of record on the standardized precipitation index calculation. *Int. J. Climatol.* **2005**, *25*, 505–520.
48. Mishra, A.K.; Singh, V.P. Drought modeling—A review. *J. Hydrol.* **2011**, *403*, 157–175.
49. Zolina, O.; Simmer, C.; Belyaev, K.; Sergey, K.; Gulev, S.; Koltermann, P. Changes in the duration of European wet and dry spells during the last 60 years. *J. Clim.* **2013**, *26*, 2022–2047.
50. Güven, O. A simplified semiempirical approach to probabilities of extreme hydrologic droughts. *Water Resour. Res.* **1983**, *19*, 441–453.
51. Mathier, L.; Perreault, L.; Bobée, B.; Ashkar, F. The use of geometric and gamma-related distributions for frequency analysis of water deficit. *Stoch. Hydrol. Hydraul.* **1992**, *6*, 239–254.
52. Sen, Z. Run-sums of annual streamflow series. *J. Hydrol.* **1977**, *35*, 311–324.
53. Sen, Z. Regional drought and flood frequency analysis theoretical consideration. *J. Hydrol.* **1980**, *46*, 265–279.
54. Sharma, T. Estimation of drought severity on independent and dependent hydrologic series. *Water Resour. Manag.* **1995**, *11*, 35–49.
55. Shiau, J.T.; Shen, H.W. Recurrence analysis of hydrologic droughts of differing severity. *J Water Resour. Plan. Manag.* **2001**, *127*, 30–40.
56. Zelenhasìc, E.; Salvai, A. A method of streamflow drought analysis. *Water Resour. Res.* **1987**, *23*, 156–168.
57. Gupta, V.; Duckstein, L. Droughts as extreme distributions from a point rainfall process. *Water Resour. Res.* **1975**, *11*, 221–228.
58. Cancelliere, A.; Salas, J.D. Drought length properties for periodic-stochastic hydrologic data. *Water Resour. Res.* **2004**, *40*, W02503.
59. Sirangelo, B.; Caloiero, T.; Coscarelli, R.; Ferrari, E. Stochastic Analysis of Long Dry Spells in Calabria (Southern Italy). Available online: http://link.springer.com/article/10.1007/s00704-015-1662-0 (accessed on 25 January 2016).
60. Wilks, D.S.; Wilby, R.L. The weather generation game: A review of stochastic weather models. *Prog. Phys. Geogr.* **1999**, *23*, 329–357.
61. Brissette, F.P.; Khalili, M.; Leconte, R. Efficient stochastic generation of multi-site synthetic precipitation data. *J. Hydrol.* **2007**, *345*, 121–133.
62. Iyengar, R.N. Stochastic modelling of monthly rainfall. *J. Hydrol.* **1982**, *57*, 375–387.
63. Semenov, M.A.; Porter, J.R. Climatic variability and the modelling of crop yields. *Agric. Forest Meteorol.* **1995**, *73*, 265–283.
64. Apipattanavis, S.; Podestà, G.; Rajagopalan, B.; Katz, R.W. A semiparametric multivariate and multisite weather generator. *Water Resour. Res.* **2007**, *43*, W11401.
65. Bardossy, A.; Pegram, G. Copula based multisite model for daily precipitation simulation. *Hydrol. Earth Syst. Sci.* **2009**, *13*, 2299–2314.
66. Bordoy, R.; Burlando, P. Stochastic downscaling of precipitation to high-resolution scenarios in orographically complex regions. Part 1: Model evaluation. *Water Resour. Res.* **2014**, *50*, 540–561.
67. Burton, A.; Kilsby, C.; Fowler, H.; Cowpertwait, P.S.; O'Connell, P. RainSim: A spatial temporal stochastic rainfall modelling system. *Environ. Model. Softw.* **2008**, *23*, 1356–1369.
68. Cowpertwait, P.S.P.; Kilsby, C.; O'Connell, P. A space-time Neyman-Scott model of rainfall: Empirical analysis of extremes. *Water Resour. Res.* **2002**, *38*, 1–14.
69. Cowpertwait, P.S.P. A spatial temporal point process model of rainfall for the Thames catchment; UK. *J. Hydrol.* **2006**, *330*, 586–595.
70. Cowpertwait, P.S.P. A spatial-temporal point process model with a continuous distribution of storm types. *Water Resour. Res.* **2010**, *46*, W12507.
71. Kang, B.; Ramırez, J.A. A coupled stochastic space-time intermittent random cascade model for rainfall downscaling. *Water Resour. Res.* **2010**, *46*, W10534.
72. Paschalis, A.; Molnar, P.; Fatichi, S.; Burlando, P. On temporal stochastic modeling of precipitation; nesting models across scales. *Adv. Water Resour.* **2014**, *63*, 152–166.
73. Peleg, N.; Morin, E. Stochastic convective rain-field simulation using a high-resolution synoptically conditioned weather generator (HiReS-WG). *Water Resour. Res.* **2014**, *50*, 2124–2139.
74. Serinaldi, F. A multisite daily rainfall generator driven by bivariate copula-based mixed distributions. *J. Geophys. Res.* **2009**, *114*, D10103.
75. Steinschneider, S.; Brown, C. A semiparametric multivariate; multisite weather generator with low-frequency variability for use in climate risk assessments. *Water Resour. Res.* **2013**, *49*, 7205–7220.
76. Hayhoe, H.N. Improvements of stochastic weather data generators for diverse climates. *Clim. Res.* **2000**, *14*, 75–87.
77. Liu, J.; Williams, J.R.; Wang, X.; Yang, H. Using MODAWEC to generate daily weather data for the EPIC model. *Environ. Model. Softw.* **2009**, *24*, 655–664.
78. Richardson, C.W. Stochastic simulation of daily

precipitation temperature and solar radiation. *Water Resour. Res.* **1981**,*17*, 182–190.

79. Sharma, A.; Lall, U. A nonparametric approach for daily rainfall simulation. *Math. Comput. Simulat.* **1999**, *48*, 361–371.

80. Srikanthan, R.; McMahon, T.A. *Stochastic Generation of Rainfall and Evaporation Data*; Technical Report 84; Australian Water Resources Council Department of Resources and Energy: Canberra, Australia, 1985.

81. Srikanthan, R.; Harrold, T.I.; Sharma, A.; McMahon, T.A. Comparison of two approaches for generation of daily rainfall data. *Stoch. Environ. Res. Risk. Assess.* **2005**, *19*, 215–226.

82. Wan, H.; Zhang, X.; Barrow, E.M. Stochastic modelling of daily precipitation for Canada. *Atmos. Ocean* **2005**, *43*, 23–32.

83. Woolhiser, D.A. Modelling daily precipitation—Progress and problems. In *Statistics in the Environmental and Earth Sciences*; Walden, A.T., Guttorp, P., Eds.; Edward Arnold: London, UK, 1992; pp. 71–89.

84. Zheng, X.; Katz, R.W. Mixture model of generalized chain-dependent processes and its application to simulation of interannual variability of daily rainfall. *J. Hydrol.* **2008**, *349*, 191–199.

85. Srikanthan, R.; McMahon, T.A. Stochastic generation of annual monthly and daily climate data: A review. *Hydrol. Earth Syst. Sci.* **2001**, *5*, 653–670.

86. McKee, T.B.; Doesken, N.J.; Kleist, J. The relationship of drought frequency and duration to time scales. In Proceedings of the 8th Conference on Applied Climatology, Anaheim, CA, USA, 17–22 January 1993; pp. 179–184.

87. Angelidis, P.; Maris, F.; Kotsovinos, N.; Hrissanthou, V. Computation of drought index SPI with Alternative Distribution Functions. *Water Resour. Manag.* **2012**, *26*, 2453–2473.

88. Stagge, J.H.; Tallaksen, L.M.; Gudmundsson, L.; van Loon, A.F.; Stahl, K. Candidate distribution for climatological drought indices (SPI and SPEI). *Int. J. Climatol.* **2015**, *35*, 4027–4040.

89. Mason, D.M.; Schuenemeyer, J.H. A modified Kolmogorov-Smirnov test sensitive to tail alternatives. *Ann. Stat.* **1983**,*11*, 933–946.

90. Crutcher, H.L. A note on the possible misuse of the Kolmogorov–Smirnov test. *J. Appl. Meteorol.* **1975**, *14*, 1600–1603.

91. Steinskog, D.J.; Tjøstheim, D.B.; Kvamstø, N.G. A cautionary note on the use of the Kolmogorov–Smirnov test for normality. *Mon. Weather Rev.* **2007**, *135*, 1151–1157.

92. Wu, H.; Svoboda, M.D.; Hayes, M.J.; Wilhite, D.A.; Wen, F. Appropriate application of the standardized precipitation index in arid locations and dry seasons. *Int. J. Climatol.* **2007**, *27*, 65–79.

93. Bryant, S.J.; Arnell, N.W.; Law, F.M. The longterm context for the current hydrological drought. In Proceedings of the IWEM Conference on the Management of Scarce Water Resources, Brighton, UK, 13–14 October 1992.

94. Blenkinsop, S.; Fowler, H.J. Changes in drought frequency, severity and duration for the British Isles projected by the PRUDENCE regional climate models. *J. Hydrol.* **2007**, *342*, 50–71.

95. Blenkinsop, S.; Fowler, H.J. Changes in European drought characteristics projected by the PRUDENCE regional climate models. *Int. J. Climatol.* **2007**, *27*, 1595–1610.

96. Bernstein, S.; Bernstein, R. *Elements of Statistics: II—Inferential Statistics*; McGraw-Hill: New York, NY, USA, 1999.

97. Priestly, M.B. *Spectral Analysis and Time Series*; Academic Press: London, UK, 1994.

98. Anderson, R.L. Distribution of serial correlation coefficient. *Ann. Math. Stat.* **1942**, *13*, 1–13.

99. Haan, C.T. *Statistical Methods in Hydrology*; Iowa State University Press: Ames, IA, USA, 1997.

100. Ferrari, E.; Caloiero, T.; Coscarelli, R. Influence of the north Atlantic Oscillation on winter rainfall in Calabria (Southern Italy). *Theor. Appl. Climatol.* **2013**, *114*, 479–494.

101. Caloiero, T.; Pasqua, A.A.; Petrucci, O. Damaging hydrogeological events: A procedure for the assessment of severity levels and an application to Calabria (Southern Italy). *Water* **2014**, *6*, 3652–3670.

102. Coscarelli, R.; Caloiero, T. Analysis of daily and monthly rainfall concentration in Southern Italy (Calabria region). *J. Hydrol.* **2012**, *416–417*, 145–156.

103. Brunetti, M.; Caloiero, T.; Coscarelli, R.; Gullà, G.; Nanni, T.; Simolo, C. Precipitation variability and change in the Calabria region (Italy) from a high resolution daily dataset. *Int. J. Climatol.* **2012**, *32*, 57–73.

104. Craddock, J.M. Methods of comparing annual rainfall records for climatic purposes. *Weather* **1979**, *34*, 332–346.

105. Press, M.H.; Teukolsky, S.A.; Vetterling, W.T.; Flannery, B.P. *Numerical Recipes in C: The Art of Scientific Computing*, 2nd ed.; Cambridge University Press: Cambridge, UK, 1997.

106. Giordano, R.; Preziosi, E.; Romano, E. Integration of local and scientific knowledge to support drought impact monitoring: Some hints from an Italian case study. *Nat. Hazards* **2013**, *69*, 523–544.

107. Wu, J.; He, B.; Lu, A.; Zhou, L.; Liu, M.; Zhao, L. Quantitative assessment and spatial characteristics analysis of agricultural drought vulnerability in China. *Nat. Hazards* **2011**, *56*, 785–801.

108. Caloiero, T.; Buttafuoco, G.; Coscarelli, R.; Ferrari, E. Spatial and temporal characterization of climate at regional scale using homogeneous monthly precipitation and air temperature data: An application in Calabria (Southern Italy).*Hydrol. Res.* **2015**, *46*, 629–646.

109. Caloiero, T.; Callegari, G.; Cantasano, N.; Coletta, V.; Pellicone, G.; Veltri, A. Bioclimatic Analysis in a Region of Southern Italy (Calabria). Available online: http://www.tandfonline.com/doi/abs/10.1080/11263504.2015.1037814(accessed on 25 January 2016).

110. Pellicone, G.; Caloiero, T.; Coletta, V.; Veltri, A.

Phytoclimatic map of Calabria (Southern Italy). *J. Maps* **2014**, *10*, 109–113.

111. Li, Y.; Gu, W.; Cui, W.; Chang, Z.; Xu, Y. Exploration of copula function use in crop meteorological drought risk analysis: A case study of winter wheat in Beijing, China. *Nat. Hazards* **2015**, *77*, 1289–1303.

112. Xu, X.; Ge, Q.; Zheng, J.; Dai, E.; Zhang, X.; He, S.; Liu, G. Agricultural drought risk analysis based on three main crops in prefecture-level cities in the monsoon region of east China. *Nat. Hazards* **2013**, *66*, 1257–1272.

CHAPTER 3

AN INDIRECT SIMULATION-OPTIMIZATION MODEL FOR DETERMINING OPTIMAL TMDL ALLOCATION UNDER UNCERTAINTY

Feng Zhou, Yanjun Dong, Jing Wu, Jiangli Zheng and Yue Zhao

Keywords: Total maximum daily load (TMDL) allocation, eutrophication, tradeoff analysis, water-quality modeling, uncertainty

INTRODUCTION

Section 303(d) of the 1972 Clean Water Act requires the development of total maximum daily load (TMDL) to restore impaired water bodies [1]. TMDL refers to the maximum loading rates of a pollutant that a water body receives to ensure compliance with water-quality standards and allocates pollutant loadings among point and nonpoint sources at different risk levels [1,2]. In most TMDL allocation analyses, load reduction scenarios are generated through trial-and-error (TAE) simulation [2], which involves repeatedly running process-oriented simulation models to derive the pollutant loadings to meet water-quality standards and allocation criteria in the water body [3,4]. However, the TAE simulation method does not necessarily generate the most cost-effective and reliable load allocations [5]. To increase the financial and technical feasibility of implementation, the U.S. Environmental Protection Agency [1] suggested that process-oriented simulation models could be directly integrated into an optimization framework to develop optimal TMDL allocations at the least cost while risk must be identified. Either traditional nonlinear optimization or modern heuristic global search algorithms can be applied in the direct simulation-optimization model (SOM) framework [6,7,8,9,10,11,12,13]; however, this direct SOM approach is rarely applied in practice, primarily due to the prohibitive computational cost and the neglect of uncertainties in both simulation modeling and the optimization process [7,14].

Most of the input parameters and model structures in both simulation and optimization procedures are uncertain, which affects the decision-making process [15]. The SOM framework under uncertainty can be grouped into four broad categories: namely stochastic optimization models, fuzzy optimization models, interval optimization models and hybrids of the above three model types [15,16,17,18,19]. However, the increasing data requirements for specifying probability density or parameter membership functions have become a big challenge for TMDL allocations [7,14]. Even when many data are available, the computational algorithms for stochastic or fuzzy optimization models, such as nonlinear programming (NP) or modern heuristic global searches, may face complex or nonlinear problems [19,20]. For example, traditional NP algorithms, including both gradient and non-gradient based ones, were limited to local optima upon solving the aforementioned SOM framework [11,21]. Although heuristic global search algorithms, including genetic algorithms, evolutionary algorithms and simulated annealing, are capable of surpassing the local optima limitations, their applications in the optimization of TMDL allocation are still restricted by their extremely high computational cost [18,22].

To mitigate the above problems, the interval linear programming (ILP) model [16], which does not require probability density or membership function, could be a

Citation: Feng Zhou, Yanjun Dong, Jing Wu, Jiangli Zheng and Yue Zhao, An Indirect Simulation-Optimization Model for Determining Optimal TMDL Allocation under Uncertainty, doi:10.3390/w7116634.

potential alternative to stochastic or fuzzy optimization models. Based on Huang's cluster analysis [16], Zhou *et al.* [23] proposed a modified interval linear programming (MILP) model to establish strict mathematical relationships between decision variables and uncertain constraint coefficients, which also provides sufficient conditions to obtain an absolutely feasible solution space. Zhou *et al.* [19] further defined a new uncertainty type, the enhanced-interval number, and developed an enhanced-interval linear programming (EILP) model, as well as a computationally-efficient solution algorithm, which provides insight into both extreme and non-extreme tradeoffs between system benefits and the different risk levels of constraint violations due to parametric uncertainties in the objective functions and constraints [15,18].

Several functional approximators have been developed for building the response function of water quality to pollutant loadings and to overcome the computation bottleneck associated with direct SOM applications. This indirect SOM framework, in place of complex process-oriented models, incorporates an approximation function into an optimization model for searching optimal TMDL allocations [7,8,24]. Generally, the framework with the approximation function needs to be sufficiently generalized to achieve optimal or approximate-optimal solutions of the original direct SOM system. This means that the regression tree model [25], the response surface [14] or the artificial neural network (ANN) [22] should truly capture the underlying functional form of the simulation model rather than just the training data used to calibrate the models. Although the ANN approach is one powerful platform for developing an approximation function, it has not yet been shown to provide explicit expressions of nonlinear stressor-response relationships and effectively quantify uncertainties in dynamic water body systems. Recently, the Bayesian recursive regression tree Version 2 (BRRT v2) model was developed for rapidly capturing the nonlinear stressor-response relationship of general aquatic systems [26], generating a series of interval linear regression equations that covered the pertinent ranges of stressors and responses reflected in the original nonlinear simulation model. Given its computational efficiency, predictive accuracy and capability of explicitly quantifying uncertainty in previous works [26], the BRRT v2 may be employed as a surrogate for a simulation model in an indirect SOM framework by replacing the process-oriented simulation model with interval linear regression equations.

The purpose of this paper is to propose an indirect SOM framework by integrating the BRRT v2 and EILP models for risk-based optimal TMDL allocation, which guarantees the approximate-optimal solutions with relatively low computational cost and to provide extreme and non-extreme tradeoff analysis for risk-based decision making. The risk-based optimal TMDL allocation aims to determine the optimal required pollutant removal rates from point and nonpoint sources to ensure compliance with water-quality standards under different risk levels. Using the real-world case study in the Swift Creek Reservoir watershed (Chesterfield County, VA), we compared the performance of the proposed BRRT-EILP model with the direct simulation-optimization model, stochastic simulation and TAE approaches in terms of computational efficiency and risk-based decision-making capability for TMDL allocation under uncertainty.

MATERIALS AND METHODS

BRRT-EILP Model

The combined BRRT-EILP model consists of two critical components: (i) the simulation model for predicting nonlinear responses to pollutant removal rates among all pollution sources; and (ii) the optimization model for achieving the optimal pollutant removal rates under uncertainty [1]. A general direct SOM model framework under uncertainty was developed to minimize load reductions as follows:

$$Min\ Z^{\pm} = \mathbf{C}^{\pm}\mathbf{X}^{\pm},\ s.t\ \mathbf{Y}^{\pm} = h(\mathbf{X}^{\pm}, \mathbf{A}^{\pm}) \leq \mathbf{Y}^{*\pm},\ \mathbf{X}^{\pm} \geq 0 \quad (1)$$

where X± represents the pollutant removal rates, Z± is the total load reductions, Yi± is the simulated concentration of pollutants, Y*± is the user-specified water-quality criteria, A± is the parameters for process-oriented models $\mathbf{Y}^{\pm} = h(\mathbf{X}^{\pm}, \mathbf{A}^{\pm})$) and $\mathbf{B}^{\pm}$ and $\mathbf{C}^{\pm}$ are the coefficients of the constraints or objective function. All parameters above are also defined as interval numbers due to the systematic uncertainty propagated from sampling collection, water quality model (i.e., structure and parameters), and coefficients in the optimization model [27,28]. Based on the experience of TMDL in the United States [1], the objective function Z± is usually defined as an acceptable measure of economic efficiency, reliability and equity. When detailed cost information is not available, minimizing load reduction could be used as an alternative to maximize the economic efficiency, since we would at least expect increased system costs for a given pollution source following increased load reduction [2]. The objective function used in this study was eventually used to determine the required minimum total load reductions; the system constraints in Equation (1) were to bring a watershed system in compliance with water-quality criteria $\mathbf{Y}^{i\pm} \leq \mathbf{Y}^{*\pm}$.

When replacing complex process-oriented models with an approximation function, Equation (1) is transferred to an indirect SOM framework. This framework integrates the BRRT-derived interval linear regressions into an EILP optimization process (defined as the BRRT-EILP model):

$$\mathbf{Y}^{i\pm} = \beta_0^i + \beta^{iT}\mathbf{X}^i \pm \varepsilon_\alpha^i,\ \text{for}\ \mathbf{X}^{i\pm} \in [\mathbf{D}^{i-}, \mathbf{D}^{i+}],\ \forall i \quad (2)$$

where *i* indicates the *i*-th interval linear regression equations, **β**i (including βi0) and εiα are the regression coefficients vector and predictive error vector under a given confidence coefficient *α*, respectively, and Di− and Di+ are the lower- and upper-bounds of X± for the interval linear regression equation *i*. To solve Equation (1) combined with Equation (2), an EILP algorithm was used to achieve the optimal TMDL allocation under uncertainty, which also provided both extreme and non-extreme tradeoffs between the benefits of TMDL allocation and the diverse risk levels violating the constraints.

Application to the Swift Creek Reservoir Watershed

The Swift Creek Reservoir (SCR) Watershed is located in the Chesterfield County, VA, the United Sates (37.4° N, 77.7° W, 165 km^2; Figure 1). Its major environmental concerns are algal blooms and hypolimnetic hypoxia in summer, the majority caused by two stressors: the loadings of inorganic nitrogen (NO_3^- and NH_4^+) and inorganic phosphorus (PO_4^{3-}) from eight sub-watersheds [28]. A hydrodynamic and water-quality model, CE-QUAL-W2 v3.1, was developed to represent such stressor-response relationships from 1998–2000 [28]. Previously, a nutrient TMDL allocation for the SCR watershed was developed through a stochastic simulation method by stochastically changing the pollutant removal rates among nonpoint sources from the eight sub-watersheds [15,18,28]. In our study, we applied the proposed BRRT-EILP model to determine optimal nutrient TMDL allocation under uncertainty for the SCR watershed. The goal of such TMDL allocation was to identify the critical load reductions from the eight sub-watersheds for the maintenance of summer (June–September) maxima of vertically averaged chlorophyll-*a* (Chl-*a*) concentrations at the reservoir outlet (defined as $y^{\pm}$) below Chl-*a*criteria. According to Equations (1) and (2), the minimum total load reduction model for inorganic phosphorus and nitrogen for the SCR watershed could be formulated as follows:

$$Min\ Z^{\pm} = \sum_{j}^{n} \phi \omega_{j1} c_{j1}^{\pm} x_{j1}^{\pm} + \sum_{j}^{n} (1-\phi) \omega_{j2} c_{j2}^{\pm} x_{j2}^{\pm} \tag{3a}$$

$$s.t\ y^{ik\pm} = \beta_0^{ik} + \sum_{l}^{L} \sum_{j}^{J} \beta_{jl}^{ik} x_{jl}^{\pm} \pm \varepsilon_{\alpha}^{ik} \le y^{*\pm},\ \text{for } x_{jl}^{\pm} \in [d_{jl}^{i-}, d_{jl}^{i+}], \forall i, k \tag{3b}$$

$$x_{jl}^{\pm} \le x_l^{max},\ \forall j, l,\ x_{jl}^{\pm} \ge x_l^{min},\ \forall j, l,\ x_{jl}^{\pm}, \omega_j, c_{jl}^{\pm}, y^{*\pm}, x_{jl}^{min\pm} \ge 0 \tag{3c}$$

where i represents the index of interval linear regression equations generated by the calibrated BRRT model, i = 1,2,…,I, I = 107, j is the index of the sub-watershed, j = 1,2,…,J, J = 8, k is the index of the compliance point, k = 1,2,…,K, K = 1 (reservoir outlet), and l is the index of the pollutant, l = 1, L, L = 2 (inorganic phosphorus and nitrogen). Z± is the total load reduction of inorganic phosphorus from the eight sub-watersheds when ϕ = 1 or the total load reduction of inorganic nitrogen from the eight sub-watersheds when ϕ = 0. ωjl is the weight for pollutant l from sub-watershed j that accounted for the relative difficulty or cost of the load reduction; whereas ωjl was assumed to be equal to 1.0 for all of the sub-watersheds, since no cost information from the previous best management practices (BMPs) implementation was available. x±jl is the pollutant removal rate for pollutant l from sub-watershed j, calculated as the difference between baseline loadings (Table 1) and maximum allowable loadings (Ajl) at river mouths divided by baseline loadings, that is $x_{jl}^{\pm} = (c_{jl}^{\pm} - A_{jl}^{\pm})/c_{jl}^{\pm}$, where $c_{jl}^{\pm}$ is the baseline load of pollutant l in sub-watershed j, computed using Hydrological Simulation Program—FORTRAN (HSPF) v11 for the period from January–December 1998 (Table 1) [18]. β_{jl}^{ik} (including β_0^{ik}) and $\varepsilon_{\alpha}^{ik}$ are the regression coefficients and predictive error of the i-th interval linear regression equation at compliance point k, respectively. To derive the interval linear regression equations, the CE-QUAL-W2 v3.1 generated the simulated Chl-a (defined as yik±) with randomly-generated pollutant removal rates from 8 sub-watersheds, resulting in 2000 stressor-response datasets. We then calibrated these coefficients and determined the associated confidence intervals using BRRT v2. Since no Chl-a criteria have been published by either the United States Environmental Protection Agency or the State of Virginia, y*± was set at 12−15 μg/L in this study based on North Carolina water-quality standards and other references [28,29]. Three Chl-a criteria scenarios had been set as [12, 15], [13, 15] and [14, 15] μg/L; xmaxl is the maximum removal level of the BMPs' performance for pollutant l, which was set at 80% according to the American Society of Civil Engineers' (ASCE) national BMP database; xminl was set as zero, indicating that no nutrient load was removed; $d_{jl}^{i-} \in \mathbf{D}^{i-}, d_{jl}^{i+} \in \mathbf{D}^{i+}$.

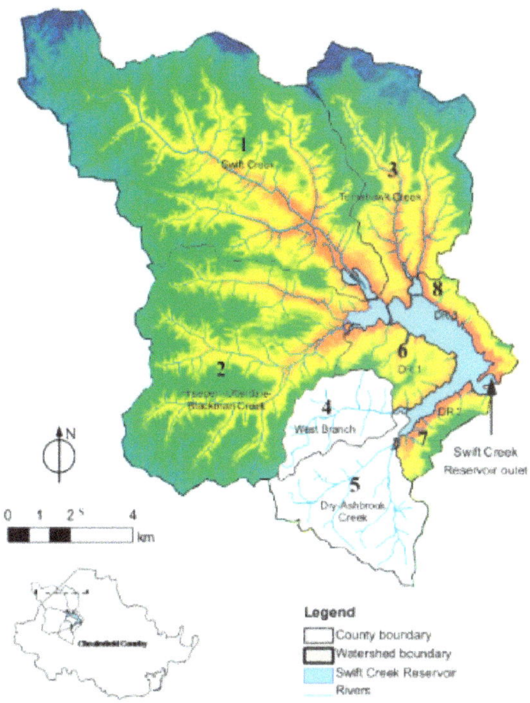

Figure 1. Swift Creek Reservoir watershed and its eight sub-watersheds.

Table 1. Baseline loadings (inorganic phosphorus and inorganic nitrogen) of Swift Creek Reservoir watershed.

Sub-Watershed *j*	Area, km²	c_{i1}, kg	c_{i2}, kg
1. Swift Creek	59.47	4451.7	157.4
2. Horsepen-Otterdale-Blackman Creek	40.47	5169.3	388.4
3. Tomahawk Creek	25.30	2677.2	126.8
4. West Branch	7.50	286.7	87.9
5. Dry-Ashbrook Creek	14.52	516.5	161.1
6. Direct runoff (1)	3.73	1916.3	114.2
7. Direct runoff (2)	6.13	1000.7	90.2
8. Direct runoff (3)	3.31	471.7	146.1
Total	160.4	16,490	1272

Algorithmic Processes

In order to explain our model in a logical and comprehensive way, we simply introduced the algorithmic processes of the BRRT-EILP model for determining optimal TMDL allocation under uncertainty.

Step 1: Generate the pollutant removal rates for pollution sources and simulate their responses at the compliance points using the water quality model, which results in a stressor-response dataset;

Step 2: Derive the interval linear regression equations by the BRRT v2 based on the stressor-response dataset (Figure 2);

Step 3: Formulate the minimum load reduction model (Equation (3)) under uncertainty by integrating interval linear regression equations into the EILP model;

Step 4: Following the EILP algorithm, solve the lower bound Z, where the optimal pollutant removal rates are x_{jopt}^- for $j = 1, 2, \ldots, k$ and x_{jopt}^+ for $j = k+1, k+2, \ldots, n$ and Z_{opt}^- is the number of x_j;

Step 5: Formulate the extra constraints ($x_j^+ \geq x_{jopt}^-$ for $j = 1, 2, \ldots, k$, $x_j^- \leq x_{jopt}^+$ for $j = k+1, k+2, \ldots, n$);

Step 6: Solve the upper bound Z^+ Equations, resulting in $Z_{opt}^{\pm}$, $E[Z_{opt}^{\pm}]$ and $\mathbf{X}_{opt}^{\pm} = \left\{x_{opt}^{\pm} = [x_{jopt}^-, x_{jopt}^+] \mid \forall j = 1, 2, \ldots, n\right\}$;

Step 7: Generate a set of pollutant removal rates or TMDL allocations under different risk levels according to the obtained optimal solutions $Z_{opt}^{\pm}$, $E[Z_{opt}^{\pm}]$, and $\mathbf{X}_{opt}^{\pm}$. Analysis of these alternatives could allow acquiring both extreme and non-extreme TMDL allocations [15].

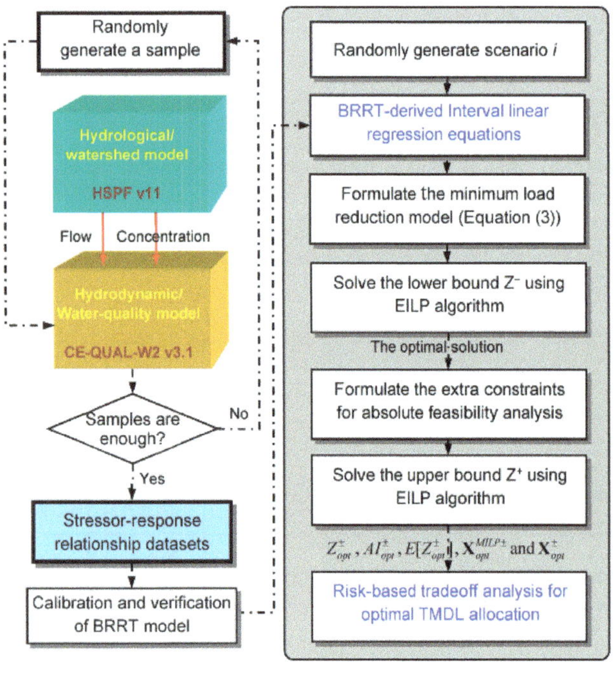

Figure 2. Algorithmic process for solving Bayesian recursive regression tree (BRRT)-enhanced-interval linear programming (EILP) model for risk-based optimal total maximum daily load (TMDL) allocation.

RESULTS

The calibrated interval linear regression equations have low biases with large coefficients of determination (0.99 for calibration and 0.97 for validation), small mean squared errors (0.105 and 0.496 μg/L) and Bayesian information criteria (−3270 and −1592). This suggests that these equations could serve as a reliable approximation function of the complex CE-QUAL-W2 v3.1 model in accurately predicting the nonlinear effects of nitrogen and phosphorous loadings on the phytoplankton response. Additionally, the calibrated interval linear regression equations were also verified by independent observations of summer maximum vertically averaged Chl-*a* concentrations at the reservoir outlet from 2004–2013. These data were obtained from Swift Creek Reservoir Reports. Although the 10-year observations are limited, the result of the verification process indicated that the calibrated interval linear regression equations could still reasonably capture the temporal variations of the summer maximum vertically-averaged Chl-*a* concentration ($R^2 = 0.73$;).

The results of the BRRT-EILP model show that the optimal required load reductions of inorganic nitrogen ($\phi = 1$) from Swift Creek were 1870 kg/year under three scenarios (Table 2), while those from Dry-Ashbrook Creek greatly increased from [0, 120] in Scenario 1 to [0, 360] and [0, 413] kg/year in Scenarios 2 and 3, respectively. Additionally, Direct Runoff (1), from one of the sub-watersheds nearest to the compliance point, was required to reduce inorganic nitrogen by [0, 238] kg/year in Scenario 3. Consequently, the total inorganic nitrogen load reductions were [1870, 1989], [1870, 2235] and [1870, 2521] kg/year, respectively. The corresponding expected values were 1929.5, 2052.5 and 2195.5 kg/year.

The optimal solutions for the inorganic phosphorus load reductions ($\phi = 0$) in Scenarios 1 and 2 indicate that it is necessary to reduce inorganic phosphorus loading only in the Swift Creek and Horsepen-Otterdale-Blackman (HOB) Creek sub-watersheds to reach a target Chl-*a* level of between 13 and 15 μg/L, with corresponding pollutant removal rates of [0.62, 0.74] and [0.62, 0.8] in the Swift Creek sub-watershed and 0.79 in the HOB Creek sub-watershed (Table 2). The total load reduction of inorganic phosphorus in Scenario 1 was [406.2, 424.2] kg/year, which is less than the optimal solution in Scenario 2 of [406.2, 433.6] kg/year. The expected load reduction values E[Z±opt] in Scenarios 1 and 2 were 415.2 and 419.9 kg/year, respectively. In Scenario 3 (c±=[12,15] μg/L), the required inorganic phosphorus load reduction from HOB Creek was slightly increased to [308, 310.8] kg compared to the Scenario 2 requirement of 308 kg/y. In addition, the Tomahawk Creek, Direct Runoff (2) and Direct Runoff (3) sub-watersheds were required to reduce inorganic phosphorus by [0, 64.8], [0, 72.2] and [0, 116.9] kg/year, respectively. Together, the total load reduction and its expected values $E[Z_{opt}^{\pm}]$ in Scenario 3 were [406.2, 690.2] and 548.2 kg/year respectively.

Based on these optimal reductions under different Chl-*a* criteria levels (*i.e.*, 12, 13, 14, 15 μg/L), vertically-averaged Chl-*a*concentrations at the reservoir outlet were further simulated and verified by CE-QUAL-W2 v3.1.

Table 2. Optimal pollution removal levels of inorganic phosphorus and inorganic nitrogen for the Swift Creek Reservoir (SCR) watershed for three scenarios.

Sub-watershed *j*	Minimum Total Load Reduction of Inorganic Nitrogen						Minimum Total Load Reduction of Inorganic Phosphorus					
	Scenario 1 (Chl-*a*: [14, 15] μg/L)		Scenario 2 (Chl-*a*: [13, 15] μg/L)		Scenario 3 (Chl-*a*: [12, 15] μg/L)		Scenario 1		Scenario 2		Scenario 3	
	x±j1 *	c±j1x±j1 *, kg	x±j1	c±j1x±j1, kg	x±j1	c±j1x±j1, kg	x±j2	c±j2x±j2, kg	x±j2	c±j2x±j2, kg	x±j2	c±j2x±j2, kg
1	[0.42, 0.42]	[1870, 1870]	[0.42, 0.42]	[1870, 1870]	[0.42, 0.42]	[1870, 1870]	[0.62, 0.74]	[98.2, 116.2]	[0.62, 0.8]	[98.2, 125.6]	[0.62, 0.8]	[98.2, 125.6]
2	[0, 0]	[0, 0]	[0, 0]	[0, 0]	[0, 0]	[0, 0]	[0.79, 0.79]	[308, 308]	[0.79, 0.79]	[308, 308]	[0.79, 0.8]	[308, 310.7]
3	[0, 0]	[0, 0]	[0, 0]	[0, 0]	[0, 0]	[0, 0]	[0, 0]	[0, 0]	[0, 0]	[0, 0]	[0, 0.51]	[0, 64.8]
4	[0, 0]	[0, 0]	[0, 0]	[0, 0]	[0, 0]	[0, 0]	[0, 0]	[0, 0]	[0, 0]	[0, 0]	[0, 0]	[0, 0]
5	[0, 0.23]	[0, 120]	[0, 0.71]	[0, 360]	[0, 0.8]	[0, 413]	[0, 0]	[0, 0]	[0, 0]	[0, 0]	[0, 0]	[0, 0]
6	[0, 0]	[0, 0]	[0, 0]	[0, 0]	[0, 0.12]	[0, 238]	[0, 0]	[0, 0]	[0, 0]	[0, 0]	[0, 0]	[0, 0]
7	[0, 0]	[0, 0]	[0, 0]	[0, 0]	[0, 0]	[0, 0]	[0, 0]	[0, 0]	[0, 0]	[0, 0]	[0, 0.8]	[0, 72.2]
8	[0, 0]	[0, 0]	[0, 0]	[0, 0]	[0, 0]	[0, 0]	[0, 0]	[0, 0]	[0, 0]	[0, 0]	[0, 0.8]	[0, 116.9]
Total load reduction		[1870, 1989]		[1870, 2235]		[1870, 2521]		[406.2, 424.2]		[406.2, 433.6]		[406.2, 690.2]
E[Z±opt]		1929.5		2052.5		2195.5		415.2		419.9		548.2

In simple terms, we used Scenario $3\ (c^{\pm} = [12, 15]\ \mu g/L)$ for inorganic nitrogen as an example to illustrate the extreme and non-extreme tradeoffs between the benefits of TMDL allocation and the diverse risk levels of violating the constraints, and the results for the other two scenarios.

Minimum Total Load Reductions of Inorganic Nitrogen

(i) The first decision alternative in the extreme tradeoff analysis was $X_1 = \{x_{jl}^{\pm} | x_{11opt} = 0.42; x_{j1opt} = 0 \text{ for other } j\}$ for other j}with the lowest total load reduction Z−opt of 1870 kg of inorganic nitrogen. This corresponded to the highest risk level of violating the constraints.

(ii) The second decision alternative in the extreme tradeoff analysis was $X_2 = \{x_{jl}^{\pm} | x_{11opt}^{\pm} = 0.42; x_{51opt}^{\pm} = 0.8; x_{61opt}^{\pm} = 0.12; x_{j1opt}^{\pm} = 0 \text{ for other } j\}$ with the highest total load reduction Z_{opt}^{+} of 2521 kg of inorganic nitrogen, corresponding to the lowest risk level of violating the constraints.

(iii) The third decision alternative in the non-extreme tradeoff analysis was $X_3 = \left\{x_{jl}^{\pm} \middle| \sum_{j}^{8} c_{jl}^{\pm} x_{jl}^{\pm} = 2195.5 \text{ where } x_{11opt}^{\pm} = 0.42; x_{51opt}^{\pm} = [0, 0.8]; x_{61opt}^{\pm} = [0, 0.12]\right.$ and $\left. x_{j1opt}^{\pm} = 0 \text{ for other } j\right\}$ with an expected value of total load reduction $E[Z_{opt}^{\pm}]$ of 2195.5 kg of inorganic nitrogen, corresponding to a moderate risk level of violating the constraints.

DISCUSSION

The optimal results in Table 2 suggest that the Swift Creek sub-watershed contributes more than 3/4 to the total nitrogen load reductions under different scenarios, though its per-area nitrogen loading ranked third from bottom (Table 1). Nevertheless, low per-area nitrogen loading does not mean that the nitrogen concentrations at the outlet of Swift Creek are lower than the other sub-watersheds. Actually, both nitrogen concentrations in this sub-watershed are highest within the simulation period [28] compared to the others. Such worse water quality is primarily caused by the urban pollution sources, such as domestic sewage and urban storm water, rather than the agricultural pollution source loss via surface runoff [28]. Additionally, both the nitrogen and phosphorous load reductions are required largely in the Swift Creek sub-watershed. Such a result is determined through the optimization processes by the BRRT-EILP model, which assigns a higher pollutant removal rate to the sub-watershed where the Chl-*a* concentration is more sensitive to the increase of load reductions. We believe that the large load reductions of both pollutants are reasonable when only considering the efficiency criteria in the optimization process, as there exists a synergy between nitrogen and phosphorous load reductions within wastewater treatment plants (WWTPs). When decision makers are interested in knowing the optimal TMDL allocation in cases where the total cost is to be minimized or the equitability among stakeholders is met, the objective functions and the associated constraints can be easily revised.

The computational cost of the BRRT-EILP model was significantly lower than that of conventional global optimization algorithms, such as genetic algorithms, evolutionary algorithms and simulated annealing (Table 3). The major computation time for the BRRT-EILP method was in the generation of the stressor-response dataset for calibrating and verifying the linear regression equations, which approximately took 2000 min, as it takes around 1 min for an Intel® Core™

2 Duo 2.4-GHz computer (CPU T8300) to complete one CE-QUAL-W2 model run. Other than that, the calibration and verification of the BRRT model takes only about 5 min, and finding the optimal solution of the BRRT-EILP model for three scenarios with 16 decision variables takes less than 1 min. Taken together, the total computational time taken to find out the approximate optimal solutions was only about 33 h. In contrast, running a traditional direct SOM framework using a GA approach might require several days (e.g., if a population size of 50 is used and the model is iterated for 300 generations, it takes over 10 days to obtain a solution) [11]. More time would be required to increase the chances of obtaining globally-approximate optimal solutions, because multiple GA runs would need to be executed for the same problem [21,22]. Moreover, the computation efficiency of the proposed BRRT-EILP model can be significantly higher than that of the traditional direct SOM approach when Equation (3) is used to evaluate solutions for different objective functions, water-quality criteria or system constraints. For example, assume that decision makers are interested in the optimal TMDL allocation in cases where the total cost, rather than the total reduction level, is to be minimized, or when water quality targets for dissolved oxygen are also considered, or when they are interested in achieving compliance at more than one location. In such cases, traditional direct SOM would involve re-running the stochastic search algorithm, which is very inefficient and can be computationally prohibitive if many scenarios need to be analyzed. In contrast, the BRRT-EILP model would require less than 1 min to compute such a new scenario, since it is not necessary to repeat the time-consuming data generation process required for BRRT calibration/verification in the scenario analysis stage.

Table 3. Method comparison of computational cost.

Method	Calibration	New Application
BRRT-EILP model	33 h	1 min
Direct SOM framework	~10 days	~10 days

The advantage of applying the BRRT-EILP model in place of stochastic simulation and TAE methods for TMDL development can be demonstrated by comparing the allocations resulting from each of the methods. First, 4000 scenarios were generated by randomly changing the inorganic nitrogen or inorganic phosphorus removal rates (no more than 80%) from eight sub-watersheds. The alternative where the sum of load reductions is minimum is eventually selected as the best allocation. Figure 3 shows that the alternatives of 3.6, 3.3, 2.6 and 2.4 × 10³ tonnes were identified as feasible inorganic nitrogen load reductions to meet Chl-*a* maximum criteria of 12, 13, 14 and 15 μg/L, respectively. Additionally, alternatives of 0.74, 0.50, 0.44 and 0.43 × 10³ tonnes were identified as feasible load reductions for inorganic phosphorus. Second, a TAE analysis of inorganic nitrogen or phosphorus load reduction was conducted by iteratively reducing loads using different reduction ratios for different sub-watersheds. In this case, we first implemented the load reductions for the three sub-watersheds (Direct Runoff (1)–(3)) closest to the outlet until a certain pre-specified maximum achievable reduction ratio was reached and then reduced loads from the five upstream sub-watersheds until compliance with Chl-*a* criteria was achieved in the three scenarios. The minimum load reductions identified by the TAE method (4.3–5.7 × 10³ tonnes for inorganic nitrogen, 0.62–0.91 × 10³ tonnes for phosphorus inorganic) are much larger than those of the BRRT-EILP method. Figure 3, Figure 4and Figure 5 show the minimum load reductions and pollution removal levels of the BRRT-EILP model compared to those of the two kinds of feasible alternatives generated by stochastic simulation and TAE methods. Specifically, the minimum inorganic phosphorus load reductions derived by the BRRT-EILP model were 7.0%, 14%, 4.8% and 5.9% less than those of the best stochastic simulation (SS) alternatives when the Chl-*a* criteria were levels below 12, 13, 14 and 15 μg/L, respectively (Figure 3a). The SS method required reducing inorganic phosphorus loads in all eight of the sub-watersheds, while the BRRT-EILP model required reductions in no more than five sub-watersheds (Figure 4). The reductions of inorganic nitrogen loads required by the BRRT-EILP solutions were 31%, 32%, 23% and 23% less than those of the best SS alternatives (Figure 3b). Additionally, the BRRT-EILP solutions only required inorganic nitrogen pollution removal in Swift Creek, Dry-Ashbrook Creek and Direct Runoff (1), while the removal rates of best SS alternatives required reductions greater than 0.3 in Sub-watersheds 1, 4, 5 and 8 (Figure 5). The plausible explanation for such sub-watershed selection by BRRT-EILP model is that the summer maximum vertically-averaged Chl-*a* concentration is more sensitive to load reductions in the selected sub-watersheds than the others. Another possible explanation is that the SS method uses random search, rather than the heuristic search applied by the BRRT-EILP model. In addition, the BRRT-EILP solutions were superior to those of the TAE method, since the latter ignores the difference in the sensitivity of Chl-*a* concentration to a unit of incremental load reduction across different sub-watersheds, while the BRRT-EILP model could rapidly identify critical sub-watersheds, resulting in the approximately most effective and reliable TMDL allocation. Therefore, the BRRT-EILP model may provide a new framework for updating the current TMDL allocation paradigm.

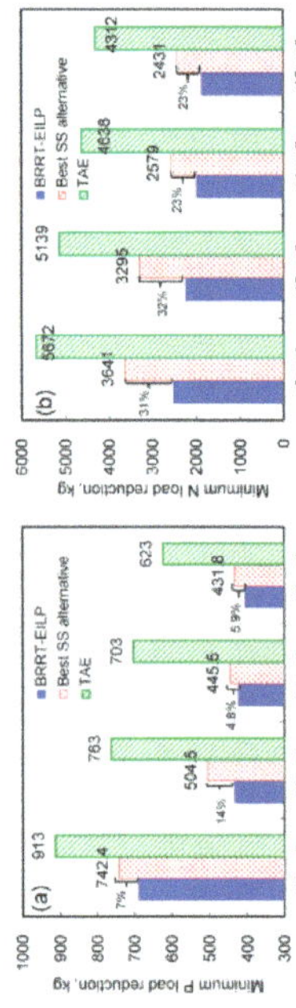

Figure 3. Minimum load reductions of the BRRT-EILP model, best stochastic simulation (SS) alternatives and trial-and-error (TAE) solutions that meet Chl-*a* criteria below 12, 13, 14 and 15 μg/L. (**a**) Inorganic phosphorus; (**b**) inorganic nitrogen. The percentage in the

panels is calculated as the difference in minimum load reductions between the BRRT-EILP model and the SS method divided by that of the SS method.

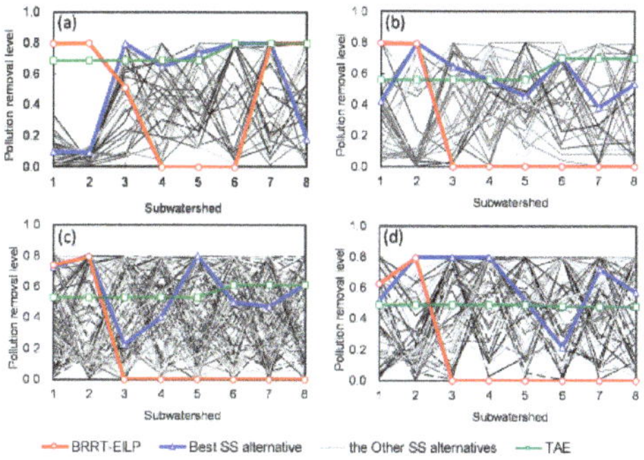

Figure 4. Pollution removal levels of inorganic phosphorus solved by the BRRT-EILP model, best stochastic simulation (SS) alternatives and the TAE method that meet diverse Chl-*a* criteria. (**a**) Chl-*a* = 12 μg/L; (**b**) Chl-*a* = 13 μg/L; (**c**) Chl-*a* = 14 μg/L and (**d**) Chl-*a* = 15 μg/L. The red, blue and green solid lines indicate the solutions of the BRRT-EILP model, the best stochastic simulation (SS) alternatives and the TAE method, while the other lines (in gray) represent the other alternatives with the SS method.

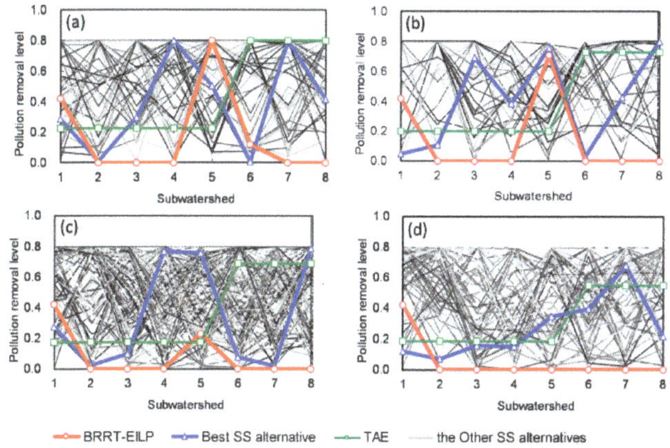

Figure 5. Pollution removal levels of inorganic nitrogen solved by the BRRT-EILP model and the TAE simulation method that meet diverse Chl-*a* criteria. (**a**) Chl-*a* = 12 μg/L; (**b**) Chl-*a* = 13 μg/L; (**c**) Chl-*a* = 14 μg/L and (**d**) Chl-*a* = 15 μg/L. The meaning of the colors is the same as in Figure 4.

Another advantage of the BRRT-EILP model is that it could provide extreme and non-extreme tradeoff analyses under uncertainty, which divides the solution space of $\mathbf{X}^{\pm}_{opt}$ based on the different risk levels of parameters. Three decision alternatives $(Z^{-}_{opt}, Z^{+}_{opt}$ and $E[Z^{\pm}_{opt}])$ could be generated corresponding to the different risk levels on both the left-hand ($yik^{\pm}$) and right-hand ($y^{*\pm}$) sides. Although the risk level could not be completely quantified, since the probability density functions of all enhanced-interval numbers were not easy to determine, the three decision alternatives with their qualitative risk levels for violating constraints represented a series of system performances of load reductions for decision makers who have a diverse tolerance of risk levels.

The optimized annual inorganic nitrogen and phosphorous load reductions could be further translated as the allowable nutrient TMDL equivalent, including individual waste load allocations (WLAs) for point sources and load allocations (LAs) for nonpoint sources. Although the permit condition in this study is written to ensure attainment of these load allocations as annual averages, it may be translated into ordinary discharge limitations with the associated risk levels. First, we suggest determining the potential load reduction by assessing the effectiveness of supporting additional BMPs for agriculture, followed by the assignment of different risk levels to account for the uncertainties of the optimal results. The sum of the WLAs needed to meet the TMDL is then calculated: $\sum$WLAs = TMDL − LAs. The next step is to translate WLAs into permits as discharge limitations. According to the USEPA guidelines [30], discharge limitations may be expressed either as numerical restrictions for continuous discharges or as average weekly and average monthly discharge limitations for WWTPs and as daily maximums and monthly averages for other dischargers, which is an additional step that can be taken in the future.

CONCLUSIONS

In this study, we developed a BRRT-EILP model to serve as an indirect SOM framework for risk-based optimal TMDL allocation. The proposed method was applied to the real-world case of the SCR watershed, and the results indicated that the BRRT-EILP model is computationally efficient and capable of risk-based decision making for TMDL allocation under uncertainty. The main conclusions are that:

(i) The interval linear regression equations derived from the BRRT approach accurately approximated the nonlinear stressor-response relationships represented in a sophisticated water quality model. The resulting BRRT-EILP optimization framework derived from interval linear regression equations could serve as an efficient decision support tool that overcomes the computational bottleneck from traditional direct SOM framework.

(ii) The BRRT-EILP model can be efficiently solved and easily acquire approximate optimal solutions at various risk levels. Furthermore, the solutions obtained from BRRT-EILP model could significantly improve TMDL allocations and the associated tradeoff analysis compared to the ones obtained by traditional stochastic simulation and TAE approaches.

(iii) Although the BRRT-EILP model was first proposed for nutrient TMDL allocation to mitigate harmful algal blooms in a reservoir, recent studies suggest that this model and its solution algorithm could also possibly be applied to new critical problems, like hypoxia [5,31] and heavy metals in other nonlinear aquatic systems under uncertainty. In particular, the objective function Equation (2a) could take various forms, including measures of reliability and equity, and the system constraints of Equation (2b)–(2e) could be modified to ensure compliance with both dissolved oxygen and heavy metal criteria. Nevertheless, the BRRT-derived equivalent needs to be further derived from a process-based model and verified by field data prior to its application in decision making.

Acknowledgments

This study was funded by the National Water Science and Technology Research Project (2013ZX07102-006), the National Natural Science Foundation of China (No.

41201077), the Research Fund for the Doctoral Program of Higher Education of China (No.20120001120129) and the 111 Project (No.B14001). We also thank Huaicheng Guo and Rui Zou for their support and suggestions on the manuscript.

Author Contributions

Feng Zhou had the original idea for the study and, with all co-authors carried out the design. Yue Zhao were responsible for recruitment and follow-up of study participants. Feng Zhou, Yanjun Dong, Jing Wu and Jiangli Zheng were responsible for data cleaning and Feng Zhou carried out the analyses. Feng Zhou drafted the manuscript, which was revised by all authors. All authors read and approved the final manuscript.

Conflicts of Interest

The authors declare no conflict of interest.

REFERENCES

1. USEPA. *Allocating Loads and Wasteloads, Technical Support Documents, Total Maximum Daily Load (TMDL)*; US EPA Press: Washington, DC, USA, 2003.
2. Jia, Y.B.; Culver, T.B. Robust optimization for total maximum daily load allocations. *Water Resour. Res.* **2006**, *42*.
3. Cerco, C.F.; Noel, M.R.; Kim, S.C. Three-dimensional management model for lake washington, part ii: Eutrophication modeling and skill assessment. *Lake Reserv. Manag.* **2006**, *22*, 115–131.
4. Cole, T.; Wells, S.A. *CE-QUAL-W2: A Two-Dimensional, Laterally Averaged, Hydrodynamic and Water Quality Model, Version 3.1*; US Army Engineering and Research Development Center: Vicksburg, MS, USA, 2003.
5. Boesch, D.F. The gulf of Mexico's dead zone. *Science* **2004**, *306*, 977–978.
6. Dhar, A.; Datta, B. Optimal operation of reservoirs for downstream water quality control using linked simulation optimization. *Hydrol. Process.* **2008**, *22*, 842–853.
7. He, L.; Huang, G.H.; Lu, H.W. Health-risk-based groundwater remediation system optimization through clusterwise linear regression. *Environ. Sci. Technol.* **2008**, *42*, 9237–9243. [PubMed]
8. He, L.; Huang, G.H.; Zeng, G.M.; Lu, H.W. An integrated simulation, inference, and optimization method for identifying groundwater remediation strategies at petroleum-contaminated aquifers in western canada. *Water Res.***2008**, *42*, 2629–2639.
9. Huang, G.H.; Huang, Y.F.; Wang, G.Q.; Xiao, H.N. Development of a forecasting system for supporting remediation design and process control based on NAPL-biodegradation simulation and stepwise-cluster analysis. *Water Resour. Res.* **2006**, *42*.
10. Rejani, R.; Jha, M.K.; Panda, S.N. Simulation-optimization modelling for sustainable groundwater management in a coastal basin of orissa, india. *Water Resour. Manag.* **2009**, *23*, 235–263.
11. Saadatpour, M.; Afshar, A. Waste load allocation modeling with fuzzy goals; simulation-optimization approach. *Water Resour. Manag.* **2007**, *21*, 1207–1224.
12. Zhou, F.; Chen, G.X.; Huang, Y.F.; Yang, J.Z.; Feng, H. An adaptive moving finite volume scheme for modeling flood inundation over dry and complex topography. *Water Resour. Res.* **2013**, *49*, 1914–1928.
13. Zhou, F.; Chen, G.X.; Noelle, S.; Guo, H.C. A well-balanced stable generalized riemann problem scheme for shallow water equations using adaptive moving unstructured triangular meshes. *Int. J. Numer. Methods Fluids* **2013**, *73*, 266–283.
14. Qin, X.S.; Huang, G.H.; Chakma, A. A stepwise-inference-based optimization system for supporting remediation of petroleum-contaminated sites. *Water Air Soil Pollut.* **2007**, *185*, 349–368.
15. Guo, H.C.; Gao, W.; Zhou, F. Three-level trade-off analysis for decision making in environmental engineering under interval uncertainty. *Eng. Optim.* **2014**, *46*, 377–392.
16. Huang, G.H. A stepwise cluster-analysis method for predicting air-quality in an urban-environment. *Atmos. Environ. B Urban Atmos.* **1992**, *26*, 349–357.
17. Sahinidis, N.V. Optimization under uncertainty: State-of-the-art and opportunities. *Comput. Chem. Eng.* **2004**, *28*, 971–983.
18. Wang, Z.; Gao, W.; Cai, Y.L.; Guo, H.C.; Zhou, F. Joint optimization of population pattern and end-of-pipe control under uncertainty for lake dianchi water-quality management. *Fresenius Environ. Bull.* **2012**, *21*, 3693–3704.
19. Zhou, F.; Huang, G.H.; Chen, G.X.; Guo, H.C. Enhanced-interval linear programming. *Eur. J. Oper. Res.* **2009**, *199*, 323–333.
20. Birge, J.R.; Louveaux, F.V. *Introduction to Stochastic Programming*; Springer: New York, NY, USA, 1997.
21. Zou, R.; Liu, Y.; Riverson, J.; Parker, A.; Carter, S. A nonlinearity interval mapping scheme for efficient waste load allocation simulation-optimization analysis. *Water Resour. Res.* **2010**, *46*.
22. Zou, R.; Lung, W.S.; Wu, J. An adaptive neural network embedded genetic algorithm approach for inverse water quality modeling. *Water Resour. Res.* **2007**, *43*.
23. Zhou, F.; Guo, H.C.; Chen, G.X.; Huang, G.H. The interval linear programming: A revisit. *J. Environ. Inform.* **2008**, *11*, 1–10.
24. Iorgulescu, I.; Beven, K.J. Nonparametric direct mapping of rainfall-runoff relationships: An alternative approach to data analysis and modeling? *Water Resour. Res.* **2004**, *40*.
25. Breiman, L.; Friedman, J.H.; Olshen, R.A.; Stone, C.G. *Classification and Regression Trees*; Wadsworth International Group: Belmont, CA, USA, 1984.
26. Zhou, F.; Shang, Z.Y.; Zeng, Z.Z.; Piao, S.L.; Ciais, P.; Raymond, P.; Wang, X.H.; Wang, R.; Chen, M.P.; Yang, C.L.; *et al.* New model for capturing the variations of fertilizer-induced emission factors of n2o. *Glob. Biogeochem. Cycles* **2015**, *29*.
27. Harmel, R.D.; Cooper, R.J.; Slade, R.M.; Haney, R.L.; Arnold, J.G. Cumulative uncertainty in measured streamflow and water quality data for small

watersheds. *Trans. ASABE* **2006**, *49*, 689–701.

28. Wu, J.; Zou, R.; Yu, S.L. Uncertainty analysis for coupled watershed and water quality modeling systems. *J. Water Resour. Plann. Manag.* **2006**, *132*, 351–361.

29. Wetzel, R.G. *Limnology—Lake and River Ecosystems*, 3rd ed.; Academic Press: New York, NY, USA, 2001.

30. U.S. Environmental Protection Agency (USEPA). *Protocol for Developing Nutrient TMDLs*; EPA 841-B-99-007; Office of Water (4503F), United States Environmental Protection Agency: Washington, DC, USA, 1999.

31. Conley, D.J.; Paerl, H.W.; Howarth, R.W.; Boesch, D.F.; Seitzinger, S.P.; Havens, K.E.; Lancelot, C.; Likens, G.E. Ecology controlling eutrophication: Nitrogen and phosphorus. *Science* **2009**, *323*, 1014–1015.

CHAPTER 4

STOCHASTIC URBAN PLUVIAL FLOOD HAZARD MAPS BASED UPON A SPATIAL-TEMPORAL RAINFALL GENERATOR

Nuno Eduardo Simões, Susana Ochoa-Rodríguez, Li-Pen Wang, Rui Daniel Pina, Alfeu Sá Marques, Christian Onof and João P. Leitão

Keywords: Urban pluvial flooding, flood hazard mapping, flood risk mapping, stochastic, rainfall radar, spatial-temporal

INTRODUCTION

Urban pluvial flooding is expected to become more frequent as a result of urbanisation and the envisaged effects of climate changes [1]. This type of flooding can happen virtually anywhere and has the potential to cause significant damage and disruption in highly urbanised areas, where the density of buildings, critical infrastructure and population is usually high. The European Directive 2007/60/EC [2] on the assessment and management of flood risks requires Member States to assess the risk of flooding. To fulfil this requirement, flood hazard maps are used to classify the likelihood of flooding and quantify its consequences.

Spatially-uniform design storms, associated to a given return period usually estimated based upon point rainfall records, have been traditionally used in the design, planning and evaluation of drainage systems, including evaluation of urban pluvial flood hazard. Although the use of such design storms is handy, the uniformly-distributed design storm concept neglects the impact of the spatial variability of rainfall fields, which has shown to have a significant impact on the hydrological response of urban catchments [3,4,5,6,7,8,9,10]. In addition, climate change may lead to more extreme rainfall events that usually have a high variability across both, temporal and spatial scales [11].

Hence, the use of uniform design storms may lead to substantial over- or underestimation of hazard which can have significant financial implications. Moreover, in most frequent studies the return period of the flood is assumed to be equal to that of the rainfall event [12,13]. However, the flood event frequency might be different from the frequency of the associated rainfall event. One of the main aims of a stormwater drainage system is to store and transport runoff in such a way that the likelihood of urban pluvial flooding is minimised. Therefore, the return period of a given flood event is generally lower than that of the rainfall event which generates the flooding. It should be noted that this assumption is only valid when the drainage system is well designed (e.g., no sewer bottlenecks) and in a good operational state (e.g., unobstructed sewer inlets) and a good structural condition (e.g., no collapsed sewer pipes).

Several studies have been published on flood risk quantification using a combination of flow/frequency, stage/damage and damage/frequency curves [14,15,16]. Ahmad and Simonovic [17] proposed a new approach to urban flood risk assessment, using fuzzy set theory to assess the spatial and temporal variability of urban flood damage, and Sun *et al.* [18] proposed a general framework for a risk-based storm sewer network design that takes future risk into account. A detailed review on flood impact assessment in urban areas presented by Hammond *et al.* [11] and Nuswantoro *et al.* [19] indicated the importance of using a probabilistic approach for deriving flood inundation maps at a regional scale.

Citation: Nuno Eduardo Simões, Susana Ochoa-Rodríguez, Li-Pen Wang, Rui Daniel Pina, Alfeu Sá Marques, Christian Onof and João P. Leitão, Stochastic Urban Pluvial Flood Hazard Maps Based upon a Spatial-Temporal Rainfall Generator, doi:10.3390/w7073396.

With the purpose of contributing towards a better estimation of flood hazard and risk in urban areas, this paper presents a methodology to stochastically assess urban pluvial flood hazard, based upon a stochastic spatial-temporal rainfall generator developed by McRobie *et al.* [20]. The proposed approach was tested in a small urban catchment (~9 km^2) in north-east London (UK) with a history of pluvial flooding. In the next section the pilot location used to demonstrate the proposed methodology is presented. Then, the rainfall generator used to produce an ensemble of spatially and temporally variable design storms is described. Afterwards, results are presented and discussed. Lastly, the main conclusions of the present study are outlined and future work is discussed.

PILOT LOCATION: THE CRANBROOK CATCHMENT

The Cranbrook catchment is located within the London Borough of Redbridge (north-east part of Greater London, UK, Figure 1) and has experienced several pluvial, fluvial and coincidental flooding in the past. It has a drainage area of approximately 9 km^2 and is predominantly urban, with a combination of residential and commercial areas, plus two off-line lakes, a couple of parks and playing fields. It has a population of approximately 41,000 inhabitants (population density ~4500 persons per km^2).

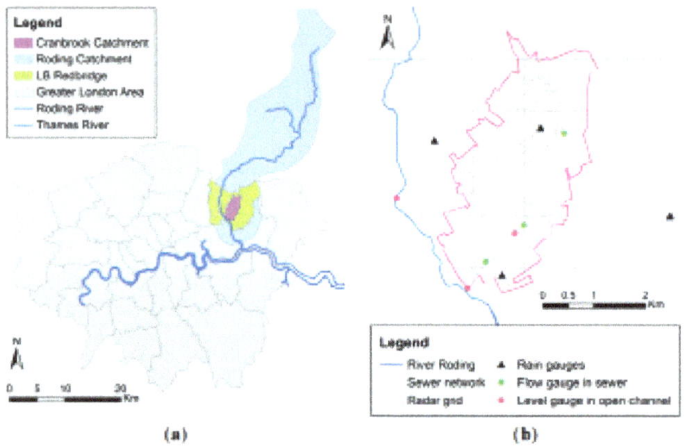

Figure 1. Cranbrook catchment: (**a**) general location; (**b**) sensor location, sewer network and radar grid over the catchment.

The main water course (the Cran Brook) is about 5.75 km long, of which 5.69 km are culverted and have become part of the stormwater drainage system, which is mainly separate. The stormwater drainage system of this catchment discharges into the River Roding, which is a tributary of the River Thames.

SPATIAL-TEMPORAL STOCHASTIC RAINFALL GENERATOR

A stochastic rainfall generator for urban-scale applications was used in this paper to generate design storms with spatial and temporal variations. This generator was developed by McRobie *et al.* [20], where storms were modelled as a Poisson clustering process and the structure of storm cells was approximated using a bivariate Gaussian model. The storm cell features represented in the model include maximum intensity (r_{max}) and spread in the direction of and perpendicular to the storm motion (s_1 and s_2 in Figure 2). This generator is an extension of the technique proposed by Willems [21]. While Willems' [21] generator's parameters were calibrated based upon historical point rainfall records from a dense network of rain gauges, the parameters of the generator used in this paper (*i.e.*, McRobie *et al.* [20]) were calibrated using radar rainfall records from 45 historical storm events which reportedly led to flooding in the Greater London area during the period 2000–2011. The statistical models employed and estimated parameters are summarised in Table 1. It is expected that more realistic spatial features of storm cells can be identified using radar images, as opposed to using rain gauge records.

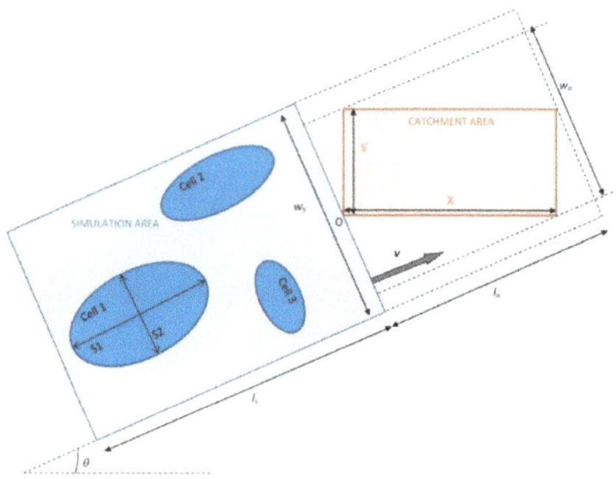

Figure 2. Schematic representation of storm movement across the catchment (adapted from Figure 1 in McRobie *et al.* [20]).

Table 1. Summary of parameter set and the fitted distributions thereof.

Parameter Set	Fitted Distribution	Distribution Parameters
Storm Duration (d)	Gamma	a (-) = 2.44, b (h) = 17.92
Number of Cells Per Square Kilometre (λ)	Normal	μ (km^{-2}) = 0.0099, σ (km^{-2}) = 0.0032
Velocity Magnitude (v)	Log Normal	μ (km/h) = 78.30, σ (km/h) = 23.64
Velocity Direction (θ)	Log Normal	μ (rads) = 0.51, σ (rads) = 1.16
Cell Spread in Direction of Motion (s_1)	Log Normal	μ = 0.700, σ = 1.260
Cell Spread in Direction Perpendicular to Motion (s_2)	Log Normal	μ = 0.704, σ = 1.201
Maximum Intensity in Cell (r_{max})	Generalised Pareto	k = 0.586, σ = 1.040, θ = 5.0 mm

With the employed rainfall generator during a storm realisation, several storm cells are initially generated (*i.e.*, sampled from the probability distributions described above) within the simulation area (see Figure 2). The overall storm simulation area then moves across the catchment at constant speed (v) and direction (θ). The number of cells to be generated within the simulation area is determined based upon the sampled cell density per km^2 (λ) and the simulation area. The dimensions of the simulation area (*i.e.*, w_s and l_s in Figure 2) are estimated based upon the sampled storm duration (d), the sampled storm speed (v) and the dimension of the catchment area (*i.e.*, w_a and l_a in Figure 2). For a detailed formulation of the generator, the reader can refer to McRobie *et al.* [20].

The aforementioned spatial-temporal rainfall generator was used to sample over 450 design storm realisations, with a spatial resolution of 1 km and temporal resolution of 5 min, with the top 45 realisations (Figure 3) being selected according to the ranking of 60-min duration rainfall maxima. The selected realisations were then used as input for hydraulic modelling. Each rainfall realisation shows a different spatial pattern (different location of rainfall cells in the simulation area; different number of rainfall cells and different speed and direction, Figure 2). The distributions of the selected storm durations and the rainfall maxima (in mm, at one radar pixel) in 15-, 30- and 60-min durations are summarised in Figure 3.

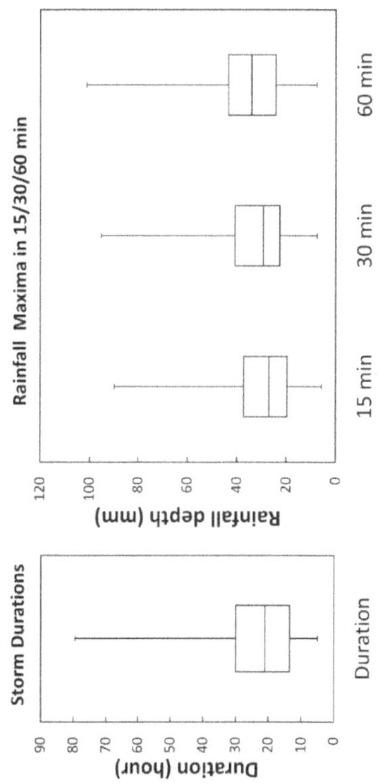

Figure 3. Boxplots of selected storms' features: (**left**) storm duration in hours; and (**right**) rainfall maxima (at one radar pixel) in 15-, 30- and 60-min time durations.

To further highlight the impact of rainfall spatial variability as well as the importance of using ensembles of storms instead of a single design storm, a spatially-uniform storm event was also generated and used as input for hydraulic modelling. Such storm was generated by averaging in space all pixel values, from the selected 45 storm realisations, at each time step. This approach leads to smoothing of local extremes (including highs and lows), but ensures preservation of the total rainfall volume falling into, thus facilitating comparison of results.

HYDRAULIC MODEL

A 1D/2D dual-drainage, physically-based model of the Cranbrook catchment was set up in InfoWorks CS 14.0. The data source used includes network and operational data for the 1D sewer model and the 2D overland surface model was created based on an available Digital Terrain Model (DTM—a LiDAR with 1 m horizontal resolution). Buildings polygons and land use data were used to characterise the model and define the surface mesh (e.g., mesh resolution, break lines, voids and boundaries).

Rainfall is applied to the model through 1765 subcatchments with areas ranging from 30 m^2 to 115,400 m^2 (subcatchment median size is 3700 m^2), slopes varying between 0.0 m/m and 0.28 m/m (average of 0.05 m/m). Runoff volumes are estimated with the NewUK model and runoff routing at subcatchments is conducted using the Wallingford model. The hydraulic model uses the dynamic wave model to solve the flow equations for the 1D sewer network and for the 2D overland surface. The 1D network consists of 1825 conduits and 1776 nodes. The conduits have a total length of 98.1 km, an average slope of 1% and most of pipes are made of concrete with circular cross sections, with diameters ranging from 100 to 1950 mm. Typical roughness coefficients were adopted for each material type. The nodes are divided in 1760 manholes, 13 storage units and three outfalls. The 2D model of the overland surface is based on an irregular mesh with 292,944 elements with areas up to 100 m^2 (average area of 25 m^2). The model was verified in 2011 using rainfall and water level data from a local monitoring system operated by the authors of the paper since 2010 [22,23].

STOCHASTIC FLOOD HAZARD MAPPING

In the proposed methodology, the spatially and temporally varying rainfall events sampled from the rainfall generator are fed into the urban pluvial flood model of the area under consideration. The resulting hydraulic outputs are statistically analysed to estimate the likelihood of a given location being flooded (*i.e.*, probabilistic flood hazard) (Figure 4).

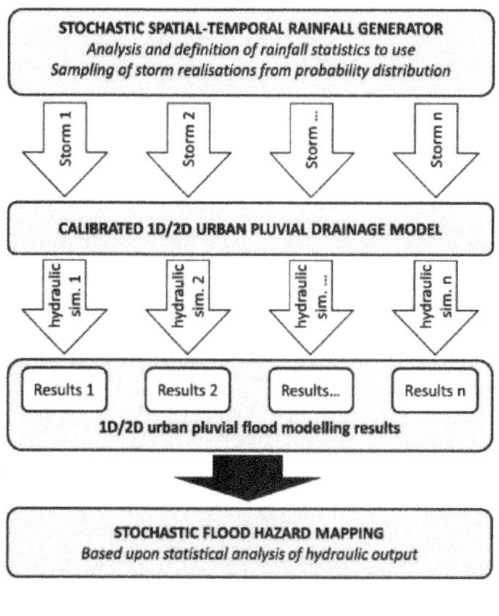

Figure 4. General approach for generation of stochastic urban pluvial flood hazard maps based upon a stochastic spatial temporal rainfall generator.

To demonstrate the proposed approach, a total of 45 hydraulic simulations were carried out, each using a storm event generated with the rainfall generator. Moreover, for the sake of comparison, a spatially-uniform storm (see description inSection 3) was also used as input for hydraulic simulation. Apart from the different rainfall inputs, no other changes were made to the hydraulic model in each of the simulations.

The hydraulic results were filtered out based upon a water depth threshold; in this case, only surface elements with water depth greater than 0.1 m were considered as flooded locations.

Taking into account the results of the 45 hydraulic simulations, the least and most extreme rainfall events correspond to storms (rainfall realisations) number 7 and 27, respectively. Table 2 presents maximum, minimum and average flooded area values for the 45 storm realisations, as well as flooded area values for the single spatially-uniform storm event.

As can be seen from Table 2, there is a large variation in flood extent for the different storm events under consideration (notice difference between maximum and minimum flood extent values). This variation is due to different rainfall characteristics, including variable spatial and temporal distribution of storm cells. The flood extent of the uniform rainfall event is close to the minimum flood extent area obtained for the stochastic storm events; this can be explained by the smoothing of rainfall peaks caused by spatial averaging of rainfall rate values.

Table 2. Statistics of flood extents associated to the storm events under consideration.

Flood Extent (m^2)			
Stochastic Realisations (45 Storm Events)			Uniform Rainfall
Minimum	Maximum	Average	
53,464	1,088,000	381,072	77,187

Figure 5a shows the maximum flood extent, corresponding to stochastic storm event number 27. In Figure 5b details of flood extents for three storm events are provided, including the maximum and minimum flood extents associated to the stochastic storm realisations, as well as the flood extent associated to the spatially-uniform storm. As can be seen in Figure 5b, the uniform rainfall event and storm 7 cause flooding in a significantly smaller area when compared with that flooded by Storm 27. The least extreme event (Storm 7) caused a flood that, in this area, is similar to that caused by the uniform average rainfall event, Figure 5b (middle and bottom).

The ultimate goal of the stochastic analysis presented in this paper is to generate stochastic flood maps, *i.e.*, to represent the catchment locations where flood is more likely to occur. This is achieved by calculating the number of times a given area is flooded (an area is the polygon of the surface mesh used in the 2D overland flow modelling). The results of this analysis are presented in Figure 6. This figure shows that some locations got flooded with every rainfall event considered in this study. In contrast, other locations were flooded only a few times. With regards to the flood extent associated to the uniform rainfall event, it can be seen that it generally matches with the high flood frequency areas identified from the stochastic analysis. However, it fails to identify other areas which may only flood under more localised storm conditions and which should be considered by flood risk managers. This highlights the importance of taking the spatial and temporal variability of the rainfall event into account when assessing urban pluvial flood hazard. Rainfall inputs, including design storms, encompass large uncertainties which hinder decision making. The stochastic approach proposed herein constitutes an alternative for dealing with these uncertainties and allows a more comprehensive assessment of flood hazard in a given area. This can help decision makers in the implementation of actions in areas where flooding is more likely to occur, thus reducing flood risk in those areas.

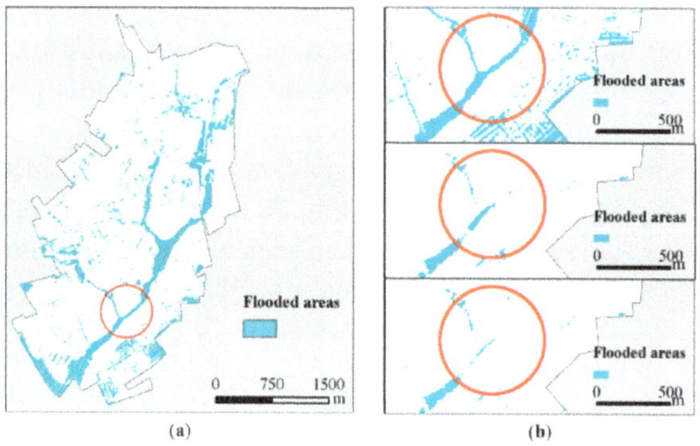

Figure 5. Simulated flood extent for three storm event realisations: storm event 27, which led to the maximum flood extent; storm event 7, which led to the minimum flood extent, and spatially-uniform storm event used for comparison. (**a**) Maximum flood extent (Storm 27); (**b**) detail of flood extents for three storm events (Storm 27, uniform rainfall and Storm 7, from top to bottom).

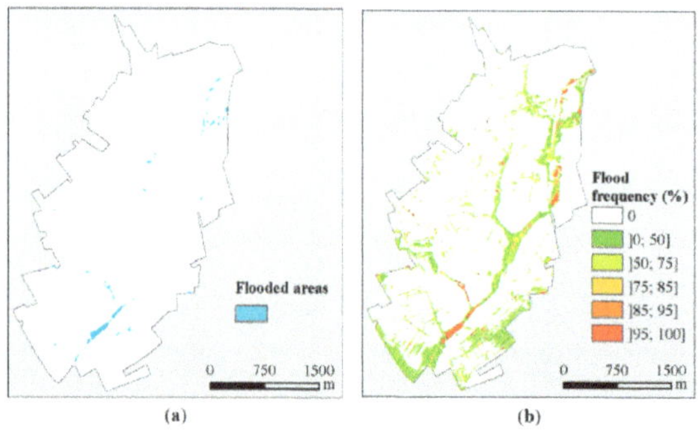

Figure 6. Modelled flood maps based on average uniform rainfall and on spatial-temporal stochastic rainfall. (**a**) Uniform rainfall flood map; (**b**) stochastic flood map.

CONCLUSIONS AND OUTLOOK

Traditionally, the return period of a flood event is associated to that of the storm event. This is however insufficient because different rainfall events with the same return period can generate flooding of different magnitudes (water depth and flood extent) and thus of different return periods. A new methodology to generate urban pluvial flood hazard maps stochastically has been presented in this paper. First, a stochastic rainfall generator for urban-scale applications was employed to generate design storms with spatial and temporal variation. Each design storm was applied to a 1D/2D drainage hydraulic model to generate flood maps. Flood hazard maps were produced based upon all the resulting hydraulic outputs (*i.e.*, water depth and flood extent).

The use of a stochastic spatial-temporal rainfall generator has enabled the generation of relatively realistic storms that vary in space and time. In turn, this has enabled a stochastic assessment of pluvial flood hazard in an urban catchment. The resulting probabilistic flood hazard map shows that some locations got flooded in almost all of the events under considerations, whereas other locations were only flooded a few times. This highlights the important role that rainfall spatial-temporal variability plays in the catchment response and the need to consider it when evaluating flood risk. Moreover, the probabilistic map provides a more comprehensive assessment of flood hazard, as compared to the flood hazard maps obtained by using single spatially-uniform design storms. Thus, the probabilistic flood hazard maps can enable better informed decisions which can result in improved flood risk management.

The proposed methodology leads to a significantly different mind-set from the current definition of design storms, where the spatial distribution is assumed to be uniform. Furthermore, this new concept creates an open question about the use of the return period to characterise spatial and temporal variable storms; this is worth further investigating. It is worth mentioning that the framework presented in this paper can be applied in other areas and with different rainfall generators and hydraulic models.

The authors suggest that future studies should involve a larger number of rainfall realisations, with the respective hydraulic simulations being conducted. Aside from the flood depth and extent, other parameters may also be included, such as velocity and velocity/water depth ratio.

The methodology presented in this paper is being further developed to generate stochastic flood risk maps, which incorporate potential consequences.

ACKNOWLEDGMENTS

The authors would like to thank Innovyze for providing the InfoWorks software. The second and third authors would like to acknowledge the support of the European Union's Interreg IVB NWE Programme to the RainGain project. The third author would like to acknowledge the support of the the Research Foundation–Flanders (FWO) and the PLURISK project for the Belgian Science Policy Office. The fourth author acknowledges the financial support from the Fundação para a Ciência e Tecnologia—Ministério para a Ciência, Tecnologia e Ensino Superior, Portugal [SFRH/BD/88532/2012]. The authors also thank Thames Water and MWH (Global) for providing historical radar data.

AUTHOR CONTRIBUTIONS

Nuno Eduardo Simões, João Paulo Leitão, Susana Ochoa, Lipen Wang and Alfeu Sá Marques had the original idea for the study. Lipen Wang and Christian Onof generated and analysed rainfall data. Nuno Eduardo Simões, Susana, João Paulo Leitão and Rui Pina build the hydraulic model and processed the hydraulic results. All authors analysed and discussed the results. The paper was written with contribution from all authors. All authors read and approved the final manuscript.

CONFLICTS OF INTEREST

The authors declare no conflict of interest.

REFERENCES

1. Overview of Climate Change Effects Which May Impact the Urban Water Cycle. Available online: http://www.prepared-fp7.eu/viewer/file.aspx?FileInfoID=439 (accessed on 24 June 2015).
2. European Parliament & Council. Directive 2007/60/EC of the European Parliament and of the Council of 23 October 2007 on the assessment and management of flood risks. *Off. J. Eur. Union* **2007**, *2007*, 27–34.
3. Aronica, G.; Cannarozzo, M. Studying the hydrological response of urban catchments using a semi-distributed linear non-linear model. *J. Hydrol.* **2000**, *238*, 35–43.
4. Berne, A.; Delrieu, G.; Creutin, J.-D.; Obled, C. Temporal and spatial resolution of rainfall measurements required for urban hydrology. *J. Hydrol.* **2004**, *299*, 166–179.
5. Gires, A.; Giangola-Murzyn, A.; Abbes, J.-B.; Tchiguirinskaia, I.; Schertzer, D.; Lovejoy, S. Impacts of small scale rainfall variability in urban areas: A case study with 1D and 1D/2D hydrological models in a multifractal framework.*Urban Water J.* **2014**, *2014*.
6. Gires, A.; Onof, C.; Maksimovic, C.; Schertzer, D.; Tchiguirinskaia, I.; Simoes, N. Quantifying the impact of small scale unmeasured rainfall variability on urban runoff through multifractal downscaling: A case study. *J. Hydrol.* **2012**, *442–443*, 117–128.
7. Leitão, J.P.; Djordjević, S.; Prodanović, D.; Maksimović, Č. Spatially distributed rainfall for surface runoff calculations in urban catchments. In Proceedings of the 8th International Workshop on Precipitation in Urban Areas, St. Moritz, Switzerland, 10–13 December 2009.
8. Schellart, A.N.A.; Shepherd, W.J.; Saul, A.J. Influence of rainfall estimation error and spatial variability on sewer flow prediction at a small urban scale. *Adv. Water Resour.* **2012**, *45*, 65–75.
9. Segond, M.L.; Neokleous, N.; Makropoulos, C.; Onof, C.; Maksimović, Č. Simulation and spatio-temporal disaggregation of multi-site rainfall data for urban drainage applications. *Hydrol. Sci. J.* **2007**, *52*, 917–935.
10. Ochoa-Rodriguez, S.; Wang, L.-P.; Gires, A.; Pina, R.D.; Reinoso-Rondinel, R.; Bruni, G.; Ichiba, A.; Gaitan, S.; Cristiano, E.; Assel, J.V.; *et al.* Impact of Spatial and Temporal Resolution of Rainfall Inputs on Urban Hydrodynamic Modelling Outputs: A Multi-Catchment Investigation. *J. Hydrol.* in press. [CrossRef]
11. Hammond, M.J.; Chen, A.S.; Djordjević, S.; Butler, D.; Mark, O. Urban flood impact assessment: A state-of-the-art review. *Urban Water J.* **2013**, *12*, 14–29.
12. Gilles, D.; Young, N.; Schroeder, H.; Piotrowski, J.; Chang, Y.-J. Inundation Mapping Initiatives of the Iowa Flood Center: Statewide Coverage and Detailed Urban Flooding Analysis. *Water* **2012**, *4*, 85–106.
13. Olsen, A.S.; Zhou, Q.; Linde, J.J.; Arnbjerg-Nielsen, K. Comparing Methods of Calculating Expected Annual Damage in Urban Pluvial Flood Risk Assessments. *Water* **2015**, *7*, 255–270.

14. Bowering, E.A.; Peck, A.M.; Simonovic, S.P. A flood risk assessment to municipal infrastructure due to changing climate part I: Methodology. *Urban Water J.* **2014**, *11*, 20–30.

15. Leitão, J.P.; do Céu Almeida, M.; Simões, N.E.; Martins, A. Methodology for qualitative urban flooding risk assessment. *Water Sci. Technol.* **2013**, *68*, 829–838.

16. Nascimento, N.; Machado, M.L.; Baptista, M.; de Paula E Silva, A. The assessment of damage caused by floods in the Brazilian context. *Urban Water J.* **2007**, *4*, 195–210.

17. Ahmad, S.S.; Simonovic, S.P. Spatial and temporal analysis of urban flood risk assessment. *Urban Water J.* **2013**, *10*, 26–49.

18. Sun, S.; Djordjević, S.; Khu, S.-T. A general framework for flood risk-based storm sewer network design. *Urban Water J.* **2011**, *8*, 13–27.

19. Nuswantoro, R.; Diermanse, F.; Molkenthin, F. Probabilistic flood hazard maps for Jakarta derived from a stochastic rain-storm generator. *J. Flood Risk Manag.* **2014**, *2014*.

20. McRobie, F.H.; Wang, L.-P.; Onof, C.; Kenney, S. A spatial-temporal rainfall generator for urban drainage design. *Water Sci. Technol.* **2013**, *68*, 240–249.

21. Willems, P. A spatial rainfall generator for small spatial scales. *J. Hydrol.* **2001**, *252*, 126–144.

22. Simoes, N.E. Urban Pluvial Flood Forecasting. Ph.D. Thesis, Imperial College London, London, UK, 2012.

23. Wang, L.-P.; Ochoa-Rodríguez, S.; Simões, N.E.; Onof, C.; Maksimović, Č. Radar-Raingauge data combination techniques: A revision and analysis of their suitability for urban hydrology. *Water Sci. Technol.* **2013**, *68*.

CHAPTER 5 STOCHASTIC FLOCCULATION MODEL FOR COHESIVE SEDIMENT SUSPENDED IN WATER

Hyun Jung Shin, Minwoo Son and Guan-hong Lee

Keywords: Cohesive sediment, flocculation process, size distribution, stochastic approach, distribution function

INTRODUCTION

The flocculation process, a series of aggregation and disaggregation (so called breakup) processes, is one of the most important characteristics of suspended cohesive sediment such as mud and organic matter. Through the flocculation process, the size and density of cohesive sediment continuously vary, resulting in variation of settling velocity and particle response time. Because the settling velocity and particle response time significantly affect the advection and diffusion process of suspended sediment, an improved understanding of the flocculation process is paramount for cohesive sediment transport. Therefore, the flocculation process has been intensively studied. The flocculation process is mainly due to eletro- or biological-chemical attraction of fine sediment (e.g., [1]). When suspended particles collide with each other, there is a chance of aggregation by attraction. Collisions of suspended sediment are caused by Brownian motion, differential settling, and turbulent shear [2,3,4]. It has been concluded in many studies that turbulent shear is the dominant mechanism of suspended sediment collisions (e.g., [5,6]). Moreover, turbulent shear is considered as the main mechanism of floc breakup (e.g., [1,7,8]). Many laboratory studies on the effects of turbulent shear have been conducted utilizing a mixing tank and Couette viscometer in order to control the intensity of turbulent shear (e.g., [9,10,11,12]). From these studies, it is apparent that the flocculation process of aggregation (mainly due to particle collision) and breakup is significantly governed by turbulent shear.

With the aim of quantitatively modeling the flocculation process, many flocculation models have been proposed. Generally, flocculation models can be classified into two groups at present; a population balance equations model (PBE) and a floc growth model (FGM). A PBE, often called a size class flocculation model, is based on discrete size classes, and typically calculates the change of floc numbers in each size class (e.g., [13,14,15]). A PBE has the advantages of considering many mechanisms of flocculation and calculating a size distribution of flocs. However, a PBE is too numerically expensive to be combined with numerical sediment transport models [16] and demands many empirical parameters, increasing uncertainties. Moreover, a discrete PBE approach is limited in terms of ensuring the mass conservation of flocs in a closed system, although many schemes have been proposed to overcome this weakness (e.g., [17,18]). A FGM, often called a Lagrangian flocculation model, is a differential equation usually composed of two terms, an aggregation term and a breakup term, representing the increase in floc size due to aggregation and the decrease due to breakup, respectively. A FGM is simple enough to be combined with numerical sediment transport models. Furthermore, a FGM ensures the mass conservation of cohesive sediment in a closed system because it calculates the average size of flocs. Based on the collisional frequency equation of Levich [19] and dimensional analysis, Winterwerp [8] proposes a semi-empirical FGM under the assumption that the fractal dimension of flocs is constant. However, many studies show that the fractal dimension of natural flocs is variable (e.g., [20,21]). Based on Levich [19] and Khelifa and Hill [20], Son and Hsu [7] improved Winterwerp's FGM [8] by adopting an empirical equation with a variable fractal dimension. However, neither Son and Hsu [7] nor Winterwerp [8] appropriately address the temporal evolution of floc

Citation: Hyun Jung Shin, Minwoo Son and Guan-hong Lee, Stochastic Flocculation Model for Cohesive Sediment Suspended in Water, doi:10.3390/w7052527.

size, although both calculate the equilibrium floc size under constant turbulence intensity conditions. Son and Hsu [1] further theoretically and mathematically derive an equation for the yield strength of flocs, which depends on the floc size and fractal dimension. The equation is similar to an empirical equation from Sonntag and Russel [22] when one empirical parameter, *r*, is set to be about 0.7 (see [1] for more details). By incorporating this equation with a FGM from Son and Hsu [7], Son and Hsu [1] propose a new FGM using a variable fractal dimension and yield strength. The model from Son and Hsu [1] has been validated with laboratory experiments by Spice *et al.* [23] and Burban *et al.* [24]. From the simulation results, it is clearly established that a new FGM from Son and Hsu [1] has the capability to replicate the temporal evolution of floc size as well as the equilibrium floc size. In addition, from comparison with results calculated by models developed by Son and Hsu [7] and Winterwerp [8], it is also concluded that both the variable fractal dimension and the variable yield strength of flocs are critical to modeling the flocculation process. Son and Hsu [25] develop a sediment transport model for cohesive sediment and incorporate it into FGMs from Winterwerp [8], Son and Hsu [7], and Son and Hsu [1]. Through validation of models with field data measured in the Ems/Dollard estuary [26], it is shown that a sediment transport model combined with Son and Hsu's model [1] calculates the characteristics observed in the Ems/Dollard estuary. Son and Hsu's model [1] is used in Son and Lee [27] again in order to understand the role of flocculation under oscillatory flow representing short waves. It is found in Son and Lee [27] that the flux of sediment under fixed short wave conditions has opposite directions depending on the flocculation process. To the best knowledge of the authors, the model from Son and Hsu [1] is one of few flocculation models with results that are in good agreement with field data when combined with a high-resolution sediment transport model which quantitatively calculates interactions between turbulence and floc dynamics.

As mentioned above, a FGM is numerically efficient and ensures mass conservation. However, compared to PBE, a FGM is disadvantaged in the sense that it does not have the capability to calculate size distribution of flocs. This is because a FGM calculates the average size of flocs using a deterministic equation. In order to simulate sediment transport, Man and Tsai [28] propose the stochastic advection-diffusion equation based on the Langevin equation, solving the equation using numerical schemes for stochastic partial differential equations [29,30,31]. In simulation results, a stochastic model showed the capability to calculate the range and variation of sediment suspension [28]. Maggi [32] proposes a FGM incorporated with stochastic approach based on Winterwerp's FGM [8]. As mentioned in the previous paragraph, Winterwerp's FGM [8] assumes that the fractal dimension is constant. To overcome this weak point, Maggi [32] adopts the concept of a variable fractal dimension. However, when the key equation for the temporal change rate of primary particles within a floc is derived, the fractal dimension is assumed to be constant again (see Equations (3) and (4) in [32]). In addition, the FGM of Maggi [33] also assumes the yield strength of flocs to be constant, whereas Son and Hsu [1] showed that the temporal evolution of floc size is reasonably replicated when a variable yield strength of flocs is assumed. This study aims to improve the existing FGM developed by Son and Hsu [1] to predict both temporal evolution of average size and size distribution of flocs by adopting a stochastic approach to flocculation modeling. A brief description of Son and Hsu's FGM [1] is given in Section 2.1. Based on this FGM, a new stochastic FGM is proposed in Section 2.2. The developed model is validated with data obtained from laboratory experiments in Section 3. Finally, the conclusions of this study are drawn in Section 4.

MATERIALS AND METHODS

Previous FGM of Son Hsu [1]

The Son and Hsu FGM [1] calculates the temporal evolution of average floc size:

$$\frac{\mathrm{d}D}{\mathrm{d}t} = \frac{Gd^{\beta}}{\beta \ln \frac{D}{d} + 1}\left[\frac{k'_A}{3}\frac{c}{\rho_s}d^{F-3}D^{-F+4-\beta} - \frac{k'_B}{3}\left(\frac{\mu G}{B_1}\right)^{q} d^{-p+\frac{2q}{3}F}D^{1-\beta+\frac{2q}{3}(3-F)}(D-d)^{p}\right] \tag{1}$$

where D is the average size of flocs; d is the size of primary particle; G is the dissipation parameter (shear rate); μ is the dynamic viscosity of the fluid; k'A and k'B are empirical coefficients for efficiency of aggregation and breakup; B1 is the coefficient for cohesive force of one primary particle; c is the mass concentration of floc; ρs is the density of primary particle; and p and q are empirical coefficients set assumed to be $p = 1$ and $q = 0.5$ in this study based on the previous study by Winterwerp [8]. The fractal dimension of flocs, F, is calculated by a power law proposed by Khelifa and Hill [20]:

$$F = \alpha\left(\frac{D}{d}\right)^{\beta} \tag{2}$$

The empirical coefficients, α and β, are specified as $\alpha = 3.0$ and $\beta = \frac{\log(F_c/3)}{\log(D_{fc}/d)}$. F_c is a characteristic fractal dimension and D_{fc} is a characteristic size of f_{loc}. The typical vales of F_c and D_{fc} are suggested to be $F_c = 2.0$ and $D_{fc} = 2000$ μm by Khelifa and Hill [20].

The first and second terms on the right hand side of Equation (1) calculate the increase and decrease of floc size due to aggregation and breakup, respectively. The Son and Hsu FGM [1] is an ordinary differential equation calculated by marching numerical methods such as Euler's method and Runge-Kutta method. In Son and Hsu's study [1], the FGM is solved by the fourth-order Runge-Kutta method. The fourth-order Runge-Kutta method is used again in this study in order to solve the stochastic FGM described in Section 2.2.

Application of Stochastic Approach to FGM

In many studies (e.g., [10,15,34,35,36,37,38,39]), it has been shown that floc size distribution is similar to a log-normal distribution containing a single peak. Contradictorily, it has also been shown from previous studies (e.g., [40,41,42,43]) that floc size distribution may be bimodal. Results from Gratiot and Manning [44] suggest that the irregular distribution of floc size result from organic matter associated with mud. When the organic content of mud is removed, the size distribution of floc is close to a log-normal distribution. Laboratory experiments using Kaolinite also show a log-normal distribution of floc size (e.g., [9,11,23,24]). In many field experiments (e.g., [15,35,36,37]), a log-normal distribution of floc size is found when the organic content of mud is less than 5%. Thus, the low organic content of mud and the log-normal distribution of floc size are assumed in this study.

In order to apply the stochastic approach to FGM, it is necessary to consider properties of the flocculation process and physical meanings of Equation (1). One of the main assumptions of the Son and Hsu FGM [1] is that a floc has the fractal structure quantified by the fractal dimension, F (see Equation (2)). That is, floc aggregation is assumed to maintain self-similarity. k′A is a coefficient accounting for the efficiency of collision and diffusion (see [8] for more details). Gibbs [45] insists that the value of k′A depends on the properties of sediment. Tsai and Hwang [11] insist that k′A represents the property of the water type (sea water and fresh water) based on the previous studies by Lick and Lick [46] and Lick *et al.*[16]. From studies by Winterwerp [8] and Mietta [47], it is also known that the aggregation process is significantly affected by physical-chemical properties such as pH and salinity and the organic compounds in the sediment. Therefore, the aggregation process is not considered to be random. The mass concentration (c), the dissipation parameter (G), and properties of primary particle such as density (ρs) and size (d) are also deterministic because they are given conditions of the flocculation process. To the best of our knowledge, there is not a general consensus about breakup processes at present [32]. Deterministic PBEs include a breakup distribution function, although the function is used to simply calculate the type of breakup such as binary and ternary fragmentations (e.g., [48,49]). In the case of FGM, k'_B representing the efficiency of floc breakup plays a similar role. k'_B determines the proportion of floc breakup due to turbulent shear. This coefficient has a close relationship with the floc yield strength (F_y), the minimum stress for floc breakup (see [1,8] for more details). However, turbulent shear exceeding F_y does not necessarily cause the breakup of flocs (e.g., [20]). Therefore, k'_B is assumed to be a random number in this study. As the size of flocs increases, the density of flocs decreases, resulting in higher probability of breakup. Thus, the distribution of floc size should be skewed positively (skewed to the right; skewness ≥ 0). Based on this idea and previous studies mentioned in the first paragraph of this section, k'_B is empirically assumed to be a random number having a log-normal distribution (see Equation

$$k'_B = \exp\left\{X\sqrt{\log\left(\frac{\sigma_l^2}{\mu_l^2}\right)+1} + \log\left(\mu_l^2/\sqrt{\sigma_l^2+\mu_l^2}\right)\right\} \tag{3}$$

To determine k'_B at each time step, a Monte Carlo simulation method is used in this study. A Monte Carlo method (or Monte Carlo experiment) is a computational algorithm using repeated random sampling in order to obtain numerical results. μ_l and σ_l are the mean and the standard deviation of log-normal distribution, respectively. In Equation (3), X is a pseudorandom number. Incorporating Equation (3) into Equation (1), the stochastic FGM of this study is developed. μ_l is equal to k'_B of the deterministic FGM. σ_l is calibrated with the distribution shape of experimental results. Compared to the previous deterministic Son and Hsu FGM [1], the new stochastic FGM has one more parameter (σ_l).However, the new FGM has a capability to calculate the size distribution of flocs ensuring the mass conservation. μ_l and σ_l are calibrated with three experimental results in Section 3.

RESULTS AND DISCUSSION

During each case of the experiment, 30 numerical simulations were conducted. This study focuses on the distribution of floc size in the equilibrium state. Therefore, the values of floc size at the last 1000 time steps (2061 total) are analyzed under the assumption that the flocculation process approaches a dynamic equilibrium state prior to 1061 time steps (see Figure 1). The stochastic FGM of this study is validated with 3 laboratory experiments by [9,23,24]. Following Spicer *et al.* [23], polystyrene particles were mixed in a stirred tank with an impeller in order to measure the size distribution of aggregates with volumetric and mass concentrations (ϕ and c) of 1.4×10−5.0 and 0.0147 kg/m³, respectively. The average value of Gproduced by the impeller is 50 s⁻¹. The size and density of polystyrene particles (c and ρs) of the experiment are 0.87 μm and 1050 kg/m³, respectively. In Biggs and Lant [9], the size distribution of activated sludge is reported under the experimental condition of G = 19.4 s⁻¹ and ϕ = 0.05. Under the assumption that ρs is 2650 kg/m³ and the density of sludge is 1300 kg/m³, the mass concentration is calculated to be 24.19 kg/m³. The activated sludge is mixed in a baffled batch vessel using a flat six-blade impeller. Burban *et al.* [24] also report the results of laboratory experiments with Detroit River sediment. c and G of the experiment are 0.8 kg/m³ and 200 s⁻¹, respectively. d and ρs of Detroit River sediment are assumed to be 4 μm and 2650 kg/m³ due to absence of further information. The detailed conditions of experiments are summarized inTable 1. The calibrated values of empirical coefficients are shown in Table 2.

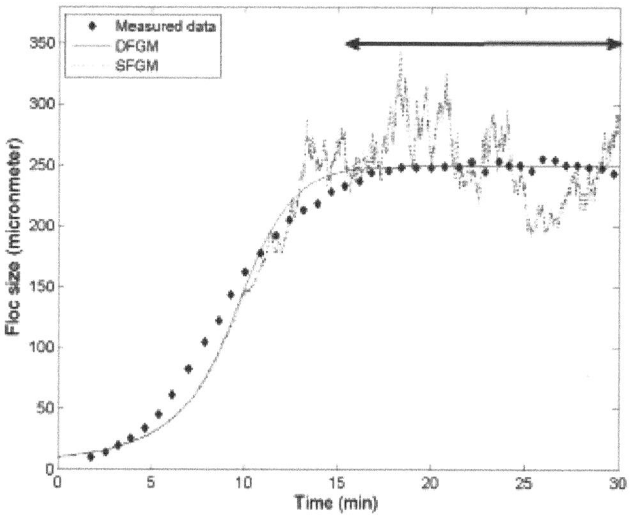

Figure 1. Example of numerical simulation and analysis data. Dots represent the experimental result of Spicer *et al.* [23]. Solid and dotted lines are the stochastic FGM (SFGM) of this study and the deterministic FGM (DFGM), respectively.

Table 1. Experimental conditions.

Experiment	ϕ	c (kg/m³)	ρ_s (kg/m³)	d (μm)	G (s⁻¹)
Spicer *et al.*, (1998)	1.4×10^{-5}	0.0147	1050	0.87	50.0
Biggs and Lant (2000)	0.05	24.19 (Assumed)	2650 (Assumed)	4.00 (Assumed)	19.4
Burban *et al.*, (1989)	N.A.	0.05	2650 (Assumed)	4.00 (Assumed)	200.0

Table 2. Calibrated coefficients of stochastic FGM.

Experiment	k'_A	μ_l	σ_l^2	B_1
Spicer *et al.*, (1998)	6.74	4.39×10^{-6}	4.47×10^{-6}	2.63×10^{-14}
Biggs and Lant (2000)	0.02	2.61×10^{-5}	5.62×10^{-5}	4.20×10^{-13}
Burban *et al.*, (1989)	0.30	3.95×10^{-5}	3.16×10^{-3}	4.10×10^{-12}

Figure 2 shows the simulation results of the stochastic FGM. The experiment from Spicer *et al.* [23] was carried out under the dilute condition (c = 0.0147 kg/m^3). The intensity of turbulent shear is not low (G = 50 s^{-1}). However, the mean size of flocs is about 200 μm. Therefore, it is clear that the polystyrene particles are easily aggregated. Under this assumption, the coefficient determining the aggregation efficiency has a relatively high value ($k'_A = 6.74$). In terms of the mean size of flocs, the simulation results are in satisfactory agreement with experimental results. The mean size is determined by calibrating k'_A and μ_l. That is, the model has the capability to calculate mean size of flocs reasonably well. It is shown in Figure 2 that the calculated size distribution is positively skewed slightly more than the measured result. The log-normal distribution function has only 2 parameters, the mean and variance. Therefore, the skewness and kurtosis of log-normal distribution is indirectly determined by the values of mean and variance. Three- or four-parameter distribution functions such as generalized extreme value distribution and Wakeby distribution have the capability to determine the skewness or kurtosis by calibrating additional parameters. However, more empirical and site-specific parameters need to be calibrated when a 3- or 4-parameter distribution function is incorporated into FGM.

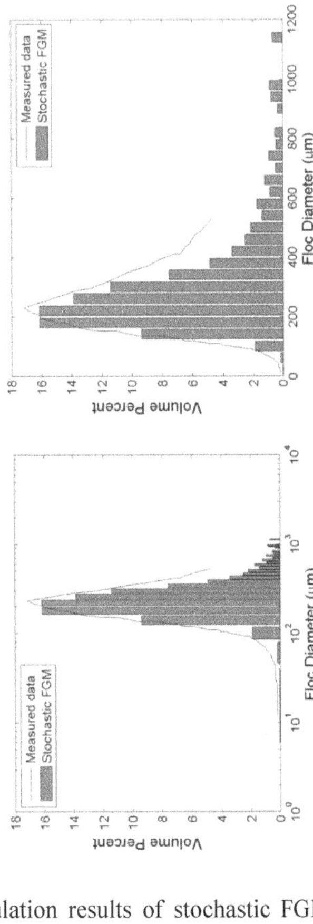

Figure 2. Simulation results of stochastic FGM and experimental results of Spicer *et al.*[23]. The bars and the solid lines represent the experimental and simulation results, respectively. (a) semi-log axis; (b) normal (Cartesian) axis.

Figure 3 represents the results of stochastic simulation and measured data from Biggs and Lant [9]. Their experiment was conducted under conditions with very high concentration (c = 24.19 kg/m^3). The intensity of turbulence is relatively low (G = 19.4 s^{-1}). Under these conditions, the mean size of flocs is not large (about 120 μm). Therefore, it is deduced that the aggregation process of activated sludge used for the experiment has a low efficiency resulting in a low value for the aggregation efficiency parameter ($k'_A = 0.02$). The simulation result is in a good agreement with laboratory measurements. Both the mean size and the distribution are reasonably replicated by the proposed stochastic FGM. The sudden increase of volume fraction around floc size 700 μm is due to the random property of stochastic simulation. Figure 4a shows the mean, maximum, and minimum floc sizes of 30 simulation cases. The dotted line is the equilibrium floc size calculated by the deterministic Son and Hsu FGM [1]. Figure 4b shows the temporal evolution of mean floc size measured by Biggs and Lant [9], a simulation result of deterministic FGM ([1]), and one case of simulation by stochastic FGM. As mentioned in the first paragraph of this section, the last 1000 values of floc size for each stochastic simulation are analyzed in this study. Therefore, the size distribution of this study is determined using 30,000 values for floc size. Figure 4 shows that a stochastic FGM calculates the floc size very randomly after the temporal evolution of floc size approaches the equilibrium state (around 15 min in Figure 4b). The maximum floc size among the values in the last 1000 time steps has a wide range (200 μm to 700 μm). Furthermore, the minimum floc size is also in the range of 10 μm to 50 μm whereas the mean values are between 70 and 120 μm in most cases.

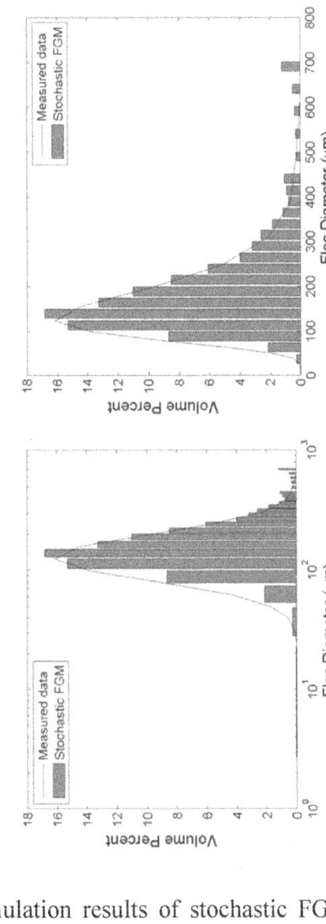

Figure 3. Simulation results of stochastic FGM and experimental results of Biggs and Lant [9]. The bars and the solid lines represent the

experimental and simulation results, respectively. (a) semi-log axis; (b) normal (Cartesian) axis.

Burban *et al.* [24] measured the size distribution of Detroit River sediments with a high dissipation parameter (G = 200 s^{-1}) and a moderate mass concentration condition (c = 0.8 kg/m^3). Due to the high turbulence intensity, a high breakup efficiency is expected. Thus, the mean of k'B is calibrated to be relatively high ($\mu_l = 3.95 \times 10^{-5}$) compared to the two previous cases ([9,23]). Turbulent shear plays two conflicting and simultaneous roles in the flocculation process. The first one is to increase the chance of particle collisions resulting in floc aggregation. Turbulence is the most important mechanism of aggregation ([6,33,50,51]). The second role is disaggregation of flocs. Based on laboratory experiments, Tsai and Hwang [38] insist that a floc usually disaggregates into two roughly equal-sized daughter flocs. From this finding, it is deduced that the main mechanism of breakup is turbulent shear rather than particle collision. Figure 5 shows the effects of concentration and turbulent shear on mean floc size. Figure 5 shows that the high turbulent shear enhances the breakup process compared to aggregation process causing reduction of floc size ([2]). Therefore, the mean floc size of Burban *et al.* [24] is relatively small (about 20 μm) resulting in a high value for μl. Figure 6 shows the results of numerical simulation by stochastic FGM and measurements from Burban *et al.* [24]. The volume fraction near the mean value is slightly overestimated by the stochastic FGM. However, the overall shape of the distribution is in good agreement with measurement results. The volume fraction slightly increases around floc size 80 μm due to the random nature of the stochastic model. A stochastic model is usually based on a random number generation technique (pseudorandom number generator in this study). Therefore, perturbations due to randomness are inevitable.

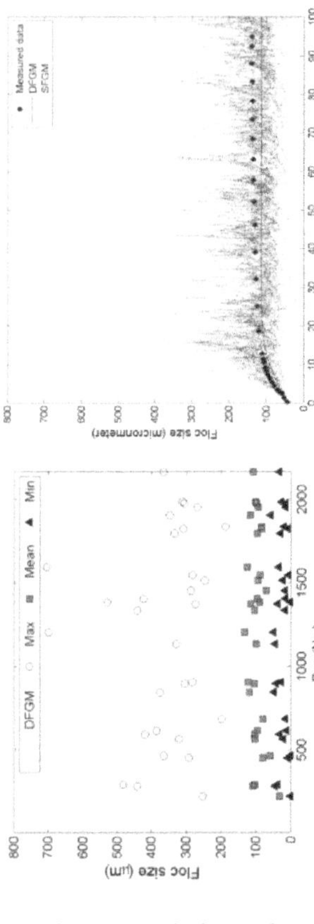

Figure 4. The random calculation of stochastic FGM. The dotted line of (a) shows the simulation results of deterministic FGM (DFGM). The circle, square, and triangle symbols represent the maximum, mean, and minimum values of each simulation, respectively; the solid and dotted lines of (b) are the results of deterministic FGM (DFGM) and stochastic FGM (SFGM), respectively; (a) mean, maximum, and minimum floc size (b) temporal evolution of floc size.

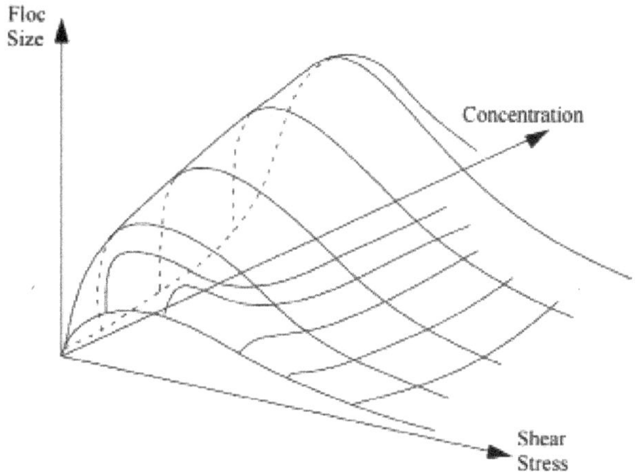

Figure 5. Conceptual diagram representing effect of turbulent shear on mean floc size.

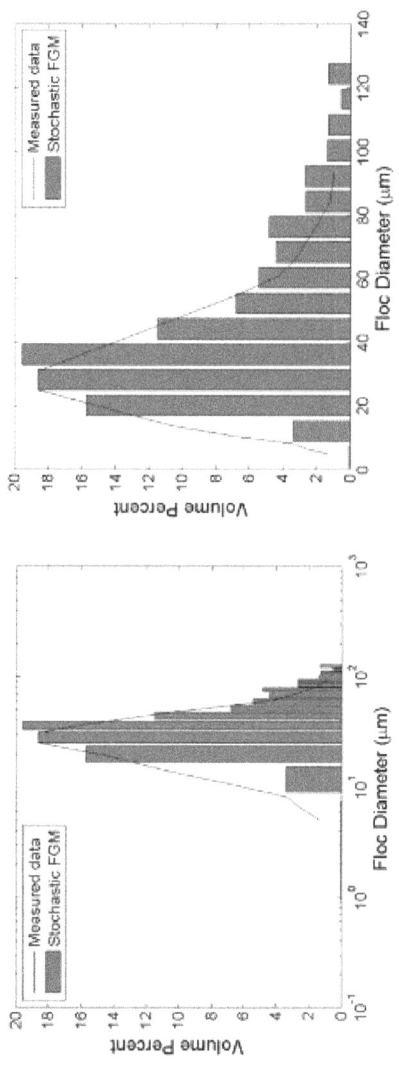

Figure 6. Simulation results of stochastic FGM and experimental results of Burban *et al.*[24]. The bars and the solid lines represent the experimental and simulation results, respectively. (a) semi-log axis; (b) normal (Cartesian) axis.

Three more probability distributions have been tested. The results of stochastic FGM using a normal distribution are drawn in Figure 7. The results are compared with Biggs and Lant [9]. A normal distribution has two parameters: mean and variance. The values of mean and variance of normal distribution are calibrated to be $2.620 \times 10^{-5.0}$ and $5.623 \times 10^{-5.0}$, respectively. Figure 8 shows the result of stochastic FGM using a Pearson type 3 distribution (Gamma distribution). A Pearson type 3 distribution needs 2 parameters: shape parameter (k_p) and scale parameter (θ). The mean and variance of Pearson type 3 distributions are indirectly determined by $k_p\theta$ and $k_p\theta^2$, respectively. Therefore, it is not simple to calibrate a stochastic FGM. The determined values of k_p and θ for the simulation are $4.780 \times 10^{-1.0}$ and, $5.248 \times 10^{-5.0}$, respectively. Compared to results of the log-normal distribution (Figure 3), FGMs using a normal distribution and a Pearson type 3 distribution overestimate the mean floc size, whereas the overall shape of size distribution is underestimated. Figure 9represents the results calculated by stochastic FGM using a generalized extreme value (GEV) distribution. A GEV distribution has 3 parameters: mean, variance, and shape parameter (ξ). In this study, the mean, variance, and ξ of the GEV distribution are set to be $1.147 \times 10^{-5.0}$, $1.0 \times 10^{-5.0}$ and 0.5, respectively. In the case of the GEV distribution, the floc size larger than the mean value is overestimated. Although a GEV distribution has one more degree of freedom (ξ) compared to log-normal distribution, no significant advantage is found.

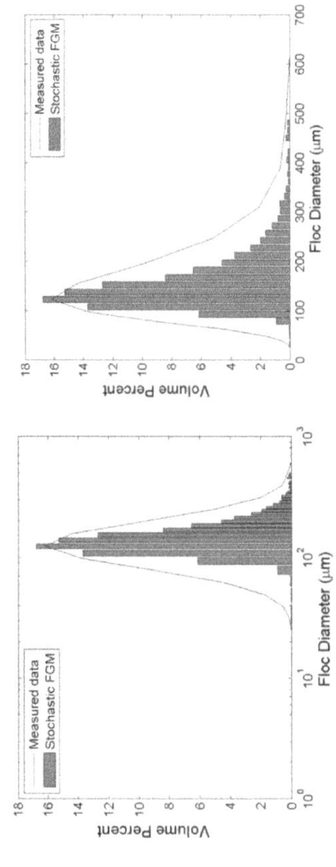

Figure 8. Simulation result of stochastic FGM using a Pearson type 3 distribution (compared to the experimental results of Biggs and Lant [9]). The bars and the solid lines represent the experimental and simulation results, respectively. (a) semi-log axis; (b) normal (Cartesian) axis.

As examined in Section 3, a stochastic FGM using a log-normal distribution has the capability to replicate the size distribution of flocs to a reasonable extent. Three different probability distributions such as normal, Pearson type 3, and GEV distributions are also tested in this study. Among them, an FGM using a log-normal distribution has the advantage of accuracy and simplicity. When a normal distribution is applied, a stochastic FGM underestimates the distribution of flocs larger than 200 μm, although the variance and mode are calculated in conformity. This finding shows that the positive skewness of $(k_p\ \theta^2)$. is needed to calculate large flocs reasonably well. The positive skewness of k'_B decreases the breakup possibility for relatively large flocs. Flocs become less dense and more fragile as the size of floc increases. As a result, large flocs have higher possibility of breakup. A stochastic FGM using a Pearson type 3 distribution shows a mean-concentrated distribution of floc size. The skewness of a Pearson type 3 distribution is inversely proportional to the shape parameter $2/\sqrt{k_p})$ whereas the variance is proportional to the shape parameter (kp θ2). Therefore, the variance decreases resulting in the mean-concentrated distribution when the skewness is increased. The result of a GEV distribution is in good agreement with measurement results. Compared to a log-normal distribution, a GEV distribution has one more degree of freedom as mentioned in Section 3. Thus, the calibration of a GEV distribution is not found to be complicated relative to a log-normal distribution.

CONCLUSIONS

This paper presents a stochastic FGM to calculate the size distribution of flocs. A stochastic approach is applied to the breakup process under the assumption that the aggregation process is significantly affected by properties of the sediment and carrier flow. For the stochastic approach, the parameter related to the breakup efficiency (k'_B) is empirically assumed to have a log-normal distribution. From many previous studies under low organic content conditions (e.g., [15,35,36,37]), it is known that the size distribution of flocs is similar to a log-normal distribution. The capacity and limitations of the developed stochastic models are validated with three experimental results from Spicer *et al.* [23], Biggs and Lant [9], and Burban *et al.* [24]. These three experiments have different experimental conditions such as sediment property, turbulence intensity, and concentration.

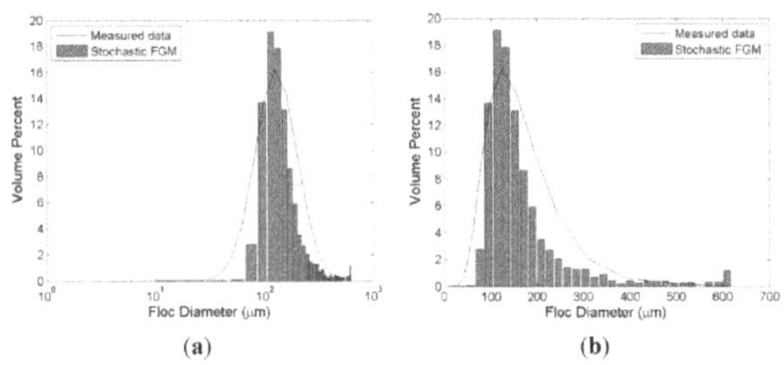

Figure 7. Simulation result of stochastic FGM using a normal distribution (compared to the experimental results of Biggs and Lant [9]). The bars and the solid lines represent the experimental and simulation results, respectively. (a) semi-log axis; (b) normal axis.

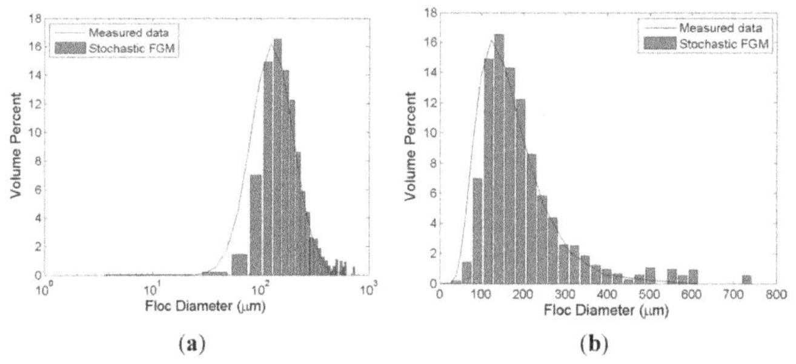

Figure 9. Simulation result of stochastic FGM using a generalized extreme value distribution (compared to the experimental results of Biggs and Lant [9]). The bars and the solid lines represent the experimental and simulation results, respectively. (a) semi-log axis; (b) normal (Cartesian) axis.

An FGM has an advantage in the sense that it ensures mass conservation in a closed system and can be easily incorporated into the numerical sediment transport model. The previous FGMs, such as those used by Son and Hsu [7] and Winterwerp [8], calculate the average size of flocs (representative floc size). Therefore, the sediment transport model considering the flocculation process also assumes the suspension of mono-sized sediment (e.g., [25]). When a stochastic FGM discussed in this study is incorporated with the sediment transport model, it has the capability to calculate the size distributions of suspended cohesive sediments (flocs) as well as the temporal and spatial variation of floc concentration. Hence, future incorporation of these results (*i.e.*, stochastic FGMs) to sediment transport models will greatly improve resolution and uncertainty, furthering our knowledge of cohesive sediment transport processes.

In this study, the mean of k'_B (μ_1 in Equation (3)) is assumed to be constant following the previous studies (e.g., [1,7,8]). k'_B is a parameter representing the probability of floc breakup. As the size of floc increases, the probability of breakup is also expected to increase because the yield stress of floc usually depends on floc size. Therefore, it is necessary to study the effect of floc size evolution through the flocculation process on calibration of k'B in the future. It is also notable that the time step size may affect the calculation of floc size distribution. In this study, the time step for the numerical calculation is automatically determined by the Dormand-Prince method. In order to examine the effect of the time step, it is necessary to calculate the stochastic FGM with different fixed time step values.

ACKNOWLEDGMENTS

This research was supported by the Basic Science Research Program through the National Research Foundation of Korea (NRF) funded by the Ministry of Education, Science and Technology (2011-0010749), the research fund of Chungnam National University in 2014, and a project, "*Development of integrated estuarine management system*", funded by the Ministry of Oceans and Fisheries of Korea.

REFERENCES

1. Son, M.; Hsu, T.-J. The effect of variable yield strength and variable fractal dimension on flocculation of cohesive sediment. *Water Res.* 2009, *43*, 3585–3592.
2. Dyer, K.R. Sediment processes in estuaries: Future research requirements. *J. Geophys. Res.* 1989, *94*, 14327–14339.
3. Dyer, K.R.; Manning, A.J. Observation of the size, settling velocity and effective density of flocs, and their fractal dimensions. *J. Sea Res.* 1999, *41*, 87–95.
4. Lick, W.; Lick, J.; Ziegler, C.K. Flocculation and its effect on the vertical transport of fine-grained sediments. In*Sed./Water Interactions, Hydrobiologia*; Hart, B.T., Sly, P.G., Eds.; Kluwer Academic Publishers: Dordrecht, The Netherlands, 1992; pp. 1–16.
5. Hunt, J.R. Particle dynamics in seawater: Implications for predicting the fate of discharged particles. *Environ. Sci. Tech.*1982, *16*, 303–309.
6. Stolzenbach, K.D.; Elimelich, M. The effect of density on collisions between sinking particles: Implications for particle aggregation in the ocean. *Deep Sea Res. Part I: Oceanogr. Res. Pap.* 1994, *41*, 469–483.
7. Son, M.; Hsu, T.-J. Flocculation model of cohesive sediment using variable fractal dimension. *Environ. Fluid Mech.*2008, *8*, 55–71.
8. Winterwerp, J.C. A simple model for turbulence induced flocculation of cohesive sediment. *J. Hydraul. Res.* 1998, *36*, 309–326.
9. Biggs, C.A.; Lant, P.A. Activated sludge flocculation: On-line determination of floc size and the effect of shear. *Water Res.* 2000, *34*, 2542–2550.
10. Bouyer, D.; Line, A.; Do-quang, Z. Experimental analysis of floc size distribution under different hydrodynamics in a mixing tank. *AIChE* 2004, *50*, 2064–2081.
11. Tsai, C.H.; Hwang, S.C. Flocculation of sediment from the Tanshui River estuary. *Mar. Freshw. Res.* 1995, *46*, 383–392.
12. Tsai, C.H.; Iacobellis, H.S.; Lick, W. Flocculation of fine-grained sediments due to a uniform shear stress. *J. Great Lakes Res.* 1987, *13*, 135–146.
13. Maggi, F.; Mietta, F.; Winterwer, J.C. Effect of variable fractal dimension on the floc size distribution suspended cohesive sediment. *J. Hydrol.* 2007, *343*, 43–55.
14. McAnally, W.H.; Mehta, A.J. Aggregation rate of fine sediment. *J. Hydraul. Eng.* 2000, *126*, 883–892.
15. Verney, R.; Lafite, R.; Brun-Cottan, J.C.; le Hir, P. Behaviour of floc population during a tidal cycle: Laboratory experiments and numerical modeling. *Cont. Shelf Res.* 2011.
16. Lick, W.; Huang, H.; Jepsen, R. Flocculation of fine-grained sediments due to differential settling. *J. Geophys. Res.* 1993,*98*, 10279–10288.
17. Hounslow, M.J.; Ryall, R.L.; Marshall, V.R. A discretized population balance for nucleation, growth and aggregation.*AIChE* 1988, *34*, 1821–1832.
18. Maerz, J.; Verney, R.; Wirtz, K.; Feudel, U. Modeling flocculation processes: Intercomparison of a size class-based model and a distribution-based model. *Continental Shelf Res.* 2011.
19. Levich, V.G. *Physicochemical Hydrodynamics*; Prentice Hall: Upper Saddle River, NJ, USA, 1962.

20. Khelifa, A.; Hill, P.S. Models for effective density and settling velocity of flocs. *J. Hydraul. Res.* 2006, *44*, 390–401.

21. Maggi, F. Variable fractal dimension: A major control for floc structure and flocculation kinematics of suspended cohesive sediment. *J. Geophys. Res.* 2007, *112*.

22. Sonntag, R.C.; Russel, W.B. Structure and breakup of flocs subjected to fluid stresses. II. Theory. *J. Colloid Interface Sci.* 1987, *115*, 378–389.

23. Spicer, P.T.; Pratsinis, S.E.; Raper, J.; Amal, R.; Bushell, G.; Meesters, G. Effect of shear schedule on particle size, density, and structure during flocculation in stirred tanks. *Power Tech.* 1998, *97*, 26–34.

24. Burban, P.Y.; Lick, W.; Lick, J. The flocculation of fine-grained sediments in estuarine waters. *J. Geophys. Res.* 1989, *94*, 8323–8330.

25. Son, M.; Hsu, T.-J. The effects of flocculation and bed erodibility on modeling cohesive sediment resuspension. *J. Geophys. Res.* 2011, *116*.

26. Van der Ham, R.; Fontijn, R.H.L.; Kranenburg, C.; Winterwerp, J.C. Turbulent exchange of fine sediments in a tidal channel in the Ems/Dollard estuary. Part I: Turbulence measurement. *Cont. Shelf Res.* 1994, *21*, 1605–1628.

27. Son, M.; Lee, G.-H. On effects of skewed and asymmetric oscillatory flows on cohesive sediment flux: Numerical study. *Water Resources Res.* 2013, *49*, 4409–4423.

28. Man, C.; Tsai, C.W. Stochastic partial differential equation-based model for suspended sediment transport in surface water flow. *J. Eng. Mech.* 2007, *133*, 422–430.

29. Davie, A.M.; Gaines, J.G. Convergence of numerical schemes for the solution of parabolic stochastic partial differential equations. *Math. Comput.* 1999, *70*, 121–134.

30. Gyongy, I. Lattice approximation for stochastic quasi-linear parabolic partial differential equations driven by space-time white noise II. *Potential Anal.* 1999, *11*, 1–37.

31. Yoo, H. Semi-discretization of stochastic partial differential equations on R^1 by a finite-difference method. *Math. Comput.* 2000, *69*, 653–666.

32. Maggi, F. Stochastic flocculation of cohesive sediment: Analysis of floc mobility within the floc size spectrum. *Water Resour. Res.* 2008, *44*.

33. McCave, I.N. Size spectra and aggregation of suspended particles in the deep ocean. *Deep Sea Res.* 1984, *31*, 329–352.

34. Hill, P.S.; Boss, E.; Newgard, J.P.; Law, B.A.; Milligan, T.G. Observations of the sensitivity of beam attenuation to particle size in a coastal bottom boundary layer. *J. Geophys. Res.* 2011, *116*, C02023.

35. Mikkelsen, O.A.; Milligan, T.G.; Hill, P.S.; Moffatt, D. INSSECT–and instrumented platform for investigating floc properties close to the seabed. *Limnol. Oceanogr. Methods* 2004, *2*, 226–236.

36. Mikkelsen, O.A.; Hill, P.S.; Milligan, T.G.; Chant, R.J. *In situ* particle size distributions and volume concentrations from a LISST-100 laser particle sizer and a digital floc camera. *Cont. Shelf Res.* 2005, *25*, 1959–1978.

37. Mikkelsen, O.A.; Hill, P.S.; Milligan, T.G. Single-grain, microfloc and macrofloc volume variations observed with a LISST-100 and a digital floc camera. *J. Sea Res.* 2006, *55*, 87–102.

38. Syvitski, J.P.M.; Hutton, E.W.H. *In situ* charateristics of suspended particles as determined by the floc camera assembly FCA. *J. Sea Res.* 1996, *36*, 131–142.

39. Xu, F.; Wang, D.-P.; Riemer, N. Modeling flocculation processes of fine-grained particles using a size-resolved method: Comparison with published laboratory experiments. *Cont. Shelf Res.* 2008, *28*, 2668–2677.

40. Manning, A.J. Observation of the properties of flocculated cohesive sediment in three western European estuaries. *J. Coastal Res.* 2002, *SI 41*, 70–81.

41. Manning, A.J.; Dyer, K.R.; Lafite, R.; Mikes, D. Flocculation measured by video based instruments in the Gironde estuary during the European Commission SWAMIEE Project. *J. Coastal Res.* 2004, *SI 41*, 58–69.

42. Manning, A.J.; Bass, S.J.; Dyer, K.R. Observations of cohesive sediment settling fluxes in high energy tidal estuarine environments. *Mar. Geol.* 2006, *235*, 177–192.

43. Manning, A.J.; Friend, P.L.; Prowse, N.; Amos, C.L. Estuarine mud flocculation properties determined using an annular mini-flume and the LabSFLOC system. *Cont. Shelf Res.* 2007, *27*, 1080–1095.

44. Gratiot, N.; Manning, A.J. An experimental investigation of floc characteristics in a diffusive turbulent flow. *J. Coast. Res.* 2004, *SI 41*, 105–113.

45. Gibbs, R.J. Coagulation rates of clay minerals and natural sediment. *J. Sediment Res.* 1983, *53*, 1193–1203.

46. Lick, W.; Lick, J. Aggregation and disaggregation of fine-grained sediments. *J. Great Lakes Res.* 1988, *14*, 514–523.

47. Mietta, F. Evolution of the Floc Size Distribution of Cohesive Sediments. Ph.D. Thesis, Delft University, Delft, The Netherlands, 2010.

48. Maggi, F. Flocculation Dynamics of Cohesive Sediment. Ph.D. Thesis, Delft University, Delft, The Netherlands, 2005.

49. Serra, T.; Casamitjana, X. Effect of the shear and volume fraction on the aggregation and break-up of particle. *Am. Instit. Chem. Eng.* 1998, *44*, 1724–1730.

50. O'Melia, C.R. Aquasols: The behavior of small particles in aquatic systems. *Env. Sci. Technol.* 1980, *14*, 1052–1060.

51. Van Leussen, W. Estuarine Macroflocs and Their Role in Fine-Grained Sediment Transport. Ph.D. Dissertation, University of Utrecht, Utrecht, The Netherlands, 1994.

CHAPTER 6 SEISMIC-RELIABILITY-BASED OPTIMAL LAYOUT OF A WATER DISTRIBUTION NETWORK

Do Guen Yoo , Donghwi Jung, Doosun Kang and Joong Hoon Kim

Keywords: Seismic reliability, water distribution system, optimal layout, Anytown network

INTRODUCTION

An earthquake occurs as a result of a sudden release of energy in the earth's crust. The released massive energy creates seismic waves, which cause deformation of the water distribution system (WDS) of which components are mainly connected beneath the ground. The damage of an earthquake is the multiple failure of system components. For example, for an earthquake occurred in Kobe, Japan in 1995, many pipes were ruptured, causing water to flow out of the system, while some of the pump stations stopped working completely due to power outage. However, the likelihood of concurrence of multiple failures within a system is low under normal conditions. Therefore, a different strategy should be adopted for WDS design of the regions with the risk of earthquake occurrence, such as Japan, Korean peninsula, and West Coast cities of the United States.

During the last two decades, many optimal design approaches have been proposed for WDSs. The early works have mainly taken into account the economic cost as a single objective [1,2,3,4,5]. Thereafter, some studies have included the system's performance index within the optimization framework. Lansey *et al.* [6] developed a chance-constrained least-cost model in which the capacity reliability is defined as the probability that stochastic pressures are equal to or higher than pressure requirement and are constrained at a certain level while minimizing cost. The nodal demands, pressure requirements, and pipe roughness coefficients were assumed to be uncertain while optimizing the pipe size. Several later studies also used chance-constrained optimization models for WDS design [7,8].

Kapelan *et al.* [9] was one of the earliest studies that proposed the multi-objective optimal design of WDS. Since then, most design approaches have adopted two objective functions to minimize economic cost and to maximize system reliability [10,11]. The various system reliability indices were suggested to reflect the uncertainties of pipe roughness and system demands when sizing the system. Component failures have been rarely considered in this stream of optimal WDS design.

Su *et al.* [12] is one of the few studies that considered component failure in WDS design. In their study, reliability was defined as the probability of water demand provision under pipe break (failure) conditions and is used as a constraint for the least-cost design of a WDS. The minimum cut-set (*i.e.*, the most critical set of pipes) was identified by closing the pipe individually for calculating the system reliability. Note that the pipe failure conditions were not considered in hydraulic simulation within the optimization framework, since the computational intensity is overwhelming.

The seismic hazard assessment models have been developed recently. HAZUS [13] was the first model to assess the economic losses of an infrastructure by earthquake. HAZUS is mainly intended to assess the seismic damage but not detailed simulation of the system behavior. Later, the Mid-America Earthquake Center developed a seismic impact assessment model and investigated interdependencies between water and power systems [14].

Early earthquake studies in the water domain focused on investigating individual components' physical behavior

Citation: Do Guen Yoo, Donghwi Jung, Doosun Kang and Joong Hoon Kim, Seismic-Reliability-Based Optimal Layout of a Water Distribution Network, doi:10.3390/w8020050.

under earthquakes rather than quantifying the system-wide performance by modeling the WDS and earthquakes [15,16,17]. In a recent study, Fragiadakis *et al.* [18] proposed a seismic reliability assessment model of a WDS using survival curves of pipes based on general seismic assessment standards and American Lifelines Alliance (ALA) [19] guidelines. However, detailed hydraulic simulations were not conducted in this paper. Later studies began proposing methodologies to evaluate seismic reliability with hydraulic simulations using well-known hydraulic solvers and seismic simulations [20,21,22,23,24,25,26,27,28,29].

The latest and most popular assessment model is the graphical iterative response analysis for flow following earthquakes (GIRAFFE [30]) developed by a research team at Cornell University. GIRAFFE can simulate various pipe leakage and breakage conditions by using EPANET [31]. The model's graphical user interface helps to visualize the model results, which is compatible with other geographic information system tools. After the initial development in 2008, the model has been improved and validated through many case studies [27,32,33,34,35,36]. However, GIRAFFE uses a controversial approach to treat negative pressure that removes the nodes' negative pressure and connected pipes. This process is repeated until the negative-pressure nodes are no longer produced. This approach can be time-consuming because the system file must be revised iteratively. Yoo *et al.* [37] developed seismic reliability assessment model under stochastic earthquake events. The model quantifies the seismic reliability of a WDS through a series of procedures: stochastic earthquake generation, seismic intensity attenuation, determination of the pipe failure status (normal, leakage, and breakage), pipe failure modeling in hydraulic simulation, and negative pressure treatment. To the authors' best knowledge, this study is the first attempt to develop an economic, cost-constrained optimal WDS design approach that takes into account seismic reliability based on detailed hydraulic simulations. The proposed model maximizes the system's seismic reliability while satisfying the constraints on economic cost and node pressure requirements. The physical impacts of a seismic wave to WDS components are simulated to determine the failure conditions. The seismic reliability is defined as the ratio of the supplied water to the required demand under stochastic earthquake events. A well-known benchmark network, the Anytown network, is used for the applications of the optimal design and layout maximizing the seismic reliability.

METHODOLOGY

In this study, a seismic reliability-based WDS optimization approach is proposed. Figure 1 shows the flowchart of the proposed WDS optimal design model which consists of two sub-models: seismic reliability estimation and optimization model. Seismic reliability estimation model (SREM) first generates stochastic earthquake events by using earthquake generation module. Here, the locations and magnitude of stochastic earthquakes are determined and peak ground acceleration (PGA) reached at each pipe is calculated (described in Section 2.1). Note that no hydraulic calculation is performed from this module. This module also determines the failure mode of pipe where each pipe is classified as normal, leakage, or breakage (Section 2.2).

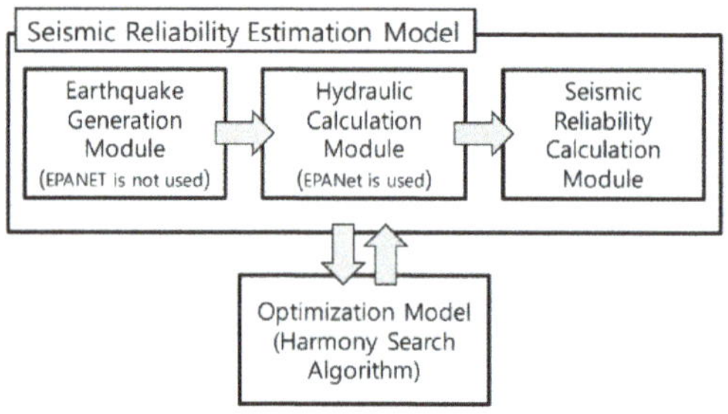

Figure 1. Flowchart of the proposed water distribution system (WDS) optimal design model.

The failure modes are transferred to hydraulic calculation module where nodal pressures under the failure conditions (*i.e.*, the boundary conditions to the WDS system equation) are calculated (Figure 1 and Section 2.3). Pipe leakage and breakage are modeled by the emitter in EPANET. In order to better simulate the conditions, a semi pressure-driven analysis (semi-PDA) is proposed. After an initial run of EPANET (a hydraulic simulator for DDA), the negative-pressure nodes' demand is set to zero, and the second run is made (Section 2.4).

The resulting hydraulics are provided to seismic reliability calculation module to quantify seismic reliability which is defined as the ratio of available supply to the required demand under stochastic earthquake events. The proposed WDS optimal design model maximizes the system's seismic reliability with constraints on economic cost and pressure requirements (Figure 1 and Section 2.5). Harmony search algorithm (HSA) is used for optimization (Figure 1) [38,39]. In the last section of the Methodology, Todini [40]'s resilience indicator is defined. This indicator is not used in the proposed model but used for a postoptimization analysis.

The following subsections describe the details of the methodology, such as earthquake simulation, objective function, optimization approach, and semi-PDA method. Note that the following subsections are in the order that Figure 1 represents.

Earthquake Intensity Attenuation

The proposed model considers the physical characteristics of the WDS's components to determine the failure modes given earthquake intensity. Therefore, supplemental information such as the types of pipes and surrounding soil characteristics should be provided. Network information such as the network layout, nodal demands and coordinates, pipe diameters and lengths, and sizes of the pump and tanks are entered from the EPANET input file. The nodal coordinates are used to calculate the location of a pipe that is used for calculating the distance from the epicenter.

Earthquake intensity is attenuated as it travels from the epicenter. In the model, three attenuation equations were adopted to quantify the damped energy of seismic waves. Kawashima *et al.* [41] relates the distance from the epicenter and earthquake magnitude to PGA (cm/s^2) based on the 90 earthquakes that have occurred in Japan as

$$PGA = 403.8 \times 10^{0.265M} \times (R + 30)^{-1.218} \quad (1)$$

where M = earthquake magnitude (e.g., M7 stands for

earthquake magnitude 7), and R = the distance from the epicenter (km).

Lee and Cho [42] presented a similar PGA function that was developed using the historical data of South Korea as

$$\log PGA = -1.83 + 0.386M - \log R - 0.0015R \quad (2)$$

Instead of using the horizontal distance from the epicenter, Baag *et al.* [43] relate the Euclidean distance from focus (Δ) and earthquake magnitude to PGA as

$$\ln PGA = 0.40 + 1.2M - 0.76\ln\Delta - 0.0094\Delta \quad (3)$$

The average of the three calculated PGAs (from Equation (1) to Equation (3)) was used in our model.

Determination of Pipe Failure Mode

The fragility of the pipes was determined based on the pipe repair rate (RR, number of pipe repairs/km) and the PGA. In this study, the RR equation suggested by Isoyama *et al.* [44] and adopted by ALA [19] was used. Isoyama *et al.* related the RR to the PGA by multiplying some correction factors denoting the pipe types, diameters and soil characteristics (Table 1) as

$$RR = C1 \times C2 \times C3 \times C4 \times 0.00187 \times PGA \quad (4)$$

where C1, C2, C3, and C4 represent the correction factors according to the pipe diameter, pipe material, topography, and soil liquefaction, respectively. Because of the lack of information, this study only considered the correction factor of the pipe diameter, C1, in Equation (4), while assuming other correction factors are equal to one. As seen in Table 1, for the smaller pipe, C1 increases and results in the higher probability of damage given the same PGA.

Table 1. Correction factors (Isoyama *et al.* [44]).

	Category	Correction Factor
Pipe diameter (mm) (C1)	D < 100	1.6
	100 ⩽ D < 200	1.0
	200 ⩽ D < 500	0.8
	500 ⩽ D	0.5
Pipe material (C2)	Asbestos-Cement Pipe	1.2
	Poly-Vinyl Chloride Pipe, Vent Pipe	1.0
	Cast Iron Pipe	1.0
	Poly-Ethylene Pipe, High Impact (3-Layer) Pipe	0.8
	Steel Pipe	0.3
	Ductile Cast Iron Pipe	0.3
Topography (C3)	Narrow Valley	3.2
	Terrace	1.5
	Disturbed Hill	1.1
	Alluvial	1.0
	Stiff Alluvial	0.4
Liquefaction (C4)	Total	2.4
	Partial	2.0
	None	1.0

The post-earthquake pipe status was classified into normal, leakage, or breakage conditions. Pipe leakage is defined as a small rupture on the pipe wall or at the joint through which continuous minor water loss occurs. Pipe breakage is defined as a complete separation of the pipe into two pieces that causes complete loss of transportation ability. ALA [19] suggested that the probability of pipe breakage (Pfbreakage,i) is a function of the RR and the length of the pipe (L_i, km) expressed as

$$P_{f_{breakage,i}} = 1 - e^{-RR \times L_i} \quad (5)$$

As a rule of thumb, the probability of pipe leakage (Pfleakage,i) is assumed to be five times higher than the probability of pipe breakage [19]. If the cumulative probability of the two conditions is less than 1.0 (e.g., $P_{fbreakage,i} = 0.1$ and $P_{fleakage,i} = 0.5$), the complementary probability is assigned for the normal condition ($P_{normal} = 0.4$). However, if the cumulative probability of the leakage and breakage is greater than 1 (e.g., $P_{fbreakage,i} = 0.18$ and $P_{fleakage,i} = 0.90$), the model assumes that the normal condition (without any damage) is not available, and the two failure probabilities are normalized to make the sum of 1.0 ($P_{fbreakage,i} = 0.18/(0.18 + 0.90) = 0.17$ and $P_{fleakage,i} = 0.90/(0.18 + 0.90) = 0.83$).

Pipe Failure Modeling

Once the status of each pipe is determined, the information is entered into a network solver, EPANET, for hydraulic simulation (*i.e.*, solving for WDS system equations consisting of conservations of mass and energy). To simulate the pipe leakage, pressure-dependent flows (Q_l) are assigned to the closest node to the damaged pipe using the following equation:

$$Q_l = C_D p^{\alpha} \quad (6)$$

where CD = the discharge coefficient (= (2gγw)αA) (the emitter coefficient in EPANET), where g = gravitational acceleration and α = the exponent of the power function; γw = the specific weight of water; A = the opening area of the damaged pipe; and p = pressure at the closest node.

Lambert [45] conducted an experimental study to investigate the accuracy of the power function model in Equation (6) and provided a guideline on using exponent α based on the pipe material and level of leakage. It was suggested that the exponent of 0.5 would be used to simulate large detectable leakages in metal pipes; the exponent of 1.0 (linear relationship between discharge and pressure) would be used if no information on the pipe material is available. Puchovsky [46] theoretically derived a discharge coefficient and validated it using sprinkler data.

The shape and area of the opening on ruptured pipe varies depending on pipe material, origin of the pipe damage, and direction of external force. Large opening area is more likely to occur in the failure of large pipes than smaller pipes. To that end, it was assumed that the total opening area is equivalent to 10% of the entire cross-sectional pipe area during leaks. In case of breakage, the entire cross-sectional area was used as the opening area. If a pipe was tagged as leaking, it was modeled in a hydraulic simulation, as illustrated in Figure 2a, and the discharge coefficient was assigned to the downstream node. The pipe breakage was modeled as shown in Figure 2b. The discharge coefficient was assigned to the upper node in flow direction. Then, the broken pipe was set to "closed" to disconnect the water flow. The demand of the node connected to a broken pipe was modified to consider the degraded delivery capacity due to disconnection. In the model, the nodal demands of both end junctions of the broken pipe were reduced by degrees of node (DoN). Here, DoN is defined as the number of connections/edges that a node has to other nodes in a network. As shown in Figure 2b, the base demand of the upstream node is reduced by 25% (= 1/DoN = 1/4), while the downstream node is reduced by 33.3% (= 1/DoN = 1/3).

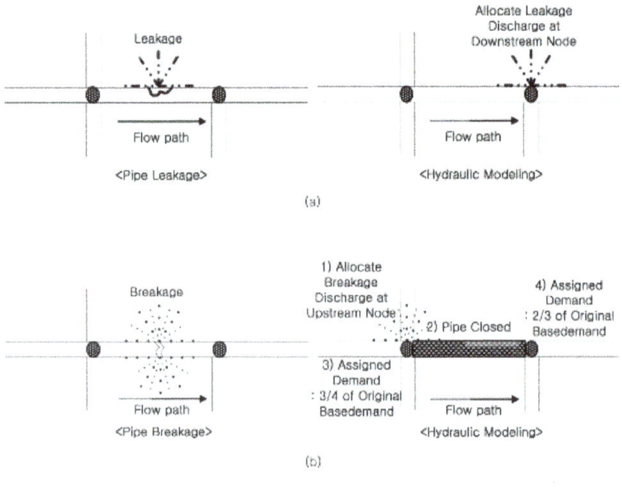

Figure 2. A schematic diagram to describe pipe damage modeling: (**a**) leakage condition and (**b**) breakage condition.

Negative Pressure Treatment

Hydraulic analysis approaches for water distribution network can be classified as a DDA and a pressure-driven analysis (PDA). DDA assumes that the demand at individual node is satisfied regardless of the associated pressure, while PDA assumes that nodal discharge is dependent on the pressure head expressed by the head-outflow relationship (HOR). DDA often generates negative pressure when simulating abnormal conditions, such as multiple pipe breaks. PDA, on the other hand, provides more realistic hydraulics avoiding negative pressures. WaterGEMS [47], WDNetXL [48], and WaterNetGen [49] are well-known programs equipped with PDA option based on HOR. However, the system-specific HOR should be provided, which is spatially and temporally inconsistent and also operational dependent. Therefore, for PDA techniques used in pipe network analyses, the analyzer should assume a HOR for the whole network. Given that hydraulic states of pipe networks can vary greatly with changes in HORs, analysis results cannot guarantee high accuracy.

Quasi-PDA methods suppress the occurrence of negative pressure through repetitive DDA analyses. In general, if negative pressure occurs from the first DDA run, better hydraulic calculation results can be derived through quasi-PDA methods by resetting the nodal demands and the components' status.

Ballantyne *et al.* [21] used the KYPIPE model [50] for hydraulic analyses. KYPIPE, a DDA program similar to EPANET, can also produce negative pressures during simulation of pipeline destruction. Ballantyne *et al.* calculated system reliability by assuming that water could not be supplied to points with negative pressures, but the hydraulic results around the negative-pressure node still did not have realistic values. The method of Shinozuka *et al.* [22] and Shi [27] completely removed negative-pressure nodes and the connected pipelines from the original network, but the process required time-consuming repetitive hydraulic analyses and regeneration of input files.

Here, a quasi-PDA approach is proposed for realistic hydraulic simulation under multiple failure conditions to avoid the occurrence of negative pressure. That is, if negative pressure occurred after DDA simulation using EPANET, the updated base demand (after modification described in Chapter 2.3) of the negative-pressure node is set to zero and DDA is repeated. If negative pressure reappears, the pressure of the relevant node is assumed to zero. Compared to Shinozuka *et al.* [22] and Shi [27] approach, the proposed approach saves overhead processing time by avoiding regeneration of the EPANET input file for the new configuration. In addition to the negative pressure treatment, pressure-dependent water supply is considered in calculating the quantity of available water at each node. The available nodal demand at node j (Qavl,j) is estimated as

$$Q_{avl,j} = \begin{cases} 0 & \text{if } P_j \leq 0 \\ Q_{new,j} \times \sqrt{\dfrac{P_j}{P_{min}}} & \text{if } 0 < P_j < P_{min} \\ Q_{new,j} & \text{if } P_j \geq P_{min} \end{cases} \quad (7)$$

where Qnew,j = the updated base demand with pipe breakage consideration and negative pressure treatment at node j; Pj= the pressure head at node j; and Pmin= the minimum pressure requirement.

Seismic Reliability Indicator

Various surrogate measures of WDS reliability have been proposed including capacity reliability [6,7], resilience [40], robustness [10,11], availability [51]. Bao and Mays [52] proposed three formulations of WDS system reliability which can be calculated from nodal reliability values: minimum nodal reliability, arithmetic mean reliability, and flow-weighted mean reliability. The first one concerns on the worst nodal reliability while the latter two values indicate system-wide reliability. In order to represent post-earthquake system performance in the water supply, a seismic reliability measure should be able to reflect system-wide water availability rather than local level of system performance.

To quantify system-wide seismic reliability, a new reliability indicator is proposed herein. The system seismic reliability (S_S) is defined as the ratio of the total available system demand to the total required system demand:

$$\begin{aligned} &\text{Maximize } F = S_s \\ &\text{s.t.} CC \leq CC_{given} \end{aligned} \quad (8)$$

where m = total number of demand nodes; and Qreq,j = the required demand at node j.

A water system planner would intend to design the most reliable network for earthquakes under a given budget condition. To that end, the proposed model here is a single-objective optimal design model that maximizes the system's seismic reliability with a constraint on economic cost. The optimization model is formulated as follows:

$$\text{Maximize } F = S_s \quad \text{s.t.} CC \leqslant CC_{given} \tag{9}$$

where CC = the pipe construction cost; CC_{given} = the available budget for pipe system.

Equation (9) is valid for each set of available commercial pipe diameters and in conditions where the minimum pressure head is guaranteed. The pipe construction cost accounting for the pipe material and installation is expressed as:

$$CC = \sum_{i=1}^{n} (uc(D_i) \times L_i) \tag{10}$$

where uc(Di) = the unit cost of the pipe with diameter D_i determined for the ith pipe (USD/m); and L_i = the length of the ith pipe.

Resilience Indicator

In this study, Todini [40]'s resilience is quantified from a range of designs of the Anytown network and compared with seismic reliability. Todini [40] introduced a resilience index to quantify the resilience of looped network. Resilience is defined as surplus power available within the network as a percentage of net input power as:

$$Resil = \frac{\sum_j^m Q_{req,j}(H_j - H_{req})}{\sum_{k=1}^{nr} Q_k H_k + \sum_{i=1}^{np} \frac{Power_i}{\gamma} - \sum_j^m Q_{req,j} H_{req}} \tag{11}$$

where Hj = total head of the jth node; Qk = flow provided by reservoir k (m^3/s); Hk = head at reservoir k; Poweri = power of the ith pump (Nm/s); γ = specific weight of water (N/m^3); nr and np = number of reservoir and pumps, respectively. Note that this indicator was used only for a postoptimization analysis.

Todini's resilience is one of the most popular and widely used surrogate measures of WDS reliability. Farmani *et al.* [53] investigated the trade-off between economic cost and the resilience for a rehabilitation problem of the Anytown network. Prasad and Park [54] proposed a multiobjective optimization approach to minimize cost and maximize modified version of Todini's resilience indicator. Recently, Gheisi and Naser [55] compared entropy-based reliability, Todini's resilience, and three modified versions of Todini's resilience in the twenty-two potential pipe layouts of a hypothetical network.

STUDY NETWORK

The proposed optimization approach is applied for optimal design of a well-known benchmark WDS, the Anytown network. The network, which was firstly published by Walski *et al.* [56], was modified by Jung *et al.* [11] for pipe only and pipe/pump designs that minimize total cost and maximize system robustness. The original benchmark network was modified by removing the two tanks and the connected riser pipes to be solely supplied by a single reservoir with a fixed source head. The fixed source head is elevated from 3 m (10 ft) to 73.2 m (240 ft). A peaking factor of 1.8 is applied to given average-based demand (2005 average daily use) to create the daily peak condition.

Other modifications of the study network were made for application purposes. First, this study suggests new pipe layouts and sizes assuming there are no existing pipes in the system. Pump and tank design are not considered. Second, in addition to 10 commercial pipe sizes (152, 203, 254, 305, 356, 406, 457, 508, 610, and 762 mm), zero diameter can be selected suggesting no pipe installation for the potential path. The unit costs of the commercial pipes are adopted from Walski *et al.* [56].

To generate random epicenters, a 2 × 2 grid was created and laid on the study network as seen in Figure 3. The rectangular boundary of the grid is defined by the four end nodes: north, south, east, and west. Total 900 earthquakes are generated and consistent number of earthquakes are assigned at each corner of the grid (marked as "x" in Figure 3). The earthquake intensity reached at the pipe is a function of the earthquake magnitude and the Euclidean distance from the epicenter. Note the earthquake magnitude of M4 was generated from each corner of the grid. The minimum pressure requirement of 28.12 m (40 psi) should be satisfied under based demand condition and is also considered in Equation (7).

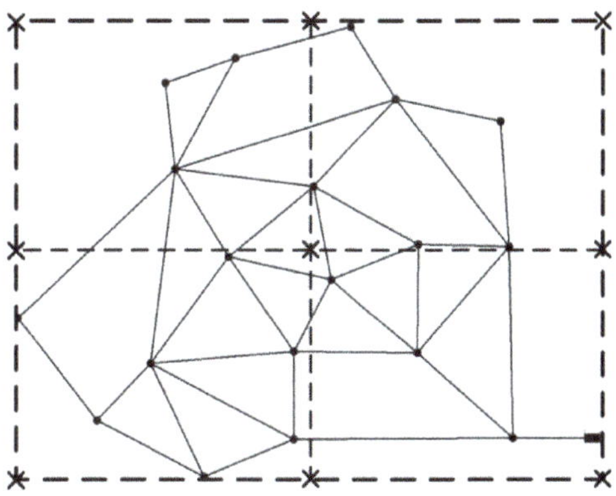

Figure 3. Study network layout (epicenter is marked as "x").

HSA was used to find an optimal solution for the pipe sizing problem. HSA was inspired by musical performance process and widely used for WDS optimizations. While the applications encompass from pipe network design to pump scheduling, HSA was proven to be generally outperform other algorithms such as genetic algorithm [57,58,59].

Several assumptions and simplifications were made in this study: (1) the seismic damages of the pump, tank, reservoir, and valve are not considered; (2) the earthquake's focal depth (tectonics) is assumed to be 10 km; (3) the coordinate of the center of the pipe is used as a reference point for calculating the distance from the epicenter; (4) Pipes are cast iron pipes; and (5) the Anytown network is laid on the alluvial plain with no liquefaction. Based on the assumption (4) and (5), all correction factors except C1 in the Equation (4) (C2-4) are equal to one.

APPLICATION RESULTS

First, we applied two different commercial pipe sets for the pipe sizing of the study network. This analysis investigates the WDS's layout changes with respect to the seismic reliability increase. The same total cost constraint was applied for the two designs and the resultant pipe layouts were compared. Then, seven designs with a fixed layout and different redundancy levels were compared for the systems' seismic reliability. Finally, discussions and suggestions on improving WDS seismic reliability were provided at the end of this section.

Different Available Pipe Sizing Options

The first application is intended to investigate the impact of seismic reliability on the network's layout. Two sets of commercial pipe sizes are assumed to be available for two design cases. In Case 1, all commercial pipes (152, 203, 254, 305, 356, 406, 457, 508, 610, and 762 mm) except the zero size option are available. In Case 2, zero diameter is available in addition to the commercial sizes in Case 1. Considering the same cost constraint (CCgiven = 18 million (M) USD) in Equation (9) provided a platform for consistent comparison of the resulting designs.

Figure 4 shows the optimal seismic reliability values of the two case designs. The corresponding optimal pipe layouts are presented in Figure 5. Contrary to expectations, the seismic reliability decreased with the availability of more pipe sizes. In Case 1, at least a 152 mm pipe should be installed because no pipe option is not available. By comparing Figure 5a,b andFigure 6, we can observe that 152 mm pipes were installed in Case 1 for the link at which no pipe was constructed in Case 2. The 152 mm pipe, which is the most vulnerable pipe to earthquakes, almost always causes failure. Being able to have no pipe instead of a 152 mm pipe increased the system's seismic reliability by 0.2 (a 20% increase in the amount of available water during an earthquake). The network layout becomes smaller from Case 2 to Case 1 (Figure 5a,b), while the overall pipe diameters also increase (Figure 6).

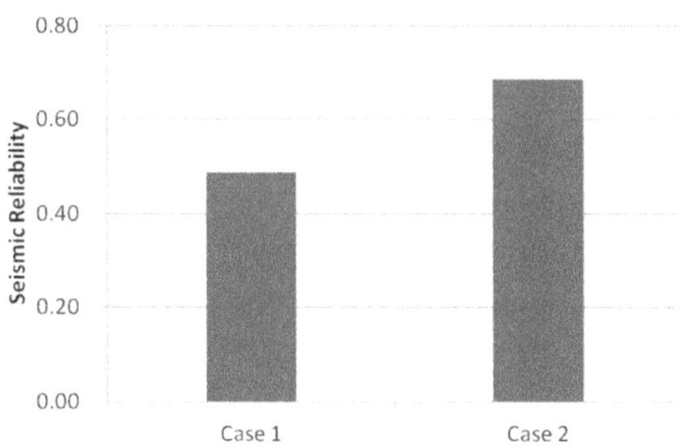

Figure 4. Maximum seismic reliability values of two case designs with different pipe sizing options.

Note that this is very different from what we observed in a traditional capacity reliability-based design. In the context of the capacity reliability field, it was believed that having additional paths and more loops would result in the increase of system reliability and redundancy [11,60]. However, the result of this study indicates that a different strategy should be available under the conditions where WDS component failures are affected by strong external forces (*i.e.*, earthquakes) and the components' physical characteristics (pipe's probability of failure as a function of RR).

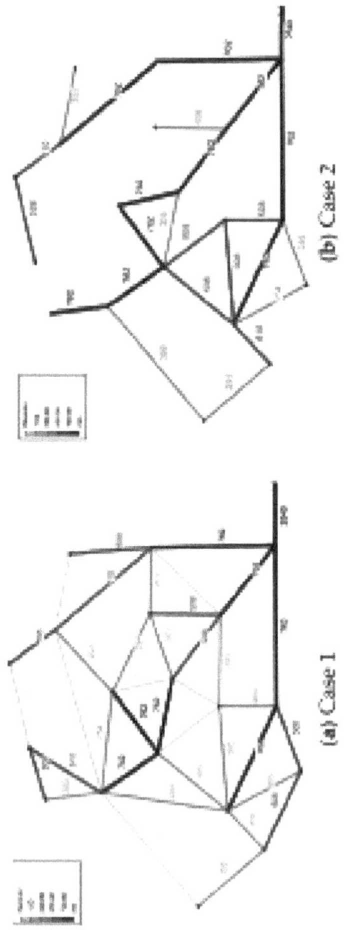

Figure 5. Pipe layout comparison for the solutions obtained from Cases 1 and 2 (Figure 4); pipe diameters are in mm; the thicker and darker pipe is larger. (**a**) Case 1; (**b**) Case 2.

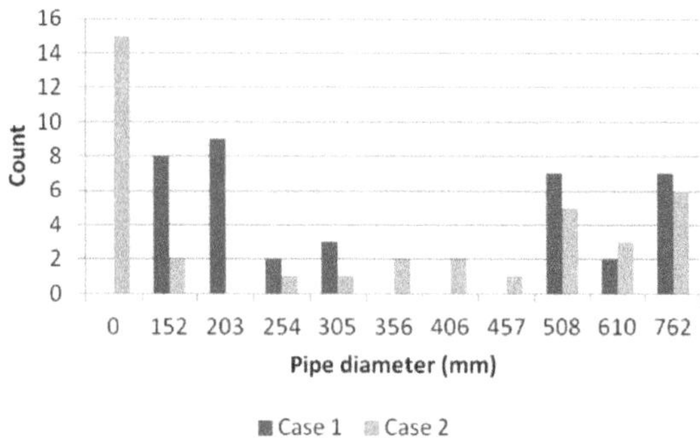

Figure 6. Histogram of pipes by the pipe diameter.

This plateau in seismic reliability was also observed in the optimal pipe designs of the Anytown network by applying different total cost constraints and using the available pipe sizes in Case 1. Figure 7 shows the Pareto optimal solutions' total cost and seismic reliability. The marginal cost becomes infinite for the designs whose cost is greater than 16.5 M USD. A reliability increase can no longer be achieved once a sufficient investment is made. Although the designs greater than 16.5 M USD have more large pipes compared to the solutions less than

16.5 M USD (the number of pipes equal to or larger than 508 mm is between 7 and 10, while the designs less than 16.5 M USD have three to five pipes), no benefit of having larger pipes was obtained with respect to seismic reliability. For effective and economical improvement of WDS seismic reliability, the threshold investment for a network should first be identified.

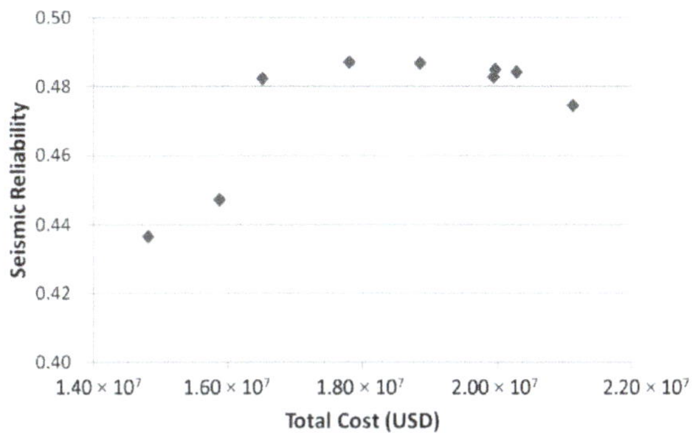

Figure 7. Trade-off relationship between total cost and seismic reliability in Case 1 where all pipes (152, 203, 254, 305, 356, 406, 457, 508, 610, and 762 mm) are available and without zero pipe option.

Constant Layout with a Single Pipe Sizing Option

The impacts of having large pipes are also investigated through the seismic reliability evaluation of seven uniform designs. The Design 1 has 305 mm for all pipes in the study network. Design 2, 3, 4, 5, 6, and 7 have 356, 406, 457, 508, 610, and 762 mm, respectively, for all pipes. The seismic reliabilities of the seven designs are shown in Figure 8. For comparison, Todini's resilience is also calculated from the seven designs and plotted in Figure 8. While there is a large increase in seismic reliability from the 457 mm design (Design 4) to the 508 mm design (Design 5), reliability decreases from the 508 mm design to the 762 mm design (Design 7). This explains why we observed a plateau in seismic reliability in Figure 7. On the other hand, resilience (a traditional reliability measure) consistently increases with increasing pipe sizes. The marginal cost of improving resilience increases substantially for a resilience value of 0.3–0.8 and stabilized for a value higher than 0.8.

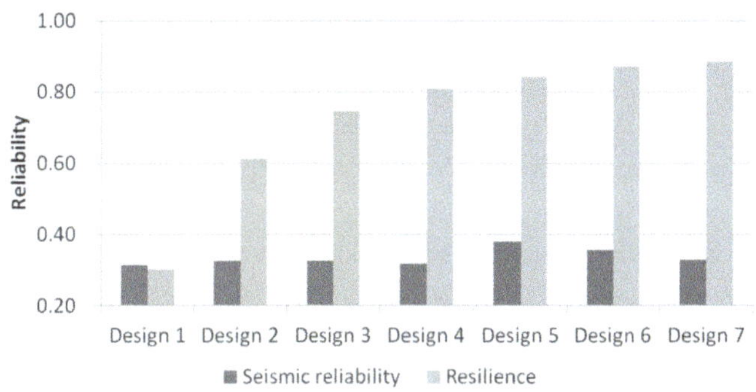

Figure 8. Seismic reliability of the seven uniform designs; all pipes are 508 mm in Case 5.

As the pipe sizes decrease, the correction factor for RR (Equation (4)) increases, finally resulting in high pipe breakage and leakage probability. However, although the pipes failed, the resulting impact is not significant to the system's seismic reliability because the calculated discharge coefficient, which is a function of the pipe's cross-sectional area (Table 2), is not large. However, as seen in Table 2, large pipes such as 508, 610, and 762 mm have a smaller failure probability compared to small pipes, but the resulting discharge coefficient is more than 10 times as big as that of small pipes. The failure effects are more significant for system seismic reliability compared to small pipes.

Table 2. Correction factors and discharge coefficients (Equation (6)) for the pipe sizes considered.

Pipe Sizes		Pipe's Cross-Sectional Area (A)	Correction Factor (C1)	Discharge Coefficient	
mm	inch			Breakage (100% of A)	Leakage (10% of A)
152	6	28	1	1074	107
203	8	50	0.8	1910	191
254	10	79	0.8	2985	298
305	12	113	0.8	4298	430
356	14	154	0.8	5850	585
406	16	201	0.8	7640	764
457	18	254	0.8	9670	967
508	20	314	0.5	11,938	1194
610	24	452	0.5	17,191	1719
762	30	707	0.5	26,861	2686

Random Designs

Finally, random designs that satisfy pressure requirements are generated from the proposed model to confirm the aforementioned conclusion. Figure 9 shows the profiles of seismic reliability and resilience of many random designs. We can clearly see that there is an inflection point around the total cost of 23 M USD and seismic reliability around 0.4, after which the overall seismic reliability decreases (Figure 9a). Because the total cost is a direct function of the pipe diameter, this plot indicates that installing large pipes does not always guarantee an increase in seismic reliability. The printed solutions are all suboptimal solutions that are dominated by the Pareto solutions found in Figure 7. On the other hand, installing large pipes resulted in an upward trend in resilience (Figure 8b).

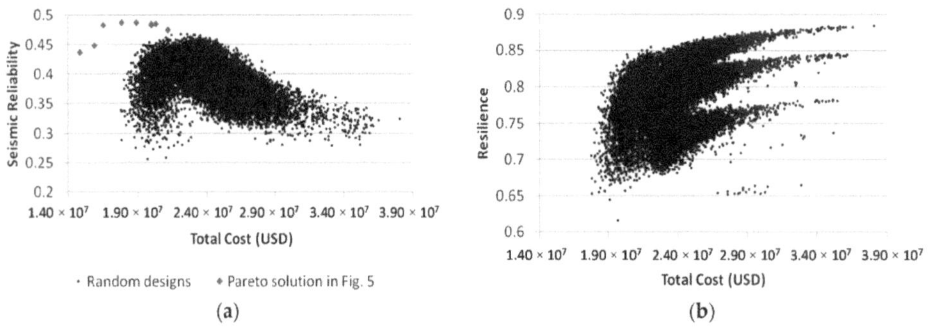

Figure 9. (**a**) Seismic reliability of randomly generated solutions (**black dot**) and Pareto solutions shown in Figure 7 (**blue diamond**) and (**b**) resilience of the same random designs.

Under normal failure condition, having more additional paths and installing large pipes throughout the system are beneficial with respect to system reliability (*i.e.*, ability to supply required quantity of water) and redundancy (*i.e.*, level of pressure redundancy). However, we observed this does not apply under earthquake failures. Earthquake deforms WDS components and their original function. Once failed under earthquakes, large pipes, which can deliver large volume of water under normal condition, help accelerate water loss out of the system. This study was the first attempt to take into account such irregular hydraulic behavior in the WDS modeling and design and highlight the need to find the optimal pipe layout considering the system's performances under the two different states.

SUMMARY AND CONCLUSIONS

Earthquakes can cause many simultaneous failures of components throughout WDSs, which result in very different and severe conditions compared to normal failure conditions that water communities usually deal with (e.g., single pipe failure). In this study, an economic, cost-constrained optimal design approach of a WDS was proposed to maximize seismic reliability. The network's seismic reliability is defined as the ratio of the available supply to the required water demand during stochastic earthquakes. The physical relationship between the earthquake intensity and the WDS components' vulnerability characteristics was defined and utilized for quantifying seismic reliability. To overcome the limitation of DDA in modeling earthquake failures, negative pressures were assumed to have realistic hydraulic calculations. Then, we investigated the seismic reliability improvement with respect to the economic investment in the system and the associated topological and pipe diameter changes. A traditional benchmark network, the Anytown network, was used to demonstrate the approach. Random earthquakes around the study network were generated for reliability quantification.

The results were quite different from what we generally observed from traditional reliability-based design (e.g., resilience-based design). First, allowing redundant small pipes in the system did not help improve seismic reliability. Those pipes cause frequent failures because of their low durability to earthquake forces. Second, having too-large pipes also degrades the system reliability during earthquakes. Compared to small pipes that are 152 mm to 305 mm, a big pipe (762 mm) has a lower failure probability; however, the failure's influence is very significant to the system once it fails. The large pipe's cross-sectional area will help release more water out of the system. Therefore, the first step to efficiently improve the WDS seismic reliability should be to identify the most appropriate pipe sizes for a system. For example, from the Anytown network, having a uniform 508 mm for all pipes provides the highest system reliability of 0.38 among seven uniform designs.

This study has several limitations that future research must address. First, this model is sensitive to correction factors for the pipe and the assumption of the opening area of the pipe under failure. Intensive sensitivity analysis should be conducted with experimental studies to simulate the most realistic failure behavior of WDS components. Second, this study only considers the pipe failures while pump, valve, tank, and reservoir failure can occur during an earthquake. Therefore, various failure types could be included in future studies. The proposed semi-PDA approach and EPANet can be replaced with a PDA-based network solver in order to simulate more realistic behavior of WDS under earthquake events. Third, while this study focuses on the system's reliability right after an earthquake occurs, post-earthquake recovery strategies should also be investigated to enhance the overall WDS's reliability/resilience. This could be found mostly in the context of operation and management. Fourth, the proposed semi-PDA approach and EPANet can be replaced with a PDA-based network solver in order to simulate more realistic behavior of WDS under earthquake events. Finally, interdependencies among multiple lifeline infrastructures (e.g., the water, power, transportation, and communication systems) can help improve and recover an individual infrastructure's reliability during an earthquake. Therefore, more efforts should be made to identify these interdependencies.

ACKNOWLEDGMENTS

This work was supported by a grant from The National Research Foundation (NRF) of Korea, funded by the Korean government (MSIP) (No. 2013R1A2A1A01013886) and the Korea Ministry of Environment as "The Eco-Innovation project (GT-11-G-02-001-2)".

AUTHOR CONTRIBUTIONS

Do Guen Yoo and Donghwi Jung carried out the survey of previous studies, analysis of proposed method, participated in the sequence alignment and drafted the manuscript. Joong Hoon Kim and Doosun Kang conceived the original idea of the study, and helped to write the final manuscript.

REFERENCES

1. Schaake, J.; Lai, D.D. *Linear Programming and Dynamic Programming Applications to Water Distribution Network Design*; Report No. 116; Department of Civil Engineering, Massachusetts Institute of Technology: Cambridge, MA, USA, 1969.
2. Alperovits, E.; Shamir, U.U. Design of optimal water distribution systems. *Water Resour. Res.* **1977**, *13*, 885–900.
3. Lansey, K.; Mays, L. Optimization model for water distribution system design. *J. Hydraul. Eng.* **1989**, *115*, 1401–1418.
4. Simpson, A.; Dandy, G.; Murphy, L. Genetic algorithms compared to other techniques for pipe optimization. *J. Water Resour. Plan. Manag.* **1994**, *120*, 423–443.
5. Savic, D.; Walters, G. Genetic algorithms for least-cost design of water distribution networks. *J. Water Resour. Plan. Manag.* **1997**, *123*, 67–77.
6. Lansey, K.; Duan, N.; Mays, L.; Tung, Y. Water distribution system design under uncertainty. *J. Water Resour. Plan. Manag.* **1989**, *115*, 630–645.
7. Xu, C.; Goulter, C. Reliability-based optimal design of water distribution networks. *J. Water Resour. Plan. Manag.* **1999**,*125*, 352–362.
8. Babayan, A.; Kapelan, Z.; Savic, D.; Walters, G. Least cost design of robust water distribution networks under demand uncertainty. *J. Water Resour. Plan. Manag.* **2005**, *131*, 375–382.
9. Kapelan, Z.; Savic, D.; Walters, G. Multiobjective design of water distribution systems under uncertainty. *Water Resour. Res.* **2005**, *41*, W11407-1–W11407-15.
10. Giustolisi, O.; Laucelli, D.; Colombo, A. Deterministic *versus* stochastic design of water distribution networks. *J. Water Resour. Plan. Manag.* **2009**, *135*, 117–127.
11. Jung, D.; Kang, D.; Kim, J.H.; Lansey, K. Robustness-based design of water distribution systems. *J. Water Resour. Plan. Manag.* **2014**, *140*.
12. Su, Y.; Mays, L.; Duan, N.; Lansey, K. Reliability-based optimization model for water distribution systems. *J. Hydraul. Eng.* **1987**, *113*, 1539–1556.
13. Federal Emergency Management Agency (FEMA). *HAZUS97 Technical Manual*; FEMA: Washington, DC, USA, 1997.
14. Kim, Y.S.; Spencer, B.F.; Song, J.; Elnashai, A.S.; Stokes, T. *Seismic Performance Assessment of Interdependent Lifeline Systems*; MAE Center: Urbana, IL, USA, 2007.
15. Hall, W.; Newmark, N. Seismic design criteria for pipelines and facilities. *J. Tech. Counc. ASCE* **1978**, *104*, 91–107.
16. Wright, J.P.; Takada, S. *Earthquake Response Characteristics of Jointed and Continuous Buried Lifelines*; Grant Report No. 15; National Science Foundation: New York, NY, USA, 1980.
17. Hwang, R.N.; Lysmer, J. Response of buried structures to traveling waves. *J. Geotech. Eng. Div.* **1981**, *107*, 183–200.
18. Fragiadakis, M.; Christodoulou, S.E.; Vamvatsikos, D. Reliability assessment of urban water distribution networks under seismic loads. *Water Resour. Manag.* **2013**, *27*, 3739–3764.
19. American Lifelines Alliance. *Seismic Fragility Formulations for Water Systems Part 1 Guideline*; American Lifeline Alliance: Washington, DC, USA, 2001.
20. Eguchi, R.T.; Taylor, C.E.; Hasselman, T.K. *Earthquake Vulnerability Models for Water Supply Components*; Technical Report No. 83-1396-2c; J.H. Wiggins Company: Redondo Beach, CA, USA, 1983.
21. Ballantyne, D.B.; Berg, E.; Kennedy, J.; Reneau, R.; Wu, D. *Earthquake Loss Estimation Modeling of the Seattlewater System*; Technical Report; Kennedy/Jenks/Chilton: Federal Way, WA, USA, 1990.
22. Shinozuka, M.; Tan, R.Y.; Toike, T. Serviceability of Water Transmission Systems under Seismic Risk. In *Lifeline Earthquake Engineering, the Current State of Knowledge*; American Society of Civil Engineers: New York, NY, USA, 1981.
23. Shinozuka, M.; Hwang, H.; Murata, M. Impact on water supply of a seismically damaged water delivery system. In*Lifeline Earthquake Engineering in the Central and Eastern U.S.*; American Society of Civil Engineers: New York, NY, USA, 1992; pp. 43–57.
24. Shinozuka, M.; Rose, A.; Eguchi, R.T. *Engineering and Socioeconomic Impacts of Earthquakes*; Monograph Series 2; Multidisciplinary Center for Earthquake Engineering Research: Buffalo, NY, USA, 1998.
25. Markov, I.; Mircea, G.; O'Rourke, T. *An Evaluation of Seismic Serviceability of Water Supply Networks with Application to the San Francisco Auxiliary Water Supply System*; Technical report NCEER 94-0001; National Center for Earthquake Engineering Research University of Buffalo, State University of New York: Buffalo, NY, USA, 1994.
26. Hwang, H.; Lin, H.; Shinozuka, M. Seismic performance assessment of water delivery systems. *J. Infrastruct. Syst.***1998**, *4*, 118–125.
27. Shi, P. Seismic Response Modeling of Water Supply Systems. Ph.D. Thesis, Cornell University, Ithaca, NY, USA, 1 January 2006.
28. Liu, G.Y.; Chung, L.L.; Yeh, C.H.; Wang, R.Z.; Chou, K.W.; Hung, H.Y.; Chen, S.A.; Chen, Z.H.; Yu, S.H. *A Study on Pipeline Seismic Performance and System Post-Earthquake Response of Water Utilities (1/2)*; Technical Report MOEA-WRA-0990095; Water Resource Agency, MOEA: Taipei, Taiwan, 2010.
29. Liu, G.Y.; Chung, L.L.; Huang, C.W.; Yeh, C.H.; Chou, K.W.; Hung, H.Y.; Chen, Z.H.; Chou, C.H.; Tsai, L.C. *A Study on Pipeline Seismic Performance and System Post-Earthquake Response of Water Utilities (2/2)*; Technical Report MOEA-WRA-1000090; Water Resource Agency, MOEA: Taipei, Taiwan, 2011.
30. GIRAFFE. *GIRAFFE User's Manual*; School of Civil and Environmental Engineering, Cornell University: Ithaca, NY, USA, 2008.

31. Rossman, L.A. *EPANET 2 User's Manual*; U.S. Environmental Protection Agency (EPA): Cincinnati, OH, USA, 2000.

32. Wang, Y. Seismic Performance Evaluation of Water Supply Systems. Ph.D. Thesis, Cornell University, Ithaca, NY, USA, 1 January 2006.

33. Shi, P.; O'Rourke, T.D.; Wang, Y. Simulation of earthquake water supply performance. In Proceedings of the 8th National Conference on Earthquake Engineering, Paper No. 8NCEE-001295, EERI, Oakland, CA, USA, 18–22 April 2006.

34. Wang, Y.; O'Rourke, T.D. Characterizations of seismic risk in Los Angeles water supply system. In Proceedings of the 5th China-Japan-US Symposium on Lifeline Earthquake Engineering, Haikou, China, 26–28 November 2007.

35. Bonneau, A.L. Water Supply Performance during Earthquakes and Extreme Events. Ph.D. Thesis, Cornell University, Ithaca, NY, USA, 2008.

36. Bonneau, A.L.; O'Rourke, T.D. *Water Supply Performance during Earthquakes and Extreme Events*; Technical Report MCEER-09-0003; University of Buffalo, State University of New York: Buffalo, NY, USA, 2009.

37. Yoo, D.G.; Jung, D.; Kang, D.; Kim, J.H.; Lansey, K. Seismic hazard assessment model for urban water supply networks. *J. Water Resour. Plan. Manag.* **2015**.

38. Geem, Z.W.; Kim, J.H.; Loganathan, G.V. A New Heuristic Optimization Algorithm: Harmony Search. *Simulation***2001**, *76*, 60–68.

39. Kim, J.H.; Geem, Z.W.; Kim, E.S. Parameter Estimation of the Nonlinear Muskingum Model Using Harmony Search. *J. Am. Water Resour. Assoc.* **2001**, *37*, 1131–1138.

40. Todini, E. Looped water distribution networks design using a resilience index based heuristic approach. *Urban Water***2000**, *2*, 115–122.

41. Kawashima, K.; Aizawa, K.; Takahashi, K. Attenuation of peak ground motion and absolute acceleration response spectra. In Proceedings of the 8th World Conference on Earthquake Engineering (WCEE), San Francisco, CA, USA, 24–28 July 1984; pp. 257–264.

42. Lee, K.; Cho, K.H. Attenuation of peak horizontal acceleration in the Sino-Korea Craton. In Proceedings of the Annual Fall Conference of Earthquake Engineering Society of Korea, Cheonan, Korea, 27 September 2002; pp. 3–10.

43. Baag, C.E.; Chang, S.J.; Jo, N.D.; Shin, J.S. Evaluation of seismic hazard in the southern part of Korea. In Proceedings of the International Symposium on Seismic Hazards and Ground Motion in the Region of Moderate Seismicity, Seoul, Korea, 1 November 1998; pp. 31–50.

44. Isoyama, R.; Ishida, E.; Yune, K.; Shirozu, T. Seismic damage estimation procedure for water supply pipelines. In Proceedings of the 12th World Conference on Earthquake Engineering (WCEE), Auckland, New Zealand, 1–4 January 2000; p. 1762.

45. Lambert, A. What do we know about pressure-leakage relationships in distribution systems? In Proceedings of the IWA Conference on Systems Approach to Leakage Control and Water Distribution System Management, Brno, Czech Republic, 16 May 2001.

46. Puchovsky, M.T. *Automatic Sprinkler Systems Handbook*; National Fire Protection Association (NFPA): Quincy, MA, USA, 1999.

47. Bentley Systems. *WaterGEMS*; Bentley Systems Incorporated: Exton, PA, USA, 2006.

48. Giustolisi, O.; Savic, D.A.; Berardi, L.; Laucelli, D. An Excel-based Solution to Bring Water Distribution Network Analysis Closer to Users. In Proceedings of the Computer and Control in Water Industry, Exeter, UK, 5–7 September 2011; Volume 3, pp. 805–810.

49. Muranho, J.; Ferreira, A.; Sousa, J.; Gomes, A.; Marques, A.S. Pressure-dependent Demand and Leakage Modeling with an EPANET Extension–WaterNetGen. *Procedia Eng.* **2014**, *89*, 632–639.

50. Wood, D. *KYPipe Reference Manual*; Civil Engineering Software Center, University of Kentucky: Lexington, KY, USA, 1995.

51. Cullinane, M.; Lansey, K.; Mays, L. Optimization-availability-based design of water-distribution networks. *J. Hydraul. Eng.* **1991**, *118*, 420–441.

52. Bao, Y.; Mays, L. Model for water distribution system reliability. *J. Hydraul. Eng.* **1990**, *116*, 1119–1137.

53. Farmani, R.; Walters, G.; Savic, D. Trade-off between total cost and reliability for Anytown water distribution network. *J. Water Resour. Plan. Manag.* **2005**, *131*, 161–171.

54. Prasad, T.D.; Park, N. Multiobjective genetic algorithms for design of water distribution networks. *J. Water Resour. Plan. Manag.* **2004**, *130*, 73–82.

55. Gheisi, A.; Naser, G. Multistate reliability of water-distribution systems: Comparison of surrogate measures. *J. Water Resour. Plan. Manag.* **2015**, *141*.

56. Walski, T.; Brill, E.; Gessler, J., Jr.; Goulter, I.; Jeppson, R.; Lansey, K.; Lee, H.; Liebman, J.; Mays, L.; Morgan, D.; Ormsbee, L. Battle of the Network Models: Epilogue. *J. Water Resour. Plan. Manag.* **1987**, *113*, 191–203.

57. Geem, Z.W. Harmony search in water pump switching problem. In *Advances in Natural Computation*; Springer: Berlin, Germany; Heidelberg, Germany, 2005; pp. 751–760.

58. Geem, Z.W. Optimal cost design of water distribution networks using harmony search. *Eng. Optim.* **2006**, *38*, 259–277.

59. Geem, Z.W. Harmony search optimisation to the pump-included water distribution network design. *Civ. Eng. Environ. Syst.* **2009**, *26*, 211–221.

60. Lansey, K. Sustainable, robust, resilient, water distribution systems. In Proceedings of the 14th Water Distribution Systems Analysis Conference, Adelaide, South Australia, 24–27 September 2012; pp. 1–18.

CHAPTER 7 ANALYSIS OF URBAN DRAINAGE NETWORKS USING GIBBS' MODEL

Yongwon Seo, Junshik Hwang and Seong Jin Noh

Keywords: Urban catchment, drainage network, flood mitigation, Gibbs' model, stochastic network model

INTRODUCTION

In 2010 and 2011, the metropolitan city of Seoul, South Korea suffered from floods that the city had not experienced before. Seoul had a record flood in 1990, which was caused by flooding of the Han River that crosses the city. Since the flood in 1990, the city government has invested heavily to reinforce the levees and pumping stations primarily to prevent a similar flood. In contrast to their predecessors, the floods in 2010 and 2011 were both caused by local heavy rainstorm converged on very small areas rather than by flooding from the Han River. In particular, the flood damage in 2011 was caused by heavy rainfall; more than 200 mm fell over a six hour periods on 27 July in Gangnam areas, which have a high-density of assets and population in the southeastern part of Seoul. Seoul has a design criteria for drainage networks to deal with rainfall amount of a 30 year-return period. The heavy rainfall in 2011 at the Gangnam area was equivalent to the rainfall-amount of a 50 year-return period, which was slightly greater than the design standards, resulting in flooding of more than four square kilometers of the city's heart near Gangnam subway station [1].

Flood disasters are the main cause of losses from natural hazards and are responsible for a large number of damaging events that threaten human safety [2]. With urbanization and climate change, many cities face increasing flooding risks playing a significant role in people's lives and the economy. Although there have been advances in the study of urban flood [3,4,5,6,7,8], most efforts have been concentrated on improving the accuracy of flood modeling in urban areas with complex geometry using cutting edge numerical methods and observations. Drainage networks are essential compartments in an urban infrastructure system for the most efficient collection and prompt drainage of flood water. On the other hand, despite their importance, the geomorphological characteristics of urban drainage networks and their impacts on flooding have not been investigated thoroughly compared to precise numerical schemes.

The relationship between the drainage network topology and shape of hydrographs has been a long standing topic in hydrology [9,10]. The configuration of the drainage network layout depends strongly on the problem size, location of the outlet, and the topography of a catchment. In contrast, the small difference in the elevation in flat areas makes the problems totally different [11]. There is no guidance, such as topography, for a designer to delineate the layout of a drainage network. In general, drainage networks in urban catchments aim at the efficient collection and prompt drainage of water as much as possible. Seo and Schmidt [12,13] examined the drainage networks of the southwestern part of Chicago, where the catchments are located at flat areas. The results revealed a wide range of network configurations in urban drainage networks in the flat areas in southwestern Chicago.

Seo and Schmidt [12] utilized the one-parameter Gibbs' model [14] to categorize the networks. As the parameter (β) tends to be zero, the resulting Gibbs' model is close to the uniform model [15,16] whereas the Gibbs' model is close to the Scheidegger model [17] as the parameter tends to infinity. The Gibbs' model is based on Gibbs' measure [18], which has maximum entropy and properties of a Markov random field [19]. As a stochastic network model, Gibbs' model is useful for categorizing an existing network, and also useful for generating an alternative layout of the drainage network. Seo and Schmidt [13] reported that Gibbs' model is efficient in terms

Citation: Yongwon Seo , Junshik Hwang, and Seong Jin Noh, Analysis of Urban Drainage Networks Using Gibbs' Model, doi:10.3390/w7084129.

of reproducing the original width function of a corresponding urban catchment. Seo and Schmidt [20,21] also, evaluated drainage networks under moving storms using Gibbs' model.

As Haghighi [11] mentioned, in steep areas, drainage networks are typically cost-effective if they are designed based on the topography. In contrast, in flat areas like Chicago, there are not only efficient and optimized networks in terms of the drainage time, but also inefficient and highly sinuous networks as reported by Seo and Schmidt [12]. Combined with the effects of network topology on the resulting hydrographs of a catchment, the configuration of drainage network layout can be a crucial point in terms of mitigating the flood risks in urban catchments.

In this regard, this study examined urban drainage networks in Seoul in terms of the network configuration and its implication for peak flows and flood mitigation. Gibb's model was used to analyze the network configuration, and the parameter, β, of Gibbs' model was used to represent the configuration. Compared to previous studies on flat areas [12,14], the methodology was evaluated for a range of urban catchments, which are hilly and have suffered from inundation in recent years. A test catchment (Seocho4) was selected among 31 catchment to demonstrate the effect of drainage network layout on the peak flows of the resulting hydrographs by comparing the hydrographs from the original network and Gibbs' model. Actual rainfall records in 2011 were used for this purpose to show how network layout affects the peak flows and floods.

METHODS

Gibbs' Model

Gibbs' model sets the probability based on Gibbs' measure, which originated from the Ising model [18]. Any positive measure with the property of a Markov random field can be regarded a Gibbs' measure with an appropriate energy function [19]. The stochastic network model based on Gibbs' measure [14] defined a Markov chain with the spanning trees S as the state space. A tree is a graph in which any two vertices are connected by precisely one simple path and a spanning tree is an acyclic tree connecting all points in the network without loops or cycles. Let s belongs to Ss and two trees, s_1 and s_2 be adjacent if one can be obtained from the other. To achieve this, a point must be selected randomly in s_1 and a new direction defined from that point will be a new spanning tree, s_2. The transition probability from s_1 to s_2, R_{s1s2}, can be defined as follows [14]:

$$R_{s_1 s_2} = \begin{cases} r^{-1}\min\{1, e^{-\beta[H(s_2)-H(s_1)]}\} & s_2 \in N(s_1) \\ 1 - \sum_{s \in N(s_1)} R_{s_1 s} & s_2 = s_1 \\ 0 & \text{otherwise} \end{cases} \tag{1}$$

where $N(s_1)$ is the set of trees adjacent to s_1, and β is a parameter that represents the extent to which the sinuosity of the network is reflected in the generation of the new spanning tree, s_2. For example, when β is equal to zero, the overall sinuosity of a network has no relationship with the transition probability and the transition probabilities are the same in all possible directions, which is identical to the uniform distribution. The maximum degree of the points in S, r, is defined as follows [14]:

$$r = \max_{s \in S} |N(s)| \tag{2}$$

The degree, r, refers to the maximum number of directions that can be selected, except for the existing direction. $H(s)$ is a measure of the sinuosity of a given spanning tree, s:

$$H(s) = \sum_{d \in D(B)} \xi_s(d) - \sum_{d \in D(B)} \xi_B(d) \tag{3}$$

where s is a spanning tree, d is a point of a finite and connected graph B, and $D(B)$ is the set of the total points of B. ξ_s is the distance to an outlet along s from v, while ξ_B is the shortest distance to the outlet from d. During every generation of adjacent spanning trees, a check is made to determine if there is any loop inside the network.

As shown in Figure 1, Gibbs' model was simulated for one of the test catchments, Seocho4, for each value of the parameter, β. The thickness of the network represents the maximum width function at that location. The width function can be converted easily to a unit hydrograh. Therefore, the thickness can be interpreted as the maximum peak flow at each point. In this regard, the magnitude of the peak flow at the outlet itself is higher on Gibbs' model with a greater value of β (e.g., when β = 10^0 in Figure 1) compared to the model with a smaller value of β (For example, when β = 10^{-4}). Figure 1shows that the maximum peak flows are observed not only at the outlet but also in other parts of the mainstream on Gibbs' model with a smaller value of β. In contrast, although the magnitude of the peak flow is higher with larger β values, the maximum peak flow is observed near the outlet as β increases. A comparison of the flow distribution of Gibbs' model indicates a close relationship between the spatial distribution of the peak flows and the drainage network, which is related directly to the risk of the corresponding drainage system.

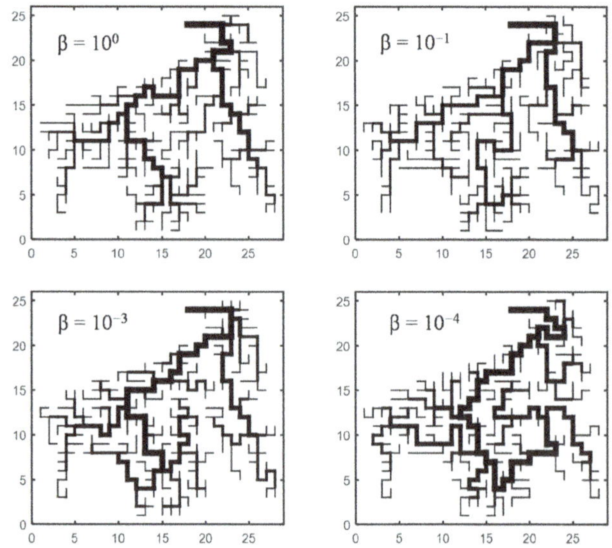

Figure 1. Gibbs' model generated for a test catchment, Seocho4

for each β with a grid size of 50 m. The horizontal and vertical axis denote distance by number of grids.

Figure 2 shows the width function of Gibbs' model (Figure 1) at the outlet of the test catchment, Seocho4. As shown inFigure 2a, the total distance from the outlet becomes shorter as β increases whereas the peak becomes larger. Figure 2b shows the maximum value of Gibbs' model averaged after one hundred simulations. Owing to the nature of stochastic network model, the averaged maximum value with β equal to 10^0 is smaller than that with β equal to 10^{-1}, but the trend shows that the maximum value of the width function at the outlet tends to increase with β. This is important because, as mentioned earlier, the width function is related directly to the hydrologic response of a drainage network. Figure 2 shows that the peak flows of a drainage network are highly dependent on the configuration of the layout.

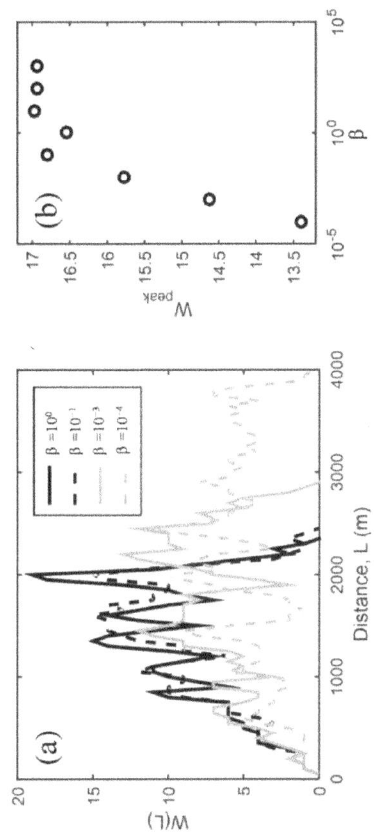

Figure 2. (**a**) The width function at the outlet of Gibbs' model depending on β and (**b**) the maximum value of the width functions averaged over 100 simulation for each β for a test catchment, Seocho4.

Study Area and Estimation of Beta

The metropolitan city of Seoul is composed of 238 catchments (Figure 3). The Han River crosses the city. The shaded area also shows the flooded areas caused by flash storms in 2010 and 2011. The storms converged on the southern part of Seoul. In particular, the damages that occurred in 2010 and 2011 were focused primarily on the southeastern (Shinweol) and southwestern areas (Gangnam), which have a high density of assets and population. Among the 238 catchments, the black-lined catchments in Figure 3 with flood damages in 2010 and 2011 were selected to estimate the value of β. Table 1 lists the test catchments selected in Seoul in this study. The areas of the selected test catchments ranges from 0.32 to 5.43 km^2 and the mean area is 1.90 km^2. The catchment slopes are rather steep, ranging from 2.92% to 19.32% with an average of 8.57%. The conduit slopes range from 0.67% to 7% with an average of 4%.

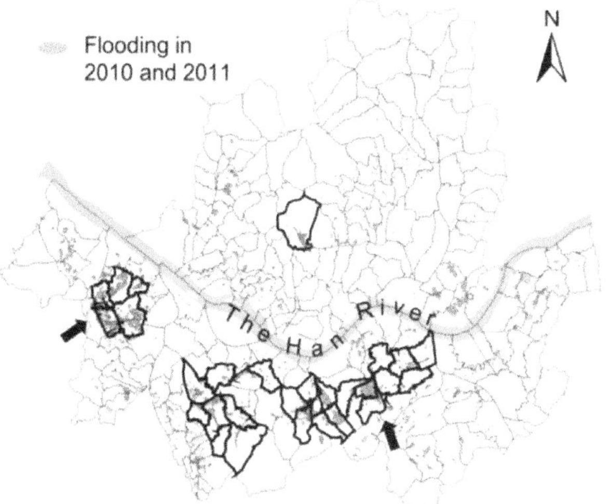

Figure 3. Test catchments in Seoul: the black-lined catchments were selected to estimate the value of β; flood damages in 2010 and 2011 primarily occurred in the southwestern areas (Shinweol) and also southeastern areas (Gangnam).

This study used the width function to estimate β for each catchment. The width function is defined as the catchment area at a distance from the outlet [22]. The width function and the area function can be defined differently based on the channelization [23] but the width function basically represents the distance-area function [24]. In this study, the width function was defined as the number of grid points for a given distance from the outlet [12,20]. The value of β that best represents the network width function of a catchment was selected. The procedure is as follows: First, a drainage network was reconstructed on a lattice, from which the actual width was calculated. Using the previously simulated Gibbs' models (one hundred simulation) for each beta, the width functions were then averaged with respect to distance from the outlet. The Nash-Sutcliffe model efficiency, E, was used to select the value of β that best respresented the original drainage network. Figure 4 presents the process of estimation for a catchment, BangBae3. The gray shaded areas are located between 25% and 75% quartiles. The dots are the averaged width function from Gibbs' model. The model efficiency, E, increased to 0.96 for a value of β equal to 10^0 for BangBae3, which is one of the catchments, where flood occured in both 2010 and 2011.

Table 1. Test catchments selected for study.

No	ID	Catchment	Area (A_w) km²	Perimeter (B_p) km	Drainage Density (D) m/m²	Conduit Slope (S_p) %	Catchment Slope (S_c) %
1	NH	Nonhyeon	1.80	8.24	0.028	5.633	6.959
2	DR	Daelim	1.25	5.92	0.024	3.565	3.721
3	DO	Daebang	2.02	8.09	0.026	3.430	4.733
4	DG	Dogok	1.82	7.33	0.022	3.377	4.330
5	DLO	Doriml	2.71	9.10	0.037	0.668	2.921
6	DCW	Deungchon2	0.87	4.57	0.022	5.638	8.461
7	BBW	Bangbae2	1.37	7.62	0.026	4.107	9.395
8	BBT	Bangbae3	1.66	7.31	0.011	6.270	17.273
9	BBF	Bangbae4	1.56	7.05	0.010	5.382	17.392
10	SD	Sadang	1.83	6.81	0.021	4.448	11.364
11	SSO	Samsung1	1.91	7.98	0.026	3.784	4.850
12	SSW	Samdung2	2.10	6.88	0.021	3.076	5.822
13	SDW	Sangdo2	1.94	7.76	0.028	3.321	9.451
14	SCO	Seocho1	1.89	8.59	0.019	4.801	7.846
15	SCT	Seocho3	1.79	8.69	0.017	3.437	6.959
16	SCF	Seocho4	1.06	6.06	0.024	3.126	3.112
17	SH	Shiheunggoji	1.60	8.61	0.022	4.604	10.007
18	SLA	Sillim1	1.98	8.84	0.033	4.489	9.503
19	SLW	Sillim2	4.35	10.56	0.012	5.670	18.232
20	SLM	Sillim3	1.36	5.82	0.019	6.996	19.319
21	SLF	Sillim4_1	0.32	2.71	0.033	1.408	3.221
22	SLF	Sillim4_2	2.23	9.08	0.025	3.342	9.953
23	SWO	Sinwol1	1.52	7.46	0.025	2.663	5.205
24	SWT	Sinwol3_1	1.20	5.77	0.035	2.345	3.014
25	SWT	Sinwol3_2	0.52	3.48	0.027	2.540	3.539
26	YS	Yeoksam	1.93	7.40	0.030	3.637	4.996
27	LM	Yimok	2.09	8.37	0.023	5.372	11.820
28	IS	Yisu	1.37	5.62	0.024	4.211	10.168
29	HGO	Hwagok1	2.28	7.22	0.025	4.404	9.820
30	HKW	Hwagok2	3.26	10.07	0.029	3.969	6.243
31	HY	Hyoja	5.43	10.97	0.016	4.263	16.030

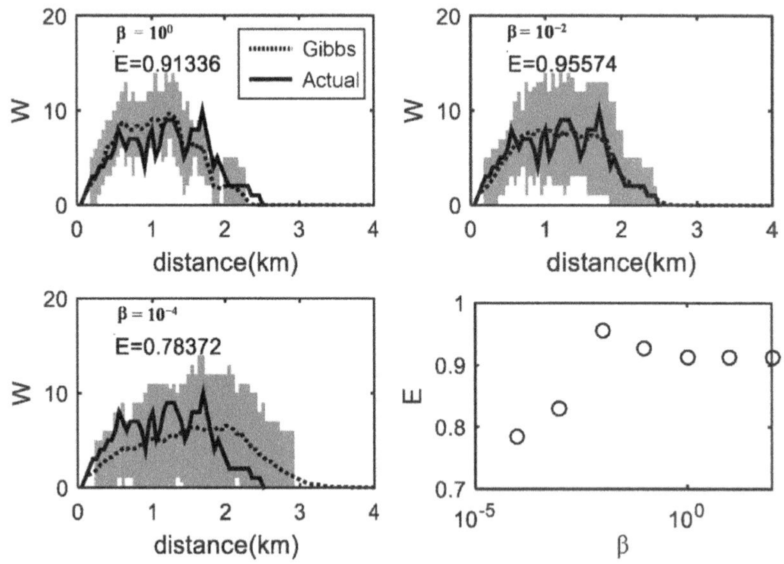

Figure 4. Comparison of the width functions from the actual and simulated drainage networks; the resulting E is plotted as a function of β for a test catchment, BangBae3; the shaded area represents interval between the 25% and 75% quartiles.

Flood in 2010 and 2011 in a Test Catchment, Seocho4

Recent rainfall events were likely to have occurred under the influence of climate change. The climate variability and change have direct consequences on global flood hazards [25]. The increased frequency of the occurrence of flood events in the world is partially attributed to the climate change-driven increase in extreme precipitation [26]. Climate change is making weather less predictable, rains more uncertain and heavy storm rainfalls more likely [27]. The quantitative estimation of the impact of climate change on small-scale and extreme rainfall events is a developing research area, and there is a large degree of uncertainty with the current estimates [26].

As mentioned previously, the floods in 2010 and 2011 were caused mainly by the insufficient conveyance capacity of drainage networks. Previous floods in Seoul were due to inundation of the Han River or rising water surface elevation of the Han River accompanied by the insufficient pumping capacity. Therefore, the city government has invested in the reinforcement of levees and pumping stations primarily to prevent flooding from the Han River. In contrast to historical floods in Seoul, the floods both in 2010 and 2011 were caused by local heavy rainfall converging on very small areas of Seoul rather than by the Han River. In particular, the flood damage in 2011 was caused by heavy rainfall, more than 200 mm, during 6 h on July 27 in Gangnam areas, which have a high-density of assets and population in the southeastern part of Seoul. Seoul has the design criteria for drainage networks to deal with the rainfall amount of a 30 year-return period. The heavy rainfall in 2011 at the Gangnam area was equivalent to the rainfall-amount of a 50 year-return period, which is slightly greater than the design standards resulting in flooding over more than four square kilometers of the city's heart near the Gangnam subway station [8]. Figure 5a presents Seocho4, which is one of the catchments around the Gangnam subway station that was flooded in both 2010 and 2011. The flood in 2010 was greater than that in 2011. Figure 5b shows the drainage network and topography of Seocho4. The arrow in Figure 5b indicates the flooded location where the conveyance capacity of the conduit was insufficient. The analysis shows that the overflowed water flows along the street and contributed to the flooding of the Gangnam subway station consequently [1].

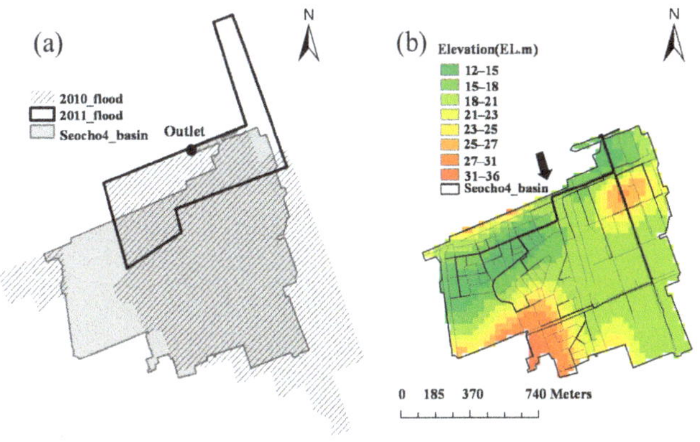

Figure 5. (**a**) Test catchment, Seocho4 with flooded areas reported; (**b**) drainage network and topography.

RESULTS AND DISCUSSION

Network Configuration of Urban Catchments in Seoul

Using the methodology described in the previous section, the network configuration of 31 drainage networks in Seoul was estimated. A constant grid size of 50 m was used to reconstruct the original network and also simulate Gibbs' model to estimate the value of β. As mentioned earlier, the catchment slopes are rather steep from 2.92% to 19.32% with an average of 8.57%. The conduit slope ranges from 0.67% to 7% with an average of 4%. Seo and Schmidt [12] examined the value of β of Chicago area and found that β ranges form 10^{-2} to 10^{2}, where the catchment slope ranges from 0.51% to 1.05% with an average catchment slope of 0.8%. Although the catchments of Chicago and Seoul have similar average catchment areas, 2.81 and 1.90 square kilometers, they have significant difference in catchment slope; the average slope (8.57%) of Seoul's catchments is more than 10 times steeper than that (0.80%) of Chicago areas.

Table 2 lists the values of β estimated for the 31 catchments in Seoul. The columns, *n* and *m* are the number of rows and columns used to reconstruct drainage networks. The goodness of fit is the Nash-Sutcliffe efficiency to determine the value of β, as shown in Figure 4. The Nash-Sutcliffe model efficiency ranges from 0.52 to 0.99 and the average is 0.83, which shows that Gibbs' model successfully simulates the original drainage network. The result shows that the value of β ranges from 10^{-4} to 10^{3}, which is even wider than that of Chicago (10^{-2}–10^{2}) reported by Seo and Schmidt [12]. Troutman and Karlinger [14] examined the value of beta in mountainous areas in Montana and found that rivers have the mean value of beta equal to 10^{0}. Seo *et al.* [28] showed that rivers in nature are close to the Scheidegger model, which is equivalent to Gibbs' model with β of infinity. These previous studies show that rivers have evolved and transformed themselves to very efficient states in terms of drainage time. In contrast, the analysis using Gibbs' model shows that urban drainage network can be less efficient compared to rivers.

As Haghighi [11] mentioned, in steep areas, the drainage networks are typically cost-effective if they were designed based on the topography. In contrast, in flat areas such as Chicago, there are not only efficient and optimized networks in terms of the drainage time or construction cost, but also inefficient and highly sinuous networks [12,13]. On the other hand, the analysis of catchments in Seoul, where the catchment slope is 10 times steeper than that of the Chicago catchments, still showed that the drainage network can be highly ineffective in terms of the drainage time.

Figure 6a shows the distribution of β for the 31 test catchments in Seoul, which shows a wide range of network configurations. As shown in Figure 2b, once $\beta > 10^{0}$, Gibbs' model revealed little difference in the increased peak of the width function. Seo and Schmidt [13] also reported that Gibbs' model does not demonstrate a difference in terms of the flow distance with $\beta > 10^{0}$. Therefore, a network with $\beta \geq 10^{0}$ can be referred to as effective in terms of the drainage time. In this regard, among the 31 drainage networks analyzed in this study, 13 catchments (shaded bar in Figure 6a), which have values of $\beta \geq 10^{0}$, are efficient in terms of the drainage time. The 31 test catchments analyzed in this study were selected among the catchments that suffered floods in 2010 and 2011. Therefore, these results suggest that an efficient network in terms of drainage time does not necessarily help protect the corresponding catchment from a flood. Obviously, there are many factors affecting the flooding condition of a catchment. The next section shows how the network configuration affects the resulting flood in the next section using the test catchment Seocho4 and historical storm simulations in 2010 and 2011 by applying different configurations of a drainage network.

Table 2. Network configuration of drainage networks in Seoul.

No	ID	Catchment	Basin Area (A_w)	Grid Size	n	m	Goodness of Fit (E)	O(β)
			km²	m				
1	NH	NonHyeon	1.80	50	43	34	0.77	10^{-4}
2	DR	DaeLim	1.25	50	24	27	0.91	10^{-2}
3	DO	DaeBang	2.02	50	34	39	0.62	10^{1}
4	DG	DoGok	1.82	50	27	42	0.70	10^{-2}
5	DLO	DoRim1	2.71	50	48	36	0.52	10^{-4}
6	DCW	DeungChon2	0.87	50	22	21	0.99	10^{-1}
7	BBW	BangBae2	1.37	50	41	27	0.91	10^{1}
8	BBT	BangBae3	1.66	50	21	31	0.96	10^{-2}
9	BBF	BangBae4	1.56	50	25	26	0.96	10^{-2}
10	SD	SaDang	1.83	50	30	23	0.82	10^{0}
11	SSO	SamSung1	1.91	50	36	38	0.87	10^{3}
12	SSW	SamDung2	2.10	50	41	33	0.87	10^{-1}
13	SDW	SangDo2	1.94	50	33	38	0.93	10^{1}
14	SCO	SeoCho1	1.89	50	47	36	0.73	10^{-1}
15	SCT	SeoCho3	1.79	50	28	40	0.79	10^{1}
16	SCF	SeoCho4	1.06	50	25	28	0.96	10^{0}
17	SH	ShiHeungGoJi	1.60	50	49	36	0.88	10^{2}
18	SLA	SilLim1	1.98	50	42	42	0.71	10^{1}
19	SLW	SilLim2	4.35	50	34	44	0.89	10^{1}
20	SLM	SilLim3	1.36	50	36	22	0.92	10^{1}
21	SLF	SilLim4_1	0.32	50	12	13	0.99	10^{-1}
22	SLF	SilLim4_2	2.23	50	46	30	0.77	10^{3}
23	SWO	Sinwol1	1.52	50	31	24	0.85	10^{-1}
24	SWT	Sinwol3_1	1.20	50	24	32	0.80	10^{-3}
25	SWT	Sinwol3_2	0.52	50	15	22	0.93	10^{-1}
26	YS	YeokSam	1.93	50	33	35	0.78	10^{-4}
27	LM	YiMok	2.09	50	39	42	0.80	10^{2}
28	IS	YiSu	1.37	50	37	24	0.73	10^{-4}
29	HGO	HwaGok1	2.28	50	38	30	0.88	10^{1}
30	HKW	HwaGok2	3.26	50	43	40	0.76	10^{-4}
31	HY	HyoJa	5.43	50	56	46	0.70	10^{1}

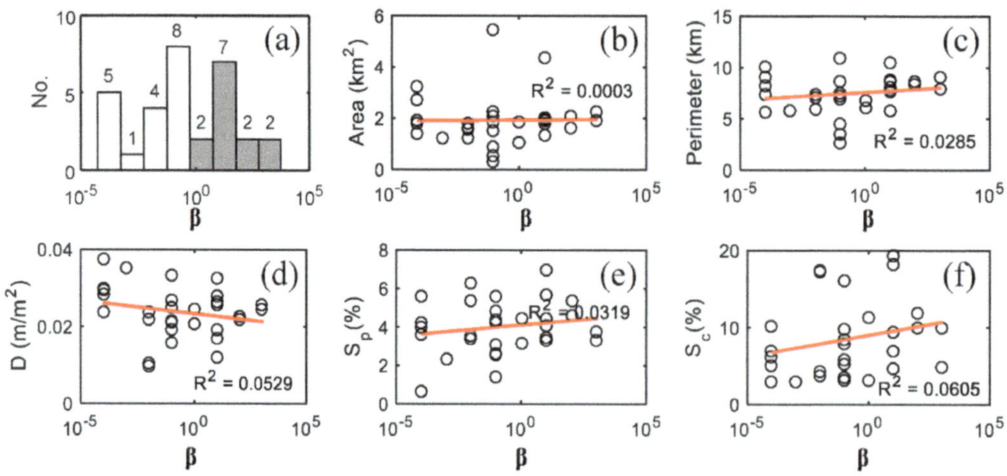

Figure 6. (**a**) Distribution of β for 31 catchments in Seoul; (**b**) the value of β as a function of catchment area; (**c**) perimeter; (**d**) drainage density (D); (**e**) conduit slope (S_p) and (**f**) catchment slope (S_c).

Figure 6b–f shows very weak relationships between β and the catchment properties, such as area, perimeter, drainage density, conduit slope, and catchment slope. In particular, Figure 6e,f shows that there is no correlation between the network configuration and conduit or catchment slopes. Therefore, the case study of urban catchments in Seoul shows that ineffective drainage network can exist in urban catchments regardless of the catchment slope. For example, the test catchment, YiSu, has a catchment slope of 10.17% but the value of β is 10^{-4}. In contrast, SamSung1 has a catchment slope of 4.85 but the value of β is 10^3.

The Effect of Drainage Network Layout

Note that the configuration of urban drainage network layout has no relationship with the catchment properties, such as catchment slope. This means that the configuration of the layout is not automatically or naturally determined by following topography. This also means that it is necessary to think of the configuration of drainage network layout intentionally because it has a relationship with flooding or peak flows of the drainage network runoff hydrographs [12]. To evaluate the effects of drainage network layout, a model was built using the US Environment Protection Agency (EPA) Stormwater Management Model (SWMM) of Seocho4 with the historical rainfall to reproduce the flooding condition in 2010 and 2011. A different layout was then provided by Gibbs' model with β = 10^{-4} instead of the original layout, of which β = 10^0. Figure 7depicts the SWMM model with the original and alternative drainage network for Seocho4. The diameters of the alternative drainage network were determined depending on the Strahler ordering. Table 3 lists the conduit information of the original network. The alternative layout (Figure 7b) was built to have the same diameter as the original network depending on the Strahler order. The impervious ratio of each grid was obtained using the land use map and the soil parameters for the Green-Ampt equation were based on the corresponding soil types [29]. Unfortunately, there is no flow record for the calibration of the SWMM model in Seocho4, but flood areas both in 2010 and 2011. The SWMM result of the original layout estimated total flooded volume of 41,328 m^3 in 2010, which is close to the flooded area (Figure 5a) assuming a flooding depth of 0.25 m.

Figure 8a,b present the simulation results for storm events in 2010 and 2011, which show the cases with flooding whereasFigure 8c shows a case with no flooding with a storm event on 23 July 2014. The flooded volume in the urban surface area is shown as white bars under the hydrographs for the actual network and the grey bars denote the alternative (Gibbs' model). The case with flooding (Figure 8a,b) showed no reduction in the peak flows. Instead, the flooded volume was reduced by 54% and 72%, respectively, compared to the actual (original) layout. In contrast, Figure 8c shows the simulation results for a rainfall event with no surcharging (flooding). The peak flow was reduced by 10.2% for the alternative layout obtained from one of Gibbs' model with β = 10^{-4}. The results with an alternative layout strongly highlight the importance of network configuration in terms of the reduction and mitigation of flood damages in urban environments.

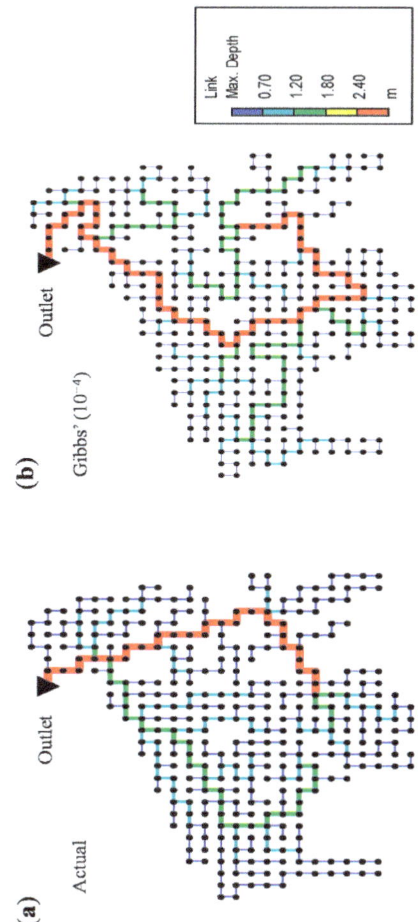

Figure 7. (**a**) Actual drainage network; (**b**) alternative drainage network of Seocho4 built with SWMM.

Table 3. Conduit information depending on Strahler ordering.

Order	Diameter (m)	S_p (%)	Length (m, Averaged)
1	0.58	2.72	37.88
2	0.80	0.19	31.37
3	1.05	2.97	30.03
4	2.11	1.02	40.09
5	2.95	0.35	31.18

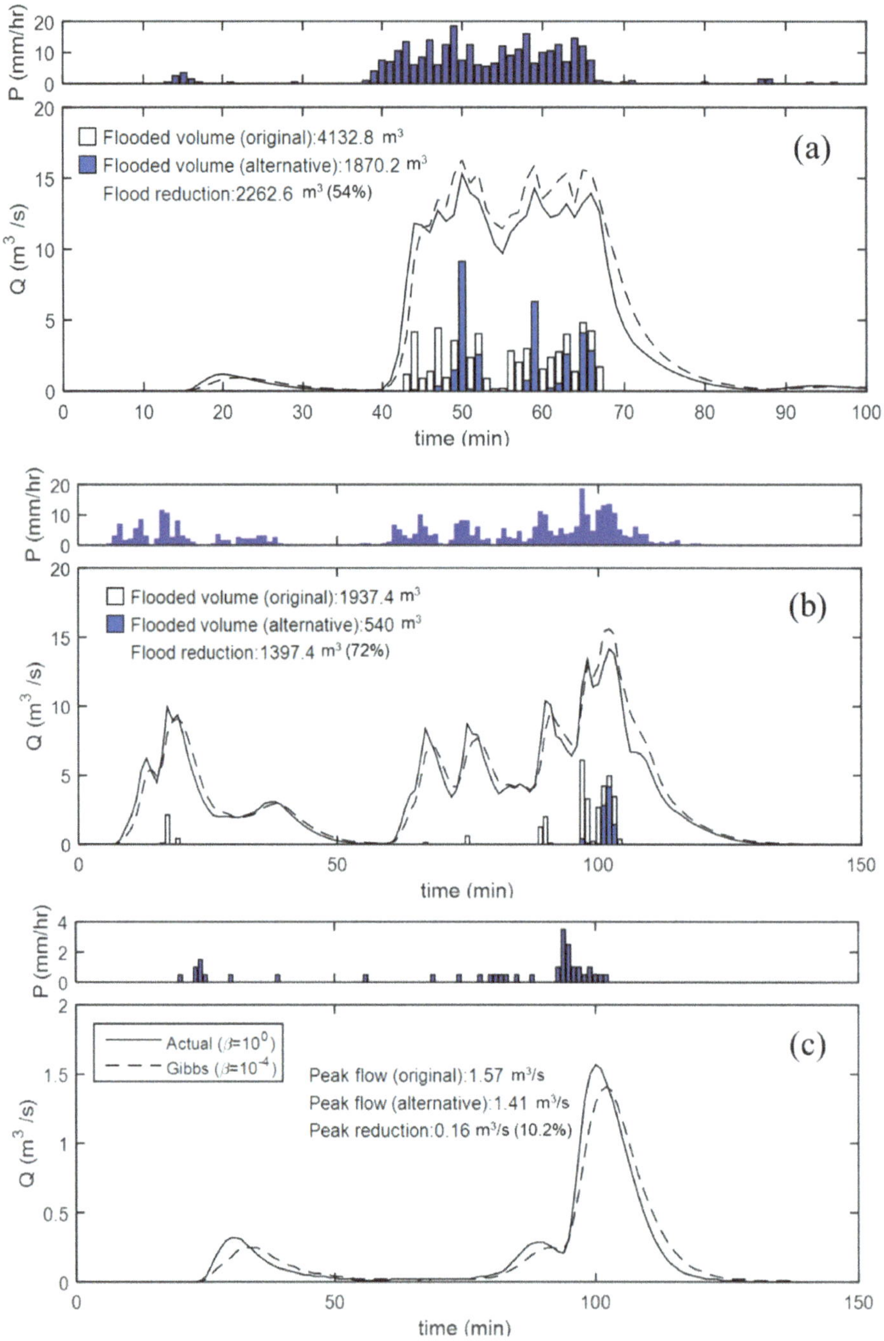

Figure 8. Simulation results with actual and alternative layout from Gibbs' model for storm events on (**a**) 21 September 2010; (**b**) 26 July 2011; (**c**) 23 July 2014.

CONCLUSIONS

This study analyzed the characteristics of urban drainage networks in Seoul using Gibbs' model. The results showed that the drainage networks in Seoul have a wide range of network configurations. Combining the results from previous studies [12,14], this study showed that the network configuration has little relation with the catchment properties, such as the catchment slope. This is important because the configuration of the layout is not automatically or naturally determined by the topography. Therefore, it is necessary to consider the configuration of the drainage network layout with intent because it does have a relationship with flooding or the peak flows of the drainage network runoff hydrographs. The case study of Seoul highlights the need to seek an alternative drainage network layout. Typically, a drainage system is designed to drain water from developed areas as promptly as possible during floods. On the other hand, the case study in Seoul shows that efficient networks can have risks of flood concentration and increase potential flood risks. This study showed that efficient networks tend to have higher peak flows at the outlet in the case of no flooding. In contrast, an alternative drainage network layout of a test catchment, which is less efficient and more sinuous, resulted in reduced peak flows in the case of no flooding and also reduced flooded volume in the case of flooding. In this regard, managing a proper drainage network layout can be an alternative measure for efficient flood mitigation in urban environments.

ACKNOWLEDGMENTS

This study was supported by a grant (14AWMP-B079364-01) from the Water Management Research Program funded by Ministry of Land, Infrastructure and Transport of Korean government.

Author Contributions

Seong Jin Noh proposed the basic idea, collected drainage networks, historical rainfall and flooding data in Seoul. Yongwon Seo designed the numerical experiment. Junshik Hwang conducted the simulation of network and SWMM models. The manuscript was written by Yongwon Seo with contribution from Junshik Hwang and Seong Jin Noh.

Conflicts of Interest

The authors declare no conflict of interest.

REFERENCES

1. Choi, S.; Kang, S.; Han, S.; Lee, D.R. Urban flooding due to heavy rainstorm with focuses on gangnam areas in Seoul.*Mag. Korea Water Resour. Assoc.* **2011**, *44*, 25–29.
2. Noh, S.J.; Tachikawa, Y.; Shiiba, M.; Kim, S. Ensemble Kalman filtering and particle filtering in a lag-time window for short-term streamflow forecasting with a distributed hydrologic model. *J. Hydrol. Eng.* **2013**, *18*, 1684–1696.
3. Brodie, I.M. Rational Monte Carlo method for flood frequency analysis in urban catchments. *J. Hydrol.* **2013**, *486*, 306–314.
4. Chang, F.J.; Chen, P.A.; Lu, Y.R.; Huang, E.; Chang, K.Y. Real-time multi-step-ahead water level forecasting by recurrent neural networks for urban flood control. *J. Hydrol.* **2014**, *517*, 836–846.
5. Chen, A.S.; Evans, B.; Djordjevic, S.; Savic, D.A. A coarse-grid approach to representing building blockage effects in 2D urban flood modelling. *J. Hydrol.* **2012**, *426*, 1–16.
6. Djordjević, S.; Prodanović, D.; Maksimović, Č. An approach to simulation of dual drainage. *Water Sci. Technol.* **1999**,*39*, 95–103.
7. Smith, M.B. A gis-based distributed parameter hydrologic model for urban areas. *Hydrol. Process.* **1993**, *7*, 45–61.
8. Zhang, S.H.; Wang, T.W.; Zhao, B.H. Calculation and visualization of flood inundation based on a topographic triangle network. *J. Hydrol.* **2014**, *509*, 406–415.
9. Rodriguez-Iturbe, I.; Valdes, J.B. Geomorphologic structure of hydrologic response. *Water Resour. Res.* **1979**, *15*, 1409–1420.
10. Saco, P.M.; Kumar, P. Kinematic dispersion in stream networks 1. Coupling hydraulic and network geometry. *Water Resour. Res.* **2002**, *38*, 1–14.
11. Haghighi, A. Loop-by-loop cutting algorithm to generate layouts for urban drainage systems. *J. Water Resour. Plan. Manag.* **2013**, *139*, 693–703.
12. Seo, Y.; Schmidt, A.R. The effect of rainstorm movement on urban drainage network runoff hydrographs. *Hydrol. Process* **2012**, *26*, 3830–3841.
13. Seo, Y.; Schmidt, A.R. Application of Gibbs' model to urban drainage networks: A case study in southwestern Chicago, USA. *Hydrol. Process* **2014**, *28*, 1148–1158.
14. Troutman, B.M.; Karlinger, M.R. Gibbs distribution on drainage networks. *Water Resour. Res.* **1992**, *28*, 563–577.
15. Leopold, L.B.; Langbein, W.B. *The Concept of Entropy in Landscape Evolution*; U.S. Government Printing: Washington, DC, USA, 1962.
16. Karlinger, M.R.; Troutman, B.M. A random spatial network model based on elementary postulates. *Water Resour. Res.***1989**, *25*, 793–798.
17. Scheidegger, A.E. A stochastic model for drainage patterns into an intramontane trench. *Int. Assoc. Sci. Hydrol. Bull.***1967**, *12*, 15–20.
18. Ising, E. Beitrag zur theorie des ferromagnetismus. *Z. für Phys. A Hadron. Nucl.* **1925**, *31*, 253–258. (In German).
19. Kindermann, R.; Snell, J.L. *Markov Random Fields and Their Applications*; American Mathematical Society: Ann Arbor, MN, USA, 1980.
20. Seo, Y.; Schmidt, A.R. Network configuration and hydrograph sensitivity to storm kinematics. *Water Resour. Res.* **2013**,*49*, 1812–1827.
21. Seo, Y.; Schmidt, A.R. Evaluation of drainage networks under moving storms utilizing the equivalent stationary storms. *Nat. Hazard.* **2014**, *70*, 803–819.
22. Moussa, R. What controls the width function shape, and can it be used for channel network comparison and regionalization? *Water Resour. Res.* **2008**, *44*, 1–19.
23. Lashermes, B.; Foufoula-Georgiou, E. Area and width functions of river networks: New results on multifractal properties. *Water Resour. Res.* **2007**, *43*.
24. Lee, M.T.; Delleur, J.W. A variable source area model of the rainfall-runoff process based on the watershed stream network. *Water Resour. Res.* **1976**, *12*, 1029–1036.
25. Milly, P.C.D.; Betancourt, J.; Falkenmark, M.; Hirsch, R.M.; Kundzewicz, Z.W.; Lettenmaier, D.P.; Stouffer, R.J. Stationarity is dead: Whither water management? *Science* **2008**, *319*, 573–574. [PubMed]
26. Pachauri, R.K.; Reisinger, A. *Contribution of Working Groups I, II and III to the Fourth Assessment Report of the Intergovernmental Panel on Climate Change*; Intergovernmental Panel on Climate Change(IPCC): Geneva, Switzerland, 2007; p. 104.
27. Huong, H.T.L.; Pathirana, A. Urbanization and climate change impacts on future urban flooding in Can Tho city, Vietnam. *Hydrol. Earth Syst. Sci.* **2013**, *17*, 379–394.
28. Seo, Y.; Schmidt, A.R.; Kang, B. Multifractal properties of the peak flow distribution on stochastic drainage networks.*Stoch. Env. Res. Risk A* **2014**, *28*, 1157–1165.
29. Mays, L.W. *Water Resources Engineering*; John Wiley: Hoboken, NJ, USA, 2011.

CHAPTER 8

BEHAVIOR OF A FULLY-LOOPED DRAINAGE NETWORK AND THE CORRESPONDING DENDRITIC NETWORKS

Yongwon Seo, Young-Ho Seo and Young-Oh Kim

Keywords: The Gibbsian model, fully looped network, dendritic network, drainage network, Storm Water Management Model (SWMM)

INTRODUCTION

From a hydraulic and hydrologic engineering point of view, urban drainage problems can be classified into two types: design purpose and prediction purpose for forecasting and operation [1]. While the design purpose problem utilizes hypothetical rainstorms with a relatively simple temporal and spatial distribution of rainfall, more precise and detailed analysis is required for the prediction purpose problem for the operation and management of a system. The Tunnel and Reservoir Plan (TARP) of Chicago in the Unites States is a good example; it requires a precise prediction of the input (runoff hydrograph) to the system in order to operate and manage the entire system for environmental and potential flood risk management purposes [2,3].

One of the recent approaches of hydrologic analysis in urban drainage problems is based on the Geomorphologic Instantaneous Unit Hydrograph (GIUH) [4,5]. Rodriguez et al. [6,7] suggest a morpho-climatic approach incorporating rainfall intensity in order to obtain the Instantaneous Unit Hydrograph (IUH) in urban catchments. Gironas et al. [8] also developed a morpho-climatic instantaneous unit hydrograph model for urban catchments based on the kinematic wave approximation. Cantone [2] and Cantone and Schmidt [9] developed the Illinois Urban Hydrologic Model (IUHM) based on the Strahler ordering scheme [10] to characterize the flow paths. They demonstrated the performance of the model in terms of predicting the shape, timing, and peak of the direct runoff hydrograph at the outlet of urban sewer systems compared with the widely-accepted model, Storm Water Management Model (SWMM); they also compared their results with observed flow hydrographs. The advantage of these approaches compared with an explicitly distributed modeling approach is that the model is applicable even in the case of data unavailability in some parts of the catchment.

In spite of the efforts of model development based on GIUH and the resulting accuracy and precision of the model prediction, one of the disadvantages of models based on GIUH is that they are not capable of analyzing a network that contains a loop; they are only applicable to a dendritic network. The fundamental idea of models based on GIUH that characterizes the flow path by the Strahler ordering scheme is only applicable to dendritic networks. Moreover, in an older city such as Chicago, it can sometimes be difficult to acquire network information such as slopes, roughness, and invert elevations of conduits required by a hydraulic model [3], thus limiting the application of the models regardless of the performances.

Recently, stochastic methodologies and algorithms are being introduced to urban drainage networks for the purpose of incomplete network delineations [11], an optimized design of sewer systems [12], and also for water supply networks [13]. Bailly et al. [11] reconstructed an entire drainage network of a cultivated landscape from disconnected reaches of the network using a spatial stochastic algorithm that satisfies two objectives about the cumulative network length and number of reconnected edges. The reconstructed networks are directed tree graphs, i.e., dendritic networks. For the purpose of generation of layouts for urban drainage systems that should be dendritic, Haghighi [14] proposed a loop-by-loop cutting algorithm, which basically converts a fully looped network into a dendritic network. The objective function reflects the layout cost in terms of the length and concave function of flow rate of each link. However, the objective function related to the

Citation: Seo, Y.; Seo, Y.-H.; Kim, Y.-O., Behavior of a Fully-Looped Drainage Network and the Corresponding Dendritic Networks. Water2015, 7, 1291-1305. doi:10.3390/w7031291.

total length of the network [11] or the construction cost [14] tends to restrict the resulting dendritic network to an efficient and optimized network.

As Haghighi [14] mentioned, in case of steep areas, drainage networks are typically cost-effective if they were designed based on the topography. However, in flat areas like Chicago, there are not only efficient and optimized networks in terms of drainage time or construction cost, but also inefficient and highly sinuous networks. In addition, it is not a straightforward problem to derive a tree type network from a looped condition. Therefore, this study seeks to obtain a dendritic network that corresponds to a looped network in terms of the shape and peak of hydrographs for a given rainstorm. For this purpose, we used the Gibbsian model [15], which generates dendritic networks based on Gibbs' measure [16]. The Gibbsian model has one parameter, β, that dictates the overall sinuosity of a network [17], thus differing from the random walk models, such as the Uniform model [18,19,20,21], which were based on an assumption of the absence of environmental controls. Therefore, the Gibbsian model enables us to generate efficient networks, as well as inefficient and highly sinuous networks, depending on the parameter values. Seo and Schmidt [22,23] showed that the network property in urban catchments represented by the parameter (β) of the Gibbsian model shows greater variability compared with that of natural river networks. They also showed that the network property can be a key link that relates the effect of rainstorm movement to the urban drainage network runoff hydrographs. Seo and Schmidt [24] also showed the applicability of the Gibbsian model in urban drainage network that the Gibbsian model can replace the actual drainage network in an urban catchment even though the network has loops inside.

As mentioned earlier, the aim of this study is to identify a dendritic network that corresponds to a looped network in terms of the shape and peak of hydrographs for a given rainstorm. This study is developed from the study of Seo and Schmidt [24] and aims at more detailed and focused study of the behavior of a fully looped network with various urban catchment characteristics such as catchment slope, impervious ratio, as well as rainfall duration. A synthetic catchment with a drainage network composed of 8 × 8 grids is introduced to investigate the behavior of the fully looped network for a given rainfall event using the Storm Water Management Model (SWMM). Dendritic networks are generated from the Gibbsian model for a given value of the parameter, β. However, because the Gibbsian model is a stochastic network model, Monte Carlo simulation, is utilized to obtain representative averaged peak flows. Therefore, the final objective is actually not to find a single dendritic network, but a value of a parameter, β, of the Gibbsian model that corresponds to the fully looped network for a given catchment slope. It should be noted that, basically, the purpose of this study is to demonstrate that an equivalent dendritic network to a fully looped network can be identified.

METHODOLOGY

Loops are easily found in urban catchments not just for water supply pipe networks, but also for stormwater drainage pipe networks. Figure 1a shows the Mokdong region in Seoul, South Korea. This area is located on a low and flat ground compared to the high water elevation of the Han River that crosses Seoul, and hence, it is prone to frequent flooding. The grey areas indicate flooded areas in 2010, and dark grey shows flooded areas in 2011. Most of the drainage network contain loops in this region as shown in Figure 1a. Figure 1b shows the downtown area of City of Chicago in USA. Urban catchments in this area have very flat topography. Drainage networks typically contains loops inside the drainage networks. As mentioned earlier, identifying a dendritic network that corresponds to a drainage network with loops need to be addressed. In flat areas like Chicago, there are not only efficient and optimized networks in terms of drainage time or construction cost, but also inefficient and highly sinuous networks. In addition, it is not a straightforward problem to derive a tree type network from a looped condition [14]. In this regard, this study introduced a square synthetic network that is fully looped. The Gibbsian model [15] is also introduced to generate dendritic networks with different degrees of sinuosity.

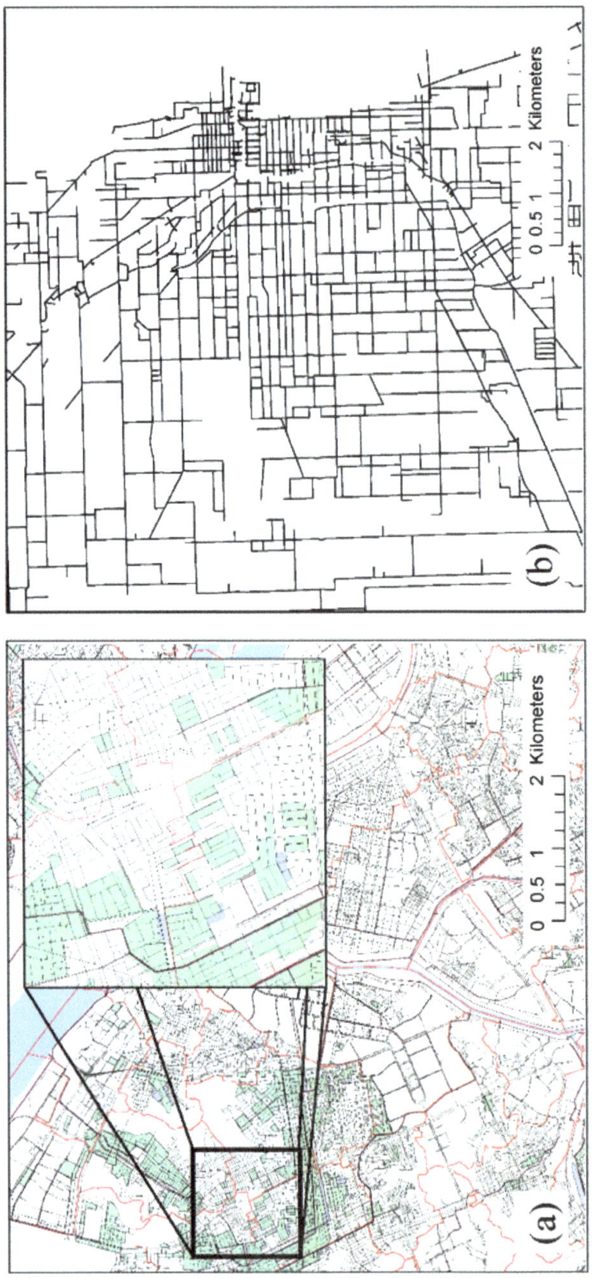

Figure 1. Drainage networks in urban catchments: (**a**) Mokdong area in Seoul, South Korea near the Han River; (**b**) downtown area of Chicago, USA near Lake Michigan.

Synthetic Catchment and the Behavior of a Fully Looped Network

In this study, a synthetic rectangular catchment is introduced with an outlet at the bottom rightmost corner in order to investigate the behavior of a fully looped network (Figure 2). The catchment is composed of 8 × 8 grids. The grid size is 500 m and each grid is connected by circular conduits with a single diameter of 6 m to avoid the surcharge condition. The catchment is assumed to have a uniform catchment slope; the bottom rightmost corner (outlet) is the lowest point in the catchment. Dendritic networks that correspond to this fully looped network in terms of the peak flow depending on the catchment slope are generated by the Gibbsian model. The sinuosity of the Gibbsian network is controlled by the value of β as shown in Figure 2. The channel width of the Gibbsian model in Figure 2 represents the maximum value of the width function. To generate a dendritic network with the Gibbsian model, a Markov chain is defined with the spanning trees of Sas the state space. Let a tree s belong to S and two trees s_1 and s_2 be adjacent. The transition probability from s_1 to s_2 can then be defined as follows [15]:

$$R_{s_1 s_2} = \begin{cases} r^{-1}\min\left\{1, e^{-\beta[H(s_2)-H(s_1)]}\right\} & s_2 \in N(s_1) \\ 1 - \sum\limits_{s \in N(s_1)} R_{s_1 s} & s_2 = s_1 \\ 0 & \text{otherwise} \end{cases} \tag{1}$$

where $N(s_1)$ is the set of trees adjacent to s_1, and β is a parameter that represents the extent to which the sinuosity of the network is reflected in the generation of the new spanning tree, s_2. For example, when β is equal to zero, the overall sinuosity of a network has no relationship to the transition probability and the transition probabilities are the same in all possible directions, which is identical to the uniform distribution. The maximum degree of the points in S, r, is defined as [15],

$$r = \max_{s \in S} |N(s)| \tag{2}$$

The degree, r refers to the maximum number of directions that can be selected, except the existing direction. H(s) is a measure of the sinuosity of a given spanning tree, s [15],

$$H(s) = \sum_{d \in D(B)} \xi_s(d) - \sum_{d \in D(B)} \xi_B(d) \tag{3}$$

where s is a spanning tree, d is a point of a finite and connected graph B, and D(B) is the set of the total points of B. ξ_s is the distance to an outlet along s from v, while ξ_B is the shortest distance to the outlet from d. The actual process to generate a network with the Gibbsian model is as follows. First, start from a network, s_1, generated by the Uniform model and randomly select a point, v, in the network and assign a new flow direction from v to generate the adjacent network s_2. Second, check whether the new network, s_2 is acyclic. If not, repeat the first step. Third, draw a random probability xbetween zero and one and check that x is greater than $e^{-\beta[\Delta H]}$ where ΔH is equal to $H(s_2)$—$H(s_1)$. If this holds, then take s_2 as a new network. In the next step, use s_2 as the starting network and repeat these steps a sufficient number of times until the resulting tree has a distribution close to the stationary Gibbs' distribution. The Uniform model [20,21] is defined on a lattice with flow allowed in each of the four possible directions with equal probability; hence, it tends to have high sinuosity. The uniform model is equivalent to the Gibbsian model with β equal to zero. Figure 2 shows a fully looped network and also dendritic networks on the same grids generated by the Gibbsian model depending on the value of β, where a smaller β produces a more sinuous network compared with a larger β.

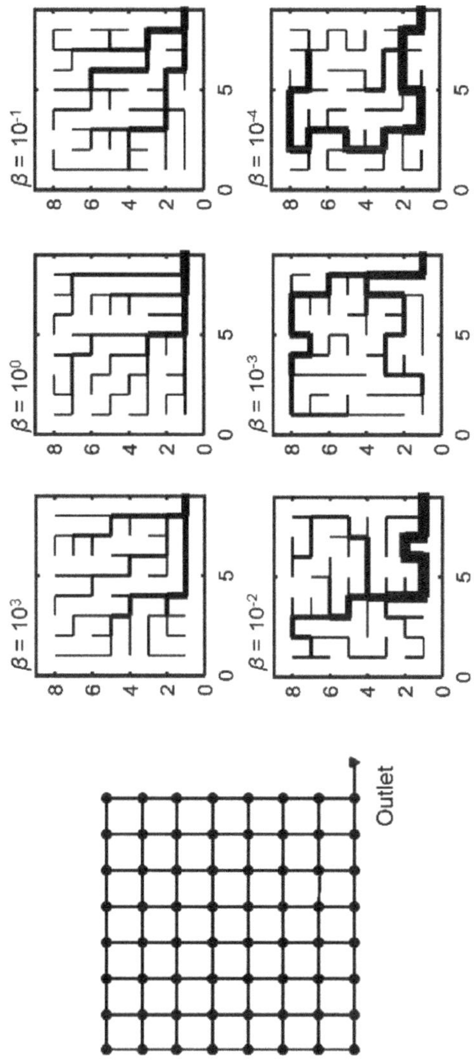

Figure 2. A synthetic rectangular catchment composed of 8 × 8 grids and dendritic networks generated by the Gibbsian model: what would behave like the fully looped network?

SWMM and Rainfall-Runoff Model Runs

Due to the nature of the Gibbsian model, which is a stochastic network model, Monte Carlo Simulation is utilized to obtain representative averaged peak flows. Therefore, in order to find dendritic networks that correspond to a fully looped network for a given catchment slope, 100 networks are generated by the Gibbsian model for a given β. Then, for each catchment slope, the shape and the peak flow of the dendritic network are tested to compare with the fully looped network. In this study, the Storm Water Management Model (SWMM) is utilized to obtain the runoff hydrograph at the outlet of the catchment for a given rainstorm. SWMM is a widely-accepted model for drainage network analysis. The dynamic wave module built in SWMM is the Extended Transport (EXTRAN) block [25,26] that solves the flow sewer by sewer by the one-sweep explicit method without the need for a simultaneous solution of the sewers of the network [1].

A SWMM model was constructed with a subcatchment, the area of which is 50 ha and catchment width is 500 m. The number of subcatchments is 64 (8 × 8). All subcatchments have the same area, width and catchment slope (0.5%) as well as the ratio of impervious area (25%). The roughness of impervious area was set to 0.01 s/m$^{1/3}$, and the roughness of pervious area was set to 0.1 s/m$^{1/3}$. The invert elevation at the outlet of the drainage network is set to zero and invert elevations of other junctions were determined based on the catchment slope and the distance from the outlet. The conduits have the same length, 500 m and the maximum depth was set to 6 m. The layout of the drainage network of the SWMM model was built depending on the Gibbsian model generated for a given value of β. The duration of the rainfall is fixed to one hour and the rainfall intensities from 4 to 30 mm were tested. In terms of the infiltration model, Horton's method was used with the same parameter values for all subcatchments. Obviously, model section and spatial variation of parameter values for the nonlinear infiltration process can affect the results of this study and finding equivalent dendritic networks for a fully looped network. We left this for a future study because the original intention of this study is focusing on the possibility of finding an equivalent dendritic network for a fully looped network.

RESULTS AND DISCUSSION

Behavior of the Fully Looped Drainage Network and a Corresponding Dendritic Network

In this study, we first focused on the catchment slope. The behavior of a fully looped network is heavily dependent on the catchment slope (S_c). Figure 3 shows the direct runoff hydrographs of the fully looped network (Figure 2) depending on slope for a 1 hour storm with hourly rainfall intensity of 15 mm. As shown in Figure 3a, the peak flow of the hydrograph sharply decreases as the catchment slope decreases. Figure 3b shows the peak flows of the fully-looped network as a function of the catchment slope. The result shows that a transition exists in the relation between the catchment slope and the peak flows of a fully looped network as shown in Figure 3b; the relation between the catchment slope and peak flows is logarithmic but the slope of peak flow with respect to the catchment slope changes near the catchment slope of 9×10^{-3}.

The behavior of a fully looped network that depends on the catchment slope implies a possibility that it can be approximated by a corresponding dendritic network for a given catchment slope. Figure 4 compares the hydrographs from a fully looped drainage network (Figure 2) and dendritic networks generated by the Gibbsian model for the slopes of 4×10^{-4}, 10^{-4}, and 4×10^{-5}. The results clearly show that the shape of the hydrograph from a fully looped network is close to the shapes of the hydrographs from dendritic networks, which are generated by the Gibbsian model depending on the catchment slope. For example, when the catchment slope is relatively steep (4×10^{-4}), the hydrograph of the fully looped network is close to that of the dendritic network with higher β (10°). In contrast, when the catchment slope is relatively mild ($S = 4 \times 10^{-5}$), the hydrograph of the fully looped network coincides with the dendritic network with smaller β (10^{-4}). The behavior of the fully looped network implies that it is possible to make a pair of dendritic networks generated by the Gibbsian model and the fully looped network depending on the catchment slope.

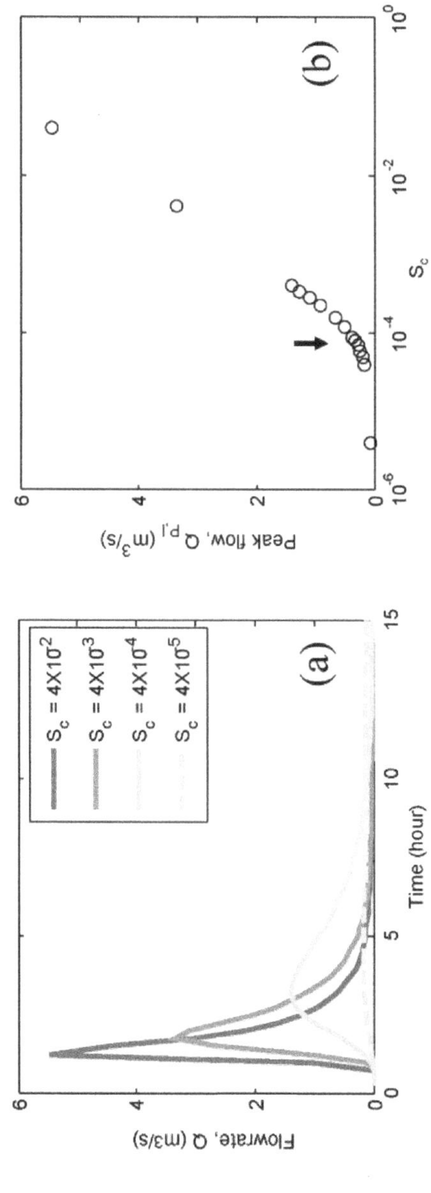

Figure 3. Behavior of a fully looped network depending on the catchment slope (S_c): **(a)** runoff hydrographs; **(b)** peak flows (rainfall intensity = 15 mm/h).

Although it is possible to identify a dendritic network (the value of β) that behaves like a fully looped network, the relation changes as the catchment slope changes. Figure 5 shows the behavior of the peak flow of a fully looped network and dendritic networks generated by the Gibbsian model for various catchment slopes. The result shows clearly that the behavior of the fully looped network in terms of the peak flow can be simulated by the corresponding dendritic networks, and is represented by β; the parameter of the Gibbsian model. Figure 5a shows that when the catchment slope is in the order of 4×10^{-5}, the peak flow of the fully looped network is close to the dendritic networks generated by the Gibbsian model with β equal to 10^{-4}.

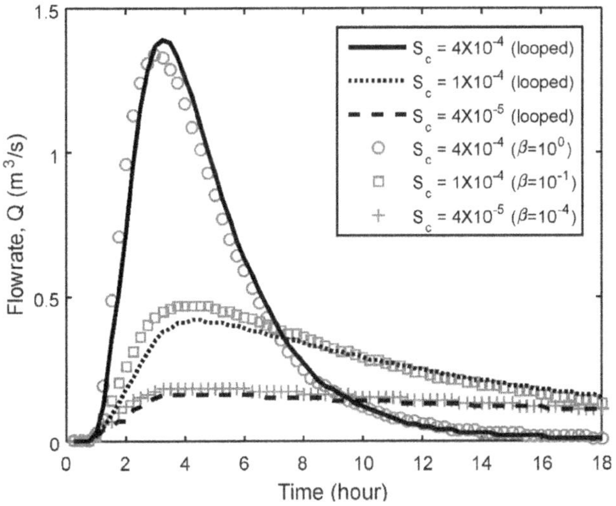

Figure 4. Comparison of the flow hydrographs from a fully looped network and dendritic networks generated by the Gibbsian model depending on the catchment slope (rainfall intensity = 15 mm/h).

In contrast, when the catchment slope becomes higher (4×10^{-3}), the closest dendritic network is generated with β higher than 10^1 (Figure 5d). Figure 5 shows that the corresponding dendritic network changes as the catchment slope changes. Table 1 lists the peak flows of the fully looped network and the dendritic networks generated by the Gibbsian model depending on slope and β. The results clearly show that β of the corresponding equivalent network to the fully looped network ($Q_p/Q_{p,l} = 1$) increases as the catchment slope increases (Figure 6). Figure 7 shows how the corresponding dendritic network (the Gibbsian model with β) changes as the catchment slope and the rainfall intensity changes. For example, the bottom left most circle indicates that, when the rainfall intensity is 30 mm/h and the catchment slope is equal to 2.93×10^{-4}, the Gibbsian model with $\beta = 6.54 \times 10^{-5}$ behaves like a fully looped drainage network.

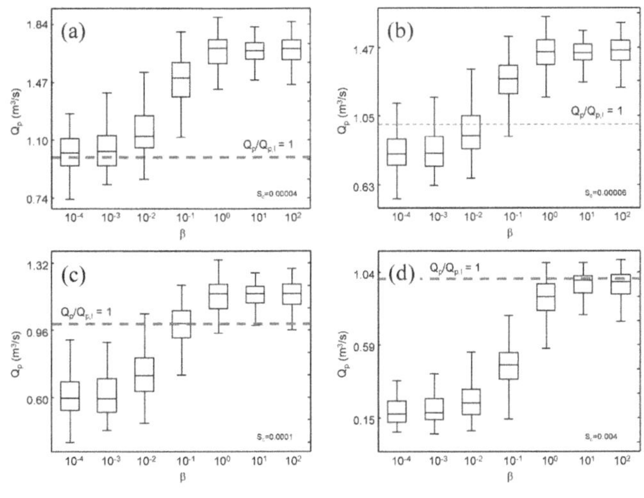

Figure 5. Behavior of the peak flow of a fully looped network depending on the catchment slope of the order of: (**a**) 4×10^{-5}; (**b**) 6×10^{-5}; and (**c**) 1×10^{-4}; 4×10^{-3}(rainfall intensity = 15 mm/h).

Table 1. Averaged peak flow ratio ($Q_p/Q_{p,l}$) of dendritic drainage networks generated with the Gibbsian model to the fully looped network depending on the catchment slope (rainfall intensity = 15 mm/h).

Catchment Slope	Gibbs Model (β)						
	10^{-4}	10^{-3}	10^{-2}	10^{-1}	10^0	10^1	10^2
4.00×10^{-5}	1.03	1.06	1.17	1.48	1.65	1.66	1.66
5.00×10^{-5}	0.93	0.96	1.06	1.37	1.54	1.54	1.55
6.00×10^{-5}	0.83	0.86	0.96	1.26	1.42	1.43	1.43
7.00×10^{-5}	0.75	0.78	0.88	1.16	1.32	1.33	1.33
8.00×10^{-5}	0.69	0.72	0.82	1.09	1.25	1.26	1.26
9.00×10^{-5}	0.65	0.67	0.77	1.04	1.19	1.20	1.20
1.00×10^{-5}	0.61	0.63	0.73	0.99	1.14	1.15	1.15
1.10×10^{-5}	0.58	0.60	0.70	0.95	1.10	1.11	1.11
1.20×10^{-5}	0.56	0.58	0.68	0.93	1.08	1.09	1.09
1.30×10^{-5}	0.54	0.56	0.66	0.91	1.06	1.07	1.07
1.40×10^{-5}	0.52	0.54	0.64	0.89	1.04	1.05	1.05
1.50×10^{-5}	0.51	0.53	0.62	0.87	1.03	1.04	1.04
1.60×10^{-5}	0.50	0.51	0.61	0.86	1.02	1.02	1.02

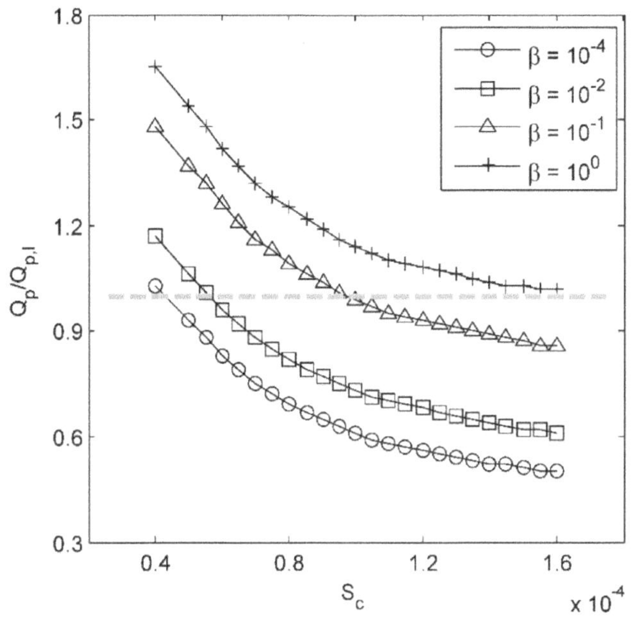

Figure 6. Averaged peak flow ratio ($Q_p/Q_{p,l}$) of dendritic drainage networks generated with the Gibbsian model to the fully looped network depending on the catchment slope (rainfall intensity = 15 mm/h).

The results show that the relation between the value of β of the corresponding dendritic network generated by the Gibbsian model and the catchment slope is dependent on the rainfall intensities. The results also indicate that the relation between β and the catchment slope can be represented by a two-step exponential relationship; one relationship for β less than 10^{-1}, the other relationship for β greater than 10^{-1} regardless of the rainfall intensities (Table 2). Figure 8 depicts the corresponding Gibbsian models (β) to a fully looped network depending on the catchment slope and rainfall intensity. As mentioned earlier, rainfall intensity is another factor affecting results: the corresponding dendritic network changes as rainfall intensity varies. However, as shown in Figure 8, the corresponding Gibbsian model (value of β) converges as rainfall intensity increases. The result also shows that the convergence exists for smaller rainfall intensity when the catchment slope is smaller.

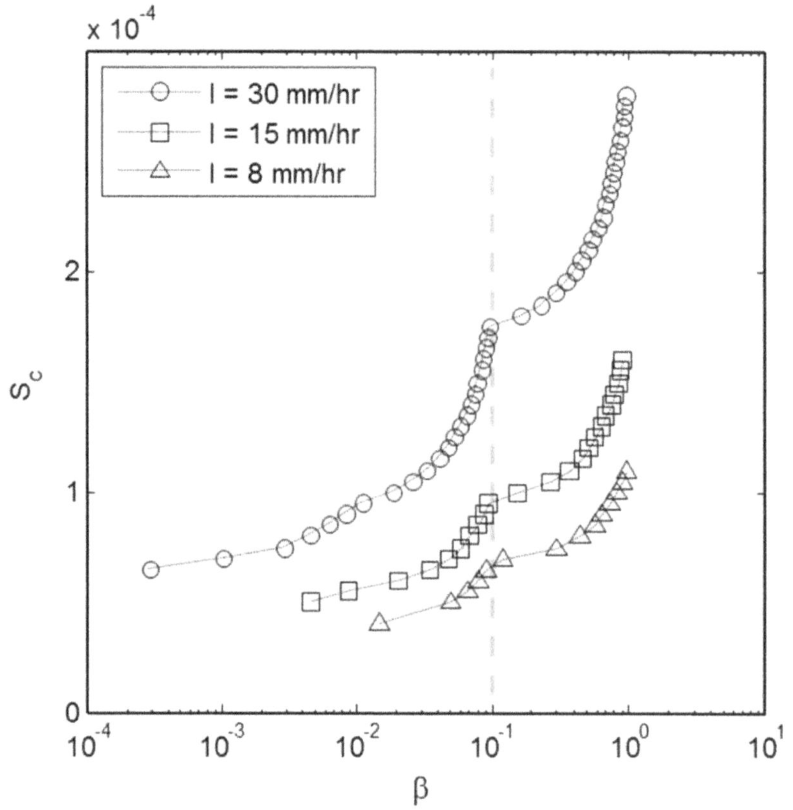

Figure 7. Behavior of the peak flow of a fully looped network depending on the catchment slope and rainfall intensity.

Table 2. Two-step regression equation between catchment slopes (S_c) and the Gibbsian model (β) (rainfall intensity = 15 mm/h).

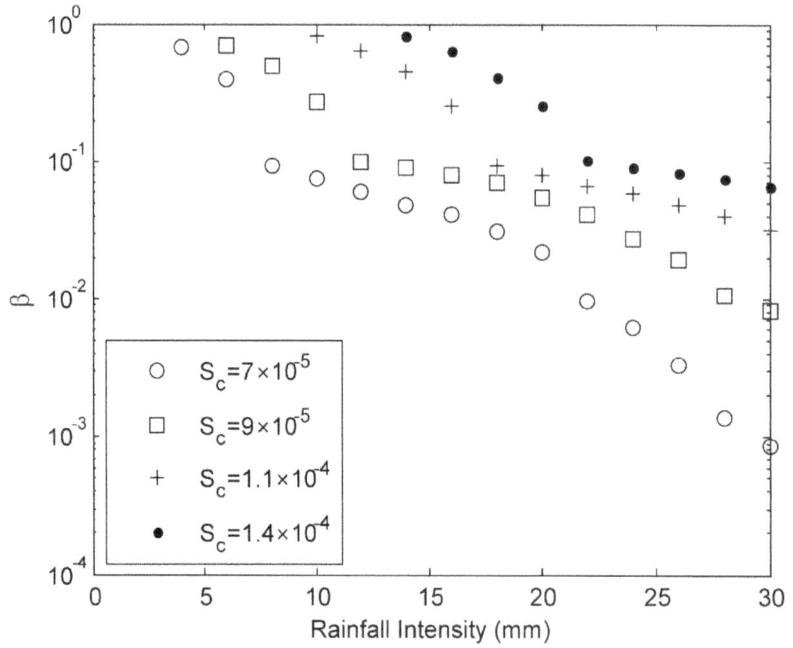

Figure 8. Corresponding Gibbsian models (β) to a fully looped network depending on the catchment slope and rainfall intensity.

Effect of the Catchment Imperviousness Ratio and Rainfall Duration on the Equivalence

It is necessary to perform a wider sensitivity analysis to show whether the association between a fully looped network and a dendritic network still holds for different conditions for rainfall duration, catchment properties, and, consequently, modeling parameters of SWMM. In this regard, we tested different catchment imperviousness ratios and evaluate the effect in terms of the equivalence. To test the effect of imperviousness ratio and temporal rainfall variation at the same time, we included various cases of rainfall variation (Table 3) and repeated the same step previously described to uncover the equivalent dendritic networks in terms of the Gibbsian parameter, β. In this case, we limit the fully looped network case to one with a catchment slope of 10^{-4}. Also, as shown in Table 3, the total rainfall amount falling on the test catchment is the same as 15 mm. Figure 9 depicts the equivalent Gibbsian networks depending on test cases and the catchment imperviousness ratio (r_{imperv}). The result is interesting in that compared with the previous results with r_{imperv} equal to 25%, the equivalent Gibbsian network shows high sensitivity to catchment imperviousness ratio. In contrast, rainfall duration shows little difference in terms of equivalent networks, which is dissimilar to the results from rainfall intensity (Figure 8). The results indicate the equivalence still can be found and the association between a fully looped network and a dendritic network can be established in different rainfall durations and imperviousness ratios. However, the equivalence can be greatly affected by the rainfall intensity and catchment characteristics.

To continue the discussion of the disadvantage of the modeling approach based on the GIUH, as outlined in the introduction section, the results of this study imply an alternative which validates the GIUH modeling approach for a drainage network that even contains loops. This study illustrates a possibility to find dendritic networks that correspond to a fully looped network in terms of the shape of the hydrograph as well as the peak flow depending on the catchment slope; this means that the modeling approach based on the GIUH in urban catchment is applicable regardless of the loops it contains. In addition, this study shows that regardless of the rainfall intensities, it is possible to obtain a two-step exponential relationship between the dendritic networks (β) and the catchment slope for two intervals; one relationship for β less than 10^{-1}, the other relationship for β greater than 10^{-1}. This relationship can be explicitly utilized to obtain the value of β depending on the catchment slopes.

Table 3. Cases of temporal rainfall variation evaluated for equivalent dendritic networks.

Range	Equation	a	b	R^2
$10^{-4} < \beta < 10^{-1}$	$S_c = a \cdot \exp(b \cdot \beta)$	5.00×10^{-5}	6.5356	0.991
$10^{-1} < \beta < 10^{1}$	$S_c = a \cdot \exp(b \cdot \beta)$	9.00×10^{-5}	0.6217	0.989

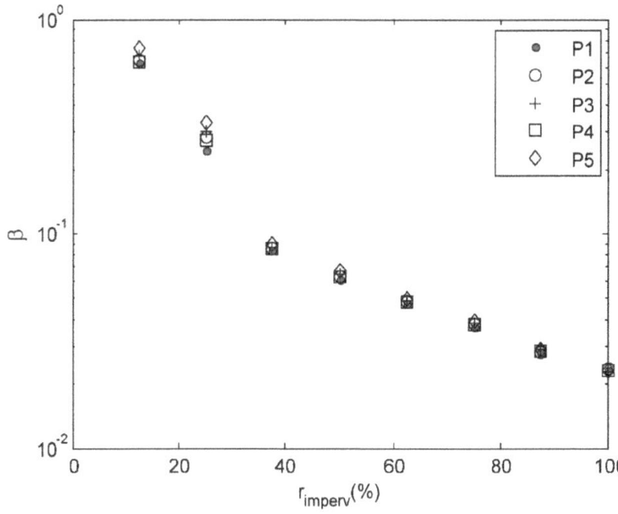

Figure 9. Corresponding Gibbsian models (β) to a fully looped network ($S_c = 10^{-4}$) depending on the imperviousness ratio and temporal rainfall variation.

The original purpose of this study was to identify an equivalent dendritic network that approximates hydrologic response of a network including loops inside it. In this regard, Figure 10 compares the hydrographs obtained using the GIUH for a dendritic networks generated by the Gibbsian model with those from a fully looped network given a catchment slopes.Figure 10a shows that when the catchment slope is relatively steep (4×10^{-4}), the hydrograph of the fully looped network is close to that of the dendritic network with higher β (10^{0}). In contrast, when the catchment slope is relatively mild ($S = 1 \times 10^{-4}$), the hydrograph of the fully looped network is close to the dendritic network with smaller β (10^{-4}), which is consistent regardless of the model (GIUH or SWMM). The behavior of the fully looped network implies that it is possible to set a pair of dendritic networks generated by the Gibbsian model and the fully looped network depending on the catchment slope. Seo and Schmidt [22] showed that the network property in urban catchments represented by the parameter (β) of the Gibbsian model shows greater variability compared with that of natural river networks, of which β is typically greater than 10°. Conversely, it can be inferred that the Gibbsian model is appropriate to represent the network properties of the drainage network in urban catchments compared to other stochastic network models, such as the Scheidegger model [18,19] and Uniform model [20,21]. Combined with the results from Seo and Schmidt [22,23,24], this study strongly suggests the application of the Gibbsian model to a hydrologic modeling approach in urban catchments, especially in the case of data unavailability.

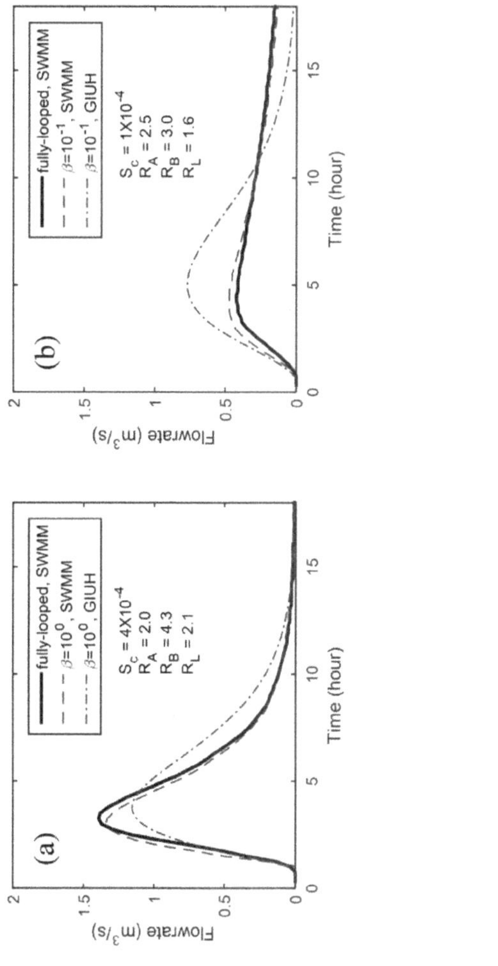

Figure 10. Comparison of the flow hydrographs from a fully looped

network using Storm Water Management Model (SWMM) and a dendritic networks generated by the Gibbsian model using Geomorphologic Instantaneous Unit Hydrograph (GIUH) and SWMM (rainfall intensity = 15 mm/h).

CONCLUSIONS

In spite of hydraulic and hydrologic modelling development and progress in urban catchments, data availability is often an important issue. Especially in older cities, it can require expensive and time-consuming processes to obtain sufficient information for modelling purposes. Moreover, combined with data unavailability, the complex loops inside drainage networks render the modelling even more difficult. Stochastic and probabilistic approaches are therefore needed in order to deal with uncertainties that arise from the hydraulic and hydrologic modelling in urban catchments. This study attempts to address the possibility of applying a stochastic network model to complement modelling approaches for this purpose. Consequently, this study aims to achieve an equivalent dendritic network that corresponds to a fully looped network in terms of the peak of the runoff hydrograph at the outlet using a stochastic network model, the Gibbsian model. A synthetic catchment with a drainage network composed of 8 × 8 grids is introduced to investigate the behaviour of a fully looped network for a given rainfall event utilizing a widely accepted sewer model, the SWMM. Dendritic networks are generated from the Gibbsian model for a given value of the parameter, β. The results indicate that the shape of the hydrograph and the peak flow of a fully looped network are heavily dependent on the catchment slope. The results also indicate the equivalent dendritic network that corresponds to the fully looped network can be identified depending on the catchment slope. Although the number of cases in this study is small and the study is limited to a synthetic square drainage network, the total runs are up to 52,000 because this study introduced a stochastic network model. This study is the very first study of this topic and, hence, focused on the possibility of identifying an equivalent dendritic network to a fully looped network. Therefore, the authors left the application to a real drainage network for future works containing loops, asymmetry and varying slopes. The study is being extended to practical application on urban drainage networks in Seoul and Chicago. The results of this study showed that, basically, a corresponding dendritic network can be identified for a fully looped network. Therefore this study implies a potential improvement of ungagged urban drainage network analyses, particularly based on the GIUH, providing a relationship between the catchment slope of a fully looped network and the corresponding equivalent dendritic networks generated by the Gibbsian model.

ACKNOWLEDGMENTS

This study was supported by Basic Science Research Program by the National Research Foundation of Korea (NRF) from the Ministry of Science, ICT and Future Planning (NRF-2013R1A1A1058964).

Author Contributions

Yongwon Seo designed the numerical experiment. Young-Ho Seo conducted the simulation of network and SWMM models. The manuscript was written by Yongwon Seo with contribution from Young-Ho Seo and Young-Oh Kim.

Conflicts of Interest

The authors declare no conflict of interest.

REFERENCES

1. Yen, B.C.; Akan, A.O. Hydraulic design of urban drainage systems. In Hydraulic Design Handbook; Mays, L.W., Ed.; McGraw-Hill: New York, NY, USA, 1999.
2. Cantone, J.P.; Schmidt, A.R.; Hollander, M.; Erickson, A.; Tang, Y.; Garcia, M.H. Application of the Illinois Urban Hydrologic Model (IUHM) to Selected Catchments in the Calumet Tunnel and Reservoir Plan System; Ven Te Chow Hydrosystems Lab, Department of Civil and Environmental Engineering, University of Illinois at Urbana-Champaign: Urbana, IL, USA, 2009; p. 94.
3. Cantone, J.P. Improved Understanding and Prediction of the Hydrologic Response of Highly Urbanized Catchments through Development of the Illinois Urban Hydrologic Model (IUHM); University of Illinois at Urbana-Champaign: Urbana, IL, USA, 2010.
4. Rodriguez-Iturbe, I.; Valdes, J.B. Geomorphologic structure of hydrologic response. Water Resour. Res. **1979**, 15, 1409–1420.
5. Gupta, V.K.; Waymire, E.; Wang, C.T. A representation of an instantaneous unit-hydrograph from geomorphology. Water Resour. Res. **1980**, 16, 855–862.
6. Rodriguez, F.; Andrieu, H.; Creutin, J.D. Surface runoff in urban catchments: Morphological identification of unit hydrographs from urban databanks. J. Hydrol. **2003**, 283, 146–168.
7. Rodriguez, F.; Cudennec, C.; Andrieu, H. Application of morphological approaches to determine unit hydrographs of urban catchments. Hydrol. Process. **2005**, 19, 1021–1035.
8. Gironas, J.; Niemann, J.D.; Roesner, L.A.; Rodriguez, F.; Andrieu, H. A morpho-climatic instantaneous unit hydrograph model for urban catchments based on the kinematic wave approximation. J. Hydrol. **2009**, 377, 317–334.
9. Cantone, J.; Schmidt, A. Improved understanding and prediction of the hydrologic response of highly urbanized catchments through development of the Illinois urban hydrologic model. Water Resour. Res. **2011**, 47.
10. Strahler, A.N. Quantitative analysis of watershed geomorphology. EOS Trans. AGU **1957**, 38, 913–920.
11. Bailly, J.S.; Levavasseur, F.; Lagacherie, P. A spatial stochastic algorithm to reconstruct artificial drainage networks from incomplete network delineations. Int. J. Appl. Earth Obs. **2011**, 13, 853–862.
12. Pan, T.C.; Kao, J.J. GA-QP model to optimize sewer system design. J. Environ. Eng. ASCE **2009**, 135, 17–24.
13. Moderl, M.; Sitzenfrei, R.; Fetz, T.; Fleischhacker, E.; Rauch, W. Systematic generation of virtual networks for

water supply. Water Resour. Res. **2011**, 47.

14. Haghighi, A. Loop-by-loop cutting algorithm to generate layouts for urban drainage systems. J. Water Resour. Plan. Manag. **2013**, 139, 693–703.
15. Troutman, B.M.; Karlinger, M.R. Gibbs distribution on drainage networks. Water Resour. Res. **1992**, 28, 563–577.
16. Kindermann, R.; Snell, J.L. Markov Random Fields and Their Applications; American Mathematical Society: Providence, RI, USA, 1980; pp. 1–142.
17. Barndorff-Nielsen, O.E. Stochastic Methods in Hydrology: Rain, Landforms, and Floods; World Scientific: River Edge, NJ, USA, 1998; pp. 1–207.
18. Scheidegger, A.E. On topology of river nets. Water Resour. Res. **1967**, 3, 103–106.
19. Scheidegger, A.E. A stochastic model for drainage patterns into an intramontane trench. Int. Assoc. Sci. Hydrol. Bull.**1967**, 12, 15–20.
20. Leopold, L.B.; Langbein, W.B. The concept of Entropy in Landscape Evolution; United States Government Printing Office: Washington, DC, USA, 1962; pp. 1–20.
21. Karlinger, M.R.; Troutman, B.M. A random spatial network model based on elementary postulates. Water Resour. Res.**1989**, 25, 793–798.
22. Seo, Y.; Schmidt, A.R. The effect of rainstorm movement on urban drainage network runoff hydrographs. Hydrol. Process. **2012**, 26, 3830–3841.
23. Seo, Y.; Schmidt, A.R. Network configuration and hydrograph sensitivity to storm kinematics. Water Resour. Res. **2013**,49, 1812–1827.
24. Seo, Y.; Schmidt, A.R. Application of Gibbs' model to urban drainage networks: A case study in southwestern Chicago, USA. Hydrol. Process. **2014**, 28, 1148–1158.
25. Roesner, L.A.; Shubinski, R.P. Improved dynamic routing model for storm drainage system. In Urban Storm Water Hydraulics and Hydrology; Yen, B.C., Ed.; Water Resources Publications: Highland Ranch, CO, USA, 1982; pp. 164–173.
26. Roesner, L.A.; Shubinski, R.P. Storm Water Management Model User's Manual Version III. Addendum I EXTRAN; Municipal Environmental Research Laboratory: Cincinnati, OH, USA, 1984.

CHAPTER 9

QUANTITATIVE ESTIMATION OF THE IMPACT OF PRECIPITATION AND LAND SURFACE CHANGE ON HYDROLOGICAL PROCESSES THROUGH STATISTICAL MODELING

Muhammad Saifullah, Zhijia Li, Qiaoling Li, Muhammad Zaman, and Sarfraz Hashim

INTRODUCTION

The runoff of some rivers in the world especially in the arid and semiarid areas has reduced remarkably due to human activities in the last decades. In addition to precipitation change, increases in land surface changes due to the growth of population such as deforestation, afforestation, and urban constructions have also altered the hydrological process, which leads to a change in the water cycle and affects the runoff [1]. Water resources managers and planners are looking for simple and easy methods to determine the contribution of precipitation and land surface change on runoff, which supports sustainable natural resources development, particularly in the catchment scale. The significant reduction in runoff is a strong indicator of influences of precipitation and land surface changes on the hydrological process of the catchment [2, 3]. Particularly in China, due to the variation in regional precipitation changes coupled with strong economic and industrial development during the past decades [4], several researchers have paid attention to understand the impacts of precipitation and land surface change on the hydrological process. The runoff reduction in the arid and semiarid areas of China has brought severe difficulties in livelihood and ecology. As a result of the recent intensification of human activities, such as agriculture activities, deforestation, hydraulic projects, and soil and water conservation measures, considerable concentration has been given to assessing the impacts of land surface changes on runoff [5]. It has been an important scientific problem in hydrology to reveal the physical characteristics of the basin and trends of runoff change and their influencing factors. Furthermore, the impacts of precipitation and land surface change impacts vary from place to place and need to be identified for local scale.

Much research has been done on the runoff variations of different river basins across the world in the previous few years. According to characteristics of the watersheds, four aspects of studies on runoff changes can be recognized: (1) arid region, for example, quantifying the effects of climate variability and human activities on runoff for Kaidu River basin [6]; (2) semiarid region, many studies discussing the influence of human activities on runoff in this region [7–10]; (3) humid region, for example, the studies on the assessment of climate change and human activities on runoff variation of the Huifa River basin [11]; (4) semihumid region, for example, separating the effect of human activities using different time scale in the Zhang River basin [3, 12].

Several methods have been used to assess the impacts of precipitation and land surface changes on the hydrological processes, but to date, no standard model has been developed. In the past, paired catchment experiment method was used to determine the impact of precipitation and land surface changes on the hydrological process. The experimental methods are time-consuming, expensive, and difficult to locate suitable controls [13–15]. On the other hand, physically based hydrological models, such as soil and water assessment tool, variable infiltration capacity model, the Xinanjiang model, and the HBV model, are always limited due to the time-consuming process of calibration and validation. These models also require extensive and complex dataset and have unpredictable output results, model parameters, and structure [16, 17].

Citation: Muhammad Saifullah, Zhijia Li, Qiaoling Li, Muhammad Zaman, and Sarfraz Hashim, "Quantitative Estimation of the Impact of Precipitation and Land Surface Change on Hydrological Processes through Statistical Modeling," Advances in Meteorology, vol. 2016, Article ID 6130179, 15 pages, 2016. doi:10.1155/2016/6130179.

According to methods, these studies could be divided into five categories: (1) time series analysis, for example, the studies on the changing trends and regime shift of streamflow in the Yellow River basin [5]; (2) change point analysis, such as Mann-Kendall rank analysis used to assess the abrupt change in runoff from the Yellow River basin [18]; (3) the rainfall-runoff relationship, for example, the assessment of runoff variation due to human activities in the middle reaches of the Yellow River [19]; (4) statistical analysis, for example, a simple linear regression and multiple linear regression methods used to quantify the human activities on runoff variation in middle reaches of the Yellow River and the Laohahe basin in northern China [20, 21]; (5) land use and land surface change impact on runoff variation assessed in West Liaohe River Basin and Yellow River basin [7, 22].

However, it is often difficult to get a quantitative estimation of land surface change to runoff variation. The studies for this region found that runoff declined during period 1990–2000 as compared to period 1980–1990 [23]. Yao et al. [23] found that land surface change leads to reduced runoff in this region. This reduced runoff can cause a severe problem for the environment, the rivers, and estuary ecosystem, and socioeconomic systems. However, the study conducted by Yao et al. [23] focused on peak flow rather than hydrography altered by land surface changes. Further, it also did not quantify the impacts of variations in precipitation and land surface changes in the Yihe River basin. More importantly, the scientific community still expresses a different opinion on how land surface and precipitation variability vary regional water resources in the Yihe River. There is a need to update the knowledge of land surface change and hydroclimatic variables for different rivers, particularly large rivers to assist global as well as regional water resources management. There is a need to study the impact of precipitation and land surface change in the Yihe River basin using simple and easily applicable method. Moreover, some new statistical methods have been developed and used to quantify the impacts of precipitation and land surface changes on the hydrological process of the basin. Before application of these methods, time series data are divided into two periods by an advanced statically method (Pettit's test) and a graphical method, that is, the double mass curve method [21].

The objectives of the present study are (1) to evaluate the new graphical method to determine the trends of low, medium, and high values of precipitation and runoff from the Yihe River basin, which is used for the first time in this region and (2) to develop the linear regression and statistical precipitation elasticity models to assess the impacts of precipitation and land surface changes on hydrological process of the Yihe River basin. This study will deepen our understanding of the runoff response to precipitation and land surface changes in this region, which is crucial for effective planning, management, and sustainable development of water resources in this regions.

MATERIALS AND METHODS

Study Area

One of largest tributaries of the Yellow River is the Yihe River and major upper tributary subbasin of the Yihe River is Dongwan basin, which exists in the semihumid climate zone. The Dongwan watershed is situated in Henan province, China, between longitudes 111° to 112°E and latitudes 33.5° to 34.5°N, with a drainage area of 2856 km2 (Figure 1). It drains the mountainous area of Fuinu located in the Qinling Mountain. It is a typical area of the semihumid basin and located in midstream of the Yellow River. The Dongwan basin controls the flood peak and volume of the downstream of Yellow River along with Luhun Reservoir, Sanmenxia Reservoir, Xiaolangdi Reservoir, and Guxian Reservoir.

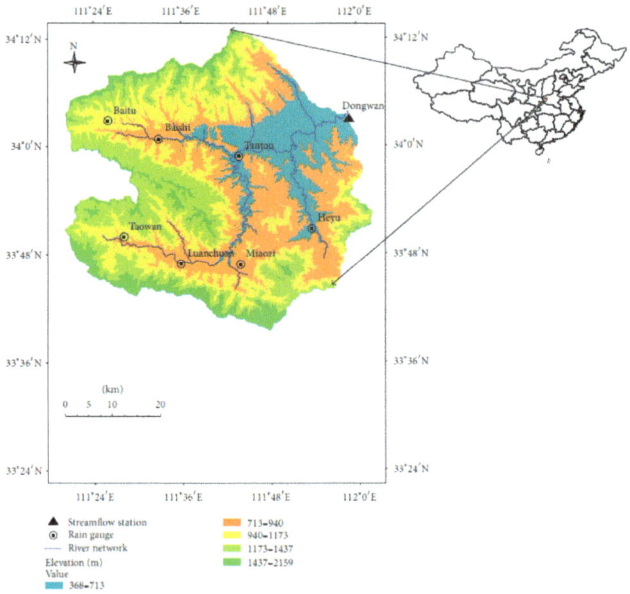

Figure 1: Location of precipitation and hydrological stations of Yihe River basin, China.

Data and Methods

Daily precipitation and streamflow data from 1961 to 2011 at the Yihe River basin were collected from the Yellow River Conservancy Commission. The locations of stations and statistics of the observed records have been presented in Table 1 for precipitation and streamflow time series, respectively. Monthly values are added to get the annual and seasonal precipitation to identify the precipitation trends. The interannual variation in rainfall is significant due to the effect of continental monsoon climate. The study area belongs to a typical semihumid climate, with an average annual precipitation of 766.11 mm and runoff 219.21 mm during the past 50 years. The rainfall in the hilly area was more than that in the plain area. Each year, the amount of precipitation for the period from July to September was more than 50% of the annual precipitation and produces the highest flooding in the period from July to August.

Land use and land cover data of 1990, 2000, and 2010 were used in this study at a scale of 1 : 100,000. The dataset which is provided by Data Center for Resources and Environmental Sciences, Chinese Academy of Sciences (RESDC) (http://www.resdc.cn/), was used to test the methods reported in this paper. LULC were classified into 6 classes, forestland, grassland, water body, urban land, agriculture land, and others [24]. In this area, grassland, forestland, and agriculture land are dominant land use type.

Table 1: Summary of basic statistics of annual precipitation, annual runoff, and land use characteristics of individual stations in the Yihe River basin.

Stations name	Latitude	Longitude	Mean (mm/year)	Median (mm/year)	Min. (mm/year)	Max. (mm/year)	Std.	C_v	Land use
Precipitation stations									
Dongwan	111.98	34.05	699.29	681.00	436.00	1059.00	149.11	0.21	FRST
Heyu	111.90	33.85	792.67	768.30	492.20	1334.00	169.41	0.21	AGL
Tantou	111.73	33.98	731.38	722.75	414.10	1155.30	171.44	0.23	FRST
Miaozi	111.73	33.78	871.36	850.85	567.40	1401.40	176.36	0.20	AGL
Luanchuan	111.60	33.78	863.81	831.20	575.60	1370.40	167.22	0.19	AGL
Baishi	111.55	34.02	732.51	717.45	468.70	1342.70	170.13	0.23	AGL
Taowan	111.47	33.83	837.61	815.50	606.90	1438.50	164.42	0.20	GRL
Baitu	111.43	34.05	714.50	700.65	420.40	1132.90	170.96	0.24	AGL
Runoff station									
Dongwan	111.98	34.05	219.21	172.50	83.26	662.62	118.93	0.54	FRST

Trend Detection

The nonparametric Mann-Kendall (MK) [25–27] test has been widely used to detect trends in hydrology and climatology [28, 29]. The MK test static (S) of variables is calculated by

$$S = \sum_{i=1}^{n-1} \sum_{j=i+1}^{n} \operatorname{sgn}\left(x_j - x_i\right) \tag{1}$$

where

$$\operatorname{sgn}\left(x_j - x_i\right) = \begin{cases} +1, & x_j > x_i, \\ 0, & x_j = x_i, \\ -1, & x_j < x_i, \end{cases} \tag{2}$$

where x_i and x_j are the data values at times i and j and n indicates the length of the dataset. While a positive value of S indicates an increasing trend, negative value of S indicates a decreasing trend. The following expression, as an assumption, is used for the series where the data length $n > 10$ and data are approximately normally distributed:

$$V(S) = \frac{n(n-1)(2n+5) - \sum_{i=1}^{n} t_i i (i-1)(2i+5)}{18} \tag{3}$$

where t_i is the number of data values. After the calculation of the variance of time series data, the standard Z value is calculated according to the following equation:

$$z = \begin{cases} \dfrac{S-1}{\sqrt{\operatorname{var}(S)}}, & S > 0, \\ 0, & S = 0, \\ \dfrac{S-1}{\sqrt{\operatorname{var}(S)}}, & S < 0. \end{cases} \tag{4}$$

The standard Z value is compared with standard normal distribution table with a two-tailed test at significance level ($\alpha = 10\%$, $\alpha = 5\%$, and $\alpha = 1\%$). If the calculated Z value is greater than $|Z| > |Z_1 - \alpha/2|$, the null hypothesis (Ho) is rejected. Therefore, the trend is statistically significant. Otherwise, the null hypothesis (Ho) is accepted. Therefore, the trend is not statistically significant. In this study, the 95% confidence level was used for the MK test.

Innovative Trend Detection Test

Sen [30] divided the time series data into two equal parts and then time series drawn on Cartesian coordinates system which was independent of the trend, preassigned significance level, magnitude of the trend, sample size, and amount of variation within time series. The observed hydrological time series is divided into two equal parts from the first date to end date. The division of hydrological time series is based on the change point, which is taken as prechange and postchange time series. Both subseries are separately arranged in ascending order. The first subseries (X_i) is then assigned to x-axis, and the other subseries (X_j) is assigned to y-axis (Figure 2) based on Cartesian coordinate system. If hydrological time series are plotted on the 1:1 (45^0) straight line, it can be said that there is a trendless time series. If data values are in below triangular area of a straight line (45^0), it can be said that there is a decreasing trend in time series. If data values are in the upper triangular area of a straight line (45^0), it can be said that there is an increasing trend in time series [31, 32]. The scatter points above or below the 1:1 line indicate increasing or decreasing monotonic trends.

Change Point Test

Pettit's test [33] is a nonparametric method to identify the occurrence of a change point. It has been commonly used to detect changes in time series data. It is rank based and distribution-free test for detecting an abrupt change in the mean of a time series when the exact time of the change is unknown. The Pettit test can be calculated from the following equation [34]:

$$U_{t,n} = \sum_{j=1}^{t}\sum_{i=1}^{n} \operatorname{sgn}\left(x_j - x_i\right), \quad (t = 1, \ldots, n) \tag{5}$$

where the maximum $U_{t,n}$ corresponds to the change point year. We will use Pettit's test to determine the change point of the runoff series as a reconfirmation of the abrupt point identified by a double mass curve.

Framework for Quantifying the Impact of Climate Change and Human Activities on Runoff

A change in observed mean annual runoff (ΔQ_{total}) may result from precipitation variability (ΔQ_p) or land surface change (ΔQ_L) [20]:

$$\Delta Q_{\text{total}} = \Delta Q_p + \Delta Q_L \tag{6}$$

To quantify the impact of precipitation and land surface, we analyzed the variations in a runoff in the prechange period and post change period. Thus, a change in average annual runoff between reference and change period is calculated as

$$\Delta Q = Q_2 - Q_1 \tag{7}$$

where ΔQ represents the change in the runoff and Q_1 and Q_2 are average annual runoff during the prechange period and postchange period, respectively.

A Simple Linear Regression

A simple linear regression between annual runoff (Q_{ref}) and averaged annual precipitation (P_{ref}) in the base period can be expressed as

$$Q_{\text{ref}} = aP_{\text{ref}} + b \tag{8}$$

where a and b are two parameters of the model. Based on (10), the runoff of catchment is calculated as

$$\Delta Q_L = Q_{\text{fit}} - Q_{\text{change}}, \quad \text{where } Q_{\text{fit}} = aP_{\text{change}} + b \tag{9}$$

$$Q_p = Q_{\text{change}} - Q_{\text{ref}} \tag{10}$$

where Q_L is the change in mean runoff due to land surface impact, Qfit calculated runoff during the change period, Pchange is rainfall during the change period, and parameters a and b are similar to (8).

Precipitation Elasticity Method

The precipitation elasticity method was widely used to assess the impact of precipitation and land surface change [8, 10]. The precipitation elasticity of runoff can be determined by the proportional change in runoff divided by the proportional change in precipitation. The precipitation elasticity of runoff is calculated as follows:

$$\varepsilon_p = \frac{dQ/Q}{dP/P} \tag{11}$$

where P and Q are precipitation and runoff, respectively. Sankarasubramanian et al. [35] presented a nonparametric approach. The precipitation elasticity of runoff can be expressed as

$$\frac{\Delta Q_i}{\overline{Q}} = \varepsilon_p \frac{\Delta P_i}{\overline{P}} \tag{12}$$

where ΔQ_i and ΔP_i are changes of annual runoff and precipitation, respectively. The value of ε_p was estimated using observed precipitation and runoff before the change point. The change in the runoff after change point could be estimated from the precipitation data during the change period using (12). The research framework is shown in Figure 3.

RESULTS

Preliminary Data Analysis

The homogeneity of the precipitation time series was assessed by applying standard normal homogeneity test (SNHT) [36, 37] and Buishand's range (BR) test [38] at 5% significant level for each station [28]. All the precipitation series were found to be homogenous. Table 2 presents the results of homogeneity tests for all precipitation stations. Figure 4 shows long-term variations in annual precipitation (P) and runoff (Q) of the Yihe River basin from 1961 to 2011. The basin exhibited a significant variation in the runoff, whereas the variations in precipitation appeared to be stationary.

Innovative Trend Pattern of Precipitation and Runoff

In this study, for the stations when the serial correlation was found, trend-free prewhitening approach had been applied. As a summary, statistically significant trends at 95% were not found in all stations according to the MK trend test. Then trends' pattern of precipitation with Şen's innovative method was used to determine the historical trends of precipitation and illustrated in Figures 5–12. These figures show variations of precipitation in eight stations, and it can be seen as increasing, decreasing, or trendless time series. At Baitu station, the low values of annual precipitation observed a decreasing trend. It is seen that the annual precipitation of Heyu and Luanchuan has no trends for medium values. However, Miaozi, Baishi, Taowan, and Baitu have a decreasing trend for medium values. All stations except Dongwan and Tantou have a downward trend for high values of annual precipitation. In the Baitu station, the low and medium values of spring season have a downward trend; there is no trend for low and medium values for remaining stations. The Tantu, Miaozi, and Baishi stations have no trend for low and medium values for the summer season. However, it is seen that while Taowan station has a decreasing trend for high values, all other stations have an increasing trend for high values of summer precipitation. The Tantou station has an increasing trend for high values and no trend for medium values; the other stations have decreasing trends for low, medium, and high values for the autumn season. All other low, medium, and high values of runoff have a decreasing trend for seasonal and annual time scale (Figure 13). As a summary of trends tests, runoff from the basin shows an obvious declining trend; the variation in precipitation and runoff shows inconsistency. Overall, downward trends have been identified in seasonal and annual, precipitation, and runoff series, indicating that the decrease in runoff may be attributed to trends in precipitation. However, runoff in the study area has reduced more as compared to precipitation. This fact may signify that the linear relationship between precipitation and runoff has altered during the past decades.

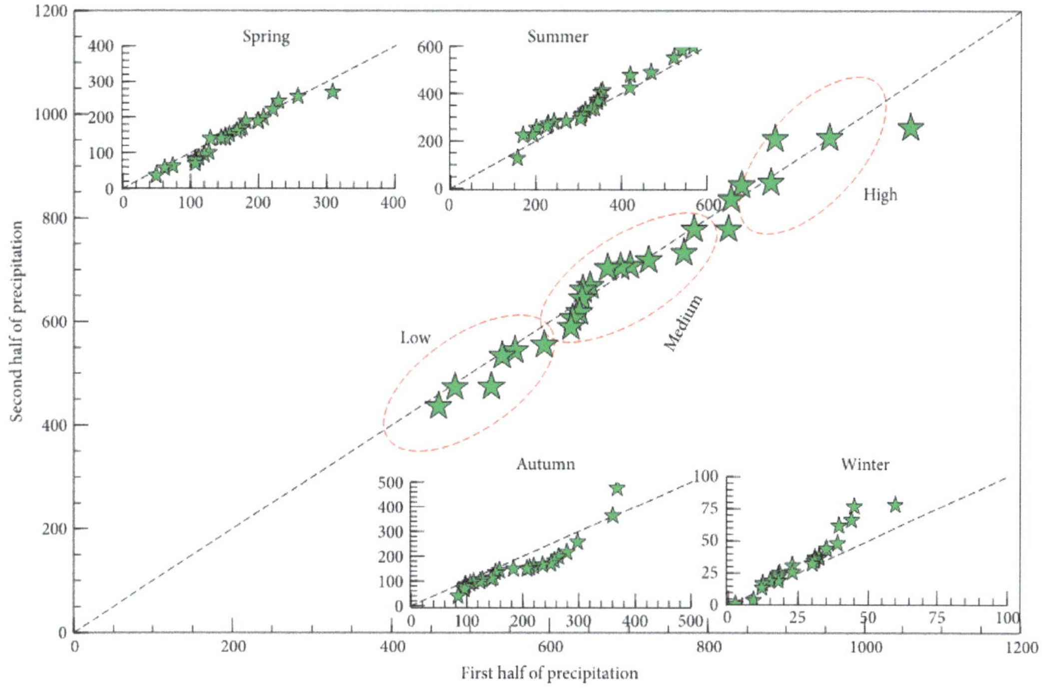

Figure 5: The precipitation pattern for Dongwan station as a result of innovative trend test.

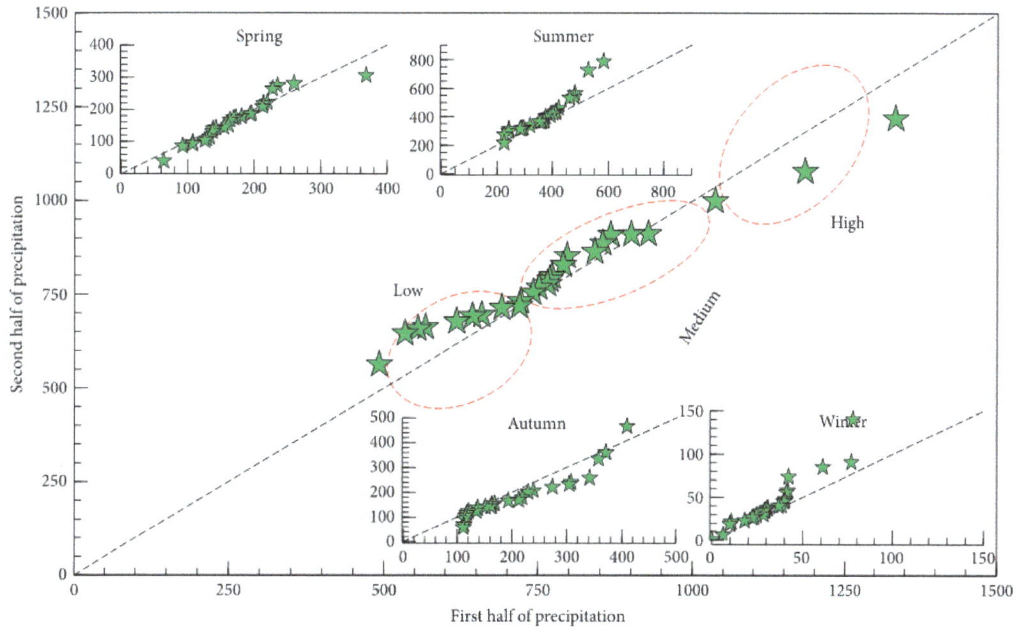

Figure 6: The precipitation pattern for Heyu station as a result of innovative trend test.

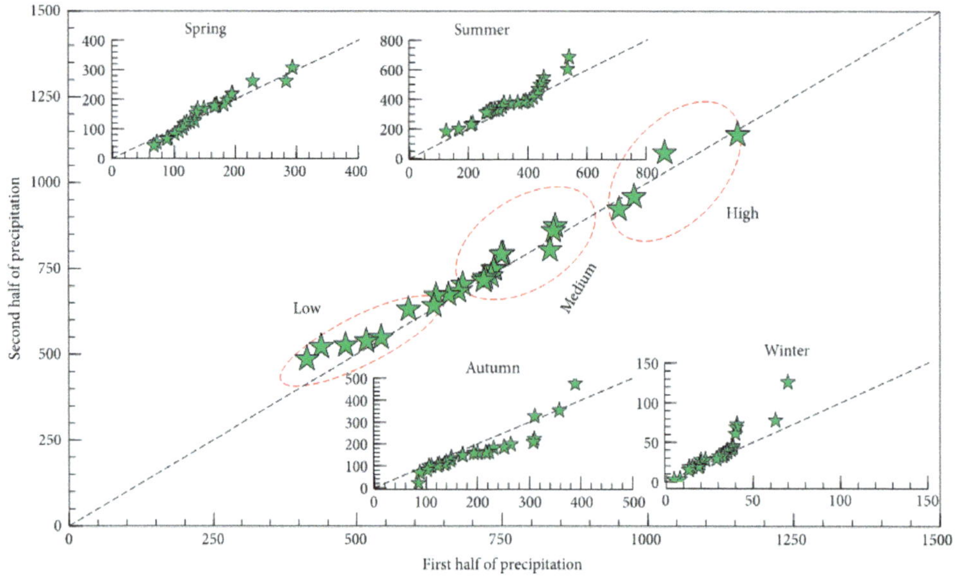

Figure 7: The precipitation pattern for Tantou station as a result of innovative trend test.

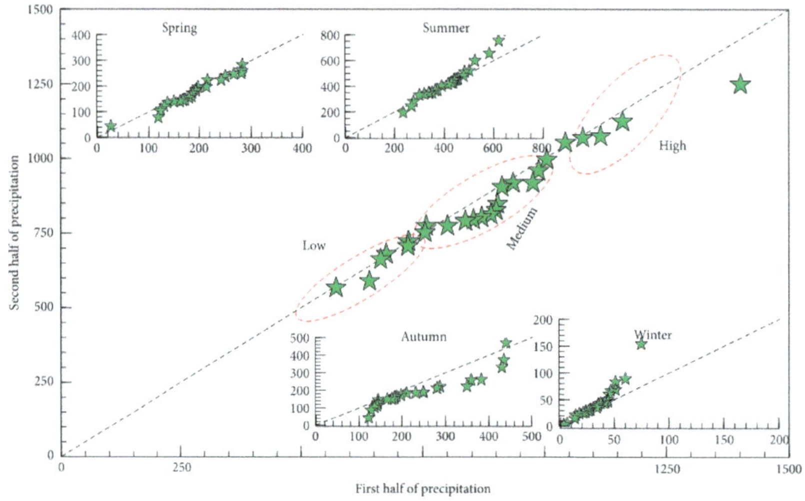

Figure 8: The precipitation pattern for Miaozi station as a result of innovative trend test.

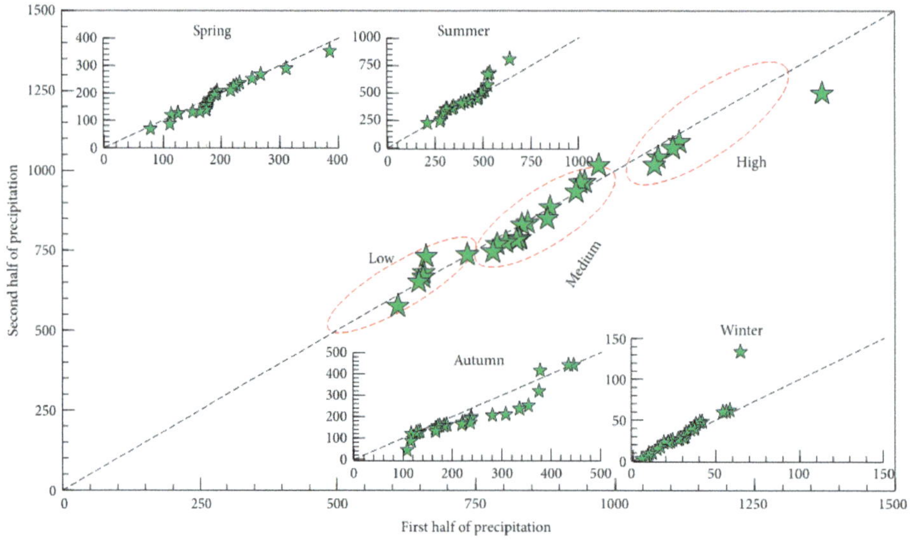

Figure 9: The precipitation pattern for Luanchuan station as a result of innovative trend test.

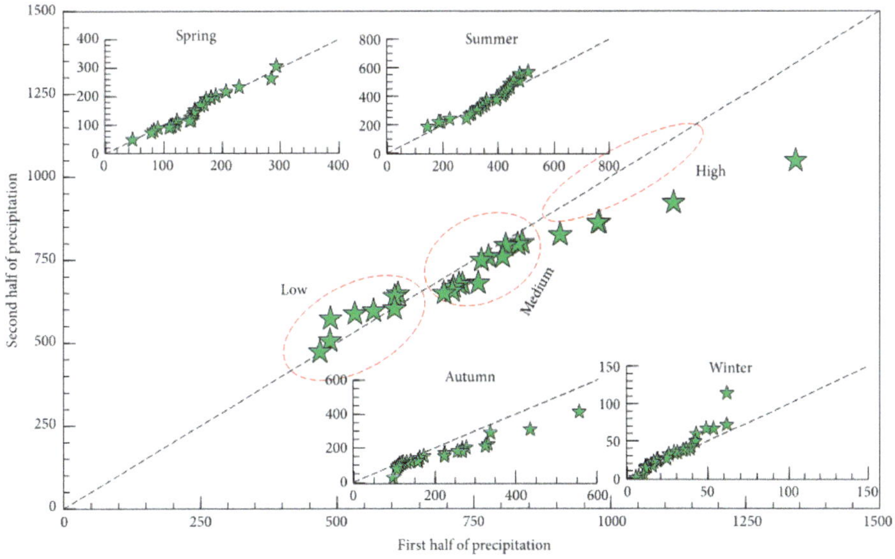

Figure 10: The precipitation pattern for Baishi station as a result of innovative trend test.

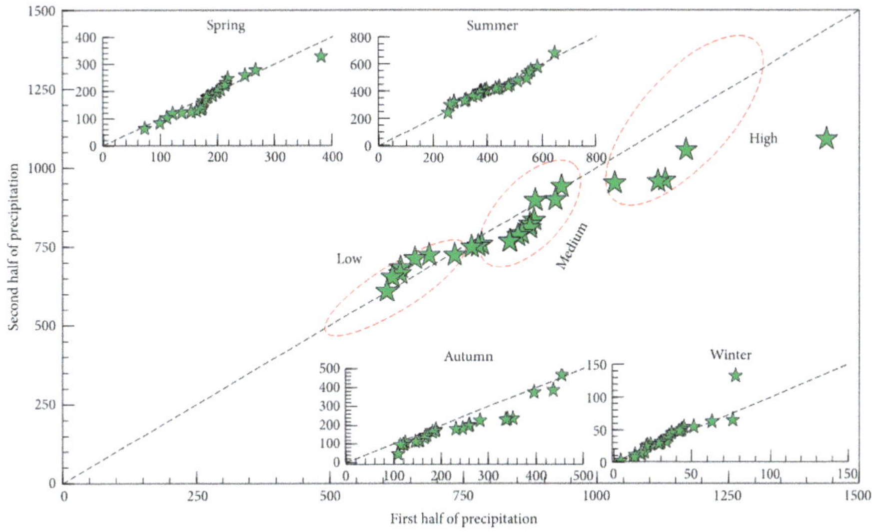

Figure 11: The precipitation pattern for Taowan station as a result of innovative trend test.

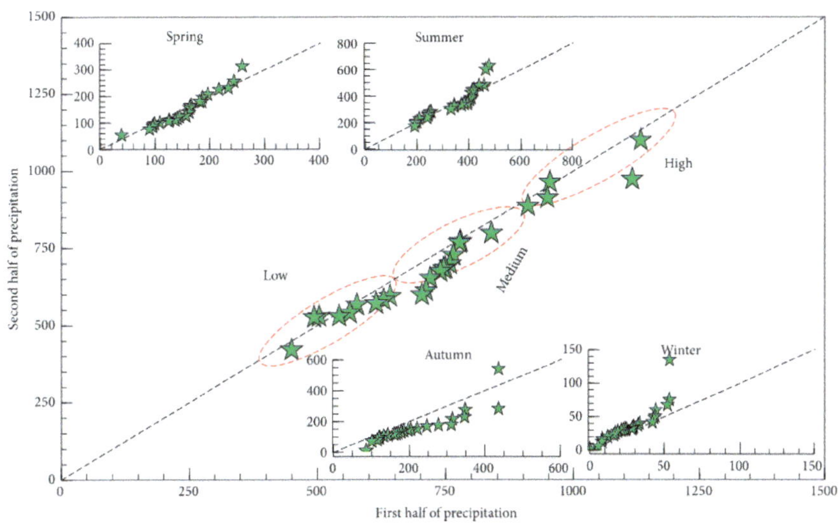

Figure 12: The precipitation pattern for Baitu station as a result of innovative trend test.

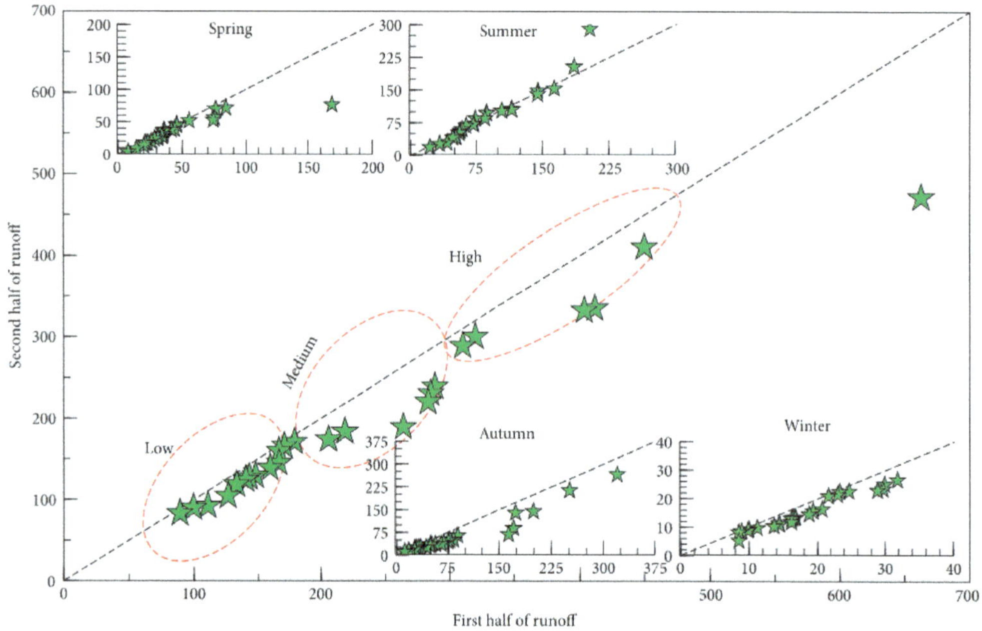

Figure 13: The runoff pattern for Dongwan station as a result of innovative trend test.

Change Point Analysis

The annual precipitation-runoff double cumulative curve is normally approximate to a straight line if the basin characteristics are stable; thus, a change in the slope of the curve may indicate that the internal relations between precipitation and runoff have altered. The change point in 1985 has only been identified in runoff series. Figure 16 presents the double cumulative curve in a nearly straight line but with different gradient before and after 1985, which may indicate that the process of the runoff production has changed.

To better understand the characteristics of the changes in runoff, the difference between the annual runoff during the prechange period and the annual runoff during postchange period was analyzed using different statistical tests. The significant difference between the means of prechange and postchange periods of slopes of two straight lines can be found at 95% confidence level using z -test. Concurrently, using F-test and t-test, there are statistically significant differences between the variances of slopes of two lines at 95% confidence level. The results of two straight lines are consistent with double mass curve lines, which indicate more variation in runoff during the postchange period. Runoff coefficient (slope of lines) of the prechange period is higher (10%) than a postchange period. The maximum value of runoff coefficient is 0.34 for the period of 1981–1985. After 1985, runoff coefficient abruptly decreases half of 1981–1985, which is evidence of the change point. The variations of runoff coefficient during the prechange period to postchange periods are 55% to 45%. The runoff coefficient of the prechange period is slightly higher than the postchange period. Recently precipitation increases and runoff coefficient is lower than the prechange period.

Hence, combining the results of the change point test and double cumulative curve, the year 1985 could be the change point representing the impact of precipitation and land surface change on runoff. In this study, the double mass curve is used to identify the change point of the runoff series as a confirmation of the change point. However, the reductions were present in average monthly runoff during the postchange period (1986–2011). The greatest decline in runoff is identified in flood seasons (July, August, and September) (Figure 17). The slightly increasing precipitation and consistent largely decreased runoff happen in July (Figures 14 and 15). The reduction in runoff during 1986–2011 might be due to land surface change. Similar findings were also identified in some previous studies [21, 39]. Figure 16 shows a correlation comparison of precipitation and runoff for the two periods. The correlation between precipitation and runoff for the prechange period (R=0.84) is stronger than that for land surface change period (R=0.78). This result again indicates that runoff should be affected by the land surface change in this region after 1985 (Table 3). Several researchers used the double mass curve method to verify the change point and found similar results [19, 34].

Table 3: Summary of annual runoff, precipitation, and runoff coefficient change point.

Period	Prechange period (1961–1985)	Postchange period (1986–2011)
Runoff		
Mean (mm)	239.66	187.44
Standard deviation	130.25	109.24
Precipitation		
Mean (mm)	781.61	749.87
Standard deviation	159.92	139.66
Runoff coefficient (slope)		
Mean	0.29	0.25
Standard deviation	0.097	0.087

Figure 14: Monthly precipitation for the prechange period and postchange period.

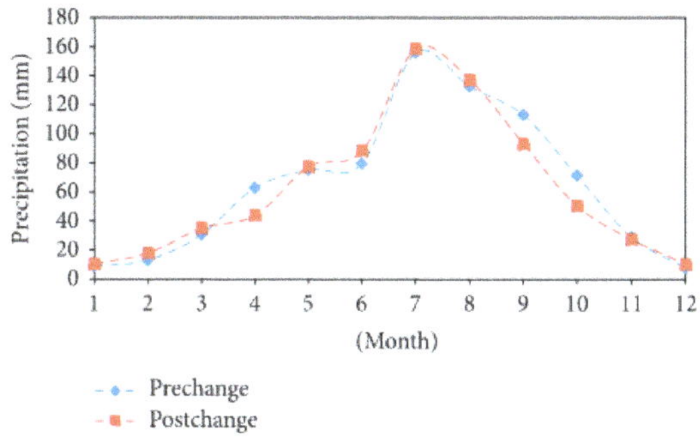

Figure 15: Monthly runoff for the prechange period and postchange period.

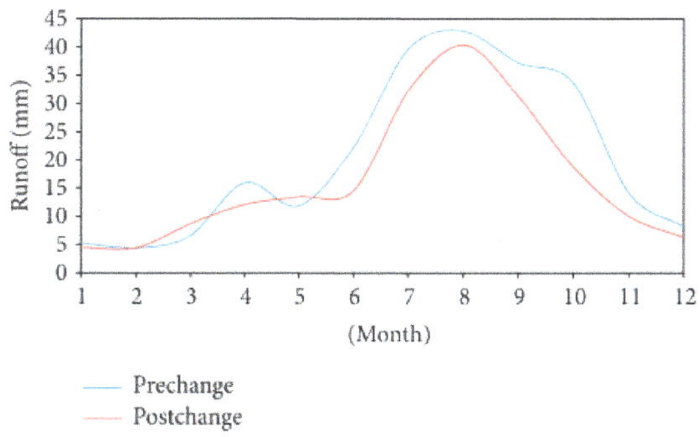

Figure 16: Correlation between precipitation and runoff for prechange period and postchange period.

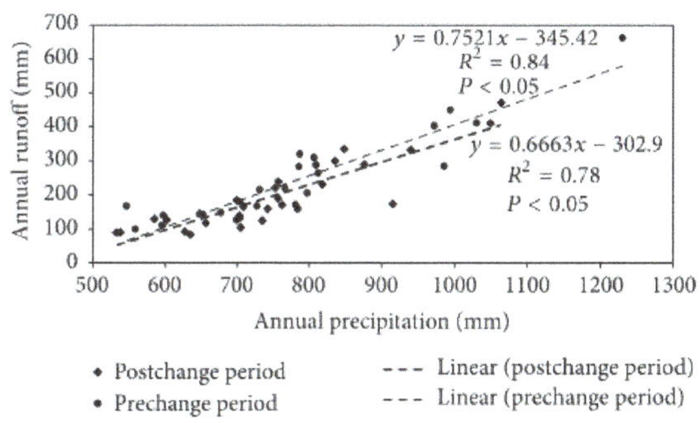

Figure 17: Double mass curve of annual precipitation and runoff.

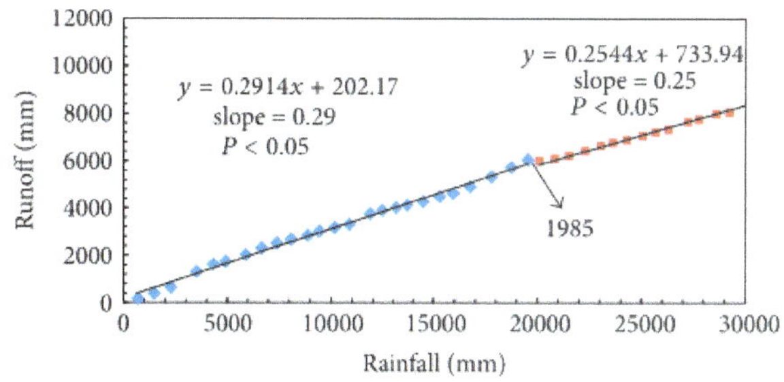

Figure 18: Land surface change of the study area from 1990, 2000, and 2010.

Impact of Precipitation and Land Surface Change on Runoff

The precipitation elasticity of runoff was estimated using (11)-(12) to assess the impact of precipitation change on runoff. The value of ε_p is 2.61. The results indicate that a 10% decrease in precipitation would result in 26.1% decrease in runoff. According to the equation, with the calculated ε_p, it can be calculated that the 31.74 mm decrease of precipitation in 1985–2011 may lessen the runoff by 45.74 mm. Therefore, a 25.79 mm decrease in annual precipitation in the study area contributes 56.38% of the total drop in the runoff, while land surface change is responsible for 43.62% of runoff decrease (Table 4). For the simple linear regression method, the difference of simulated runoff before and after the change point is assessed for the impact of precipitation change. Annual runoff in the study area had reduced from 242.46 mm in 1961–1985 to 196.72 mm in 1986–2011. The 45.74 mm of runoff indicates that precipitation change could have accounted for 67.68% of the runoff reduction. Therefore, land surface change may have resulted in 21.84 mm decrease in annual runoff, estimating 32.32% of the runoff reduction. After 1990, land surface change was pointed out in Dongwan basin. The main land cover types of the study area include agriculture, forest, urban, water body, grassland, and other forms (saline, wetland, and barren). The agriculture land, grassland, and forest land are predominant land use types in this study area. From 1990 to 2010, there have been occurring land changes in the basin, such as an increase in urban land, water body, and other types (barren, saline, and wetland) of land and decrease in agriculture land. The area of agriculture decreased from 45.9% to 44.2%, while other uses of land increased from 0.07% to 0.17%. From 1990 to 2000, urban land increases from 4.26% to 5.26%; meanwhile other uses of land type and water body also increased. During 2000–2010, the area of urban land increased from 4.57% to 5.26%, while it was only 4.2% during 1990–2000. **Table 4**: Quantifying the effects of climate variability and human activities on runoff.

	Precipitation elasticity	Simple linear regression
Q_P	25.79	45.74
Q_L	19.95	21.84
Q_P (%)	56.38	67.68
Q_L (%)	43.62	32.32

The agriculture land decreased from 45.95% to 45.82% in the first period. Meanwhile urban land increased by 4.26% to 5.26% in the whole study period (Figures 18 and 19).

The land surface change in basin suggested that agriculture land has decreased in the 21st century, accompanied with the fast urbanization process and serious desertification.

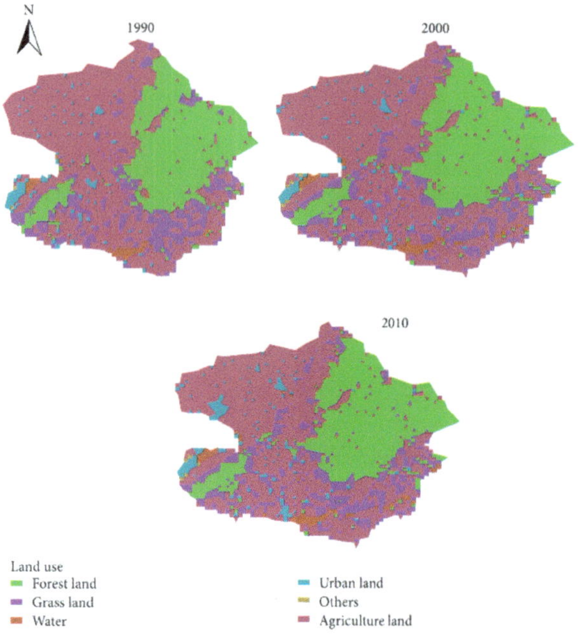

Figure 19: Variations in land cover of Yihe River basin during the study period.

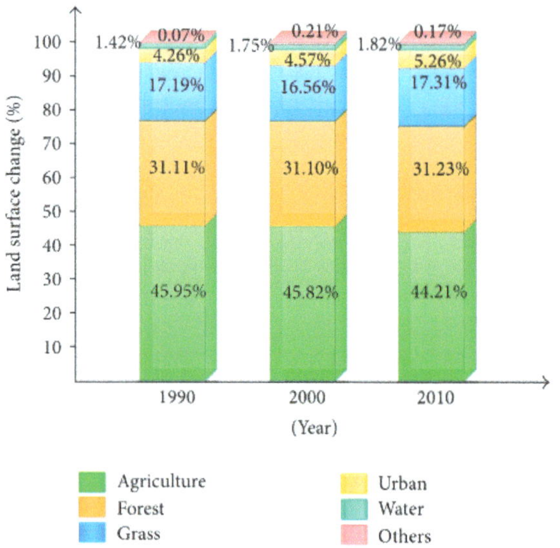

For the Dongwan basin, the precipitation elasticity and simple linear regression method provided consistent results. Similar results were found for the Chaohe River [40]. Similar results were also found in some earlier studies in another area of China [20, 21].

DISCUSSIONS

In this study, two different approaches, that is, linear regression and precipitation elasticity methods, have been used to assess the impacts of the precipitation and land surface changes on runoff. In a comparison of the two approaches, the precipitation elasticity method is relatively simple and can be easily applied to other areas, and it gives a natural runoff change with lesser data and parameters. The simple linear regression method, on the other hand, is a statistical method.

These two approaches were applied independently at the same time scale and showed highly consistent results. The main outcomes of this study agree with the finding of other studies [3, 6, 39, 41]. Yao et al. [23] found that there was an increase in land surface change after 1990 with a reduction in the runoff. Wang et al. [42] concluded that climate change was the main driving force for runoff decrease in midregion of the Yellow River. Over the decades, runoff due to human activities has continuously increased from 11% to 44%. The land surface change increased due to the start of the project "Conservation of Cropland to Forest and Grassland Program (CCFGP)," which increased the area from 0.09(10^4 hm^2) to 3.41(10^4 hm^2) from 2000 to 2009 [43]. The construction of Dragon Island National Forest Park and other tourist activities has reduced the runoff of the basin [23].

Uncertainty arises from the assumption of the method that precipitation and land surface changes are independent of each other [22]. The precipitation variability and land surface changes are strongly related to each other and cannot be separable. The uncertainty may still exist, which overestimates or underestimates the runoff. Therefore, both methods have some uncertainty on the results to a certain extent. The further studies should be needed in the future to improve the quantitative estimation of precipitation variability and land surface impact on runoff with considering the uncertainties.

CONCLUSIONS

It is important to quantify the impacts of precipitation variability and land surface changes on runoff in the Yihe River basin. This study investigates the characteristics of changes in precipitation and runoff, during 1961–2011, in the Yihe River basin. Downward trends were found for precipitation and runoff using new method, that is, innovative trend pattern. Runoff series in the study area presented greater decline compared with the precipitation time series. An abrupt change in runoff can be found in 1985. The consistent results from precipitation elasticity and linear regression methods to quantify the impacts of precipitation variability and land surface changes in runoff were identified. Precipitation variability comprised up to 56.38 and 67.68% of the difference in annual runoff between the two periods by the two methods, that is, precipitation elasticity and linear regression, indicating that precipitation variability played main driving force in the runoff decrease. The role of land surface changes is increasing during the recent decades in runoff reduction due to the rapid growth of urbanization. The results obtained in this study would provide more evidence and useful reference for water resources planning and management in this region.

Competing Interests

The authors declare that they have no competing interests.

ACKNOWLEDGMENTS

This work was supported by the National Natural Science Foundation of China (Grant nos. 41130639, 51179045, and

41201028) and the Non-Profit Industry Financial Program of Ministry of Water Resources of China (201501022 and 201301068) with combined funding from Higher Education Commission of Pakistan, the special program for scholars for Engineering Faculty Development.

REFERENCES

1. Y. He, F. Wang, X. Mu, H. Yan, and G. Zhao, "An assessment of human versus climatic impacts on Jing River Basin, Loess Plateau, China," Advances in Meteorology, vol. 2015, Article ID 478739, 13 pages, 2015. · ·
2. X. Zhou, Y. Zhang, and Y. Yang, "Comparison of two approaches for estimating precipitation elasticity of streamflow in China's Main River Basins," Advances in Meteorology, vol. 2015, Article ID 924572, 8 pages, 2015. ·
3. Z. Bao, J. Zhang, G. Wang et al., "Attribution for decreasing streamflow of the Haihe River basin, northern China: climate variability or human activities?" Journal of Hydrology, vol. 460-461, pp. 117–129, 2012. · ·
4. S. Zeng, C. Zhan, F. Sun, H. Du, and F. Wang, "Effects of climate change and human activities on surface runoff in the luan river basin," Advances in Meteorology, vol. 2015, Article ID 740239, 12 pages, 2015. · ·
5. G. Zhao, E. Li, X. Mu, Z. Wen, S. Rayburg, and P. Tian, "Changing trends and regime shift of streamflow in the Yellow River basin," Stochastic Environmental Research and Risk Assessment, vol. 29, no. 5, pp. 1331–1343, 2015. · ·
6. H. Chen, C.-Y. Xu, and S. Guo, "Comparison and evaluation of multiple GCMs, statistical downscaling and hydrological models in the study of climate change impacts on runoff," Journal of Hydrology, vol. 434-435, pp. 36–45, 2012. · ·
7. X. Yang, L. Ren, Y. Liu, D. Jiao, and S. Jiang, "Hydrological response to land use and land cover changes in a sub-watershed of West Liaohe River Basin, China," Journal of Arid Land, vol. 6, no. 6, pp. 678–689, 2014. · ·
8. S. Wang, Y. Yan, M. Yan, and X. Zhao, "Quantitative estimation of the impact of precipitation and human activities on runoff change of the Huangfuchuan River Basin," Journal of Geographical Sciences, vol. 22, no. 5, pp. 906–918, 2012. · ·
9. J. Fan, F. Tian, Y. Yang, S. Han, and G. Qiu, "Quantifying the magnitude of the impact of climate change and human activity on runoff decline in Mian River Basin, China," Water Science & Technology, vol. 62, no. 4, pp. 783–791, 2010. · ·
10. Q. Li, X. Yu, Z. Xin, and Y. Sun, "Modeling the effects of climate change and human activities on the hydrological processes in a semiarid watershed of Loess Plateau," Journal of Hydrologic Engineering, vol. 18, no. 4, pp. 401–412, 2013. · ·
11. A. Zhang, C. Zhang, G. Fu, B. Wang, Z. Bao, and H. Zheng, "Assessments of impacts of climate change and human activities on runoff with SWAT for the Huifa River Basin, Northeast China," Water Resources Management, vol. 26, no. 8, pp. 2199–2217, 2012. · ·
12. S. Zeng, J. Xia, and H. Du, "Separating the effects of climate change and human activities on runoff over different time scales in the Zhang River basin," Stochastic Environmental Research and Risk Assessment, vol. 28, no. 2, pp. 401–413, 2014. · ·
13. F. Zhao, L. Zhang, Z. Xu, and D. F. Scott, "Evaluation of methods for estimating the effects of vegetation change and climate variability on streamflow," Water Resources Research, vol. 46, no. 3, Article ID W03505, 2010. · ·
14. N. Fohrer, S. Haverkamp, and H.-G. Frede, "Assessment of the effects of land use patterns on hydrologic landscape functions: development of sustainable land use concepts for low mountain range areas,"Hydrological Processes, vol. 19, no. 3, pp. 659–672, 2005. · ·
15. N. Zégre, A. E. Skaugset, N. A. Som, J. J. McDonnell, and L. M. Ganio, "In lieu of the paired catchment approach: hydrologic model change detection at the catchment scale," Water Resources Research, vol. 46, no. 11, Article ID W11544, 2010. · ·
16. X. Li, Z. Li, and J. Dong, "Application of SWAT model in runoff simulation in upper reaches of Yihe River," Journal of Hohai University: Natural Sciences, vol. 37, no. 1, pp. 23–26, 2009 (Chinese).
17. P. Huang, Z. Li, C. Yao, Q. Li, and M. Yan, "Spatial combination modeling framework of saturation-excess and infiltration-excess runoff for semihumid watersheds," Advances in Meteorology, vol. 2016, Article ID 5173984, 15 pages, 2016. ·
18. B. He, C. Miao, and W. Shi, "Trend, abrupt change, and periodicity of streamflow in the mainstream of Yellow River," Environmental Monitoring and Assessment, vol. 185, no. 7, pp. 6187–6199, 2013. · ·
19. P. Gao, X.-M. Mu, F. Wang, and R. Li, "Changes in streamflow and sediment discharge and the response to human activities in the middle reaches of the Yellow River," Hydrology and Earth System Sciences, vol. 15, no. 1, pp. 1–10, 2011. · ·
20. G. Zhao, P. Tian, X. Mu, J. Jiao, F. Wang, and P. Gao, "Quantifying the impact of climate variability and human activities on streamflow in the middle reaches of the Yellow River basin, China," Journal of Hydrology, vol. 519, pp. 387–398, 2014. · ·
21. S. Jiang, L. Ren, B. Yong, V. P. Singh, X. Yang, and F. Yuan, "Quantifying the effects of climate variability and human activities on runoff from the Laohahe basin in northern China using three different methods," Hydrological Processes, vol. 25, no. 16, pp. 2492–2505, 2011. · ·
22. H. Zheng, L. Zhang, R. Zhu, C. Liu, Y. Sato, and Y. Fukushima, "Responses of streamflow to climate and land surface change in the headwaters of the Yellow River Basin," Water Resources Research, vol. 45, no. 7, Article ID W00A19, 2009. · ·
23. C. Yao, L. Chang, J. Ding, Z. Li, D. An, and Y. Zhang, "Evaluation of the effects of underlying surface change on catchment hydrological response using the HEC-HMS model," in Proceedings of the International Association

of Hydrological Sciences (ICWRS '14), vol. 364, pp. 145–150, Bologna, Italy, June 2014.

24. J. Liu, W. Kuang, Z. Zhang et al., "Spatiotemporal characteristics, patterns, and causes of land-use changes in China since the late 1980s," Journal of Geographical Sciences, vol. 24, no. 2, pp. 195–210, 2014. · ·

25. H. B. Mann, "Nonparametric tests against trend," Econometrica, vol. 13, no. 3, pp. 245–259, 1945. · · View at MathSciNet

26. M. G. Kendall, Rank Correlation Methods, Griffin, London, UK, 1975.

27. R. Sneyers, Sur L'analyse Statistique des Séries D'observations, Secrétariat de l'Organisation Météorologique Mondiale, 1975.

28. I. Ahmad, D. Tang, T. Wang, M. Wang, and B. Wagan, "Precipitation trends over time using mann-kendall and Spearman's rho tests in Swat River Basin, Pakistan," Advances in Meteorology, vol. 2015, Article ID 431860, 15 pages, 2015. ·

29. M. Zaman, G. Fang, K. Mehmood, and M. Saifullah, "Trend change study of climate variables in Xin'anjiang-Fuchunjiang watershed, China," Advances in Meteorology, vol. 2015, Article ID 507936, 13 pages, 2015. · ·

30. Z. Şen, "Innovative trend analysis methodology," Journal of Hydrologic Engineering, vol. 17, no. 9, pp. 1042–1046, 2012. ·

31. K. Saplioglu, M. Kilit, and B. K. Yavuz, "Trend analysis of streams in the Western Mediterranean Basin of Turkey," Fresenius Environmental Bulletin, vol. 23, no. 1, pp. 313–324, 2014.

32. Z. Şen, "Trend identification simulation and application," Journal of Hydrologic Engineering, vol. 19, no. 3, pp. 635–642, 2014. · ·

33. A. N. Pettitt, "A nonparametric approach to the change-point problem," Journal of the Royal Statistical Society, Series C: Applied Statistics, vol. 28, no. 2, pp. 126–135, 1979. · · View at MathSciNet

34. Y. Guo, Z. Li, M. Amo-Boateng, P. Deng, and P. Huang, "Quantitative assessment of the impact of climate variability and human activities on runoff changes for the upper reaches of Weihe River," Stochastic Environmental Research and Risk Assessment, vol. 28, no. 2, pp. 333–346, 2014. · ·

35. A. Sankarasubramanian, R. M. Vogel, and J. F. Limbrunner, "Climate elasticity of streamflow in the United States," Water Resources Research, vol. 37, no. 6, pp. 1771–1781, 2001. · ·

36. H. Alexandersson and A. Moberg, "Homogenization of Swedish temperature data. Part I: homogeneity test for linear trends," International Journal of Climatology, vol. 17, no. 1, pp. 25–34, 1997.

37. H. Alexandersson, "A homogeneity test applied to precipitation data," Journal of Climatology, vol. 6, no. 6, pp. 661–675, 1986. · ·

38. T. A. Buishand, "Some methods for testing the homogeneity of rainfall records," Journal of Hydrology, vol. 58, no. 1-2, pp. 11–27, 1982. · ·

39. S. Peng, W. Liu, W. Wang et al., "Estimating the effects of climatic variability and human activities on streamflow in the Hutuo River Basin, China," Journal of Hydrologic Engineering, vol. 18, no. 4, pp. 422–430, 2013. · ·

40. W. Wang, Q. Shao, T. Yang et al., "Quantitative assessment of the impact of climate variability and human activities on runoff changes: a case study in four catchments of the Haihe River basin, China,"Hydrological Processes, vol. 27, no. 8, pp. 1158–1174, 2013. · ·

41. H. Ma, D. Yang, S. K. Tan, B. Gao, and Q. Hu, "Impact of climate variability and human activity on streamflow decrease in the Miyun Reservoir catchment," Journal of Hydrology, vol. 389, no. 3-4, pp. 317–324, 2010. · ·

42. J. Wang, Y. Hong, J. Gourley, P. Adhikari, L. Li, and F. Su, "Quantitative assessment of climate change and human impacts on long-term hydrologic response: a case study in a sub-basin of the Yellow River, China," International Journal of Climatology, vol. 30, no. 14, pp. 2130–2137, 2010. ··

43. Y. Qiu, Sustainable Land Use Change in the North West Provinces of China Research Reports, School of Economics and Government, Australian National University, 2007.

CHAPTER 10 THE INFLUENCE OF STATISTICAL UNCERTAINTY IN THE HYDRAULIC BOUNDARY CONDITIONS ON THE PROBABILISTICALLY COMPUTED HIGH WATER LEVEL FREQUENCY CURVE IN THE RHINE DELTA

Hua Zhong, Pieter van Gelder, Wen Wang, Gaoxu Wang, Yongzhi Liu and Shuai Niu

Keywords: Statistical uncertainty, the bootstrap method, high water level frequency, Lower Rhine Delta

INTRODUCTION

The hydrodynamic characteristics of a delta or estuary are mainly governed by discharges of rivers and water levels at sea (or lake) boundaries. The joint probability approach is widely used to quantify the high water level frequency in deltas [1,2,3,4,5,6,7]. In the Rhine Delta, the relevant hydrodynamic loading variables at boundaries, namely the astronomical tides, the wind induced storm surges and the river flows, are jointly investigated and their joint probability distribution is estimated from the historical flood events [7]).

The high water level frequency assessment consists of two steps: first, a large number of stochastic scenarios of extreme boundary conditions are generated from the joint probability distribution; second, these generated scenarios are used as inputs to drive a deterministic hydrodynamic model to result in the peak water levels at locations of interest in the delta. The resulting peak water levels can be converted to the high water level frequency curve. The probabilistically computed high water level frequency is critically influenced by the joint probability distribution, time evolution of the relevant loading variables and accuracy of the hydrodynamic model.

In the joint probability approach, to get the marginal distributions (Rhine flood, wind induced storm surges), the annual maximal value during the observed year is determined and then a distribution to this series of annual maxima is fitted. From the obtained distribution, a flood event with a certain return period can be estimated. In general two different sources of uncertainty exist in this process [8]. Natural uncertainty refers to the variability of the flood events, mostly due to the limited number of flood events. For example, to estimate a 1000-year flood event based on a few dozen years flood records. Epistemic uncertainty consists of two parts: the uncertainty in distribution type chosen and in the parameter estimation. In this article, we focus on this natural uncertainty.

Statistical uncertainty exists in the marginal distributions of the joint probability distribution, which refers to the uncertainty in the parameters of the marginal distributions caused by estimating them from a limited number of flood events.And this is what we have to work with. Hence, better assessment of the statistical uncertainty due to insufficient data would be helpful.

The nonparametric bootstrap method is that from the given sample, new samples are generated by using resampling with replacement [9,10,11]. It relies on re-sampling with replacements from the given observations and providing estimates of uncertainty of distribution variables and quantiles. It is employed to quantify the natural uncertainty in the distributions as it is simple to present and easy to implement [12,13,14]

Citation: Hua Zhong, Pieter van Gelder , Wen Wang, Gaoxu Wang, Yongzhi Liu and Shuai Niu, The Influence of Statistical Uncertainty in the Hydraulic Boundary Conditions on the Probabilistically Computed High Water Level Frequency Curve in the Rhine Delta, doi:10.3390/w8040147.

The statistical uncertainty can be incorporated in the marginal distributions to form new marginal distributions by a Monte Carlo integration method. It is expected that the new distributions will increase the probability of extreme load values because of the uncertainty in the low quantiles of the marginal distributions. In the joint probability distribution, the marginal distributions with/without incorporating the statistical uncertainty are applied to probabilistically compute the high water level frequency curve. Then the impact of the statistical uncertainty of the marginal distributions on the high water level frequency curve can be investigated and quantified.

This study briefly introduces the joint probability approach to assess the high water level frequency in the Lower Rhine Delta, and aims to quantify the statistical uncertainty in the marginal distributions, and further evaluate the impact on the high water level frequency. This article will specifically focus on the case of the Lower Rhine Delta. It is organized as follows: Section 2 presents the method; the case is shown in Section 3; followed by the conclusion in Section 4.

METHODS

The outline of the method is illustrated in Figure 1. First, the bootstrap method is applied to estimate statistical uncertainty in the marginal distributions. Second, the statistical uncertainty is incorporated into the marginal distributions to form into new marginal distributions. Third, the new marginal distributions are applied to compute the high water level frequency in the Lower Rhine Delta, and to assess the impact of statistical uncertainty on the high water level frequency.

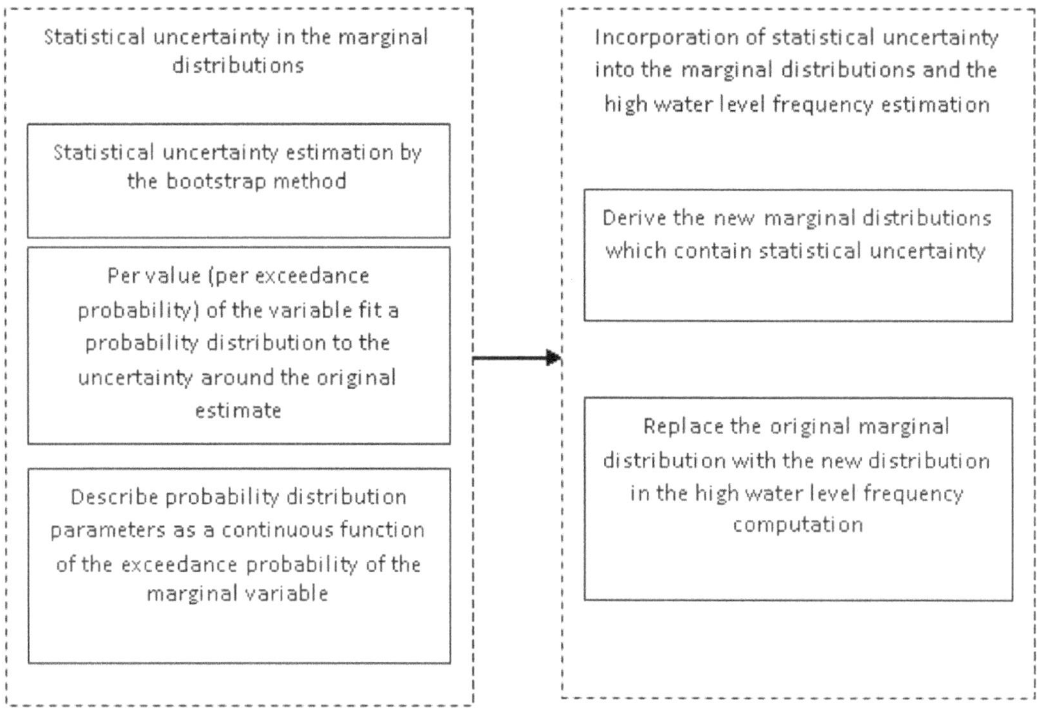

Figure 1. Outline of the method.

Statistical Uncertainty in a Distribution

The statistical uncertainty in a distribution can be estimated using the nonparametric bootstrap method, and subsequently parameterized as a probability distribution function of the exceedance probability of the variable.

The nonparametric bootstrap method generates a predetermined N samples of the variable by randomly resampling with replacement from the original sample. Each generated sample has the same size of the original sample. Each sample fits to a same probability distribution, and then there are N distributions. For a given exceedance probability of the variable, around the original estimate there are N values derived from the N distributions. These N values can fit to a distribution. For example, the log-normal distribution, a simple skewed distribution can be chosen.

The probability density function of a log-normal distribution is:

$$f_x(x;u,\sigma) = \frac{1}{x\sigma\sqrt{2\pi}} e^{-\frac{(\ln x - u)^2}{2\sigma^2}} \tag{1}$$

u and σ are the location parameter and the scale parameter on a logarithmic scale respectively.

The log-normal parameters can be made continuous functions of the exceedance probability of the variable. The parameter σ can be modeled as a function of the exceedance probability P by fitting a polynomial.

The mode of the log-normal distribution is the value with maximum probability density. The original estimate of the variable x is assumed to serve as the mode of the log-normal distribution, see Equation (2), and therefore the other log-normal parameter u can be estimated from Equation (3):

$$x = mode(x) = \exp(u - \sigma^2) \tag{2}$$

$$u(P) = \ln(x(P)) + \sigma(P)^2 \tag{3}$$

here P is the exceedance probability of the marginal variable.

The statistical uncertainty in the distribution can be estimated by the Log-normal distribution: given a value of x, the corresponding Log-normal distribution, in terms of the parameters u and σ, around that x is known.

Uncertainty-Incorporated Distribution

The statistical uncertainty can be incorporated into the distribution according to Equation (4) by the Monte Carlo Integration method. In Equation (4), x is the original variable, ε is the uncertainty value and x_{un} is the uncertainty incorporated variable, where xun=x+ε.

$$F_{un}(x_{un}) = \int_x F_{\log n}((x_{un}); u(x), \sigma(x)) f_x(x) dx \tag{4}$$

here $F_{un}(x_{un})$ is the uncertainty incorporated distribution of the variable; f_x is the original distribution of the variable;F_{logn} is the Log-normal (statistical uncertainty) distribution conditioned on x.

Impact of the Uncertainty on the High Water Level Frequency in the Rhine Delta

The hydrodynamic characteristics of the Rhine Delta are mainly governed by discharges of the Rhine and Meuse and the water level at the North Sea boundaries, as can be seen in Figure 2 and Figure 3. A joint probability approach using a 1-D hydrodynamic model was applied to assess the high water level frequency in the Lower Rhine Delta [7]. The operational control of the existing flexible hydraulic structures (red points in Figure 4) was taken into account.

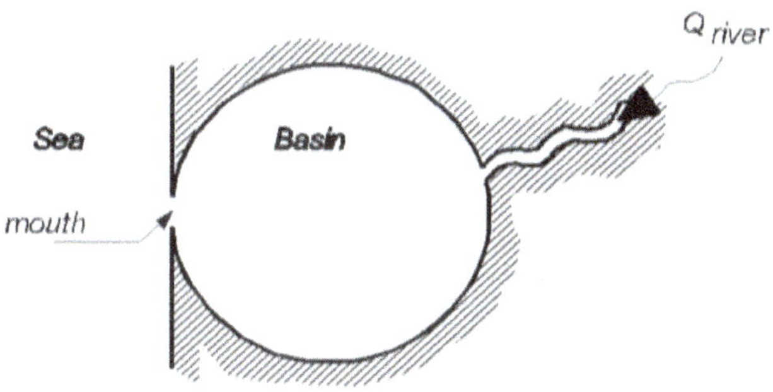

Figure 2. The conceptual model of the open delta.

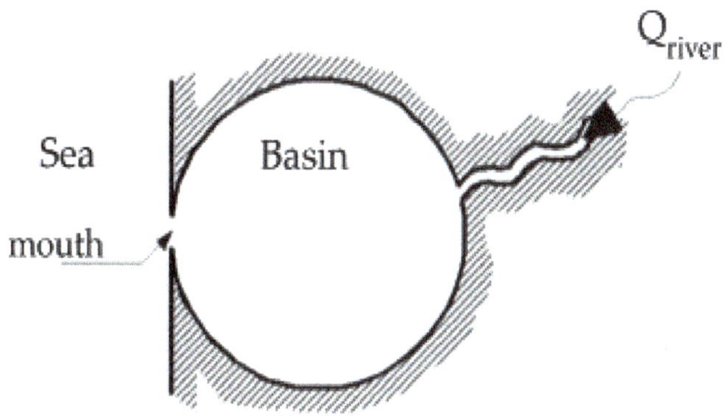

Figure 3. The conceptual model of the closable delta.

Figure 4. The Rhine Delta with operated hydraulic structures [19,20].

To reduce the computation burden, A conceptual model, the "Equal Level Curves" [15,16] is preferred to be applied to examine the interaction of sea level, fluvial flows and infrastructure operations to produce water levels at locations of interest in the Rhine Delta. Equal Level Curves are a simple steady state function, which can simulate the highest water level at Rotterdam by boundary conditions during one tidal period.

Two states of Equal Level Curves are introduced, see Figure 2 and Figure 3: one for the open delta where all dams and barriers are open, and the other for the closable delta where all dams and barriers along the coast are closed. This model has advantages: first, less information needed and a strong reduction in computation time; second, convenient to combine with the present operation control of the Maeslant storm surge barrier at the mouth of the delta. More information is in Appendix A.

According to previous studies [7,16,17], particular attention is paid to the simultaneous occurrence of storm surges and Rhine floods. Facing this kind of flood event, the Haringvliet dam gates and the Maeslant Storm Surge Barrier with the Hartel Storm Surge Barrier should be closed to protect high sea levels from propagating into the delta [18,19,20], but unfortunately the simultaneous high Rhine flow accumulating in a long closure duration can result in extreme high water level behind the dams and barriers.

In the joint probability of this kind of flood events, it is commonly assumed that the magnitude of the Rhine flow is independent of the magnitude of the storm surge [7,21,22,23] found no significant dependence of simultaneous occurrence of storm surges and Rhine floods. The marginal distributions were estimated from the selected flood events (Peak over Threshold values) with the parametric distributions (for example, generalized Pareto distribution). All parameters in the marginal distributions were estimated by the Maximum likelihood method. The joint probability distribution of this kind of event and its marginal distributions are shown in the Appendix B. The statistical uncertainty of three critical marginal distributions in terms of the wind induced storm surge peak level h_{smax}, the wind induced storm surge duration T_s and the high Rhine flow Q_r, are discussed.

The flood risk map of the Netherlands indicated the urbanized areas Rotterdam and Dordrecht are more hazardous and vulnerable than other places in the Netherlands, and with the higher fatalities [24]. As a result, Rotterdam is taken as the study areas of interest for the high water level frequency estimation.

In the joint probability distribution, three marginal distributions with/without incorporating the statistical uncertainties are applied to compute the high water level frequency in Rotterdam. Thus, the impact of the statistical uncertainty on the high water level frequency can be investigated and quantified.

RESULTS

Statistical Uncertainty

For each marginal variable, 1000 bootstrap samples are generated and fitted to 1000 bootstrap curves. Of which 41 curves (by order 25th, 50th ..., 475th, 525th ..., 950th, 975th, as well as 5th and 995th) are selected, shown as blue lines in Figure 5. The exceedance probability curve estimated by the original samples is shown in the red line in Figure 5. For a given exceedance probability of a marginal variable, 41 bootstrap estimates around the original estimate are fitted to a Log-normal distribution, as can be seen in Figure 6. Here only the figures with regard to the wind induce surge peak (h_{smax}) are given.

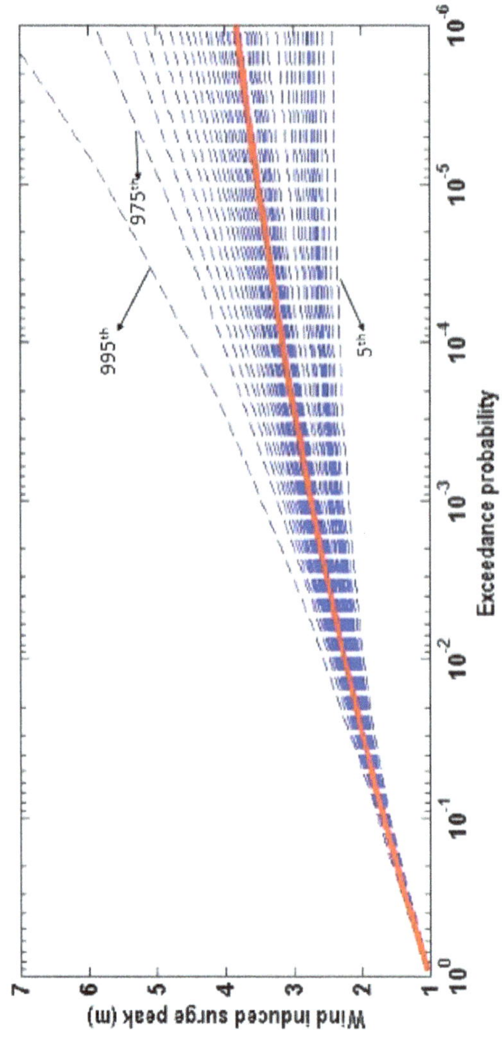

Figure 5. Selected bootstrap quantiles for wind induced surge h_{s-max}.

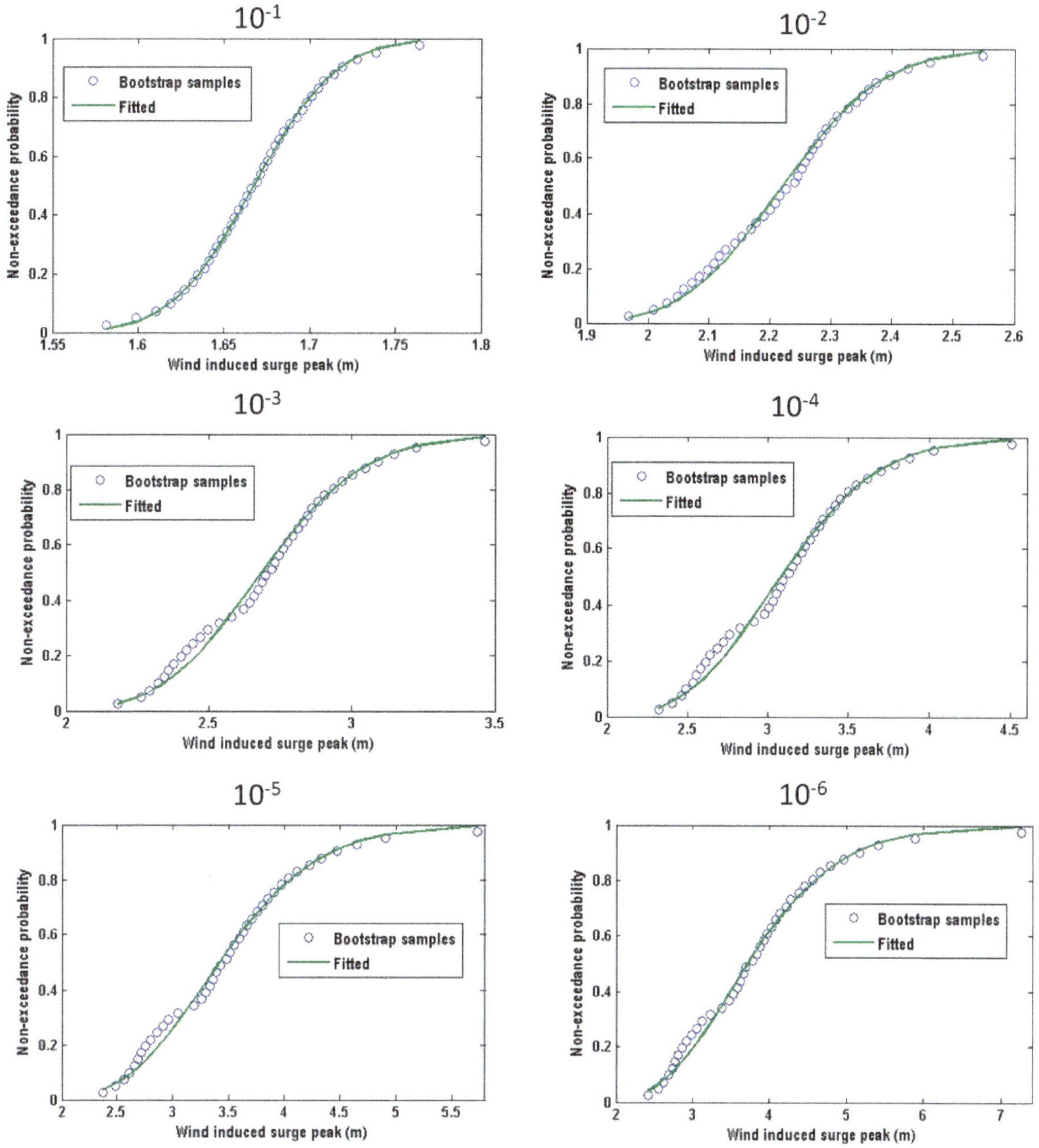

Figure 6. The fitted Log-normal distributions at the exceedance probability of 10^{-1}, 10^{-2}, 10^{-3}, 10^{-4}, 10^{-5}, 10^{-6} of wind induced surge h_{smax}.

The Log-normal parameters can be made continuous functions of the exceedance probability of the marginal variables. The estimated log-normal parameter σ for the exceedance probability is shown in Figure 7. The parameter σ can be modeled as a third-degree polynomial of the exceedance probability in Equation (5) in the region shown.

$$\sigma(P) = a_1 \log_{10}(P)^3 + a_2 \log_{10}(P)^2 + a_3 \log_{10}(P) + a_4 \quad (5)$$

here P is the exceedance probability of the marginal variable, a_1, a_2, a_3, a_4 are the parameters estimated by fitting a polynomial with the method of least squares. The estimates of a_1, a_2, a_3 and a_4 for each marginal variable are shown in Table 1.

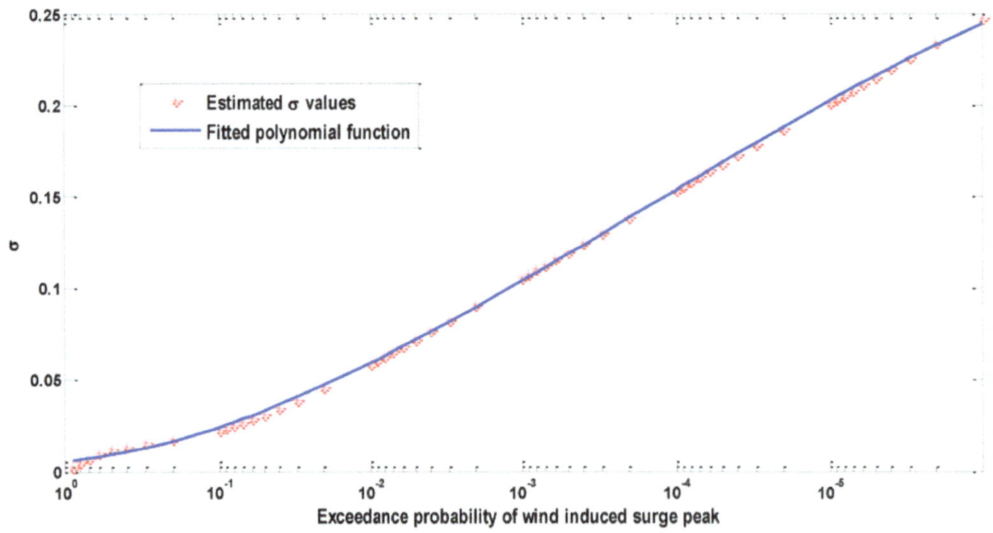

Figure 7. The Log-normal distribution's parameter σ is modeled as a function of the exceedance probability of the wind induced surge h_{smax}.

Table 1. The estimates of a_1, a_2, a_3 and a_4 for the exceedance probabilities of each marginal variable.

σ	a_1	a_2	a_3	a_4
$\sigma(P(h_{smax}))$	0.0010	0.011	-0.0078	0.0065
$\sigma(P(T_s))$	-0.0001	-0.0021	-0.0141	0.0189
$\sigma(P(Q_r))$	0.0017	0.0247	0.0069	0.0189

The other Log-normal parameter u can be estimated based on Equation (3).

Uncertainty-Incorporated Marginal Distributions

Through application of Equation (4), the statistical uncertainties can be incorporated into the marginal distributions. The results are shown in Figure 8, Figure 9 and Figure 10.

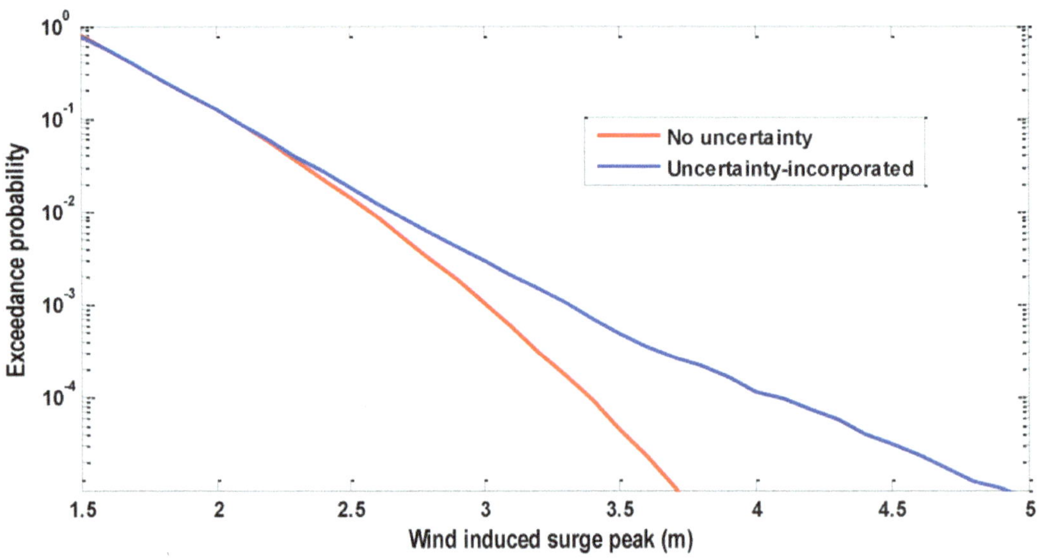

Figure 8. The uncertainty-incorporated marginal distribution of the wind induced surge h_{smax}.

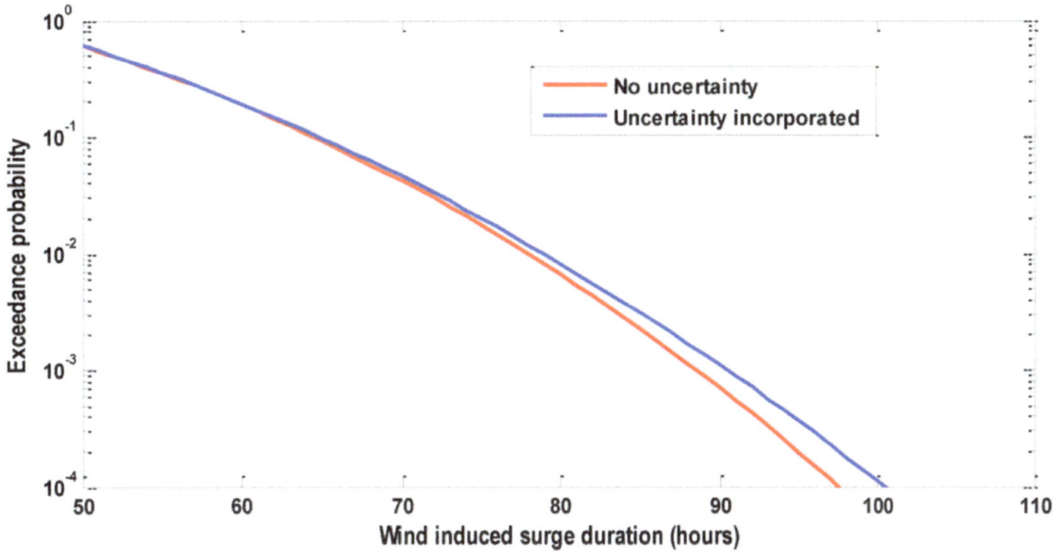

Figure 9. The uncertainty-incorporated marginal distribution of the wind induced storm surge duration Ts.

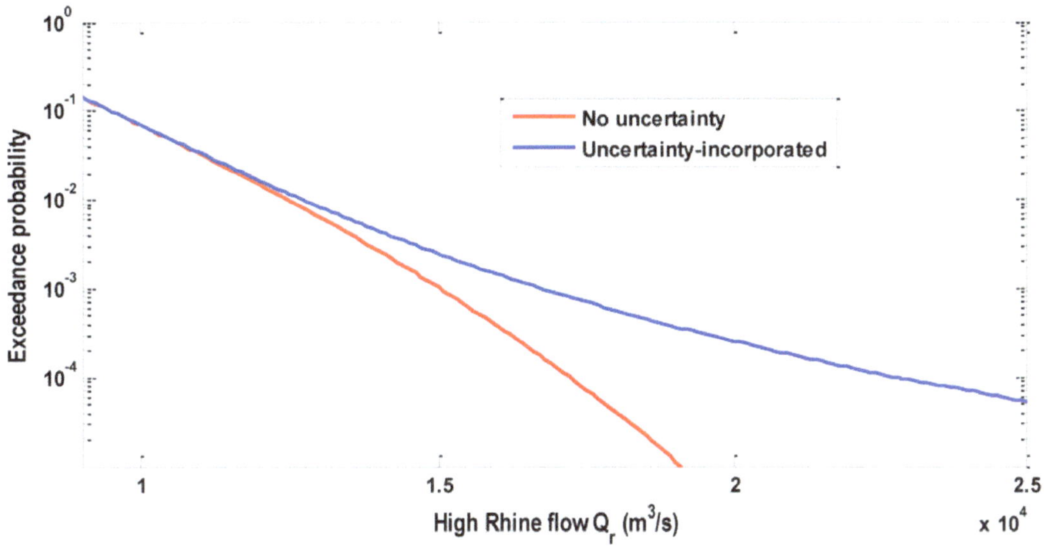

Figure 10. The uncertainty-incorporated marginal distribution of the high Rhine flowQr.

For higher exceedance probabilities (lower return periods), the uncertainty has a negligible effect on the uncertainty-incorporated distribution, mainly due to the small variance of the uncertainty for higher exceedance probabilities (seeFigure 7). For lower exceedance probabilities (higher return periods), the variance increases, and the effect on the distributions becomes more substantial.

The results indicate that uncertainty incorporated marginal distributions result in more extreme values of the marginal variables at low exceedance probabilities.

However, considering the physical conditions the statistical uncertainty in the marginal distributions needs to be constrained at extreme return periods in order to avoid unreal situations.

Impact on the High Water Level Frequency

The impact of the statistical uncertainty on the high water level frequency can be addressed by the differences between the high water level frequency curves derived from two types of marginal distributions, as indicated in Section 2.3. The results are shown in Figure 11, Figure 12 and Figure 13 respectively.

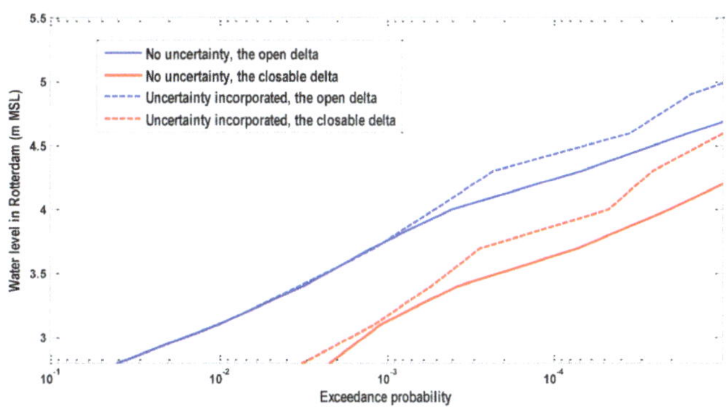

Figure 11. The high water level frequency curve considering the statistical uncertainty in the marginal distribution of the wind induced surge h_{smax}.

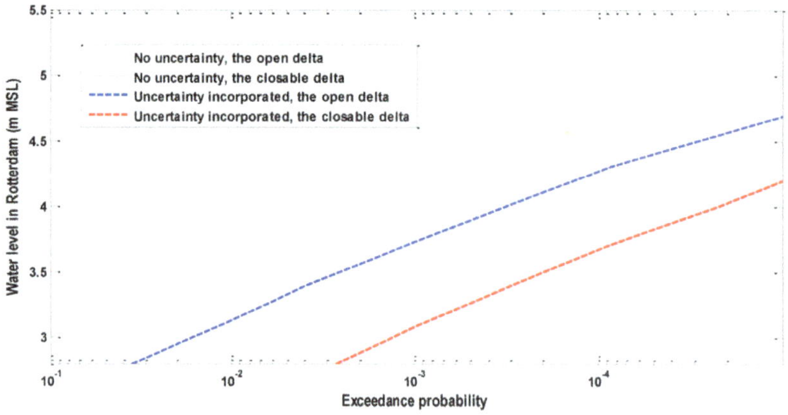

Figure 12. The high water level frequency curve considering the statistical uncertainty in the marginal distribution of the wind induced storm surge duration Ts.

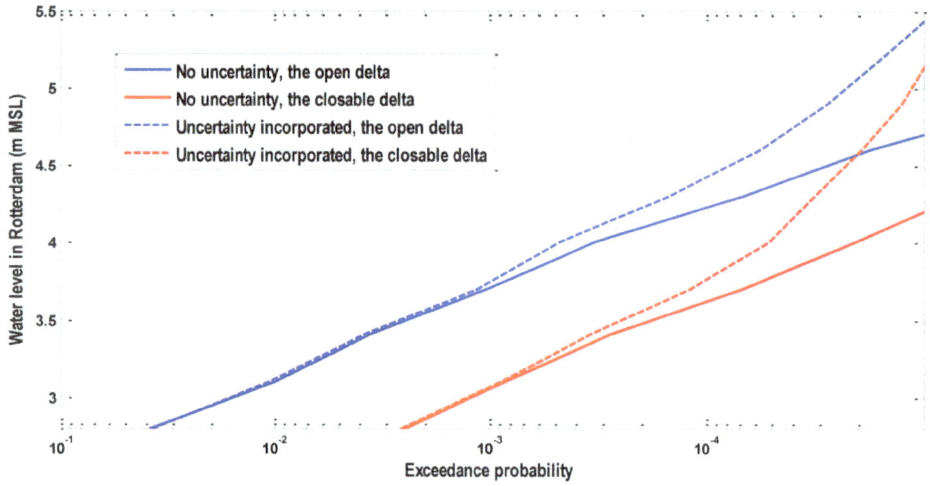

Figure 13. The high water level frequency considering the statistical uncertainty in the above two marginal distributions.

In Figure 11, the high water level frequency curves in Rotterdam are estimated from two different marginal distributions of h_{smax} respectively. Incorporating the statistical uncertainty in the marginal distribution of h_{smax} significantly increases the high water level frequency in Rotterdam for the exceedance probabilities lower than 10^{-3}.

In Figure 12, incorporating the statistical uncertainty in the marginal distribution of Ts does not affect the high water level frequency in Rotterdam.

In conclusion, the statistical uncertainty in each marginal distribution differs and its impact on the high water level frequency also varies.

In Figure 13, incorporating the statistical uncertainty in two marginal distributions significantly increases the high water level frequency in Rotterdam. Generally for higher exceedance probabilities (lower return periods), the statistical uncertainty has a negligible effect on the high water level frequency, while for lower exceedance probabilities (higher return periods), the effect on the high water level frequency becomes more substantial.

The design water level in Rotterdam is regarded as the water level with an exceedance frequency of 1/10,000; and its present value is 3.60 m above mean sea level (in brief MSL) [25]. As can be seen in Figure 12, the design water level of Rotterdam corresponds to 3.60 m MSL without considering the statistical uncertainty in three marginal distributions, while it corresponds to 3.75 m with considering the statistical uncertainty in three marginal distributions.

CONCLUSIONS AND RECOMMENDATIONS

Quantifying high water level frequency is critical but complex in deltas or estuaries. In this article, the joint probability approach using a deterministic hydrodynamic model is applied to estimate the high water level frequency. This study aims at investigating statistical uncertainty of marginal distributions of the joint probability distribution and its impact on the high water level frequency. In the Rhine Delta, the results show that incorporating the statistical uncertainty in the marginal distributions will increase the high water level frequency because the probability of extreme hydraulic boundary conditions increases. The statistical uncertainty in each marginal distribution differs and its impact on the high water level frequency also varies.

Generally only a limited amount of data available when calculating the high water level frequency or flood frequency, however, this is what we have to work with. Hence, better assessment of the statistical uncertainty due to insufficient data would be helpful. In condition, considering real physical conditions, this statistical uncertainty should be constrained. For example, Rhine discharges in Lobith and sea levels in Hook of Holland have an physical upper limit. This constrain should be elaborated in future research.

ACKNOWLEDGMENTS

We appreciate financial supports from National Natural Science Foundation of China (51479222, 41471427, 41101510) , the 111 Project (B08048).

Author Contributions

Hua Zhong and Pieter van Gelder designed this study and wrote this article; Wen Wang contributed to the statistical uncertainty analysis; Gaoxu Wang, Yongzhi Liu and Shuai Niu assised data analysis and drew the conclusions . All authors have read and approved the final manuscript.

Conflicts of Interest

The authors declare no conflict of interest

APPENDIX A

The Conceptual Model of the Rhine Delta

Considering the characteristics of the Rhine Delta, the conceptual model is illustrated. Two states of Equal Level Curves are introduced: one for the open delta where all dams and barriers are open, and the other for the closable delta where all dams and barriers along the coast are closed, as can be seen in Figure A1 and Figure A2. The closable delta can be open to the sea except that the delta can be closed with the help of hydraulic structures during the extreme weather conditions (storm surges). The conceptual model of the Lower Rhine Delta was introduced in Zhong et al. [16].

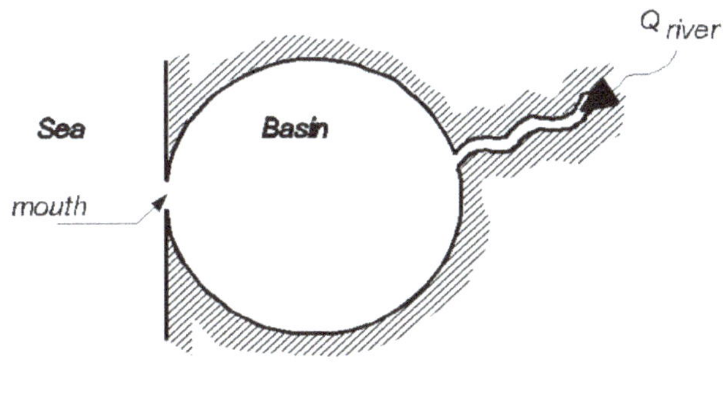

Figure A1. The conceptual model of the open delta.

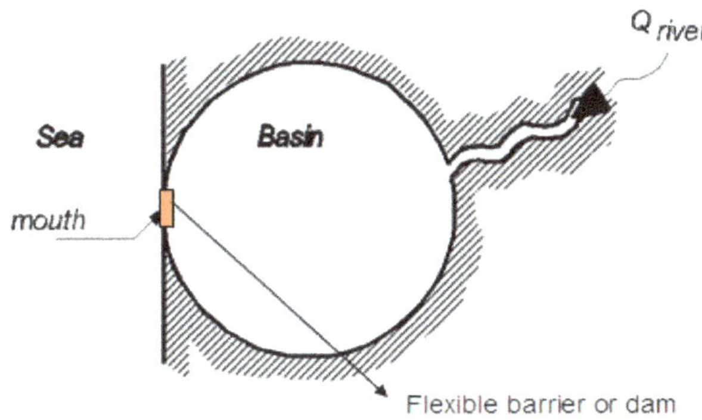

Figure A2. The conceptual model of the closable delta.

Equal Level Curves are a simple steady state function, which can simulate the highest water level at Rotterdam by boundary conditions during one tidal period. For the open delta, Rotterdam water levels can be modeled by the Equal Level Curves with the boundary conditions of the Rhine flow at Lobith and the sea level at Hook of Holland, see Equation (A1):

$$h_R = h_{hvh} + (\frac{Q_r + Q_m}{\mu A})^2 \cdot \frac{1}{2 \cdot g}$$

(A1)

here h_R is the water level at Rotterdam; while h_{hvh} stands for the sea water level at Hook of Holland. Q_r is the Rhine flow at Lobith, and Q_r is the Meuse flow at Borgharen. μ is the discharge coefficient, A stands for the surface area of the cross section in Hook of Holland and g is the gravitational acceleration. The parameters of μ and A can be estimated by the linear regression method with the selected historical flood event, and μ·A is estimated to be 3620 m^2.

$$h_{hvh} = h_s + h_a + h_0$$

(A2)

here h_0 is the mean sea level, h_a is the astronomical tide level, h_s is the wind induced surge level. From a statistical point of view, the occurrence of the astronomical tide component is independent of the occurrence of the wind induced storm surge component at the mouth of the Rhine Delta. However, these two components can interact with each other when they propagate into the delta. Their nonlinear interaction generally increases the surge height at a rising astronomical tide and decreases the surge height at a high astronomical tide. Quantifying the nonlinear effect is beyond the scope of this study. For the sake of convenience, it can be assumed that the wind induced storm surge is independent of the astronomical tide.

Equal Level Curves with the open delta are shown in Figure A3.

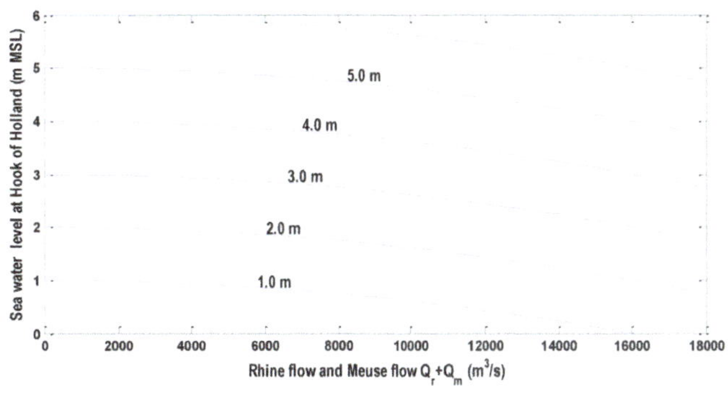

Figure A3. Equal Level Curves with the open delta at Rotterdam.

At present, the Rhine delta can be kept open always via the New Waterway and be closed by closing the Maeslant Barrier and the Haringvliet dams when facing storm surges.

Being a closable delta, there are two states for the Equal Level Curves: the state of open can be described by Equation (A1), the state of closure can be described by Equation (A3). After the closure of the delta, the water level behind the barrier rises because the Rhine flow cannot be discharged into the

North Sea and will accumulate. In this process the heights of the surrounding dikes are assumed to be infinite high and no dike breaches occur.

$$h_R = h_{R,c} + \frac{(Q_r + Q_m) \cdot \Delta T}{B} \tag{A3}$$

where h_R is Rotterdam water level after the closure duration ΔT; $h_{R,c}$ is the average water level behind the Maeslant barrier at the closure time; Q_r is the Rhine flow at Lobith; Q_m is the Meuse flow at Borgharen; B is the surface area of the delta where water can be stored.

$h_{r,c}$ can be estimated from the average water level of four locations (Rotterdam, Goidschalxoord, Dordrecht, and Moerdijk) at the moment of the Maeslant barrier closing.

The water levels at these four locations at time t depend on the sea water level of a few hours ago at Hook of Holland and on the Rhine and Meuse discharges of 24 hours ago at Lobith and Borgharen. The functions are shown in Equations (A4)–(A7). The parameters of the below equations are estimated from system identification of historical measurements. The time unit of t is an hour.

$$h_R^t = 0.9735 \cdot h_{hvh}^{t-1} + 7.781 \cdot 10^{-9} \cdot (Q_r^{t-24} + Q_m^{t-24})^2 \tag{A4}$$

$$h_G^t = 0.7335 \cdot h_{hvh}^{t-1} + 1.013 \cdot 10^{-8} \cdot (Q_r^{t-24} + Q_m^{t-24})^2 \tag{A5}$$

$$h_D^t = 0.6065 \cdot h_{hvh}^{t-2} + 1.737 \cdot 10^{-8} \cdot (Q_r^{t-24} + Q_m^{t-24})^2 \tag{A6}$$

$$h_M^t = 0.3753 \cdot h_{hvh}^{t-3} + 1.458 \cdot 10^{-8} \cdot (Q_r^{t-24} + Q_m^{t-24})^2 \tag{A7}$$

The parameter B can be estimated by inverting Equation (A3) into:

$$B = \frac{(\frac{8}{9}Q_r + Q_m) \cdot \Delta T}{h_R - h_{R,c}} = \frac{(\frac{8}{9}Q_r + Q_m) \cdot \Delta T}{\Delta h} \tag{A8}$$

The factor 8/9 comes from the distribution of the Rhine River inflow of which 1/9 flows north towards the IJsselmeer. The water level can rise by Δh after the closure time ΔT. In the closure event of 2007, the Rhine discharge Q_r was 1171 m^3/s, the Meuse discharge Q_m was 148 m^3/s and the water level at Rotterdam rose from 0.70 to 1.12 m after 15 hours closure. In Figure A4, the derivative is constant and the value of B is estimated to be 152 km^2.

The closure duration ΔT depends on the operational control of closing and opening of the delta. Two hydraulic structures: the Maeslant storm surge barrier and the Haringvliet dams are mainly responsible for closing the Rhine delta. When a high Rhine flow coincides with a storm surge, the present operational control of the Haringvliet dams keep the Rhine delta partly open, depending on the quantity of the Rhine flow in Lobith. However, to keep the analysis simple, it is assumed that when the Maeslant Barrier closes, the Haringvliet dams fully close. This assumption will overestimate the water level when a storm surge coinciding with a high Rhine flow occurs. Therefore, the closure duration ΔT depends on the operational control of the Maeslant Barrier responding to the hydraulic boundary conditions.

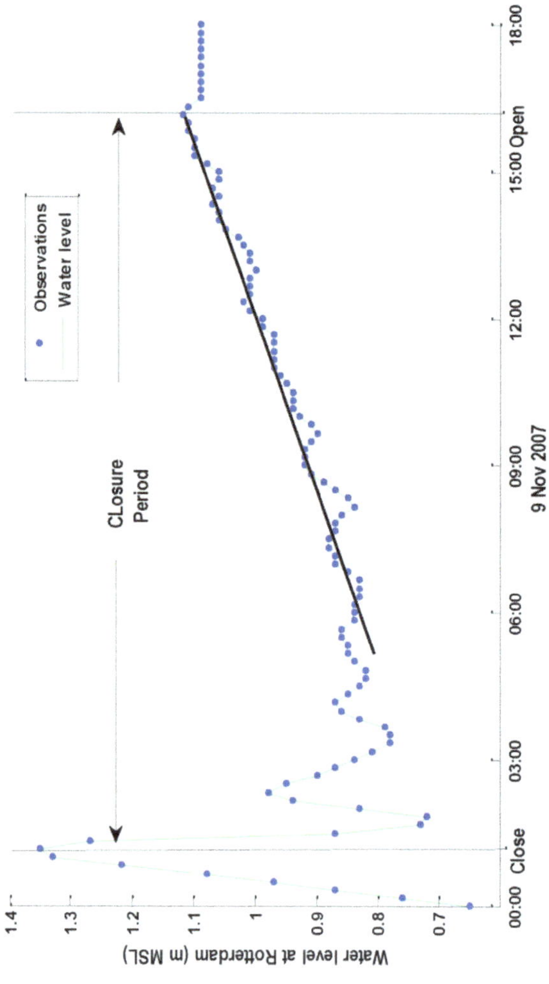

Figure A4. Rotterdam water level during the first closure event of 9 November 2007.

The control system of the Maeslant Barrier (named BOS, in Dutch: Beslissing & Ondersteunend Systeem; in English, Decision and Support System) has the responsibility to close the barrier completely autonomously [18]. To keep the analysis simple, it is assumed that only one control parameter is considered: the closing decision level H_d, in the operational control of the barrier. When Rotterdam water level is predicted to exceed the closing decision level H_d (3.0 m MSL in Rotterdam), the barrier is assumed to close at two hour before. In reality, the barrier closes at the moment the current starts to change from seaward direction to landward direction, however, to calculate this moment needs extra information which is not the main concern in this article. The water level in Hook of Holland drops after the storm surge, and when the water level in Rotterdam is higher than the water level in Hook of Holland, the barrier re-opens. Considering the time the procedure of close and re-open takes, the minimum closure duration is 6 hours. The closure decision making of the Maeslant Barrier is illustrated in Figure A5.

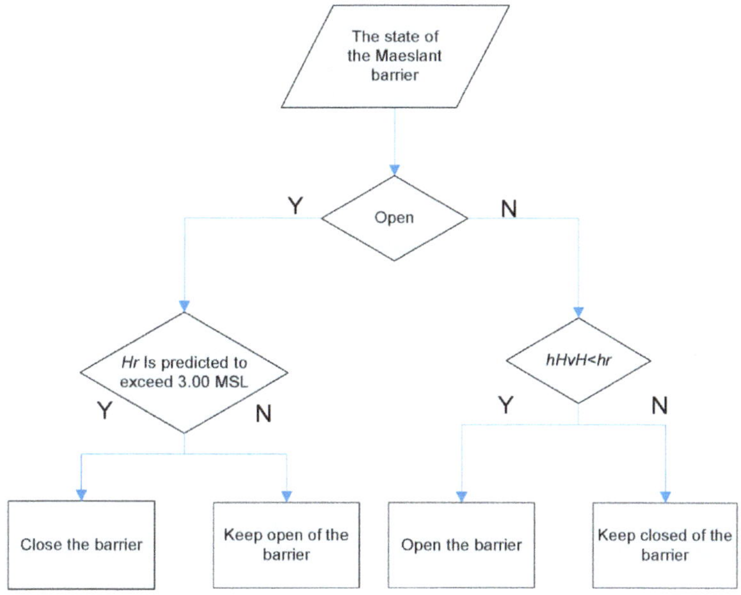

Figure A5. The closure decision making of the Maeslant Barrier.

Equal Level Curves for the closable delta are shown in Figure A6 in which Equal Level Curves in the shaded area are presented in Figure A7.

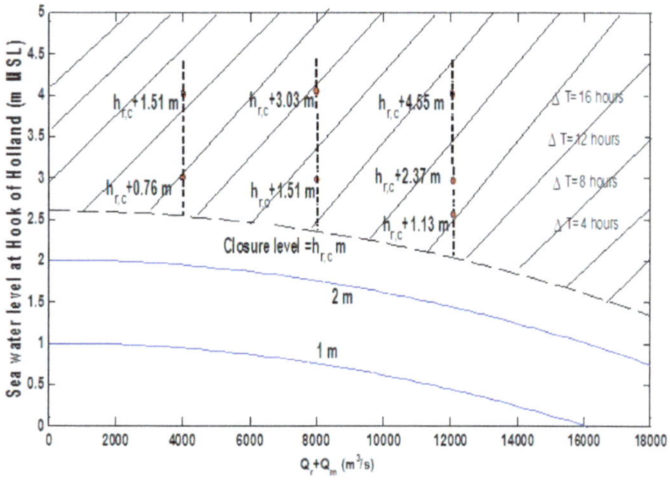

Figure A6. Equal Level Curves with the closable delta at Rotterdam.

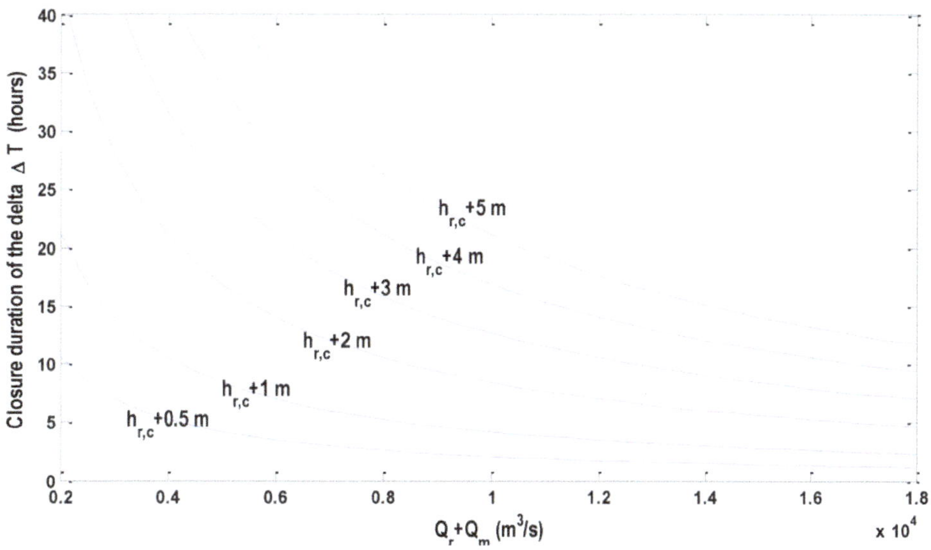

Figure A7. Equal Level Curves with the closable delta at Rotterdam after the delta closes at $h_{r,c}$.

APPENDIX B

Probability Distributions

The wind induced storm surge peak level h_{smax} fits the generalized Pareto distribution (GPD):

$$f(h_{s\max}) = \frac{1}{\sigma}(1+\xi\frac{h_{s\max}-\mu}{\sigma})^{-(\frac{1}{\xi}+1)}$$

(B1)

In this equation the shape parameter ξ is −0.0677; the scale parameter σ is 0.3140 m; the location parameter u is 1.0 m.

The wind induced storm surge duration T_s fits the Weibull distribution

$$f(T_s) = \frac{k}{\lambda}(\frac{T_s}{\lambda})^{k-1} e^{-(\frac{T_s}{\lambda})^k}$$

(B2)

In this equation, $T_s > 0$, k is the shape parameter, 2.5237; λ is the scale parameter, 38.0887 hours.

The high Rhine discharge Q_r fits the generalized Pareto distribution:

$$f(Q_r) = \frac{1}{\sigma}(1+\xi\frac{Q_r-\mu}{\sigma})^{(-\frac{1}{\xi})-1}$$

(B3)

In this equation, ξ is the shape parameter; σ is the scale parameter; u is the location parameter; and the parameters' values are -0.0667, 1629.7 m^3/s and 6000 m^3/s respectively.

The high Meuse discharge Q_m fits the Log-normal distribution:

$$f(Q_m) = \frac{1}{\sigma \cdot Q_m \cdot \sqrt{2\pi}} e^{-\frac{(\ln Q_m - \mu)}{2\cdot\sigma^2}}$$

(B4)

In this equation, u is the mean value, 6.8667; σ is the stand deviation value, 0.3752.

The joint cumulative probability distribution of high Rhine flow and high Meuse flow fits a Gumbel Copula function:

$$F_{Q_r,Q_m}(Q_r,Q_m) = C_\alpha(u,v) = \exp\{-[(-\ln u)^\alpha] + (-\ln v)^\alpha]^{\frac{1}{\alpha}}\}$$

(B5)

where α is estimated as 1.7158; F_r is the marginal distribution of high Rhine flow; F_m is the marginal distribution of the associated Meuse flow.

The joint probability distribution of storm surges and high Rhine flows

$$P(h_R^*) = \frac{9}{70} \cdot \iiint \iiint I(*)p(h_{s\max})p(h_{HW})p(T_s)p(u)p(Q_r,Q_m)dh_{s\max}dh_{HW}dT_s du dQ_r dQ_m$$
$$I = 1 : h_R^* <= h_R(h_{s\max}, h_{HW}, T_s, u, Q_r, Q_m)$$
$$I = 0 : h_R^* > h_R(h_{s\max}, h_{HW}, T_s, u, Q_r, Q_m)$$

(B6)

here I is an indicator function and h_R is the highest Rotterdam water level calculated from the specific input variables using the 1-D model; 9/70 is the occurrence probability per year for the combination event of storm surges and high Rhine flows; h_{smax} stands for the wind induced storm surge peak, and its unit is m; T_s is the wind induced storm surge duration, and its unit is hours; h_{HW} is the high tide level and h_{LW} is the low tide level, their unit is m MSL; u is the time shift between peaks of tide and surge residual; Q_r is Rhine discharge and Q_m is Meuse discharge, their unit is m^3/s; p(Qr,Qm) is the joint probability density distribution of Rhine discharge and Meuse discharge. More detailed information is in Zhong et al. [7].

REFERENCES

1. Mantz, P.A.; Wakeling, H.L. Forecasting flood levels for jointevents of rainfall and tidal surge flooding using extreme valuestatistics. Proc. Institut. Civil Eng. Res. Theor. **1979**, 67, 31–50.
2. Acreman, M.C. Assessing the joint probability of fluvial and tidalfloods in the River Roding. J. Inst. Water Environ. Manage. **1994**, 8, 490–496.
3. Gorji-Bandpy, M. The joint probability method of determining theflood return period of a tidally affected pond, Iran.J. Sci. Technol. **2001**, 25, 599–610.
4. Samuels, P.G.; Burt, N. A new joint probability appraisal offlood risk. Proc. Institut. Civil Eng. Water Maritime Eng.**2002**, 154, 109–115.
5. Adib, A.; Vaghefi, M.; Labibzadeh, M.; Rezaeian, A.; Tagavifar, A.; Foladfar, H. Predicting extreme water surface elevation intidal river reaches by different joint probability methods. J. FoodAgric. Environ. **2010**, 8, 988–991.
6. Lian, J.J.; Xu, K.; Ma, C. Joint impact of rainfall and tidal level on flood risk in a coastal city with a complex river network: A case study for Fuzhou city, China. Hydrol. Earth Syst. Sci. Discuss. **2012**, 9, 7475–7505.
7. Zhong, H.; van Overloop, P.J.; van Gelder, P. A joint probability approach using a 1-d hydrodynamic model for estimating high water level frequencies in the lower Rhine delta. Nat. Hazard. Earth Syst. Sci. **2013**, 13, 1841–1852.
8. Bruno, M.; Annegret, H.T. Separating natural and epistemic uncertainty in flood frequency analysis. J. Hydrol. **2005**,309, 114–132.
9. Efron, B. Bootstrap methods: Another look at the jackknife. Ann. Statist. **1979**, 7, 1–26.
10. Efron, B.; Tibschirani, R.J. An Introduction to the Bootstrap; Chapman&Hall: New York, NY, USA, 1993.
11. Davison, A.C.; Hinkley, D.V. Bootstrap Methods and Their Application; Cambridge University Press: Cambridge, UK, 1997.
12. Dunn, P.K. Bootstrap confidence intervals for predicted rainfall quantiles. Int. J. Climatol. **2001**, 21, 89–94.
13. Kysely, J. Coverage probability of bootstrap confidence intervals in heavy-tailed frequency models, with application to precipitation data. Theor. Appl. Climatol. **2010**, 101, 345–361.
14. Roscoe, K.L.; Diermanse, F. Effect of surge uncertainty on probabilistically computed dune erosion. Coast. Eng. **2011**,58, 1023–1033.
15. Vrijling, J.K.; van Gelder, P.H.A.J.M. Probabilistic

Design in Hydraulic Engineering; Delft University of Technology: Delft, The Netherlands, 1996.

16. Zhong, H.; van Overloop, P.J.; van Gelder, P.; Rijcken, T. Influence of a storm surge barrier's operation on the high water level frequency in the Rhine delta area. Water **2012**, 4, 474–493.
17. van der Made, J. Design levels in the transition zone between the tidal reach and the river regime reach, hydrology of deltas. In Proceedings of the Bucharest Symosium, Bucharest, Rumania, 6–14 May 1969; pp. 246–257.
18. Bol, R. Operation of the "Maeslant Barrier": Storm surge barrier in the Rotterdam new waterway. In Flooding and Environmental Challenges for Venice and Its Lagoon: State of Knowledge; Cambridge University Press: Cambridge, UK, 2005; pp. 311–316.
19. van Overloop, P.J. Operational Water Management of the Main Waters in the Netherlands; Delft Unversity of Technology: Delft, The Netherlands, 2009.
20. van Overloop, P.J. Optimization of "Everything", Prediction and Control of the entire Delta and River System of The Netherlands; Technical Report for Water INNovation (WINN); Delft University of Technology: Delft, The Netherlands, 2011.
21. Dantzig, D.; Hemelrijk, J.; Kriens, J.; Lauwerier, H. Rapport Deltacommissie; Staatsdrukkerij En Uitgeverijbedrijf: 's-Gravenhage, The Netherlands, 1960. (In Dutch)
22. Jorigny, M.; Diermanse, F.; Hassan, R.; van Gelder, P. Correlation analysis of water levels along dike-ring areas.Develop. Water Sci. **2002**, 47, 1677–1684.
23. Klerk, W.J.; Winsemius, H.C.; van Verseveld, W.J.; Bakker, A.M.R.; Diermanse, F.D. The co-incidence of storm surges and extreme discharges within the Rhine-Meuse Delta. Environ. Res. Lett. **2015**, 10, 035005.
24. De Bruijn, K.M.; Klijn, F. Risky places in the netherlands: A first approximation for floods. J. Flood Risk Manag. **2009**, 2, 58–67.
25. Ministerie van Verkeer en Waterstaat. Hydraulische Randvoorwaarden Primaire Waterkeringen, Voor de Derde Toetsronde 2006–2011 (hr 2006); Ministerie van Verkeer en Waterstaat: Den Haag, The Netherlands, 2007. (In Dutch)

CHAPTER 11

FUZZY-BOUNDARY INTERVALS COMBINED WITH MULTI-STAGE STOCHASTIC PROGRAMMING MODEL

Eleni Bekri, Markus Disse and Panayotis Yannopoulos

Keywords: Optimal water allocation, fuzzy, interval bounds, uncertainty, stochastic hydrologic scenarios, future scenarios

INTRODUCTION

Optimal water allocation of a river basin poses great challenges for engineers due to various uncertainties associated with the hydrosystem, its parameters and its impact factors as well as their interactions. These uncertainties are often associated with various complexities in terms of information quality [1]. The random characteristics of natural processes (*i.e.*, precipitation and climate change) and stream conditions (*i.e.*, stream inflow, water supply, storage capacity, and river-quality requirement), the errors in estimated modeling parameters (*i.e.*, benefit and cost parameters), and the vagueness of system objectives and constraints are all possible sources of uncertainties. These uncertainties may exist in both left- and right-hand sides of the constraints as well as coefficients of the objective function. Some uncertainties may be expressed as random variables. At the same time, some random events can only be quantified as discrete intervals with fuzzy boundaries, leading to multiple uncertainties presented as different formats in the system's components [2]. Traditional optimization techniques can embody various characteristics but only as deterministic values. In various real-world problems, results generated by these traditional optimization techniques could be rendered highly questionable if the modeling inputs could not be expressed with precision [1,3,4]. For these reasons conventional deterministic optimization approaches have given their place to stochastic (SP), fuzzy (FP) and interval-parameter programming (IPP) approaches and their hybrid combinations in order to face up these difficulties. Various methodologies have been developed and proposed [4,5,6,7,8,9,10,11,12,13,14,15,16] in order to embody in optimal water allocation uncertainties of various influencing factors and hydrosystem characteristics.

SP can handle uncertainties expressed as random variables with known probability distributions and at the same time connect efficiently the pre-regulated policies and the associated economic implications caused by improper policies [2]. Reference [17] introduced a stochastic dynamic programming model for a single reservoir deriving optimal reservoir operating policies subject to reliability constraints. Reference [18] presented a stochastic optimization approach for a multi-reservoir planning with hydropower system under uncertainty. They assigned a given probability to each of a range of inputs occurring at different stages of an optimization horizon. A common SP method is the two-stage stochastic programming (TSP). It is based on the concept of recourse, expressed as the ability to take corrective actions after a random event has occurred. The initial action is called the first-stage decision, and the corrective one is named the recourse decision. The first-stage decisions have to be made before further information of initial system uncertainties is revealed, whereas the recourse decisions are allowed to adapt to this information [19]. Reference [6] developed an inexact two-stage stochastic programming method for water resources management, dealing with uncertainties expressed as both probability distributions and intervals and accounting for economic penalties due to infeasibility. Reference [20] proposed a scenario-based multistage stochastic programming model for water supplies planning from highland lakes. A number of inflow scenarios are explicitly taken into account in order to determine a contract for water delivery in the coming year. In general, multistage stochastic programming (MSP) approach permitted modified decisions in each time

Citation: Bekri, E.; Disse, M.; Yannopoulos, P. Fuzzy-Boundary Intervals Combined with Multi-Stage Stochastic Programming Model. Water 2015, 7, 6427-6466. doi:10.3390/w7116427.

stage based on the real-time realizations of uncertain system conditions [8,21].

SP cannot handle randomness in the right-hand-side parameters. However, chance-constrained programming (CCP) method can deal with this type of uncertainty. It can reflect the reliability of satisfying (or risk of violating) system constraints under uncertainty [22,23]. Reference [24] used a CCP model for reservoir systems planning of irrigation districts. Reference [23] developed an inexact CCP method for assessing risk of violating system constraints, in which uncertainties are expressed as probabilities and intervals. Reference [25] presented a mathematical programming model for the reservoir capacity planning under random stream inflows. Based on the CCP method it considers a special target-priority policy based on given system reliabilities. Reference [26] developed a stochastic model for optimal multi-period operation of a multi-reservoir system for a basin operating under a conjunctive use of ground and surface water framework, with uncertainties in the inflows dealt using CCP method. However, TSP and CCP have difficulties in dealing with uncertain parameters when their probabilistic distributions are not available [13]. Moreover, the increased data requirements for specifying the parameters' probability distributions may affect their practical applicability [1].

Uncertainties may be also related to the incompleteness or impreciseness of observed information [27]. This type of uncertainty, expressed as fuzziness, cannot be handled by SP but by fuzzy programming (FZ). FZ can deal with decision problems under fuzzy goal and constraints and ambiguous and vague coefficients not only in the objective function but also in the constraints [28,29]. Reference [30] used FZ method in order to optimize a multi-reservoir system, expressing the uncertainties in reservoir inflows as fuzzy sets. Reference [31] introduced a fuzzy compromise approach for water resources planning under imprecision uncertainty. Reference [32] proposed an interactive fuzzy approach for planning a stream water resources management system including vague and imprecise information. Reference [1] proposed a multistage fuzzy-stochastic programming model for water-resources allocation and management with uncertainties expressed as probability distributions and fuzzy sets.

Interval parameter programming (IPP) can handle uncertain parameters expressed as intervals with known lower- and upper-bounds, without any distributional information that is always required in fuzzy and stochastic programming [33]. However, in many real-world problems, the lower- and upper-bounds of some interval parameters can rarely be acquired as deterministic values [10,11]. Instead, they may often be provided as subjective information and therefore defined as fuzzy sets. This drives to dual uncertainties that cannot be addressed through the conventional IPP and FP methods. Hybrid approaches that link IPP with FP have been proposed for handling this combined type of complexities. However, these combined approaches have difficulties in tackling uncertainties expressed as random variables [34]. Additionally, a linkage to economic consequences of violated policies preregulated by authorities through taking recourse actions in order to correct any infeasibility is missing.

Therefore, in case of multiple uncertainties expressed at various and complex formats, one possible approach is to build hybrid modeling techniques combining IPP, FP and SP. In [12], a violation analysis approach has been developed for planning water resources management system associated with uncertain information, based on a fuzzy multistage stochastic integer programming model within a scenario-based frame. However, by using such a scenario-based approach, the resulting multistage programming model could become too large when all water-availability scenarios are considered. In [1], a multistage fuzzy-stochastic programming method has been developed dealing with uncertainties presented as fuzzy sets and probability distributions by employing vertex analysis and generating of a set of representative scenarios within a multistage context. The same problem due to the scenario-based approach is also here identified. Reference [13] developed a two-stage fuzzy-stochastic programming method for planning water resources allocation of agricultural irrigation systems. Also in this case a scenario-based approach sets limitations when the study system is very large and complicated. In reference [14], a method is developed for tackling multiple uncertainties through integration of stochastic dynamic programming, fuzzy-Markov chain, vertex analysis and factorial analysis techniques. It may have, though, computational (among others) difficulties to handle many other uncertain parameters (such as interval or dual-probabilities) that exist in large-scale practical problems. In [15], an optimal water allocation method is proposed incorporating techniques of interval-parameter programming and fuzzy vertex analysis within a fixed-mix stochastic programming framework to deal with uncertainties presented as probability distributions and dual intervals. In this study, only one reservoir is considered for all subareas and crops, in order to enable the use of linear programming method. Reference [16] presented an interval-fuzzy De Novo programming method for planning water-resources management systems under uncertainty, mainly useful for designing an optimal system rather than optimizing a given system.

In the present work, an optimal water allocation method under uncertain system conditions is explored for the Alfeios river basin in Greece. Alfeios is a water resources system of great natural, ecological, social and economic importance for Western Greece, since it has the longest watercourse and highest flow rate in the Peloponnese region [35,36,37]. The main water user in the basin is irrigation, playing a vital social, economic and environmental role associated among others with agricultural income and with water, food and energy efficiency. The other water-use sectors are hydropower generation and drinking water supply. As analyzed in [38], in Alfeios river basin, as in most Mediterranean countries, water resources management has been focused up to now on an essentially supply-driven approach. Authority responsibility relationships are fragmented and law enforcement and policy implementation are weak, facts that lead to the difficulty of gathering the necessary data for water resources management or even worst to data loss. In some cases, river monitoring, which is crucial for water quantity and quality assessment, if present, is either inefficient with intermittent periods with no measurements, or due to low financial means the monitoring programs are short and undertaken by a small number of personnel leading to unreliable or/and short-term data. In this case some sources of obtaining hydrologic, technical, economic and environmental data required for water resources management is by making additional periodic measuring expeditions, indirectly by expert knowledge, by informal knowledge by the local population, or by more general data concerning a wider geographical location (*i.e.*, country level) from national, European or international databases. Data of this type with a high degree of uncertainty may be easily defined as fluctuation ranges, and therefore, simulated as intervals with lower- and upper-bounds, either as deterministic or fuzzy, without the need of any distributional or probabilistic information.

This work is the second part of two papers, both aiming at analyzing and applying two similar in terms of their basic concepts optimization techniques for optimal water allocation under uncertain system conditions in a real and complex multi-tributary and multi-period water resources system, the Alfeios river basin. The first method is an inexact two-stage stochastic programming (ITSP) as developed by [6]. The second methodology, described and discussed in this paper, extends the ITSP in order to take into account fuzzy instead of deterministic boundaries for the variables, which are expressed as intervals, since some intervals are fuzzy in nature. This fuzzy-boundary interval-stochastic programming (FBISP) method proposed by [2] is selected. This algorithmic process is advanced including two different solution methods in order to take into account different risk attitudes of decision makers concerning system uncertainties. In [2], the uncertain random information of the water inflow is modeled through a multi-layer scenario tree having the limitation of resulting in too large mathematical problem to be applied to large-scale real-world problems. Additionally, this approach is not capable to incorporate the persistence in hydrological records and to take into consideration conditional probabilities for quantifying water availability, which are important in many real-world cases. In this work, in order to overcome these difficulties, the system dynamics related to random water inflows are reflected through the consideration of the various equal-probability hydrologic scenarios that have been stochastically generated simultaneously at multiple sites of the river basin. A thorough description of this proposed change in the methodology of [2] is provided in Section 2.1.

The results obtained from this methodology include (a) the optimized water-allocation target with a minimized risk of economic penalty from shortages and opportunity loss from spills; and (b) an optimized water-allocation plan (identification of water allocation and shortages based on the optimized water allocation targets) with a maximized system benefit over a multi-period planning horizon. These types of results are derived as fuzzy-boundary intervals. The total net system benefits and the benefits and penalties of each main water uses for the Alfeios are studied and analyzed based on the application of the FBISP method for a baseline scenario and four water and agricultural future scenarios developed within the Sustainability of European Irrigated Agriculture under Water Framework Directive and Agenda 2000 (WADI) project [39,40,41,42]. These future scenarios cover various possible technical, environmental and socio-economics aspects of the future space for different European Union (EU) water and agricultural policies, having an impact mainly on agriculture and also on water resources management. Changes on crop patterns, yields, subsidies, farmer income, variable input costs, market prices per agricultural product, fertilizers, and water and hydropower prices are some of the variables described into the narratives of these scenarios, which in turn serve as inputs into the optimization algorithm for the evaluation and the estimation of their effect on the water allocation pattern and the system benefits. Finally, for applying the above mentioned optimal water allocation methodology, benefit analysis of each water use identifying the unit benefit and unit penalties of each m^3 of water allocated to each one of the water uses has been undertaken for the Alfeios River in [38].

The paper is organized as follows: Section 2 includes the presentation of the mathematical background of the FBISP as provided in [2]. In Section 2.1, the limitations of the applied methodology and the proposed by this study changes are analyzed in details. Section 3 describes shortly the Alfeios river basin and the formulation of optimization problem for the Alfeios hydrosystem. The steps of this process and the used software programs are presented schematically in the form of a flow chart in Section 3.2. Section 4 analyzes in brief the WADI future scenarios. Section 5 presents the results and their interpretation and finally, Section 6 discussion and conclusions.

MATHEMATICAL FORMULATION OF THE FBISP METHOD

In the present work, a FBISP methodology as developed by [2,43,44] is employed for optimizing water allocation under uncertain system conditions in the Alfeios river basin in Greece. The used methodology is based on the combination of three optimization techniques: (a) the multistage-stochastic programming; (b) the fuzzy programming (employing the vertex analysis for fuzzy sets) and (c) the interval parameter programming. Each technique has a unique contribution in enhancing the model's capability of incorporating uncertainty presented as multiple formats. Its theoretical and mathematical background of the model and its parameters is presented below based on [2], but for simplicity reason the terms, referring to the variables with negative coefficients, are not included into all equations and inequalities, since they are absent from the examined application to Alfeios river basin. The complete mathematical model including also these terms with negative coefficients can be found in [2].

In the FBISP model, assume that there is no intersection between the fuzzy sets at the two bounds (e.g., let $\widetilde{b_{rt}^{\pm}} = [\widetilde{b_{rt}^{-}}, \widetilde{b_{rt}^{+}}] = \left|[b_{rt}^{-}, \overline{b_{rt}^{-}}], [b_{rt}^{+}, \overline{b_{rt}^{+}}]\right]$, where $\widetilde{b_{rt}^{-}}$ and $\widetilde{b_{rt}^{+}}$ are fuzzy lower- and upper-bounds of $\widetilde{b_{rt}^{\pm}}$, b_{rt}^{-} and $\overline{b_{rt}^{-}}$ are the lower- and upper-boundary of $\widetilde{b_{rt}^{-}}$; b_{rt}^{+} *and* $\overline{b_{rt}^{+}}$ are the lower-and upper-boundary of b+rt˜). This is due to satisfy the definition of an interval value that its lower-bound should not be larger than its upper-bound [5]. Secondly, interval numbers are used to express uncertainties without distribution information. If the fuzzy sets of an interval's lower- and upper- bounds intersect, then the so-called "interval" is actually described by fuzzy membership functions, such that the interval representation becomes unnecessary [45]. Thirdly, if the fuzzy sets of an interval's lower- and upper-bound intersect, the interactive algorithm for solving the interval-parameter programming problem cannot be used for solving such a FBISP model. Then, two solution methods are proposed for solving the FBISP model, which are based on an optimistic and pessimistic approach of the uncertainties by the decision makers, respectively. In the first solution methods (*i.e.*, optimistic or risk-prone), a set of submodels corresponding to f+~ can be first formulated based on the interactive algorithm; for each f+~ submodel, take one end point from each of the fuzzy intervals (*i.e.*, $[\underline{c_{jt}^{+}}, \overline{c_{jt}^{+}}]$, $[\underline{d_{jtk}^{-}}, \overline{d_{jtk}^{-}}]$ and $[\underline{b_{rt}^{+}}, \overline{b_{rt}^{+}}]$); then, the obtained end points can be combined into an *n*-array, leading to 2n combinations for n fuzzy sets [9,46]. Through solving 2n problems, a set of upper-bound objective-function values $(f_1^{+}, f_2^{+}, \ldots, f_{2^n}^{+})$ can be obtained. In detail, for each α-cut level, a set of $\widetilde{f^{+}}$ submodels can be formulated as follows (assume that the right-hand sides and objective are both greater than zero):

$$Max\ \widetilde{f^{+}} = \sum_{t=1}^{T}\left\{\sum_{j=1}^{j_1}\left[\left[\underline{c_{jt}^{+}}, \overline{c_{jt}^{+}}\right] + c_{jt}^{+}\right] x_{jt}^{+}\right\} - \sum_{t=1}^{T}\sum_{k=1}^{K_r} p_{tk}\left\{\sum_{j=1}^{j_2}\left[\left[\underline{d_{jtk}^{-}}, \overline{d_{jtk}^{-}}\right] + d_{jtk}^{-}\right] y_{jtk}^{-}\right\} \quad (1)$$

subject to:

$$\sum_{j=1}^{j_1}\left[\left|\underline{a_{rjt}^{-}}\right| Sign\left(\underline{a_{rjt}^{-}}\right), \left|\overline{a_{rjt}^{-}}\right| Sign\left(\overline{a_{rjt}^{-}}\right)\right] x_{jt}^{+} \le [\underline{b_{rt}^{+}}, \overline{b_{rt}^{+}}], \forall r, t \quad (2)$$

$$\sum_{j=1}^{j_1} |a_{ijt}^{-}| Sign(a_{ijt}^{-}) x_{jt}^{+} + \sum_{j=1}^{j_2}\left[\left|\underline{a'^{+}_{ijtk}}\right| Sign\left(\underline{a'^{+}_{ijtk}}\right), \left|\overline{a'^{+}_{ijtk}}\right| Sign\left(\overline{a'^{+}_{ijtk}}\right)\right] y_{jtk}^{-} \ge \left[\underline{w_{itk}^{-}}, \overline{w_{itk}^{-}}\right]$$

$$\forall t, k = 1, 2, \ldots, K_t$$

$$x_{jt}^{+}, y_{jtk}^{-} \ge 0 \quad (3)$$

where x_{jt}^{+} $(j = 1, 2, \ldots, j_1)$ are upper-bounds of the first-stage decision variables $(x_{jt}^{\pm})$, and y_{jtk}^{-} $(k = 1, 2, \ldots, K_t$ and $j = 1, 2, \ldots, j_2)$ are lower-bounds of the recourse decision variables $(y_{jtk}^{\pm})$.

Through solving 2^n submodels, a set of values $(f_1^{+}, f_2^{+}, \ldots, f_{2^n}^{+})$ can be obtained. Let f_{opt}^{+} be the minimum value of the upper-bound (for the objective-function value) with $\underline{f_{opt}^{+}} = min(f_1^{+}, f_2^{+}, \ldots, f_{2^n}^{+})$, and $\overline{f_{opt}^{+}}$ be the maximum value of the upper-bound with $\overline{f_{opt}^{+}} = min(f_1^{+}, f_2^{+}, \ldots, f_{2^n}^{+})$. Then, the optimized upper-bound interval for the objective function value (under an α-cut level) can be identified as follows:

$$\left[\underline{f_{opt}^{+}}, \overline{f_{opt}^{+}}\right]^{a} = [min(f_1^{+}, f_2^{+}, \ldots, f_{2^n}^{+}), max(f_1^{+}, f_2^{+}, \ldots, f_{2^n}^{+})]^{a} \quad (4)$$

Based on the solutions from the first set of submodels, submodels corresponding to $\widetilde{f^{-}}$ can be formulated as:

$$Max\ \widetilde{f^{-}} = \sum_{t=1}^{T}\left\{\sum_{j=1}^{j_1}\left[\left[\underline{c_{jt}^{-}}, \overline{c_{jt}^{-}}\right] + c'^{-}_{jt}\right] x_{jt}^{-}\right\} - \sum_{t=1}^{T}\sum_{k=1}^{K_r} p_{tk}\left\{\sum_{j=1}^{j_2}\left[\left[\underline{d_{jtk}^{+}}, \overline{d_{jtk}^{+}}\right] + d'^{+}_{jtk}\right] y_{jtk}^{+}\right\} \quad (5)$$

subject to:

$$\sum_{j=1}^{j_r}\left[\left|\underline{a_{rjt}^{\pm}}\right| Sign\left(\underline{a_{rjt}^{\pm}}\right), \left|\overline{a_{rjt}^{\pm}}\right| Sign\left(\overline{a_{rjt}^{\pm}}\right)\right] x_{jt}^{-} \leq [\underline{b_{rt}^{\pm}}, \overline{b_{rt}^{\pm}}], \forall r,t$$

$$\sum_{j=1}^{j_1}|a_{ijt}^{\pm}|Sign(a_{ijt}^{\pm})x_{jt}^{-} + \sum_{j=1}^{j2}\left[\left|\underline{a'^{\pm}_{ijtk}}\right| Sign\left(\underline{a'^{\pm}_{ijtk}}\right), \left|\overline{a'^{\pm}_{ijtk}}\right| Sign\left(\overline{a'^{\pm}_{ijtk}}\right)\right] y_{jtk}^{+} \geq \left[\underline{w_{itk}^{\pm}}, \overline{w_{itk}^{\pm}}\right], \forall i,t; k = 1,2,\ldots,K_t \quad (6)$$

$$0 \leq x_{jt}^{-} \leq X_{jtopt}^{+}, \forall t; j = 1,2,\ldots,j_{t1}$$

$$y_{jtk}^{+} \geq y_{jtkopt}^{-}, \forall t; j = 1,2,\ldots,j_2, k = 1,2,\ldots,K_t \quad (7)$$

Where

$$x_{jtopt}^{+}\ (j = 1,2,\ldots,j_1),\ x_{jtopt}^{-}\ (j = j_1+1, j_1+2,\ldots,j_1+n),\ y_{jtkopt}^{-}\ (j = 1,2,\ldots,j_2)$$

and

$$y_{jtkopt}^{+}\ (j = j_2+1,\ j_2+2,\ldots,j_2+n_2)$$

are solutions corresponding to $\underline{f_{opt}^{+}}$. Through solving 2ⁿ deterministic problems, a set of values $(f_1^{-}, f_2^{-}, \ldots, f_{2^n}^{-})$ can be obtained. The optimized lower-bound interval for the objective-function value (under an α-cut level) can be identified as follows:

$$\left[\underline{f_{opt}^{-}}, \overline{f_{opt}^{-}}\right]^a = [min(f_1^{-}, f_2^{-}, \ldots, f_{2^n}^{-}), max(f_1^{-}, f_2^{-}, \ldots, f_{2n}^{-})]^a \quad (8)$$

where $\underline{f_{opt}^{-}}$ is the minimum value of the lower-bound (for the objective function value) with $\underline{f_{opt}^{-}} = min(f_1^{-}, f_2^{-}, \ldots, f_{2^n}^{-})$; $\overline{f_{opt}^{-}}$ is the maximum value of the lower-bound with $\overline{f_{opt}^{-}} = \max(f_1^{-}, f_2^{-}, \ldots, f_{2n}^{-})$.

Then, through integrating the computational results of the two sets of submodels, the solution for the objective function value (under an α-cut level) can be obtained. Iteratively, the computational process can be repeated with the other α-cut levels.

The above optimistic solution method identifies the solutions for the first-stage and recourse decisions variables by first solving the best-case submodel (*i.e.*, upper-bound objective function value when the problem is to be maximized). This includes the upper-bound system solution for total benefit, which is associated with more advantageous (more favorable) conditions. In example, this is related to the upper-bound benefit coefficients, lower-bound cost coefficients, upper-bound reservoir capacities, lower-bound reserved storage requirements, *etc*. The resulting solution can provide intervals for the objective function value and decision variables, and can be easily interpreted for generating decision alternatives.

However, this solution method may provide a wide-ranging objective function value because significant (and costly) first-stage and recourse decisions are required under unfavorable conditions (represented by worst-case parameter values) [47]. Consequently, another solution method based on risk adverse is proposed for solving the FBISP model to reduce the interval width of the objective-function value, in which the worst-case submodel (*i.e.*, corresponding to the lower-bound objective function) can be first solved to identify a more appropriate set of first-stage and recourse decision variables. In this case, the interval for the objective function value is narrower, but it may lead to increased opportunity loss, being incapable of achieving the highest benefit under advantageous conditions. Thus, we have:

$$Max\ \tilde{f}^{-} = \sum_{t=1}^{T}\left\{\sum_{j=1}^{j_1}\left[\left[\underline{c_{jt}^{-}}, \overline{c_{jt}^{-}}\right] + c'^{-}_{jt}\right] x_{jt}^{-}\right\} - \sum_{t=1}^{T}\sum_{k=1}^{K_r} p_{tk}\left\{\sum_{j=1}^{j_2}\left[\left[\underline{d_{jtk}^{+}}, \overline{d_{jtk}^{+}}\right] + d'^{+}_{jtk}\right] y_{jtk}^{+}\right\} \quad (9)$$

subject to:

$$\sum_{j=1}^{j_r}\left[\left|\underline{a^{+}_{rjt}}\right| Sign\left(\underline{a^{+}_{rjt}}\right), \left|\overline{a^{+}_{rjt}}\right| Sign\left(\overline{a^{+}_{rjt}}\right)\right] x^{-}_{jt} \le [\underline{b^{-}_{rt}}, \overline{b^{-}_{rt}}], \forall r,t \tag{10}$$

$$\sum_{j=1}^{j_1}|a^{+}_{ijt}|Sign(a^{+}_{ijt})x^{-}_{jt} + \sum_{j=1}^{j2}\left[\left|\underline{a'^{-}_{ijtk}}\right| Sign\left(\underline{a'^{-}_{ijtk}}\right), \left|\overline{a'^{-}_{ijtk}}\right| Sign\left(\overline{a'^{-}_{ijtk}}\right)\right] y^{+}_{jtk} \ge \left[\underline{w^{+}_{itk}}, \overline{w^{+}_{itk}}\right], \forall i,t; k = 1,2,\ldots,K_t \tag{11}$$

Through solving 2^n deterministic problems, a set of values $(f^{-}_{1}, f^{-}_{2}, \ldots, f^{-}_{2n})$ can be obtained. The optimized lower-bound interval for the objective-function value (under an α-cut level) can be identified as follows:

$$\left[\underline{f^{-}_{opt}}, \overline{f^{-}_{opt}}\right]^{a} = [min(f^{-}_{1}, f^{-}_{2}, \ldots, f^{-}_{2^n}), max(f^{-}_{1}, f^{-}_{2}, \ldots, f^{-}_{2n})]^{a} \tag{12}$$

In the second solution method, characterized as risk adverse and corresponding to the worst-case solution, the previously described process is reversed so that the lower-bound submodels are solved first, and their solution is integrated into the upper-bound formulation of the problem, which is solved in the second step. Based on the solutions from the first set of submodels, submodels corresponding to $\widetilde{f^{+}}$ can be formulated as:

$$Max\ \widetilde{f^{+}} = \sum_{t=1}^{T}\left\{\sum_{j=1}^{j_1}\left[\left[\underline{c^{+}_{jt}}, \overline{c^{+}_{jt}}\right] + c^{+}_{jt}\right] x^{+}_{jt}\right\} - \sum_{t=1}^{T}\sum_{k=1}^{K_r} p_{tk}\left\{\sum_{j=1}^{j_2}\left[\left[\underline{d^{-}_{jtk}}, \overline{d^{-}_{jtk}}\right] + d^{-}_{jtk}\right] y^{-}_{jtk}\right\} \tag{13}$$

subject to:

$$\sum_{j=1}^{j_1}\left[\left|\underline{a^{-}_{rjt}}\right| Sign\left(\underline{a^{-}_{rjt}}\right), \left|\overline{a^{-}_{rjt}}\right| Sign\left(\overline{a^{-}_{rjt}}\right)\right] x^{+}_{jt} \le [\underline{b^{+}_{rt}}, \overline{b^{+}_{rt}}], \forall r,t \tag{14}$$

$$\sum_{j=1}^{j_1}|a^{-}_{ijt}|Sign(a^{-}_{ijt})x^{+}_{jt} + \sum_{j=1}^{j_2}\left[\left|\underline{a'^{+}_{ijtk}}\right| Sign\left(\underline{a'^{+}_{ijtk}}\right), \left|\overline{a'^{+}_{ijtk}}\right| Sign\left(\overline{a'^{+}_{ijtk}}\right)\right] y^{-}_{jtk} \ge \left[\underline{w^{-}_{itk}}, \overline{w^{-}_{itk}}\right], \forall t, k = 1,2,\ldots,K_t$$

$$x^{+}_{jt}, y^{-}_{jtk} \ge 0 \tag{15}$$

$$x^{+}_{jt} \ge x^{-}_{jtopt}, \forall t; j = 1,2,\ldots,j_1$$

$$0 \le y^{-}_{jtk} \le y^{+}_{jtkopt}, \forall t; j = 1,2,\ldots,j_2, k = 1,2,\ldots,K_t$$

Where $x^{-}_{jtopt}\ (j = 1,2,\ldots,j_1), y^{+}_{jtkopt}\ (j = 1,2,\ldots,j_2)$ are solutions corresponding f^{-}_{opt}.

Through solving 2^n submodels, a set of values $(f^{+}_{1}, f^{+}_{2}, \ldots, f^{+}_{2^n})$ can be obtained. Then, the optimized upper-bound interval for the objective function value (under an α-cut level) can be identified as follows:

$$\left[\underline{f^{+}_{opt}}, \overline{f^{+}_{opt}}\right]^{a} = [min(f^{+}_{1}, f^{+}_{2}, \ldots, f^{+}_{2^n}), max(f^{+}_{1}, f^{+}_{2}, \ldots, f^{+}_{2^n})]^{a} \tag{16}$$

Limitations of the Applied Methodology and Corresponding Changes

The main limitations of the chosen FBISP methodology as described in [2] are the following. The uncertain random information of the water inflow is modeled through a multi-layer scenario tree which is representative for the universe of water-availability conditions for the relative simple application of [2] including a hydro network with two tributaries and two reservoirs. With such a scenario-based approach, the resulting mathematical problem can become too large to be applied to large-scale real-world problems. The same problem has been mentioned among others in [12,13,14,15]. Moreover, the random variables (mainly the water inflows) are assumed to take on discrete distributions, such that the study can be solved through linear programming method. However, when water resources management problems are complicated by the need to take adequate account of persistence in hydrological records, conditional probabilities may need to be handled for quantifying water availability. This may lead to non-linearity in system responses and raise major problems for the linear assumption in the developed model. In order to understand the above-mentioned limitations, some theoretical information of the TSP are considered. The TSP problem with recourse, originating in [48,49], is generally nonlinear, and the set of feasible constraints is convex only under some particular distributions. However, the problem can be equivalently formulated as a linear programming model by assuming discrete distributions for the uncertain parameters [6,50]. Then, the random vector is assumed to have a finite number of possible realizations, called scenarios, *i.e.*, $s_1, s_2, \ldots, s_n$ with respective probability masses $p_1, p_2, \ldots, p_n$ with $p_i > 0$ and $\sum_{i=1}^{n} p_i = 1$. The expected value of the second-stage optimization problem can be written as the summation of the products of the values of each scenario with its probability mass. Based on this transformation, the TSP problem is expressed as a large linear programming problem forming the deterministic equivalent of the original problem. This approach has been further advanced and various techniques have been suggested in order to enable its efficient numerical solution [51,52].

The number of the constructed scenarios should be relatively modest so that the obtained deterministic equivalent can be solved with reasonable computational effort. Let's assume that *K*independent random component are contained in the optimization problem (*i.e.*, *K* water inflow sites) and each has three possible realizations (*i.e.*, low, medium and high), then the total number of scenarios is 3^K. The number of realizations/scenarios of the random variables (or in case of continuous distribution the number of discretization points) typically grows exponentially with the dimensionality of the variables and therefore, this number can quickly become prohibiting for the computational capacities of modern computers. As analyzed in [53], a common technique for reducing the number of scenario set to a manageable size is by using Monte Carlo simulation through generation of a sample $x_1, x_2, .., x_N$ of replications *N* of the random variable. Given a sample $x_1, x_2, .., x_N$ of replications *N*, the expectation function is approximated by the sample average. By the Law of Large Numbers this average value converges pointwise to the expected value as $N \rightarrow \infty$. This approach is called Sample Average Approximation (SAA) method. The SAA problem is a function of the considered sample and in that sense is random. For a given sample $x_1, x_2, .., x_N$, the SAA problem is of the same form as a two-stage stochastic linear programming problem with the scenarios $s_1, s_2, \ldots, s_n$ each taken with the same probability equal to $1/N$.

In this work, for the application of the FBISP methodology in the Alfeios River Basin a different approach for embodying the stochastic uncertainty of multiple water inflows has been adapted based on the Monte Carlo sampling and the SAA method in order to overcome the limitations associated with the scenario-based approach. The proposed modification is based on the generation of stochastic equal-probability hydrologic realizations/scenarios as thoroughly described in [38] using the stochastic software of CASTALIA [54,55,56]. CASTALIA is a system for the stochastic simulation and forecast of hydrologic variables, including (a) multivariate analysis (for many hydrologic processes, such as rain, temperature and discharge, and geographical correlated locations) and (b) multiple time scales (monthly and yearly) in a disaggregation framework. It enables the preservation of essential marginal statistics up to third order (skewness) and joint second order statistics (auto- and cross-correlations), and the reproduction of long-term persistence (Hurst phenomenon) and periodicity. More specifically, an original two-level multivariate scheme was introduced, appropriate for preserving the most important statistics of the historical time series and reproducing characteristics peculiarities of hydrologic processes, such as persistence, periodicity and skewness. At the first stage, the annual synthetic values are generated based on the alternative expression of the backward moving average algorithm [57] from [54] resulting to the symmetric moving average (SMA). This modified version extends the stochastic synthesis not only backward but also forward using the condition of symmetry for the corresponding backward and forward parameters ($a_j = a_{-j}$). This model reproduces the long-term persistence, and has been further generalized for application to simultaneous generation of stochastically dependent multiple variables. This is achieved by generating correlated (multivariable) white noise. At the second stage, the monthly synthetic values are generated posing emphasis on the reproduction of periodicity. A periodic first-order autoregression, abbreviated as PAR(1), model is used, which has been also generalized for multi-variable simulation. The final step is the coupling of the two time scales through a linear disaggregation model [55]. A brief description of the process for the generation of fifty short-time equal-probability scenarios simultaneously for the monthly rain and temperature variables and the corresponding hydrologic simulation for the computation of the discharges at the main four subcatchments of the Alfeios river basin is included in Section 3.2 and the detailed description is given in [38].

It is worth mentioning that the examined water resources management problem in [2] includes a relative simple hydronetwork with two tributaries with two reservoirs and three stages, ending up with a scenario-tree composed of 258 scenarios. In Alfeios River Basin, the simplified schematization of the river network, as presented in Section 3, includes 5 streams and therefore, by considering one year with monthly time step (12 stages), this results in a much more complex scenario-tree (*i.e.*, taking into account only 6 stages of the 12 stages (6 months): 2.8×10^{11} scenarios).

FORMULATION OF OPTIMIZATION PROBLEM FOR THE ALFEIOS RIVER BASIN

Description of the Alfeios River Basin

The Alfeios river basin has been extensively described in the past [35,36,37]. This section describes briefly the information for the Alfeios required for the application of the FBISP method, since the thorough analysis of the hydrosystem is included in [38].

The Alfeios River is a water resources system of great natural, ecological, social and economic importance for Western Greece, since it is the longest watercourse (with a length of 112 km) and has the highest flow-rate (absolute maximum and minimum values recorded 2380 and 13 m^3/s) in the Peloponnese region in Greece. The main river and its six tributaries represent a significant source of water supply for the region, aiming at delivering and satisfying the expected demands from a variety of water users, including irrigation, drinking water supply, hydropower production and to a smaller extend recreation. A plethora of environmental stresses were exerted on its river basin the last decades. It drains an area of 3658 km^2. Its annual water yield is estimated to be 2100 $\times 10^6$ m^3. The main water uses in the basin include: (a) the hydropower production at Ladhon hydropower station (HPS) linked with the Ladhon Dam and reservoir situated in the middle mountainous Alfeios; (b) the agricultural demand of the Flokas scheme linked with the diversion Flokas Dam situated almost 20 km before the discharge of Alfeios into the Kyparissiakos Gulf and very close to Ancient Olympia; (c) the hydropower production at the small HPS at Flokas Dam; and (d) the drinking water supply to the Region of Pyrgos and the neighboring communities from Alfeios tributary, Erymanthos.

Beginning with the Ladhon river which is the most important tributary of Alfeios river basin draining almost one third of its total basin area, the operation goals of Ladhon reservoir are specified firstly, as the satisfaction of irrigation demand at the Flokas irrigation scheme, and secondly, as the hydropower production at HPS Ladhon, situated at around 8 km downstream the Ladhon Dam. The general operation rule is to start at the beginning of January with a minimum level of the Ladhon lake and then filling up the lake till May, in order to satisfy mainly irrigation demand. For the application of the FBISP method the upper-and lower-bounds of the optimized hydropower production target $T^{\pm}$ (in MWh) at Ladhon (Table 1) are required. These bounds are approximated from the statistical analysis of the monthly time series of hydropower production at Ladhon from 1985 to 2011. The ranges between the mean value of the historical timeseries minus its standard deviation (lower-bound) and its mean value plus its standard deviation (upper-bound) are considered as analyzed in [38].

Proceeding to the Flokas irrigation region, the irrigation scheme is connected to the diversion Flokas Dam, draining an area of 3600 km^2. The irrigation water demand extends officially from May to September, whereas in most years could be further extended from April up to October due to dry climatic conditions. The present monthly irrigation water demand is composed of two parts: (a) a regulated and stable irrigation demand pattern, referring only to the required water volume releases from Ladhon Reservoir, which is derived from the official agreement between Hellenic Public Power Corporation and General Irrigation Organization for the Flokas Irrigation Area; and (b) an extra uncertain irrigation demand at Flokas Dam site based on the actual crop patterns and the water inflows at this position. The total irrigation requirements for the crop pattern of Flokas are estimated for each stochastic hydrologic scenario using the FAO software CROPWAT 8.0. For the application of the FBISP methodology the upper- and lower-bounds of the water allocation targets for irrigation in EUR/m^3 are required. The optimized water allocation target for irrigation (Table 2) is explored, assuming that the irrigation demand can vary between the maximum demand of the present crop pattern and the maximum demand given in the study of the small HPS at Flokas. Based on this assumption, the lower-bound of the optimized water allocation target is set equal to the maximum of all sets of irrigation water requirements for the fifty hydrologic scenarios as computed by CROPWAT for the present irrigated area and crop pattern.

The small Flokas HPS is situated directly after the diversion of water from the Flokas Dam, and is operated automatically based on the upstream water level. In this way when the river flow rate is between 9 and 90 m^3/s, the entire part of the river flow passes through the Flokas HPS, maintaining the water level of Dam at a stable level. When river flow rate exceeds 90 m^3/s then the surplus flows over the spillways of Flokas Dam. Whereas for flood water volumes exceeding 300 m^3/s the HPS Flokas closes for security reasons and the flood volume passes through the spillways of the dam and the opened gateways. For the application of the FBISP method the upper- and lower-bounds of the optimized hydropower production target $T\pm$ (in MWh) at Flokas small HPS (Table 3) are required. These bounds are approximated also in this case, from the statistical analysis of the monthly timeseries of hydropower production at Flokas from 2011 to 2015. The ranges between the mean value of the historical timeseries minus its standard deviation (lower-bound) and its mean value plus its standard deviation (upper-bound) are taken into account as specified in [38].

Finally, a monthly water flow rate of 0.6 m^3/s for the drinking water supply system for the north and central part of the Region of Hleias is diverted from Erymanthos to the water treatment plant and then to the neighboring communities extending up to the city of Pyrgos. Due to the short operation period (starting in 2013), this water use is not incorporated in the optimization process as a variable but as a steady and known water abstraction demand.

The schematization of the Alfeios river network is depicted as shown in Figure 1, including the main five water inflow locations, where historical timeseries (rain, temperature and river discharge) are available and the main water users as described above.

Table 1: Upper- (THydroLadhon$^+$) and lower- (THydroLadhon$^-$) bounds of optimized target for hydropower production at HPS at Ladhon

Bounds of	Target Limits for Hydropower Production at Ladhon HPS (MWh)												
Optimized Target	January	February	March	April	May	June	July	August	September	October	November	December	Annual
THydroLadhon$^-$	11,857	12,553	11,810	11,046	11,081	8965	9077	7613	5925	7387	9427	8540	115,282
THydroLadhon$^+$	37,353	38,947	48,311	35,391	23,237	15,868	15,598	14,233	13,642	17,062	17,971	24,276	301,890

Table 2: Upper- and lower-water allocation targets for irrigation in EUR/m^3

Time Stages	Irrigation Water Demand (m³/s)	
	Lower-Bound of Optimized Allocation Target	Upper-Bound of Optimized Allocation Target
	Tirrigation⁻	Tirrigation⁺
t = 1—January	0	0
t = 2—February	0	0
t = 3—March	0	6
t = 4—April	2.0	6
t = 5—May	5.0	6
t = 6—June	8.9	12
t = 7—July	11.5	12
t = 8—August	9.2	12
t = 9—September	2.7	6
t = 10—October	1.2	6
t = 11—November	0	0
t = 12—December	0	0
Annual (m³)	108,756,934	174,700,800

Table 3: Upper- (THydroFlokas⁺) and lower- (THydroFlokas⁻) bounds of optimized target for hydropower production at HPS at Flokas

Bounds of	Target Limits for Hydropower Production at Flokas HPS (MWh)												
Optimized Target	January	February	March	April	May	June	July	August	September	October	November	December	Annual
THydroFlokas⁻	1244	1740	2450	2045	1574	437	219	218	232	395	299	1129	11,982
THydroFlokas⁺	2379	2894	3435	2840	1861	773	251	255	571	1111	1397	2097	19,865

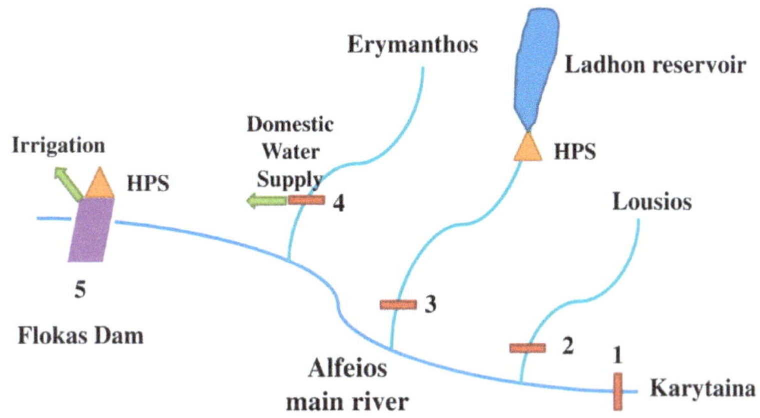

Figure 1: The simplified schematics of the Alfeios river basin

Optimization Problem of the Alfeios Hydrosystem

The goal of this optimization problem is to identify an optimal water allocation target with a maximized economic benefit over the planning period for the Alfeios river basin. Different water allocation targets are related not only to different policies for water resources management, but also to different economic implications (probabilistic penalty and opportunity loss). The penalty is associated with the nonproper water allocation/management, and therefore, resulting to shortages and spills for hydropower and for water shortage for irrigation. The optimization problem is structured as follows:

$$Max\ f^{\pm} = \sum_{i=1}^{3}\sum_{t=1}^{T} \widehat{NB_{ti}^{\pm}} T_{ti}^{\pm} - \sum_{i=1}^{3}\sum_{t=1}^{T}\sum_{k_1=1}^{K_1^t} p_{tk_1} PE_{ti}^{\pm} Y_{tk_1 i}^{\pm} - \sum_{t=1}^{T}\sum_{k_1=1}^{K_1^t} p_{tk_1} PE_{ti=1}^{\pm} SP_{tk_1 i=1}^{\pm} \quad (17)$$

subject to:

Constraints of water-mass balance for the Ladhon reservoir:

$$R_{tk_1}^{\pm} + SP_{tk_1}^{\pm} = S_{tk_1}^{\pm} + Q_{tk_1 j=1}^{\pm} - E_{1t}^{\pm} - S_{(t+1)k_1}^{\pm},\ \forall t;\ k_1 = 1, 2, \ldots, K_1^t \quad (18)$$

$$E_{1t}^{\pm} = e_{1t}^{\pm}\left(\frac{A_{tk_1}^{\pm} + A_{(t+1)k_1}^{\pm}}{2}\right) \quad (19)$$

$$A_{tk_1}^{\pm} = a_1 S_{tk_1}^{\pm} + b_1 \quad (20)$$

Constraints for the minimum and maximum water volume released by turbines at HPS Ladhon:

$$R_{tk_1}^{\pm} \leq TLU_1^{\pm}, \forall t;\ k_1 = 1, 2, \ldots, K_1^t \quad (21)$$

$$R_{tk_1}^{\pm} \geq TLL_1^{\pm}, \forall t;\ k_1 = 1, 2, \ldots, K_1^t \quad (22)$$

Constraints of Ladhon reservoir capacity:

$$S_{tk_1}^{\pm} \leq RSC_1^{\pm}, \forall t;\ k_1 = 1, 2, \ldots, K_1^t \quad (23)$$

$$S_{tk_1}^{\pm} \geq RSL_1^{\pm}, \forall t;\ k_1 = 1, 2, \ldots, K_1^t \quad (24)$$

Constraints for target of hydropower production at Ladhon:

$$T_{t\,i=1}^{\pm} - Y_{tk_1\,i=1}^{\pm} = HP_{tk_1\,i=1}^{\pm}, \forall t;\ k_1 = 1, 2, \ldots, K_1^t \quad (25)$$

$$HP_{tk_1\,i=1}^{\pm} = d_1 R_{tk_1}^{\pm} + e_1, \forall t;\ k_1 = 1, 2, \ldots, K_1^t \quad (26)$$

$$THP_{k_1\,i=1}^{\pm} = \sum_{t=1}^{12}\left(HP_{tk_1\,i=1}^{\pm}\right), \forall k_1 = 1, 2, \ldots, K_1^t \quad (27)$$

Constraints of water-mass balance for the Flokas Dam:

Water availability at Flokas dam based on the upstream water inflows:

$$Q_{tk_1 5}^{\pm} = \sum_{j=2}^{4} Q_{tk_1 j}^{\pm}\, c_j + R_{tk_1}^{\pm} + SP_{tk_1}^{\pm}, \forall t;\ k_1 = 1, 2, \ldots, K_1^t \quad (28)$$

Water-mass balance at Flokas Dam:

$$Q_{tk_1 5}^{\pm} = T_{t\,i=3}^{\pm} - WIB_{tk_1}^{\pm} + WIO_{tk_1}^{\pm} + WFL_{tk_1}^{\pm}, \forall t;\ k_1 = 1, 2, \ldots, K_1^t \quad (29)$$

Water-mass balance at the small HPS at Flokas:

$$WIO_{tk_1}^{\pm} = WVF_{tk_1}^{\pm} + WVFO_{tk_1}^{\pm} +, \forall t;\ k_1 = 1, 2, \ldots, K_1^t \quad (30)$$

Constraints for target of hydropower production at Flokas:

$$T_{t\,i=2}^{\pm} - Y_{tk_1\,i=2}^{\pm} = HP_{tk_1\,i=2}^{\pm}, \forall t;\ k_1 = 1, 2, \ldots, K_1^t \quad (31)$$

$$HP_{tk_1\,i=2}^{\pm} = f_1 R_{tk_1}^{\pm} + g_1, \forall t;\ k_1 = 1, 2, \ldots, K_1^t \quad (32)$$

$$THP_{k_1\,i=2}^{\pm} = \sum_{t=1}^{12}\left(HP_{tk_1\,i=2}^{\pm}\right), \forall k_1 = 1, 2, \ldots, K_1^t \quad (33)$$

Constraints for the minimum and maximum water volume released by turbines at HPS at Flokas:

$$WVF_{tk_1}^{\pm} \leq TFU_1^{\pm}, \forall t;\ k_1 = 1, 2, \ldots, K_1^t \quad (34)$$

$$WVF_{tk_1}^{\pm} \geq TFL_1^{\pm}, \forall t;\ k_1 = 1, 2, \ldots, K_1^t \quad (35)$$

$$Y_{tk_1\,i=3}^{\pm} = WIB_{tk_1}^{\pm}, \forall t;\ k_1 = 1, 2, \ldots, K_1^t \quad (36)$$

Constraints of upper- and lower-bounds for allocation targets:

$$\widetilde{T_{ti}^{min}} \leq T_i^{\pm} \leq \widetilde{T_{ti}^{max}, \forall t} \text{ and } i \quad (37)$$

Non-negative and technical constrains

$$T_{ti}^{\pm} \geq Y_{tk_1\,i}^{\pm}, \forall t;\ k_1 = 1, 2, \ldots, K_1^t;\ i = 1 \ldots 3 \quad (38)$$

where

$f^{\pm}$ = net system benefit over the planning horizon (EUR);

t = time period, and $t = 1, 2, \ldots, T$ ($T = 12$);

A_{tk1} = surface area of Ladhon reservoir in period t under scenarios k_1 (m^2);

a_1 = slope coefficient from linear regression between surface area of Ladhon reservoir and storage volume;

b_1 = intercept coefficient from linear regression between surface area of Ladhon reservoir and storage volume;

$HP_{tk_1\,i}^{\pm}$ = monthly hydropower production for $t = 1, 2, \ldots, T$; in period t under scenarios k_1 for the water user i with $i = 1, 2$ corresponding to Ladhon and Flokas, respectively;

$THP_{k_1\,i}^{\pm}$ = annual hydropower production in period t under scenarios k_1 for the water user i with $i = 1, 2$ corresponding to Ladhon and Flokas, respectively;

d_1 = slope coefficient from linear regression between hydropower production of Ladhon reservoir and water volume released through the turbines;

e_1 = slope coefficient from linear regression between hydropower production of Ladhon reservoir and water volume released through the turbines;

c_j = area-based conversion factor multiplied with each stream discharge $Q_{tk_1 j}^{\pm}$ to add the contribution of water inflows from intermittent drainage areas;

f_1 = slope coefficient from linear regression between hydropower production of Flokas and water volume released through the turbines;

g_1 = slope coefficient from linear regression between hydropower production of Flokas HPS and water volume released through the turbines;

$\widehat{T^{min}_{t_i}}$ = lower-bound of the optimized target for the water user i in period t ((m^3) for irrigation and (MWh) for hydropower;

$\widehat{T^{max}_{t_i}}$ = upper-bound of the optimized target for the water user i in period t ((m^3) for irrigation and (MWh) for hydropower);

$e^{\pm}_{1t}$ = average evaporation rate for Ladhon reservoir in period t (m);

$E^{\pm}_{1t}$ = evaporation loss of Ladhon reservoir in period t (m^3);

K^t_1 = number of flow scenarios in period t;

$NB^{\pm}_i$ = net benefit per unit of water allocated for each water user i—(EUR/m^3) for irrigation and (EUR/MWh) for hydropower;

$PE^{\pm}_i$ = penalty per unit of water not delivered (EUR/m^3) for each water user i—for irrigation and (EUR/MWh) for hydropower; and $PE_t > NB_t$;

p_{tk_1} = probability of occurrence of scenario k1 in period t, with $p_{tk_1} > 0$ and $\sum_{k_1=1}^{K^t_1} p_{tk_1} = 1$;

$Q^{\pm}_{tk_1 j}$ = water inflow level into stream j in period t under scenario k_1 (m^3);

$i = 1, 2, 3$ for the water users being hydropower production at Ladhon, hydropower production at Flokas and irrigation at Flokas;

$j = 1, 2, 3, 4, 5$ stream index for the river flows at Karytaina, Lousios, Ladhon, Erymanthos and Flokas;

$R^{\pm}_{tk_1}$ = release flow from the turbines of Ladhon reservoir in period t under scenario k_1 (m^3);

$SP^{\pm}_{tk_1}$ = spill volume over Ladhon Dam in period t under scenario k_1 (m^3);

$RSC^{\pm}_1$ = maximum storage capacity of Ladhon reservoir (m^3);

$RSL^{\pm}_1$ = minimum storage capacity of Ladhon reservoir (m^3);

$TLU^{\pm}_1$ = maximum capacity of turbines at Ladhon HPS (m^3);

$TLL^{\pm}_1$ = minimum capacity of turbines at Ladhon HPS (m^3);

$TFU^{\pm}_1$ = maximum capacity of turbines at Flokas HPS (m^3);

$TFL^{\pm}_1$ = minimum capacity of turbines at Flokas HPS (m^3);

$S^{\pm}_{tk_1}$ = storage level in Ladhon reservoir in period t under scenario k_1 (m^3);

$T^{\pm}_{ti}$ = water allocation target that is promised to the user i in period t (m^3);

$Y^{\pm}_{tk_1 i}$ = shortage level by which the water-allocation target is not met in period t under scenariosk1 for the water user i, which is associated with probability of p_{tk1}—(m^3) for irrigation and MWh for hydropower;

$WIB^{\pm}_{tk_1}$ = irrigation shortage volume in period t under scenarios k_1 (m^3);

$WIO^{\pm}_{tk_1}$ = water volume left at Flokas after having allocated the irrigation water in period $t$under scenarios k1 (m3);

$WFL^{\pm}_{tk_1}$ = water volume flowing through the fish ladder at Flokas dam in period t under scenarios k_1(m^3);

$WVF^{\pm}_{tk_1}$ = water volume flowing through the turbines at Flokas HPS in period t under scenarios k_1 (m^3);

$WVFO^{\pm}_{tk_1}$ = spill volume at Flokas dam in period t under scenarios k_1 (m^3);

CASTALIA model is applied for the simultaneous stochastic generation of fifty equal-probability scenarios for the hydrologic variables of rain and temperature at the four considered subcatchments (Karytaina, Lousios, Ladhon and Erymanthos) having a time length of ten years (since the future WADI water and agriculture scenarios are projected ten years after the baseline scenario) and monthly time step. The stochastically simulated rain and temperature timeseries are introduced into the calibrated simple lumped conceptual river basin model ZYGOS [58,59] for the four subcatchments in order to compute the mean monthly discharges for this ten years period. The uncertainty from the hydrologic model structure and the parametrization is taken into account through the computation of the standard error between the measured and the simulated water discharge timeseries. Based on this standard error, upper-bound water inflows timeseries for all the hydrologic scenarios (which are used in the f^+ model), and lower-bound (which are used in the f^-model) are created. The steps of this process and the software programs used are presented schematically in the form of a flow chart in Figure 2. The last year of each of the fifty stochastic monthly discharge scenarios (since the future scenarios refer to ten years after the baseline) serves as input inflows into the optimization model for the optimal water allocation of Alfeios river basin. The monthly discharge at Flokas Dam, which is of interest for the optimization process, since at this position the available water is diverted to the irrigation canal, is computed as the sum of the four subcatchments as described in [38].

In the optimization problem, there are some nonlinear equations, such as the relationship between water flowing through the turbines and the hydropower energy produced. In order to introduce them into the linear programming algorithm their linear regression equations are considered. The uncertainty resulting from this simplification has not been considered in the process, but it is worth mentioning that in all cases the R^2 takes values ≥ 0.9.

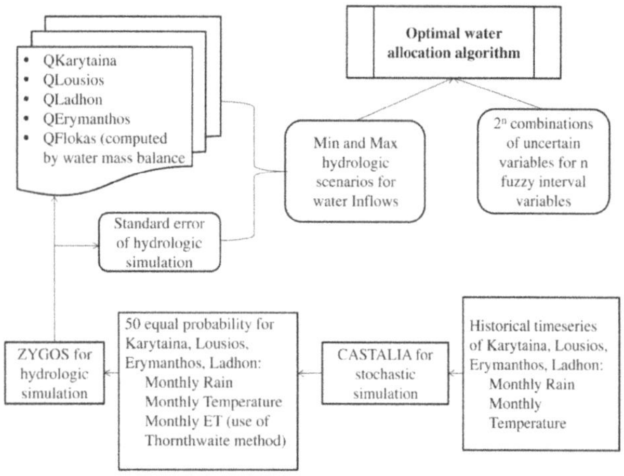

Figure 2: Methodological framework for optimal water allocation of Alfeios river basin.

The uncertain variables in this case are firstly, the coefficients of the objective function including (a) the unit benefits and penalties from hydropower production of Ladhon (EUR/MWh); (b) the unit benefits and penalties from the hydropower of Flokas (EUR/MWh); and (c) the unit benefits and penalties from Flokas irrigation (EUR/m^3), and secondly, the initial water level of Ladhon reservoir at stage zero (m^3) which is expressed as deterministic-boundary interval (12,362,644.01, 26,783,729.12). A detailed analysis of the estimation of the prementioned unit benefits and penalties is provided in [38]. Briefly, based on the estimations of the Chief Engineer at Ladhon hydropower station, the upper- and lower- fuzzy-boundary intervals for the unit benefit and penalty of Ladhon are defined as given in Table 4. The shadow penalty for the hydropower production at Ladhon is composed of two parts: (a) the penalty due to hydropower shortage in comparison to the hydropower production target and (b) the penalty for the water spilled from the Ladhon Dam (if any) and is not available for hydropower production, which intends to express the opportunity loss of hydropower energy production. Based on the monthly selling price data of Flokas HPS, its unit benefit is approximated as a single deterministic interval (and not as a fuzzy-boundary interval) as presented in Table 4. The unit shadow penalty is approximated as a single deterministic interval and is taken equal to the upper-bound solution of the unit penalty of Flokas HPS, as shown in Table 4. The unit benefit for water allocated to irrigation is interpreted as the probable net income from agricultural production of the Flokas crop pattern, taking into account the brutto farmer, the cost of production, the cost of the irrigation canal (associated to the water charge to the farmers from the general irrigation organization) and the organizational structure of local irrigation organizations (the charges of the local irrigation organizations). In Table 5, the lower- and upper- fuzzy-boundaries of the unit benefit of water allocated to the Flokas irrigation scheme is provided for the baseline scenario and the WADI future scenarios. Finally, the unit penalty from the irrigation water deficits is based on the crop yield reduction and the corresponding net farmer income loss. In Table 6 the lower- and upper- fuzzy-boundaries of the unit penalties for irrigation water deficits of the Flokas irrigation scheme is provided for the baseline scenario and the WADI future scenarios (where as presented in Section 4 the World agricultural markets scenario is denoted as Future Scenario 1 (FS1), the Global agricultural sustainability scenario as Future Scenario 2 (FS2), the Provincial agriculture scenario as (Future Scenario 3-FS3) and the Local community agriculture as (Future Scenario 4-FS4).

Table 4: Lower- and upper- fuzzy-boundary intervals for the unit benefit and unit penalty for hydropower production EUR/MWh at Ladhon and at Flokas

Variables	NBHP Ladhon EUR/MWh	NBHP Flokas EUR/MWh	CHP Ladhon EUR/MWh	CHP Flokas EUR/MWh
Lower-Bound—Minimum	40	87.5	120	140
Lower-Bound—Maximum	55	–	130	150
Upper-Bound—Minimum	60	80	140	140
Upper-Bound—Maximum	75	–	150	150

Table 5: Lower- and upper- fuzzy-boundary intervals for the unit benefit from irrigation for the baseline and the WADI future scenarios for Flokas irrigation scheme EUR/m^3

Fuzzy-Boundary Intervals		NBIrrigationFlokas EUR/m^3				
		Baseline	FS 1	FS 2	FS 3	FS 4
Upper-Bound	Min	0.166	0.127	0.189	0.191	0.221
	Max	0.175	0.136	0.265	0.276	0.294
Lower-Bound	Min	0.187	0.190	0.266	0.277	0.295
	Max	0.205	0.234	0.269	0.314	0.431

FS1: Future Scenario 1; FS2: Future Scenario 2; FS3: Future Scenario 3; FS4: Future Scenario 4.

Table 6: Lower- and upper- fuzzy-boundary intervals for the unit penalties for water deficits to irrigation EUR/m^3 for the baseline and the future scenarios

Fuzzy-Boundary Intervals		PEIrrigationFlokas EUR/m^3				
		Baseline	FS 1	FS 2	FS 3	FS 4
Upper-Bound	Min	0.989	0.748	1.052	1.035	1.043
	Max	1.051	1.159	1.075	1.073	1.070
Lower-Bound	Min	1.715	3.361	1.537	1.552	2.184
	Max	1.812	3.410	1.891	1.871	2.279

FS1: Future Scenario 1; FS2: Future Scenario 2; FS3: Future Scenario 3; FS4: Future Scenario 4.

WADI WATER AND AGRICULTURE FUTURE SCENARIOS

The Sustainability of European Irrigated Agriculture under Water Framework Directive and Agenda 2000 project (WADI) focuses on changes in EU agricultural and water policy as they affect the economic, social and environmental performance of irrigation in the partner countries [39]. Its aim was to investigate the impacts of policy change on the irrigation sector in Spain, Greece, Italy and the UK with a particular focus on Water Framework Directive (WFD) and the reform of Common Agricultural Policy (CAP). The agricultural policy regime in place in 2001, as determined by CAP at that time is taken as baseline scenario. This 2001 baseline year is used to provide a relative reference point for the definition of future scenarios. This scenario is extrapolated to 10 years after the reference year (thus from 2001 to 2010) based on predictions (rather than possibilities) of agricultural markets and prices from EU, the Organization for Economic Co-operation and Development (OCDE) and other sources. The estimates of the main parameters (Table 7), determined for each future scenario, are used within this paper as inputs in the developed optimal water allocation model based on [2] under uncertain and vague water system conditions in order to enable the assessment of the impact of the EU agricultural and water policy changes on the hydrosystem's benefits of the baseline scenario.

Table 7: Analysis of the Foresight scenarios based on the regional analysis in [39,40]—expressed as a percentage of the baseline year at constant values

Crop Prices	Baseline	World Agricultural Markets		Global Agricultural Sustainability		Provincial Agriculture		Local Community Agriculture	
Crops selling prices perceived by the farmers	–	Min	Max	Min	Max	Min	Max	Min	Max
Maize	100	85	95	95	105	100	110	100	110
Maize area subsidy	100	0	–	75	85	90	100	85	95
Set aside quota	100	0	–	95	–	100	–	105	–
Tomato	100	85	95	110	120	100	110	120	130
Potato	100	85	95	95	105	105	115	120	130
Watermelons	100	85	95	95	105	105	115	120	130
Cotton	100	80	90	90	100	85	95	110	120
Cotton subsidy	100	0	–	85	–	90	–	105	–
Olive Trees	100	80	90	85	95	90	100	100	110
Olive trees area subsidy	100	0	–	95	–	95	–	105	–
Alfalfa	100	80	90	90	100	100	110	110	120
Citrus	100	85	95	95	105	100	110	120	130
Input prices	–	Min	Max	Min	Max	Min	Max	Min	Max
Fertilizers	100	85	100	140	150	100	110	150	160
Pesticides	100	110	120	100	105	105	115	95	100
Energy	100	85	95	120	130	100	110	130	140
Seeds	100	100	110	110	120	120	130	130	140
Machinery	100	100	115	115	135	100	115	120	140
Contractor services	100	130	135	120	130	130	140	110	120
Water prices	100	100	110	115	130	100	110	120	140
Irrigation infrastructure	100	100	110	120	130	115	125	130	150
Labour	100	90	100	100	110	95	105	110	120
Land	100	110	120	110	125	100	110	85	95
Other inputs	100	85	95	125	135	85	95	130	140
Crop yield changes due to technology	100	110	120	100	115	100	105	85	105
Restriction on chemical use	100	130	140	120	130	110	120	100	110

To provide a better understanding of the content and the philosophy of each future scenario, the four WADI agricultural and water scenarios are briefly described based on [39,40]. *World Markets*scenario is related to private consumption and a highly developed and integrated world trading system. *Global Sustainability* scenario poses emphasis on social and ecological values associated with global institutions and trading systems. In comparison to the first scenario, slow but more equally distributed growth is considered. Active public policy and international co-operation within the European Union and at a global level are central. *Provincial Enterprise* scenario emphasizes private consumption within national and regional level to depict local priorities and interests. The dominance of market values is noticed within the national/regional boundaries.*Provincial agricultural markets* scenario is also characterized by protectionist regimes similar to that under pre-reform Common Agricultural Policy. People aspire to personal independence and material wealth within a nationally rooted cultural identity. *Local Stewardship* scenario is focused on strong local or regional governments with emphasis on social values, self-reliance, self-sufficiency, and conservation of natural resources and the environment. *Local community agricul-*

ture scenario emphasizes sustainability at a local level. In the following tables the *World agricultural markets*scenario is denoted as Future Scenario 1 (FS1), the *Global agricultural sustainability* scenario as Future Scenario 2 (FS2), the *Provincial agriculture* scenario as (Future Scenario 3-FS3) and the *Local community agriculture* as (Future Scenario 4-FS4).

RESULTS

In the proposed methodology, two types of solution methods, an optimistic (1st solution method, also called as risk-prone by [2]) and a pessimistic (2nd solution method, also called as risk-adverse by [2]), have been incorporated in order to compute the optimized water allocation targets under uncertain system conditions. The term of "risk", used to characterize these two different solution approaches, does not imply the measuring of risk with its strict mathematical definition, but the willingness of the decision makers to take the risk or not of economic penalty or of opportunity loss associated with their attitude concerning the uncertainties' values of the optimization problem as described below. According to [2], in general, the first solution method could identify the highest system benefit, but may be, however, associated with higher risk since it is based on an optimistic anticipation for the system components. The second solution method could assist to compute a narrower interval for the system benefit with a lower risk, since it is based on a conservative anticipation for system components and constraints, but the system might lose the opportunity of achieving the highest benefit value.

The solution in terms of the objective function value (meaning the total net benefit) results to a fuzzy-boundary interval for each α-cut level and each solution method $[(\underline{f_{opt}^-}), (\underline{f_{opt}^+}), (f_{opt}^-, f_{opt}^+)]^{\alpha}$ composed of four options of maximized system benefits in combination with minimized probabilistic penalties corresponding to different system conditions. These four options for each solution method correspond to lower-min $\underline{f_{opt}^-}$ (and in tables and figures written as min (f^-)), lower-max $\underline{f_{opt}^+}$ (and in tables and figures written as max (f^-)), upper-min $\overline{f_{opt}^-}$ (and in tables and figures written as min (f^+)), upper-max $\overline{f_{opt}^+}$ (and in tables and figures written as max (f^+)). These results (four prementioned options), however, do not necessarily construct a set of stable intervals [2].

For the Alfeios river basin considering five uncertain variables (as analyzed in Section 3.2) with fuzzy-boundary intervals and two uncertain variables with deterministic-boundary intervals (Table 4, Table 5 and Table 6), $2^5 = 32$ possible combinations of the uncertain variable values/runs of the algorithm for each WADI future scenario (32 × 5 = 160 runs for all examined scenarios in total) has been undertaken based on the FBISP algorithm as proposed by [2].

For each examined case/run (1) the optimized water allocation target for each time stage (the twelve months of the examined year), as well as for total annual (as the summation of the values of the twelve time stages); (2) the probabilistic shortages and allocations for each one of the fifty hydrologic scenarios and for each of the three water users for each time stage (the twelve months of the examined year), as well as for the total annual (as the summation of the values of the twelve time stages) and (3) the total benefits and the benefits and penalties for each of the three water users are derived. The analysis of these results is divided in two parts: (a) the analysis for the results of the baseline scenario, enabling the understanding of the optimization technique used and (b) the comparison of the results between the baseline and the future WADI water and agriculture scenarios in order to assess the effect of the EU water and agricultural policy changes in the system benefits and penalties on the baseline scenario.

Results Analysis for the Baseline Scenario

The total annual maximized net benefits of the Alfeios hydrosystem for the two solution methods are presented in Table 8. The first solution method estimates a wider range for the objective function value (meaning the total net benefits) equal to [(96,192,950, 102,180,847), (127,801,604, 135,950,230)] in EUR, but at the same time it embodies a higher uncertainty or risk. The second solution process, on the other hand, results to a narrower interval for the objective function value method [(104,523,859, 109,324,450), (128,786,579, 134,978,247)], being related to a more conservative view of the uncertainties of the hydrosystem and to the incapability to derive higher benefits under favorable conditions. The interval solution can be easily interpreted for generating decision alternatives, where upper-bound (first solution method) system benefits is associated with more advantageous conditions (e.g., associated with upper-bound inflows, upper-bound benefit coefficients, lower-bound cost coefficients, upper-bound reservoirs capacities, lower-bound reserved storage requirements) while the lower-bound (second solution method) one corresponds to the demanding conditions [1].

Table 8: Total annual net benefit (EUR) for all water uses

WADI Scenarios	Total Annual Net Benefit (EUR)							
	1st Solution Method				2nd Solution Method			
	Min (f^-)	Max (f^-)	Min (f^+)	Max (f^+)	Min (f^-)	Max (f^-)	Min (f^+)	Max (f^+)
Baseline	96,192,950	102,180,847	127,801,604	135,950,230	104,523,859	109,324,450	128,786,579	134,978,247
FS1	76,667,389	81,865,504	113,991,044	126,002,601	84,858,261	89,043,374	114,966,693	125,209,412
FS2	127,853,717	147,770,816	179,855,724	187,230,205	140,082,873	157,717,364	181,108,510	185,941,332
FS3	100,450,661	119,991,812	143,479,720	154,964,308	108,428,316	126,072,379	144,472,397	154,016,575
FS4	139,941,260	159,235,851	194,409,861	225,532,763	152,798,364	169,543,642	195,726,034	224,148,365

FS1: Future Scenario 1; FS2: Future Scenario 2; FS3: Future Scenario 3; FS4: Future Scenario 4.

The total annual benefits and penalties for the irrigation at Flokas, for the hydropower production at Ladhon HPS and for the hydropower production at Flokas HPS are given in Table 9,Table 10 and Table 11. From these tables, it is shown that for the second (pessimistic) solution method the benefits are lower but also the penalties are lower in comparison to the corresponding results from the first (optimistic) solution method.

Table 9: Total annual benefit and penalties (EUR) for irrigation at Flokas

WADI Scenarios	Total Annual Benefits (EUR)							
	1st Solution Method				2nd Solution Method			
	Min (f^-)	Max (f^-)	Min (f^+)	Max (f^+)	Min (f^-)	Max (f^-)	Min (f^+)	Max (f^+)
Baseline	29,064,003	30,643,655	32,706,273	35,881,125	26,641,243	28,170,265	32,706,273	35,881,125
FS1	20,626,258	22,151,608	33,275,869	40,903,212	19,075,077	20,587,187	33,275,869	40,903,212
FS2	32,942,354	46,373,935	46,548,636	46,988,116	30,191,272	44,417,804	46,548,636	46,988,116
FS3	33,406,833	48,219,300	48,394,001	54,915,529	30,710,414	46,185,328	48,394,001	54,915,529
FS4	38,614,195	51,370,227	51,544,928	75,236,645	35,390,899	47,823,846	51,544,928	75,236,645
WADI Scenarios	**Total Annual Penalties (EUR)**							
Baseline	2,470,856	2,611,203	432,349	459,697	2,257,577	2,385,810	118,146	581,477
FS1	3,245,851	3,292,577	326,846	506,598	2,452,467	2,586,774	54,110	382,445
FS2	2,214,386	2,724,846	459,945	469,950	2,298,278	3,527,484	125,687	594,446
FS3	2,235,536	2,696,322	452,675	469,307	2,485,131	3,675,421	123,700	593,634
FS4	3,146,448	3,283,159	456,063	467,608	2,874,068	3,650,108	124,626	591,484

FS1: Future Scenario 1; FS2: Future Scenario 2; FS3: Future Scenario 3; FS4: Future Scenario 4.

Table 10: Total annual benefit and penalties (EUR) for hydropower production at Ladhon HPS

WADI Scenarios	Total Annual Benefits (EUR)							
	1st Solution Method				2nd Solution Method			
	Min (f^-)	Max (f^-)	Min (f^+)	Max (f^+)	Min (f^-)	Max (f^-)	Min (f^+)	Max (f^+)
Baseline	11,027,398	15,924,814	16,711,593	21,878,560	7,346,406	10,657,147	11,202,510	14,950,226
FS1	9,373,289	13,536,092	14,204,854	18,596,776	6,161,067	9,059,977	9,513,407	12,692,143
FS2	15,438,358	22,294,739	23,396,230	30,629,985	10,261,288	14,920,853	15,683,515	20,954,784
FS3	11,027,398	15,924,814	16,711,593	21,878,560	7,346,676	10,657,666	11,203,456	14,967,703
FS4	16,541,098	23,887,221	25,067,389	32,817,841	11,000,389	15,986,499	16,803,766	22,427,814
WADI Scenarios	**Total Annual Penalties (EUR)**							
Baseline	18,854,936	20,501,510	7,598,174	8,375,765	4,805,658	5,564,719	1,655,029	2,279,614
FS1	16,003,268	17,401,182	6,458,448	7,119,400	4,065,699	4,711,214	1,401,002	1,930,522
FS2	26,396,911	28,702,114	10,637,443	11,726,070	6,727,922	7,830,586	2,341,889	3,210,311
FS3	18,854,936	20,501,510	7,598,174	8,375,765	4,807,341	5,593,869	1,655,633	2,293,079
FS4	28,282,404	30,752,265	11,397,261	12,563,647	7,208,488	8,354,489	2,689,422	3,425,385

FS1: Future Scenario 1; FS2: Future Scenario 2; FS3: Future Scenario 3; FS4: Future Scenario 4.

In Table 12 the optimized targets of annual water volume for irrigation (m^3) and of annual hydropower production at HPS Ladhon and Flokas (MWh) for the two different solution processes of the model are presented for the baseline scenario and the four future WADI scenarios. For the irrigation at Flokas the range of the optimized target of the annual water volume for the upper-bound model (f^+) is the same for both solution methods and is equal to its maximum possible value of 174,700,800 m^3. For the lower-bound model (f^-) higher maximum allocation targets are computed by the first solution method (160,079,569, 174,700,800) with much wider ranges between the minimum and the maximum value (14,621,231 m^3) compared to the ones from the second solution method (160,574,718, 160,599,895) with corresponding range (25,177 m^3). By comparing the corresponding results of the FBISP method in this paper with the ITSP as presented in [38], it is worth noticing that the monthly optimized water allocation target values for irrigation at Flokas are equal to the maximum possible allocation for the upper- and the lower-bound solution of the ITSP. This shows that the incorporation of the fuzzy nature of the uncertainties in the FBISP results in lower optimized water allocation target values for the lower-bound (second solution method) solution, reflecting a more analytic and fine approximation of the effect of the uncertainties on the minimum and maximum values of the boundaries of the results.

Table 11: Total annual benefit and penalties (EUR) for hydropower production at Flokas HPS

WADI scenarios	Total Annual Benefits (EUR)							
	1st Solution Method				2nd Solution Method			
	Min (f^-)	Max (f^-)	Min (f^+)	Max (f^+)	Min (f^-)	Max (f^-)	Min (f^+)	Max (f^+)
Baseline	79,312,739	79,312,739	86,996,160	86,996,160	79,312,739	79,312,739	86,996,160	86,996,160
FS1	67,415,828	67,415,828	73,946,736	73,946,736	67,415,828	67,415,828	73,946,736	73,946,736
FS2	111,037,834	111,037,834	121,794,624	121,794,624	111,037,834	111,037,834	121,794,624	121,794,624
FS3	79,312,739	79,312,739	86,996,160	86,996,160	79,312,739	79,312,739	86,996,160	86,996,160
FS4	118,969,108	118,969,108	130,494,240	130,494,240	118,969,108	118,969,108	130,494,240	130,494,240
WADI scenarios	**Total Annual Penalties (EUR)**							
Baseline	1,294,736	1,295,664	235,104	244,521	1,279,939	1,284,407	196,202	199,798
FS1	1,071,046	1,071,835	199,839	207,843	1,064,610	1,070,039	168,108	172,620
FS2	1,812,630	1,813,930	329,146	342,330	1,791,914	1,821,487	272,398	279,717
FS3	1,294,736	1,295,664	235,104	244,521	1,281,003	1,301,062	194,570	199,372
FS4	1,942,104	1,943,496	352,657	366,782	1,919,908	1,934,281	292,242	299,697

FS1: Future Scenario 1; FS2: Future Scenario 2; FS3: Future Scenario 3; FS4: Future Scenario 4.

Table 12: Optimized target for total annual water volumes for irrigation (m^3)

Irrigation (m^3)	Optimized Target for Total Annual Water Volumes for Irrigation (m^3)							
	1st Solution Method				2nd Solution Method			
	Min (f^-)	Max (f^-)	Min (f^+)	Max (f^+)	Min (f^-)	Max (f^-)	Min (f^+)	Max (f^+)
Baseline	160,079,569	174,700,800	174,700,800	174,700,800	160,574,718	160,599,895	174,700,800	174,700,800
FS1	150,527,097	162,767,927	174,700,800	174,700,800	151,272,706	151,272,706	174,700,800	174,700,800
FS2	160,079,569	174,700,800	174,700,800	174,700,800	167,331,623	167,331,623	174,700,800	174,700,800
FS3	160,541,634	174,700,800	174,700,800	174,700,800	167,331,623	167,331,623	174,700,800	174,700,800
FS4	160,079,569	174,700,800	174,700,800	174,700,800	162,606,691	162,640,203	174,700,800	174,700,800

FS1: Future Scenario 1; FS2: Future Scenario 2; FS3: Future Scenario 3; FS4: Future Scenario 4.

Table 13: Annual Shortage for irrigation ($m^3 \times 10^6$)

Hydroscenarios	Annual Shortage for Irrigation ($m^3 \times 10^6$)											
	Optimistic Solution Method				Pessimistic Solution Method				Best/Worst Case Solution			
(Exceedance Probability %)	Min (f^+)	Max (f^+)	Min (f^-)	Max (f^-)	Min (f^-)	Max (f^-)	Min (f^+)	Max (f^+)	Min (f^-)	Max (f^-)	Min (f^+)	Max (f^+)
1 (68.1%)	0.00	0.00	0.00	0.00	0.00	0.45	0.00	0.00	0.00	0.00	0.00	0.45
2–18	0.00	0.00	0.00	0.00	0.00	0.00	0.00	0.00	0.00	0.00	0.00	0.00
19 (96.9%)	21.86	21.86	60.22	60.22	45.84	46.30	5.97	27.65	5.97	27.65	45.84	60.22
20–28	0.00	0.00	0.00	0.00	0.00	0.00	0.00	0.00	0.00	0.00	0.00	0.00
29 (79.8%)	0.00	0.00	0.00	0.00	9.23	9.68	0.00	0.00	0.00	0.00	0.00	9.68
30–40	0.00	0.00	0.00	0.00	0.00	0.00	0.00	0.00	0.00	0.00	0.00	0.00
41 (93.5%)	0.00	0.00	11.83	11.83	8.67	9.11	0.00	0.00	0.00	0.00	8.67	11.83
42–43	0.00	0.00	0.00	0.00	0.00	0.00	0.00	0.00	0.00	0.00	0.00	0.00
44 (92.9%)	0.00	0.00	0.00	0.00	2.08	2.53	0.00	0.00	0.00	0.00	0.00	2.53
45–50	0.00	0.00	0.00	0.00	0.00	0.00	0.00	0.00	0.00	0.00	0.00	0.00

Generally, the shortage of a water user (irrigation or hydropower production) is the result of the nonsatisfaction of the predefined target for the examined water use and is expressed as the difference between target value and available water. Therefore, the water allocation is given by the difference between the target value and the probabilistic shortage. The results for the annual shortage (Table 13) and the annual allocation (Table 14) for irrigation, as computed by the optimization algorithm from the two solution methods and for the fifty hydrologic equal-probability scenarios are presented. As shown in Table 13, the hydrologic scenario 19 (with exceedance probability value = 96.9%) is the worst-shortage condition. For this hydrologic scenario, the annual shortage is [(21,860,787, 21,860,787), (60,217,736, 60,217,737)] in m^3 from the first solution method and [(45,840,807, 46,302,872), (5,973,779, 27,652,026),] in m^3 from the second solution method.

Table 14: Annual Allocation for irrigation ($m^3 \times 10^6$)

Hydroscenarios (Exceedance Probability %)	Optimistic Solution Method Min (f^+)	Max (f^+)	Min (f^-)	Max (f^-)	Pessimistic Solution Method Min (f^-)	Max (f^-)	Min (f^+)	Max (f^+)	Best/Worst Case Solution Min (f^-)	Max (f^-)	Min (f^+)	Max (f^+)
1 (68.1%)	1.75	1.75	1.75	1.75	1.60	1.60	1.75	1.75	1.75	1.75	1.60	1.75
2–18	1.75	1.75	1.75	1.75	1.60	1.61	1.75	1.75	1.75	1.75	1.60	1.75
19 (96.9%)	1.53	1.53	1.14	1.14	1.14	1.14	1.47	1.69	1.46	1.69	1.14	1.14
20–28	1.75	1.75	1.75	1.75	1.60	1.61	1.75	1.75	1.75	1.75	1.60	1.75
29 (79.8%)	1.75	1.75	1.75	1.75	1.51	1.51	1.75	1.75	1.75	1.75	1.51	1.75
30–40	1.75	1.75	1.75	1.75	1.60	1.61	1.75	1.75	1.75	1.75	1.60	1.75
41 (93.5%)	1.75	1.75	1.63	1.63	1.51	1.51	1.75	1.75	1.75	1.75	1.51	1.63
42–43	1.75	1.75	1.75	1.75	1.60	1.61	1.75	1.75	1.75	1.75	1.60	1.75
44 (92.9%)	1.75	1.75	1.75	1.75	1.58	1.58	1.75	1.75	1.75	1.75	1.58	1.75
45–50	1.75	1.75	1.75	1.75	1.60	1.61	1.75	1.75	1.75	1.75	1.60	1.75

In Table 15, presenting together the annual water allocation, shortage and preregulated targets for this scenario, the shortage to target ratio (%) is also introduced. From this ratio, it is indicated that the shortage is serious ranging from 12.5% to 37.6% of the corresponding target for the first solution method and from 3.4% to 28.5% for the second solution. In this case, the farmers should find an alternative water source such as pumping water from groundwater. If the farmers do not have an alternative water source (such as pumping water from groundwater or wastewater reuse), a yield reduction is highly possible, which is introduced into the objective function as a penalty for irrigation. The solutions of water shortage and allocation for the other hydrologic scenarios can be accordingly interpreted.

From the Table 15, it is verified that for the first solution method the benefits and the optimized targets are higher with wider ranges, but the penalties are also higher and wider in comparison to the corresponding results from the second solution method, since the two solution methods are associated with different risk attitudes of the decision makers considering system uncertainties. They drive the results of the optimization algorithm to different solutions in terms of target, shortage and allocation. The shortage intervals can be low under favorable system conditions and high and critical under demanding conditions as in the case of irrigation for the hydrologic scenario 19. Additionally, in Table 13 and Table 14 the best/worst case results are provided in order to enable an evaluation of the system capacity to fulfill the preregulated goals. These results, however, do not necessarily construct a set of stable intervals [2].

Table 15: Annual target, shortage and allocation for irrigation (m^3) for the hydrologic scenario 19

Baseline	Annual Water Volumes for Irrigation (m^3) for the Hydrologic Scenario 19							
	1st Solution Method				2nd Solution Method			
	Min (f^-)	Max (f^-)	Min (f^+)	Max (f^+)	Min (f^-)	Max (f^-)	Min (f^+)	Max (f^+)
Target	174,700,800	160,079,569	174,700,800	174,700,800	160,574,718	160,599,895	174,700,800	174,700,800
Shortage	60,217,737	60,217,737	21,860,788	21,860,788	45,840,807	46,302,872	5,973,779	27,652,026
Allocation	114,483,063	114,483,063	152,840,012	152,840,012	114,297,023	114,297,023	147,048,774	168,727,021
Shortage/target	34.5%	37.6%	12.5%	12.5%	28.5%	28.8%	3.4%	15.8%

In Table 16 the optimized annual target of the hydropower production at the small hydropower station at Flokas Dam are provided. It is worth noticing that this optimized annual target remains unchanged and equal to 19,828 MWh for all WADI scenarios as well as for both solution methods, constituting a robust value. By comparing the corresponding results of the FBISP method in this paper with the ITSP as presented in [38], it is worth noticing that the optimized annual hydropower target at Flokas HPS are the same for both optimization methodologies.

Table 16: Optimized target for total annual hydropower production at HPS Flokas (MWh)

HP Flokas (MWh)	Optimized Target for Total Annual Hydropower Production at HPS Flokas (MWh)							
	1st Solution Method				2nd Solution Method			
	Min (f^-)	Max (f^-)	Min (f^+)	Max (f^+)	Min (f^-)	Max (f^-)	Min (f^+)	Max (f^+)
Baseline	19,828	19,828	19,828	19,828	19,828	19,828	19,828	19,828
FS1	19,828	19,828	19,828	19,828	19,828	19,828	19,828	19,828
FS2	19,828	19,828	19,828	19,828	19,828	19,828	19,828	19,828
FS3	19,828	19,828	19,828	19,828	19,828	19,828	19,828	19,828
FS4	19,828	19,828	19,828	19,828	19,828	19,828	19,828	19,828

FS1: Future Scenario 1; FS2: Future Scenario 2; FS3: Future Scenario 3; FS4: Future Scenario 4.

The corresponding monthly optimized targets of the hydropower production at Flokas Dam in comparison to the maximum allowable ones are given in Table 17. These optimized target values are the same for both solution methods. From this table it is observed that for all months the optimized targets are equal to the maximum ones except of May and June, for which the optimized target values are 1.8% and 0.5%, respectively, below the maximum values. Similar results are derived also from the ITSP method with the difference that the corresponding ratios are higher with 11% and 1% for May and June, respectively.

From Table 18 the optimized targets of annual hydropower production at HPS Ladhon (MWh) for the two different solution processes of the model are presented for the baseline scenario and the four future WADI scenarios. For the first solution method the optimized target is [(275,685, 289,542), (278,527, 291,714)] in MWh and for the second solution method [(179,672, 179,743), (186,709, 199,336)] in MWh for the baseline scenario. The maximum annual allowable target value for hydropower production at Ladhon has been set equal to 301,890 MWh. Based on this and by comparing the results from the two solution methods, it is concluded that the optimistic solution method results into a much higher optimized target interval (ranging from 91% to 97% of the maximum allowable value = 301,890 MWh), and the pessimistic solution method results into a significant lower target interval (ranging from 60% to 66% of the maximum allowable value).

In Table 19, the corresponding minimum and maximum monthly optimized targets of the hydropower production at Ladhon HPS in comparison to the maximum allowable ones, as derived from the first solution method, are provided. From this table it is observed that for the hydropower production at Ladhon, the monthly optimized hydropower production target values are equal to the maximum allowable values, as derived from the first solution method, for all months except for July (from 83% up to 93% of its maximum value), September (from 45% to 93% of its maximum value), October (from 43% up to 73% of its maximum value) and November (from 80% to 84% of its maximum value). In Table 20, the corresponding minimum and maximum monthly optimized targets of the hydropower production at Ladhon HPS in comparison to the maximum allowable ones, as derived from the second solution method, are provided. From this table it is observed that for the hydropower production at Ladhon, the monthly optimized hydropower production target values deviate from the maximum allowable values for all months within a range of 47% up to 94% of their maximum values. Therefore, it is obvious that when the lower-bound water inflows (see the description for the incorporation of the uncertainty of the rainfall-runoff model in Section 3.2) are used at the first step of the algorithm, which corresponds to the process of the second solution (pessimistic) method, representing the unfavorable (demanding) hydrologic conditions, the optimized targets for the monthly hydropower production at Ladhon are significantly lower in comparison to the ones resulting from the first solution method.

By comparing the corresponding results of the FBISP method in this paper with the ITSP as presented in [38], it is worth noticing that from the ITSP method the optimized monthly hydropower target at Ladhon HPS are equal to the maximum allowable values as derived from the first solution method for all months except for July (60% of its maximum value), September (67% of its maximum value), October (from 23% of its maximum value) and November (from 58% of its maximum value). The four months with the deviations from the maximum allowable values are the same as the ones derived by the first solution method of the FBISP. The only difference is that the optimized targets from the ITSP are lower. Also in this case, the FBISP method provides a more detailed overview of the intervals for the optimized targets.

From the optimized targets of the three main users, as analyzed above, it can be concluded that the highest priority for water allocation is set to irrigation, since it has the highest unit benefit, but at the same time also the highest unit penalty. The next priorities are given to hydropower production at Flokas and finally to the hydropower production at Ladhon, which has the smallest unit benefit.

In Figure 3 and in Table 21, the interplay between the optimized total targets in m^3, which are derived from the addition of the optimum target water volumes allocated to the three water users (hydropower production at Ladhon, hydropower production at Flokas and irrigation at Flokas), and the system's net benefits for these four options for both types of solution methods is shown.

Table 17: Maximum allowable (THydroFlokasPlus) and Optimized (Optimized THydroFlokas) monthly targets of hydropower production at Flokas HPS (MWh)

Monthly Hydropower Targets	Maximum and Optimized Monthly Targets of Hydropower Production at Flokas HPS (MWh)											
	January	February	March	April	May	June	July	August	September	October	November	December
THydroFlokasPlus	2379	2894	3435	2840	1861	773	251	255	571	1111	1397	2097
Optimized THydroFlokas	2379	2894	3435	2840	1828	770	251	255	571	1111	1397	2097

Table 18: Optimized annual target for hydropower production at HPS Ladhon (MWh)

WADI Scenarios	Optimized Annual Target for Hydropower Production at HPS Ladhon (MWh)							
	1st Solution Method				2nd Solution Method			
	Min (f^-)	Max (f^-)	Min (f^+)	Max (f^+)	Min (f^-)	Max (f^-)	Min (f^+)	Max (f^+)
Baseline	275,685	289,542	278,527	291,714	179,672	179,743	186,709	199,336
FS1	275,685	289,542	278,527	291,714	179,486	179,507	186,537	199,092
FS2	275,685	289,542	278,527	291,714	179,672	180,006	186,709	199,569
FS3	275,685	289,542	278,527	291,714	179,688	180,006	186,724	199,569
FS4	275,685	289,542	278,527	291,714	179,672	179,790	186,709	199,358

FS1: Future Scenario 1; FS2: Future Scenario 2; FS3: Future Scenario 3; FS4: Future Scenario 4.

Table 19: Maximum allowable (THydro-LadhonPlus) and minimum (MinOptimized THydroFlokas) and maximum (MaxOptimized THydroFlokas) optimized monthly targets of hydropower production at Flokas HPS (MWh) and their ratios in (%) for the first solution method (optimistic)

Monthly Hydropower Targets	Target for Hydropower Production at Ladon HPS (MWh) from the First Solution Method (Optimistic)											
	January	February	March	April	May	June	July	August	September	October	November	December
(1) Thydro-Ladhon Plus	37,353	38,947	48,311	35,391	23,237	15,868	15,598	14,233	13,642	17,062	17,971	24,276
(2) MinOptimized ThydroFlokas	37,353	38,947	48,311	35,391	23,237	15,868	12,997	14,233	6132	7387	14,395	24,276
(3) MaxOptimized ThydroFlokas	37,353	38,947	48,311	35,391	23,237	15,868	14,431	14,233	12,638	11,980	15,048	24,276
(2)/(1) %	100%	100%	100%	100%	100%	100%	93%	100%	93%	70%	84%	100%
(3)/(1) %	100%	100%	100%	100%	100%	100%	83%	100%	45%	43%	80%	100%

Table 20: Maximum allowable (THydro-LadhonPlus) and minimum (MinOptimized THydroFlokas) and maximum (MaxOptimized THydroFlokas) optimized monthly targets of hydropower production at Flokas HPS (MWh) and their ratios in (%) for the second solution method (pessimistic)

Monthly Hydropower Targets	Target for Hydropower Production at Ladon HPS (MWh) from the Second Solution Method (Pessimistic)											
	January	February	March	April	May	June	July	August	September	October	November	December
(1) Thydro-LadhonPlus	20,519	30,385	27,639	17,727	11,081	10,513	13,635	11,413	5925	7387	9427	13,358
(2) Min Optimized ThydroFlokas	20,519	30,385	27,639	18,390	11,081	11,247	13,635	11,413	5925	7387	9427	13,358
(3)Max Optimized ThydroFlokas	37,353	38,947	48,311	35,391	23,237	15,868	14,431	14,233	12638	11,980	15,048	24,276
(2)/(1)-%	55%	78%	57%	50%	48%	66%	94%	80%	47%	62%	63%	55%
(3)/(1)-%	55%	78%	57%	52%	48%	71%	94%	80%	47%	62%	63%	55%

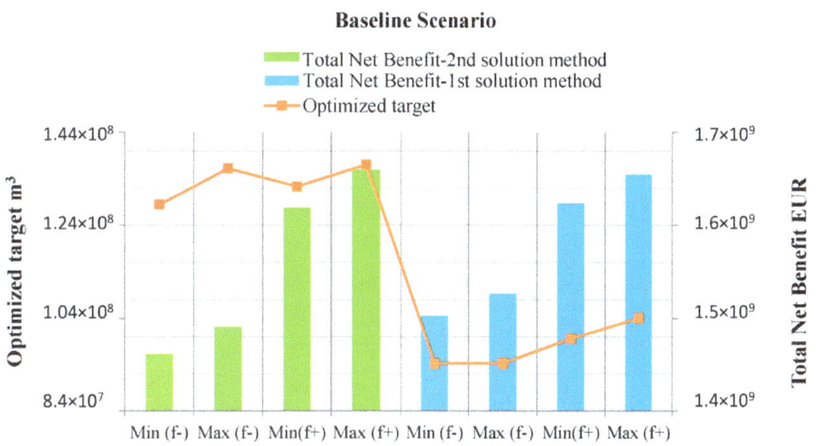

Figure 3: Interconnections between total net benefit and optimized total target for the four options and for both solution methods.

Table 21: Interconnections between total net benefit and optimized total target for the four options and for both solution methods

Baseline	1st Solution Method				2nd Solution Method			
	Min (f^-)	Max (f^-)	Min (f^+)	Max (f^+)	Min (f^-)	Max (f^-)	Min (f^+)	Max (f^+)
Total Net Benefit EUR	96,192,950	102,180,847	127,801,604	135,950,230	104,523,859	109,324,450	128,786,579	134,978,247
Optimized target m³	1,621,866,797	1,661,184,353	1,641,552,336	1,665,055,450	1,451,246,461	1,451,397,999	1,477,912,994	1,500,418,501

Results Analysis for the Baseline and the Four Future Scenarios

The four WADI future scenarios represent four different possible water and agricultural policy alternatives in comparison to the baseline scenario, which may have an impact on the optimal water allocation. The differences between the future scenarios include among others, changes of hydropower energy prices, water prices, selling prices of the agricultural products, yield functions, subsidies, farmer income variable costs, labor, and fertilizers. Since the uncertainties of water availability, water demand, benefits and costs associated with probabilistic water allocations and shortages are incorporated in the optimization algorithm, it is very interesting to investigate also the effect of the various water and agricultural policies on the water allocation targets and on the corresponding maximized system benefits.

The optimized total annual water allocation target, derived by the addition of the water allocation targets of the three examined water uses, for all WADI scenarios is given in Table 22. For comparison reasons the ratio of these targets for each future scenario to the baseline target (%) is also shown. From these ratios (99.3% up to 100.5%) it is concluded that the optimized total annual water allocation targets for the various alternative water and agricultural policies compared to the baseline are only slightly affected. By applying the ITSP method in [38], the optimized total annual water allocation targets are exactly the same for all four WADI future scenarios. Even though the quantitative changes of these target values are relatively low, in the case of the application of the FBISP method the consideration of the fuzzy nature for some of the uncertain variables results in different water allocation targets revealing a more complicated structure of the results.

The total annual maximized net benefits of the hydrosystem in EUR and their ratios of the four future scenarios to the baseline are presented in Table 8 and Table 23, respectively, for all scenarios and for the first (optimistic) and second (pessimistic) solution method. It is obvious that the highest increase of these benefits is observed for the Local Stewardship scenario (FS4) ranging for the first solution method from 45.5% up to 65.9% and for the second solution method from 46.2% to 66.1%. The only decrease of the net benefits compared to the benefits of the baseline scenario occurs for the World Market scenario (FS1) for both upper-(f^+) and lower-(f^-) intervals of both solution methods up to 20%.

The statistical analysis of these optimized total annual net benefits from the various runs (combinations of the uncertain variables values) is provided in Figure 4 for the baseline and the four future scenarios through the use of box plots for Alfeios River Basin in Western Greece. It is worth mentioning that for the Local Stewardship scenario (FS4) not only the highest total net benefits are derived, but also the widest intervals ranges. On the other hand, the baseline scenario has the narrowest upper- and lower-bound intervals.

Since the WADI scenarios focus mainly on the effects of the agricultural policy changes on the irrigation sector, it is worth examining separately the changes between the scenarios of the annual net benefits of water allocated to irrigation, which represent the agricultural income. From Table 24the highest reduction of the agricultural income (net benefit of irrigation) compared to the baseline scenario is observed in the lower-bound (f^-) interval for both solution methods of the World Market scenario (FS1) (from 65.5% up to 69.80% farmer income reduction compared to the baseline scenario). The corresponding upper-bound intervals for this scenario drive to a small increase of agricultural income ranging from 1.2% to 11.4% compared to the one for the baseline scenario. These results are similar and compatible with the corresponding ones from the application of the ITSP method in [38]. More precisely, the highest increase of the total system benefits is also observed for the *Local Stewardship* scenario ranging from 52% to 59% and the only decrease occurs for the *World Market* scenario (9%–24%). The above mentioned reduction of the agricultural income can be explained by the fact that in the World Market scenario the highest decrease of selling prices and a significant increase of the prices of most of input variables for agriculture (pesticides, seeds, water price, *etc.*) in comparison to the other scenarios is noticed. Moreover, for most of the crops cultivated at Flokas Irrigation scheme, the presence of area subsidy plays a balancing role for the positive sign (profit) of the agricultural income. In this scenario, no subsidies are provided to the farmers. This fact in combination with the existence of mainly small agricultural units, mainly family farms, renders this agricultural region and Greece in general into weak competitor to bigger and stronger economically agricultural markets, such as U.S.A. or Brazil. Through this analysis, the importance of a balancing area subsidy for economically sensitive agricultural products for the Greek market is verified. The globalization and the liberalization of the agricultural market in combination with the different orientation of the new CAP reform 2014–2020 pose great challenges for the Greek farmers for modernization and increased agricultural and technical expertise.

Finally, as already mentioned in the general analysis of the four future scenarios, moving from*Global Sustainability* towards *Local Stewardship* the agricultural income and the net benefits in general, increase (Figure 5). In *Local Stewardship* scenario, where the focus is on strong local or regional governments with emphasis on social values, self-reliance, self-sufficiency, and conservation of natural resources and the environment, the highest agricultural income is derived and more specifically an increase of 33.5% to 111% compared to the baseline scenario is noticed.

Table 22: Optimized total annual water allocation target of the four future scenarios as ratio of the baseline (%)

WADI Scenarios	Optimized Total Annual Water Allocation Target (m^3)							
	1st Solution Method				2nd Solution Method			
	Min (f^-)	Max (f^-)	Min (f^+)	Max (f^+)	Min (f^-)	Max (f^-)	Min (f^+)	Max (f^+)
Baseline	1,621,866,797	1,661,184,353	1,641,552,336	1,665,055,450	1,451,246,461	1,451,397,999	1,477,912,994	1,500,418,501
FS1	1,612,314,325	1,649,251,480	1,641,552,336	1,665,055,450	1,441,612,922	1,441,650,080	1,477,608,041	1,499,983,803
FS2	1,621,866,797	1,661,184,353	1,641,552,336	1,665,055,450	1,458,003,367	1,458,598,441	1,477,912,994	1,500,833,807
FS3	1,622,328,862	1,661,184,353	1,641,552,336	1,665,055,450	1,458,031,456	1,458,598,441	1,477,941,083	1,500,833,807
FS4	1,621,866,797	1,661,184,353	1,641,552,336	1,665,055,450	1,453,278,435	1,453,522,032	1,477,912,994	1,500,457,705
FS1/Baseline	99.4%	99.3%	100.0%	100.0%	99.3%	99.3%	100.0%	100.0%
FS2/Baseline	100.0%	100.0%	100.0%	100.0%	100.5%	100.5%	100.0%	100.0%
FS3/Baseline	100.0%	100.0%	100.0%	100.0%	100.5%	100.5%	100.0%	100.0%
FS4/Baseline	100.0%	100.0%	100.0%	100.0%	100.1%	100.1%	100.0%	100.0%

Table 23: Total annual net benefit (EUR) of the four future scenarios as ratio of the baseline (%)

WADI scenarios	Annual Total Net Benefit of Future Scenarios as Ratio of the Baseline (%)							
	1st Solution Method				2nd Solution Method			
	Min (f^-)	Max (f^-)	Min (f^+)	Max (f^+)	Min (f^-)	Max (f^-)	Min (f^+)	Max (f^+)
FS1/Baseline	79.7%	80.1%	89.2%	92.7%	81.2%	81.4%	89.3%	92.8%
FS2/Baseline	132.9%	144.6%	140.7%	137.7%	134.0%	144.3%	140.6%	137.8%
FS3/Baseline	104.4%	117.4%	112.3%	114.0%	103.7%	115.3%	112.2%	114.1%
FS4/Baseline	145.5%	155.8%	152.1%	165.9%	146.2%	155.1%	152.0%	166.1%

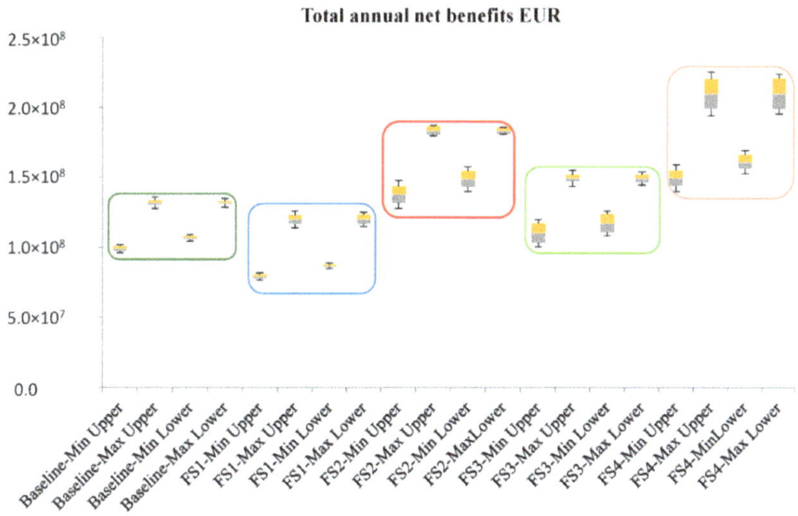

Figure 4: Box plots for the four options of total net optimized benefits in EUR for the baseline and the four future scenarios (FS1: Future Scenario 1; FS2: Future Scenario 2; FS3: Future Scenario 3; FS4: Future Scenario 4).

Table 24: Annual net benefit (EUR) for irrigation and ratios (%) of annual net benefit of the four future scenarios compared to baseline

WADI Scenarios	Annual Net Benefit for Irrigation (EUR)							
	1st Solution Method				2nd Solution Method			
	Min (f^-)	Max (f^-)	Min (f^+)	Max (f^+)	Min (f^-)	Max (f^-)	Min (f^+)	Max (f^+)
Baseline	26,452,801	28,172,800	32,246,577	35,448,777	24,255,433	25,835,452	32,124,796	35,762,980
FS1	17,333,682	18,905,757	32,769,271	40,576,366	16,587,306	18,037,122	32,904,418	40,849,102
FS2	30,217,508	44,159,549	46,078,686	46,528,170	27,702,908	40,891,389	45,954,190	46,862,429
FS3	30,710,512	45,983,764	47,924,693	54,462,854	28,162,550	42,625,668	47,800,367	54,791,829
FS4	35,331,036	48,223,779	51,077,320	74,780,583	32,392,287	44,325,729	50,953,444	75,112,020
FS1/Baseline	65.5%	67.1%	101.6%	114.5%	68.4%	69.8%	102.4%	114.2%
FS2/Baseline	114.2%	156.7%	142.9%	131.3%	114.2%	158.3%	143.0%	131.0%
FS3/Baseline	116.1%	163.2%	148.6%	153.6%	116.1%	165.0%	148.8%	153.2%
FS4/Baseline	133.6%	171.2%	158.4%	211.0%	133.5%	171.6%	158.6%	210.0%

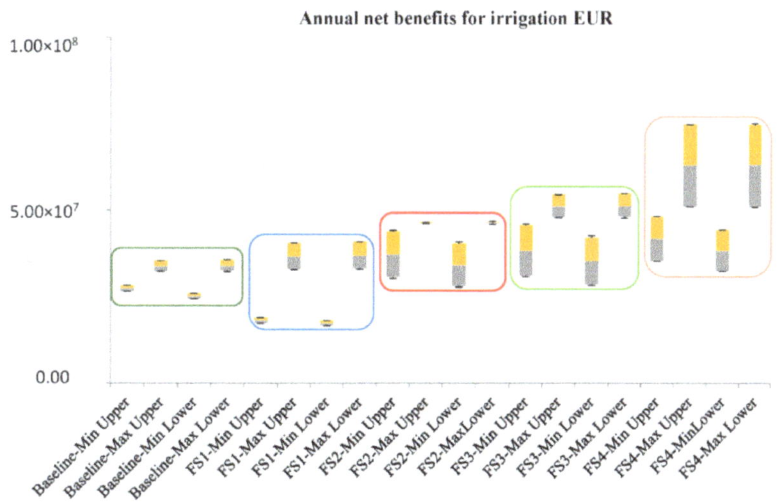

Figure 5: Box plots for the four options of annual net optimized benefits for irrigation in EUR for the baseline and the four future scenarios (FS1: Future Scenario 1; FS2: Future Scenario 2; FS3: Future Scenario 3; FS4: Future Scenario 4).

DISCUSSION AND CONCLUSIONS

In this present study, an optimization technique named fuzzy-boundary interval stochastic programming and developed by [2], which incorporates the most important types of uncertainty (possibilistic, probabilistic and interval), is chosen and applied for optimal water allocation under vague and fuzzy conditions within the Alfeios river basin in Greece. After the analysis of the methodology for water resources management under uncertainty as developed by [6], where an inexact two-stage stochastic programming with deterministic-boundary intervals has been applied into Alfeios [38], the optimization algorithm by [2] has been chosen because it proceeds a step forward by embodying interval-parameter programming, fuzzy vertex analysis, and multistage stochastic programming techniques. More precisely, it can handle uncertainties expressed as probability distributions and fuzzy-boundary intervals, since the lower- and upper-bounds of some intervals may rarely be defined as deterministic values, and they may be fuzzy in nature. The related probability and possibility information can also be included in the solutions for the objective function value and decision variables. The risk attitude of the decision-maker is considered by [2] solving the algorithm through two different processes for a risk-adverse (pessimistic) and a risk-prone (optimistic) attitude of the decision makers. The term "risk", used to characterize these two different solution approaches, does not imply the measuring of risk with its strict mathematical definition, but the willingness of the decision makers to take the risk or not of paying higher penalties in case of selecting the optimistic solution under demanding (unfavorable) conditions or receiving lower benefits in case of selecting the pessimistic solution under favorable conditions.

To the best of our knowledge, this application in the Alfeios river basin is the first application of the proposed methodology by [2] to a real and complex multi-tributary and multi-period water resources system for optimal water allocation, although other hybrid methods with similar concepts have been applied to real-world hydrosystems (*i.e.*, [13,15]). The Alfeios river basin in Greece has been selected because it is characterized by uncertain and limited data required for optimal water allocation, which is also a common problem in many countries including the Mediterranean countries. Authority responsibility relationships are fragmented, fact that leads to the difficulty of gathering the necessary data or even worst to data loss. In some cases, river monitoring, if present, is either inefficient with intermittent periods with no measurements, or due to low financial means the monitoring programs are short and with small number of personnel leading to unreliable or/and short-term data. In this case the only sources of obtaining hydrologic, technical, economic, and environmental data required for water resources management is by periodic measuring expeditions, indirectly by expert knowledge or by informal knowledge by local population, or by more general data concerning a wider geographical location (*i.e.*, country level) from national, European or international databases. Data of this type with a high degree of uncertainty can be defined as fluctuation ranges and therefore simulated as intervals with lower- and upper-bounds either as deterministic values or as fuzzy without the need of distributional or probabilistic information.

The benefits and penalties of the main water users are studied and analyzed through investigation of technical, environmental and socio-economics aspects within the framework of the four WADI water and agricultural future scenarios. Consideration of the hydropower energy market of Greece, crop patterns, yield functions, subsidies, farmer income variable costs, market prices per agricultural product and fertilizers are taken into account for the valuation and the estimation of the hydropower energy and irrigation benefits.

According to [2], the proposed methodology handles uncertainties through constructing a set of scenarios (scenario-tree) that are representative for the universe of water-availability conditions for two tributaries. With such a scenario-based approach, the resulting mathematical programming model could become too large to be applied to large-scale real-world problems. Moreover, the random variables (*i.e.*, water inflows from two tributaries) are assumed to take on discrete distributions and to be mutually independent, such that the study problem can be solved through linear programming method. However, conditional probabilities need to be handled for quantifying water availability, particularly for a multi-stream and multi-reservoir system. This may lead to non-linearity in system responses and raise a main challenge for identifying global optimal solution.

An alternative approach to these limitations of the FBISP methodology is proposed by incorporating the water inflow uncertainty through the simultaneous generation of stochastic equal-probability hydrologic scenarios for stochastically dependent multiple variables at various locations of water inflows in the river basin. This is enabled by using CASTALIA software for stochastic simulation and forecast of hydrologic variables, combining not only multivariate analysis (for many hydrologic processes and geographical correlated locations) as well as multiple time scales (monthly and yearly) in a disaggregation framework. This software permits the preservation of essential marginal statistics up to third order (skewness) and joint second order statistics (auto- and cross-correlations), and the reproduction of long-term persistence (Hurst phenomenon) and periodicity. In this application twelve time stages, one for each month of the examined year have been used (whereas in [2] only three stages have been considered) and fifty equal-probability hydrologic scenarios (whereas in [2] 258 scenarios should be taken into account for only three stages). By increasing the number of the generated equal-probability scenarios, the accuracy of the results also increases. But it is worth mentioning that an increase of the time stages to more than 12 (*i.e.*, in 24 stages for a 2 years analysis), would mean that $24 \times 50 = 1200$ probabilistic values for shortages and allocations should be analyzed. This would make the analysis of the results very complicated, setting also a matter of use of this methodology to a more complex time horizon. From the analysis of the results, it is clear that due to the space limitations, the monthly results cannot be presented in tabular form and analyzed as thoroughly

Finally, in terms of the results from this methodology, the goal of this technique is from one side to identify the optimized water-allocation target with a minimized risk of economic penalty and opportunity loss, and from the other side to determine an optimized water-allocation plan with a maximized system benefit over a multi-period planning horizon. Fuzzy upper- and lower-bound intervals (expressing the effect of the embodied uncertainties) for the optimal water allocation targets and the probabilistic water allocations and shortages as well as the total benefits for the main water uses are identified. The results show that variations in water-allocation targets could express different strategies for water resources management and thus produce varied economic implications under uncertainty. The major results through the application of the FBISP method to optimal water resources allocation in the Alfeios River Basin are the following:

(1) The monthly optimized water allocation target values are: (i) for irrigation for the upper-bound model (f^+) the same for both solution methods and equal to its

maximum possible value; for the lower-bound model (f^-) are higher for the first solution method with much wider ranges between the minimum and the maximum value compared to the ones from the second solution method; (ii) for the hydropower production at Ladhon equal to the maximum allowable values for all months except for July, September-November for the first solution method; deviate from the maximum allowable values for all months for the second solution method; and (iii) the maximum possible allocation for all months except May and June for the hydropower production at Flokas. From the optimized targets of the three main users, as analyzed above, it can be concluded that the highest priority for water allocation is set to irrigation, since it has the highest unit benefit, but at the same time also the highest unit penalty. The next priorities are given to hydropower production at Flokas and finally to the hydropower production at Ladhon, which has the smallest unit benefit.

(2) The optimized total annual water allocation targets for the various alternative water and agricultural WADI policies compared to the baseline are only slightly affected, since the main impact of these scenarios is on the net system benefits. Based on the comparison of the total system benefits from the four future scenarios to the baseline, the highest increase is observed for the *Local Stewardship* scenario and the only decrease for the *World Warket* scenario.

(3) For irrigation, in most hydrologic scenarios, annual water shortages are zero, since the water allocation is equal to the optimized water allocation target. There are only a few hydrologic scenarios with nonzero shortages, for which, if the farmers do not have an alternative water source, a yield reduction is highly possible. These shortages occur in August and September, which can be justified by the low flow rate at Flokas Dam for these two months in combination with the increased irrigation demand. On the other hand, the hydropower production at Ladhon and Flokas in most hydrologic scenarios deviates from the optimized target, therefore resulting in nonzero annual shortages for both hydropower stations. For the hydropower production at Ladhon, the highest shortages take place from January–April (with the highest in March), since in order to satisfy completely the most important water use, that being irrigation (starting mainly from May), the water volume flowing into the Ladhon Reservoir from December–April should be stored and not released. A conflict between the two uses for this time period is observed. For the hydropower production at Flokas, the highest shortages occur during the irrigation period from June–October (with the highest in October), showing a conflict between the two uses. The small HPS at Flokas is only set in operation after the satisfaction of irrigation demand, driving toward water shortages for these months if the available water at Flokas Dam is not adequate.

(4) By comparing the corresponding results of the FBISP method in this paper with the ITSP as presented in [38], it is worth noticing that the results are consistent and compatible but it can be concluded that the incorporation of the fuzzy nature of the uncertainties in the FBISP results in a more analytic and fine approximation of the effect of the uncertainties on the minimum and maximum values of the boundaries of the results providing also a more complicated structure of the results.

ACKNOWLEDGMENTS

This research has been co-financed by the European Union (European Social Fund—ESF) and Greek national funds through the Operational Program "Education and Lifelong Learning" of the National Strategic Reference Framework (NSRF)—Research Funding Program: Heracleitus II: Investing in knowledge society through the European Social Fund. We would like to express our gratitude firstly, to Dimitris Demetracopoulos, Ioannis Argyrakis, Ioannis Mavros and Ioannis Stathas from the Hellenic Public Power Corporation for providing valuable operational data for Ladhon HPS, secondly, to the Technische Universität München (Germany) and to HYDROCRITES University Network for their support and, thirdly, to the anonymous reviewers for their insightful and helpful comments.

Author Contributions

Eleni Bekri was responsible for literature work, data preparation, selection of the methodology, optimization model development and analysis, and paper writing, having as principal author the highest contribution to the manuscript. Markus Disse and Panayotis Yannopoulos enhanced and enriched the overall concept of this research topic, being also responsible for the guidance and consultations about model analysis, paper writing and manuscript modification.

Conflicts of Interest

The authors declare no conflict of interest.

REFERENCES

1. Li, Y.P.; Huang, G.H.; Huang, Y.F.; Zhou, H.D. A multistage fuzzy-stochastic programming model for supporting sustainable water-resources allocation and management. *Environ. Model. Softw.* 2009, *24*, 786–797.
2. Li, Y.P.; Huang, G.H.; Nie, S.L. Planning water resources management systems using a fuzzy-boundary interval-stochastic programming method. *Adv. Water Resour.* 2010, *33*, 1105–1117.
3. Fan, Y.R.; Huang, G.H. A robust two-step method for solving interval linear programming problems within an environmental management context. *J. Environ. Inf.* 2012, *19*, 1–12.
4. Suo, M.Q.; Li, Y.P.; Huang, G.H.; Deng, T.L.; Li, Y.F. Electric power system planning under uncertainty using inexact inventory nonlinear programming method. *J. Environ. Inf.* 2013, *22*, 49–67.
5. Huang, G.H.; Baetz, B.W.; Patry, G.G. A grey linear programming approach for municipal solid waste management planning under uncertainty. *Civ. Eng. Syst.* 1992, *9*, 319–335.
6. Huang, G.H.; Loucks, D.P. An inexact two-stage

stochastic programming model for water resources management under uncertainty. *Civ. Eng. Environ. Syst.* 2000, *17*, 95–118.

7. Maqsood, I.; Huang, G.H.; Yeomans, J.S. An interval-parameter fuzzy two-stage stochastic program for water resources management under uncertainty. *Eur. J. Oper. Res.* 2005, *167*, 208–225.
8. Li, Y.P.; Huang, G.H.; Nie, S.L. An interval-parameter multi-stage stochastic programming model for water resources management under uncertainty. *Adv. Water Resour.* 2006, *29*, 776–789.
9. Nie, X.H.; Huang, G.H.; Li, Y.P.; Liu, L. IFRP: A hybrid interval-parameter fuzzy robust programming approach for municipal solid waste management planning under uncertainty.*J. Environ. Manag.* 2007, *84*, 1–11. [PubMed]
10. Li, Y.P.; Huang, G.H.; Yang, Z.F.; Nie, S.L. Interval-fuzzy multistage programming for water resources management under uncertainty. *Resour. Conserv. Recycl.* 2008, *52*, 800–812.
11. Yeomans, J.S. Applications of simulation-optimization methods in environmental policy planning under uncertainty. *J. Environ. Inform.* 2008, *12*, 174–186.
12. Li, Y.P.; Huang, G.H. Fuzzy-stochastic-based violation analysis method for planning water resources management systems with uncertain information. *Inf. Sci.* 2009, *179*, 4261–4276.
13. Li, Y.; Huang, G. Planning agricultural water resources system associated with fuzzy and random features. *J. Am. Water Resour. Assoc.* 2011, *47*, 841–860.
14. Fu, D.Z.; Li, Y.P.; Huang, G.H. A Factorial-based Dynamic Analysis Method for Reservoir Operation Under Fuzzy-stochastic Uncertainties. *Water Resour. Manag.* 2013, *27*, 4591–4610.
15. Liu, J.; Li, Y.P.; Huang, G.H.; Zeng, X.T. A dual-interval fixed-mix stochastic programming method for water resources management under uncertainty. *Resour. Conserv. Recycl.* 2014, *88*, 50–66.
16. Miao, D.Y.; Huang, W.W.; Li, Y.P.; Yang, Z.F. Planning water resources systems under uncertainty using an interval-fuzzy de novo programming method. *J. Environ. Inform.* 2014,*24*, 11–23.
17. Stedinger, S.; Loucks, D.P. Stochastic dynamic programming models for reservoir operation optimatization. *Water Resour. Res.* 1984, *20*, 1499–1505.
18. Pereira, M.; Pinto, L. Stochastic optimization of a multireservoir hydroelectric system: A decomposition approach. *Water Resour. Res.* 1985, *6*, 779–792.
19. Dupacova, J. Application of stochastic programming: Achievements and questions. *Eur. J. Oper. Res.* 2002, *140*, 281–290.
20. Watkins, D.W., Jr.; McKinney, D.C.; Lasdon, L.S.; Nielsen, S.S.; Martin, Q.W. A scenario-based stochastic programming model for water supplies from the highland lakes. *Int. Trans. Oper. Res.* 2000, *7*, 211–230.
21. Birge, J.R. Decomposition and partitioning methods for multistage stochastic linear programs. *Oper. Res.* 1985, *33*, 989–1007.
22. Charnes, A.; Cooper, W.W. Response to Decision problems under risk and chance constrained programming: Dilemmas in the transitions. *Manag. Sci.* 1983, *29*, 750–753.
23. Huang, G.H. A hybrid inexact-stochastic water management model. *Eur. J. Oper. Res.* 1998,*107*, 137–158.
24. Abrishamchi, A.; Marino, M.A.; Afshar, A. Reservoir planning for irrigation district. *J. Water Resour. Plan. Manag.* 1991, *117*, 74–85.
25. Edirisinghe, N.C.P.; Patterson, E.I.; Saadouli, N. Capacity Planning Model for a Multipurpose Water Reservoir with Target-Priority Operation. *Ann. Oper. Res.* 2000, *100*, 273–303.
26. Azaiez, M.N.; Hariga, M.; Al-Harkan, I. A Chance-Constrained Multi-period Model for a Special Multi-reservoir System. *Comput. Oper. Res.* 2005, *32*, 1337–1351.
27. Freeze, R.A.; Massmann, J.; Smith, L.; Sperling, J.; James, B. Hydrogeological decision analysis, 1. a framework. *Ground Water* 1990, *28*, 738–766.
28. Dubois, D.; Prade, H. Operations on fuzzy number. *Int. J. Syst. Sci.* 1978, *9*, 613–626.
29. Zimmermann, H.J. *Fuzzy Set Theory and Its Applications*; Kluwer Academic Publishers: Dordrecht, The Netherlands, 1995.
30. Jairaj, P.G.; Vedula, S. Multireservoir system optimization using fuzzy mathematical programming. *Water Res. Manag.* 2000, *14*, 457–472.
31. Bender, M.J.; Simonovic, S.P. A fuzzy compromise approach to water resources planning under uncertainty. *Fuzzy Sets Syst.* 2000, *115*, 35–44.
32. Lee, C.S.; Chang, S.P. Interactive fuzzy optimization for an economic and environmental balance in a river system. *Water Res.* 2005, *39*, 221–231. [PubMed]
33. Huang, G.H. IPWM: An interval parameter water quality management model. *Eng. Optim.*1996, *26*, 79–103.
34. Inuiguchi, M.; Ramik, J. Possibilistic linear programming: A brief review of fuzzy mathematical programming and a comparison with stochastic programming in portofolio selection problem. *Fuzzy Sets Syst.* 2000, *111*, 3–28.
35. Bekri, E.S.; Yannopoulos, P.C. The interplay between the Alfeios river basin components and the exerted environmental stresses: A critical review. *Water Air Soil Pollut.* 2012, *223*, 3783–3806.
36. Manariotis, I.D.; Yannopoulos, P.C. Adverse effects on Alfeios river basin and an integrated management framework based on sustainability. *Environ. Manag.* 2004, *34*, 261–269.
37. Podimata, M.; Yannopoulos, P.C. Evaluating challenges and priorities of a trans-regional river basin in Greece by using a hybrid SWOT scheme and a stakeholders' competency overview. *Int. J. River Basin Manag.* 2013, *11*, 93–110.
38. Bekri, E.S.; Yannopoulos, P.C.; Disse, M. Optimizing water allocation under uncertain system conditions in Alfeios river basin (Greece)—Part A: Two-stage stochastic programming model with deterministic-

boundary intervals. *Water* 2015, *7*, 5305–5344.

39. Sustainability of European Irrigated Agriculture under Water Framework Directive and Agenda 2000—WADI. Available online: http://www.lu.lv/materiali/biblioteka/es/pilnieteksti/vide/Sustainability%20of%20European%20Irrigated%20Agriculture%20under%20Water%20Framework%20Directive%20and%20Agenda%202000.pdf (accessed on 20 February 2015).
40. Manos, B.; Bournaris, T.; Kamruzzaman, M.; Begum, A.A.; Papathanasiou, J. The regional impact of irrigation water pricing in Greece under alternative scenarios of European policy: A multicriteria analysis. *Reg. Stud.* 2006, *40*, 1055–1068.
41. Berkhout, F.; Hertin, J. *Foresight Futures Scenarios Developing and Applying a Participative Strategic Planning Tool*; Greenleaf Publishing: Sheffield, UK, 2002.
42. Department of Trade and Industry. *Foresight Futures 2020 Revised Scenarios and Guidance*; Department of Trade and Industry; HMSO: London, UK, 2002.
43. A Combined Linear Optimisation Methodology for Water Resources Allocation in an Alfeios River SubBasin (Greece) under Uncertain and Vague System Conditions. Available online: http://meetingorganizer.copernicus.org/EGU2013/EGU2013-1753.pdf (accessed on 20 February 2015).
44. Investigation and Incorporation of Water Inflow Uncertainties through Stochastic Modelling in a Combined Optimization Methodology for Water Allocation in Alfeios River (Greece). Available online: http://meetingorganizer.copernicus.org/EGU2014/EGU2014-8657.pdf(accessed on 20 February 2015).
45. Chen, H.K.; Hsu, W.K.; Chiang, W.L. A comparison of vertex method with JHE method.*Fuzzy Sets Syst.* 1998, *95*, 201–214.
46. Dong, W.; Shah, H.C. Vertex method for computing functions of fuzzy variables. *Fuzzy Sets Syst.* 1987, *24*, 65–78.
47. Rosenberg, D.E. Shades of grey: A critical review of grey-number optimization. *Eng. Optim.*2009, *41*, 573–592.
48. Beale, E. On minimizing a convex function subject to linear inequalities. *J. R. Stat. Soc. Ser. B*1955, *17*, 173–184.
49. Dantzig, G. Linear programming under uncertainty. *Manag. Sci.* 1955, *1*, 197–206.
50. Birge, J.R.; Louveaux, F.V. A Multicut Algorithm for Two-Stage Stochastic Linear Programs.*Eur. J. Oper. Res.* 1988, *34*, 384–392.
51. Kall, P.; Wallace, S.W. *Stochastic Programming*; Wiley: Chichester, UK, 1994.
52. Birge, J.R.; Louveaux, F. *Introduction to Stochastic Programming*; Springer: New York, NY, USA, 1997.
53. Stochastic Programming by Monte Carlo Simulation Methods. Available online: http://edoc.hu-berlin.de/series/speps/2000-3/PDF/3.pdf (accessed on 6 October 2015).
54. Koutsoyiannis, D. A generalized mathematical framework for stochastic simulation and forecast of hydrologic time series. *Water Resour. Res.* 2000, *36*, 1519–1534.
55. Koutsoyiannis, D. Coupling stochastic models of different time scales. *Water Resour. Res.*2001, *37*, 379–392.
56. Efstratiadis, A.; Koutsoyiannis, D.; Kozanis, S. *Theoretical Documentation of the Model of Stochastic Simulation of Hydrologic Parameters Castalia*; Odysseus Program: Athens, Greece, 2005.
57. Box, G.E.P.; Jenkins, G.M. *Time Series Analysis Forecasting and Control*; Holden-Day: San Francisco, CA, USA, 1970.
58. Kozanis, S.; Efstratiadis, A. Zygos: A basin process simulation model. In Proceedings of the 21st European Conference for ESRI Users, Athens, Greece, 6–8 November 2006.
59. Kozanis, S.; Christoforides, A.; Efstratiadis, A. *Scientific Documentation of Hydrognomon Software (Version 4). Development of Database and Software Application in a Web Platform for the "National Database and Meterological Information"*; Department of Water Resources and Environmental Engineering-National Technical University of Athens: Athens, Greece, 2010. (In Greek)

CHAPTER 12

ASSESSMENT OF GROUNDWATER CHEMISTRY AND STATUS IN A HEAVILY USED SEMI-ARID REGION WITH MULTIVARIATE STATISTICAL ANALYSIS

Xuedi Zhang, Hui Qian, Jie Chen and Liang Qiao

Keywords: Multivariate analysis, principal component analysis, cluster analysis, ground water, water quality, Yinchuan region, drought conditions

INTRODUCTION

Groundwater resources are a central source of drinking water around the world, because they are often good quality, difficult to pollute, widely distributed, and adjust throughout the year [1,2,3,4]. China is no exception, and relies greatly on these water resources. In the country's northwest, water for drinking, industrial uses, and irrigation comes mainly from groundwater, which thus plays an essential role in development [5]. With the rapid development of China's economy, the gap between water supply and demand in the northwest is widening each year. In many regions, growing water shortages and the pollution of shallow groundwater seriously restrict economic development [1,3,4], especially in the Yinchuan plain of Ningxia Province. Given the presence of few rivers at a low density, the region relies on artificial irrigation, via diversion of water from the Yellow River through a network of canals and ditches [5]. As contaminants and ions contained in soils dissolve into irrigation water, they infiltrate phreatic water, causing water quality to decline in some regions [3,5].

Groundwater hydrochemistry can be affected by multiple natural factors, such as chemical reactions between water and soil or sediments, biochemical reactions, and surface water–groundwater interactions, as well as human activities [1,2,3]. The industrial and municipal wastewater, whether treated or not, is a constant source of pollution that influences groundwater hydrochemistry. Additionally, groundwater may be affected by non-point sources of pollution, caused by surface and subsurface runoff from irrigation water and waste treatment plants in urban areas [3,4]. These human impacts on groundwater chemistry have been documented extensively [5,6,7]. Phenomena such as increasing industrialization, urbanization, agricultural irrigation, water pollution, and changes in irrigation practices (e.g., the channel line project and dredging of discharge ditches) greatly influence the phreatic water resources of Yinchuan region and may therefore alter water hydrochemistry.

Phreatic water is the main source for domestic and irrigation uses in the Yinchuan region, especially in villages. The over-populated area has traditionally depended on agriculture, with activities such as vegetable and greenhouse cultivation continuing year round [5]. Along with the aforementioned factors, the pumping of water from the Yellow River adds a further pressure to groundwater management. Water levels vary dramatically, influencing the amount that can be diverted from the River. Phreatic water in the region is characterized by a variable regime of recharging with annual irrigation. In these systems, artificial perturbations from irrigation (i.e., fluctuations of the amount and quality of diverted water), population expansion, excess exploitation of groundwater, and other human activities have greatly changed and/or influenced the evolution of the water chemistry and hydrochemical characteristics of natural water in the Yinchuan region. Human intervention has resulted in complicated spatial-temporal changes in the quantity and quality of phreatic water, giving rise to a unique artificial ecosystem. These pressures make

Citation: Zhang, X.; Qian, H.; Chen, J.; Qiao, L. Assessment of Groundwater Chemistry and Status in a Heavily Used Semi-Arid Region with Multivariate Statistical Analysis. Water 2014, 6, 2212-2232. doi:10.3390/w6082212.

understanding the dynamics of phreatic water hydrochemistry in the region relatively more complex.

The goal of this study was to determine the factors influencing phreatic water quality in Yinchuan, and measure its current status, to facilitate sound water management in the region. We used multivariate statistical techniques as they have proven useful for summarizing water-quality data [4,8]. Our specific objectives were to: (1) assess the chemical characteristics of phreatic water in order to identify the factors that affect the region's phreatic water chemistry; (2) use principal component analysis (PCA) to identify the main drivers of phreatic water hydrochemistry; and (3) use cluster analysis (CA) to identify distinct water groups and assess different characteristics and the factors that influence them.

STUDY AREA

Location

The study area is located in the middle of the Yinchuan plain, in northwest China, near the upper reaches of the Yellow River. It is bounded by Helan Mountain in the west, and the Yellow River's western bank in the east. It measures 42–52 km wide from east to west and 38 km long from south to north, with a total area of 1836 km^2 (Figure 1).

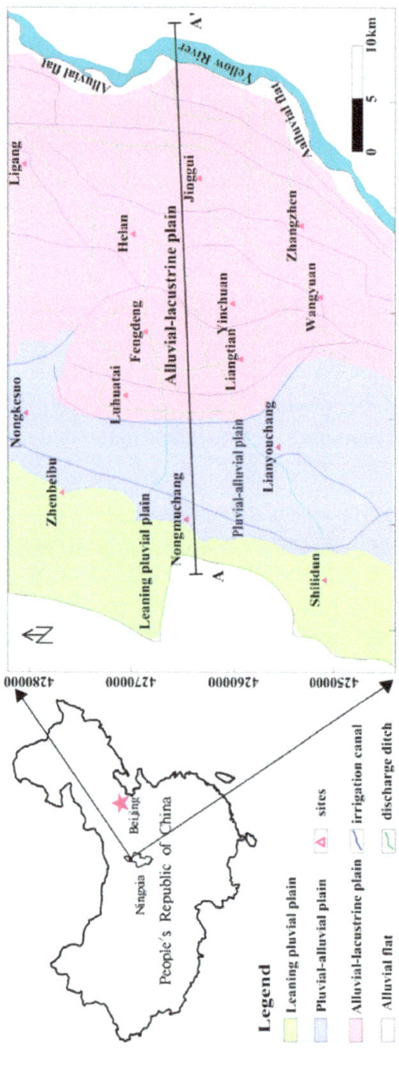

Figure 1. Location of the study area and landforms in Yinchuan, Ningxia Province, China.

The Yinchuan region has an arid to semi-arid climate within the north temperature zone, with a long winter, short summer, less rain and more frequent droughts, ample sunshine, high wind and evaporation, and dramatic temperature changes. According to monitoring data from 1951 to 2010, annual precipitation is 191.9 mm, with 70% of it concentrated from July to September, and the average annual evaporation is 1582.8 mm, more than 8 times the amount of precipitation [9].

Hydrogeology

The elevation of the study area is higher in the west and lower in the east, ranging from approximately 1100 m to 1200 m above mean sea level, the lowest area in Ningxia. Landforms show a zonal distribution, sloping from a leaning pluvial plain at the foot of Helan Mountain in the west, to the pluvial-alluvial plain of the Yellow River, and alluvial-lacustrine plain and alluvial flat in the east (Figure 2). The Yinchuan Plain was formed by deposition of river, lake and flood sediments over a Cenozoic fault basin. It is a graben basin based on the evolution of the Helan tectonic belt and the Helan tectonic belt went through the tension and closing of the structure evolution many times from Mesoproterozoic Era to Cenozoic era, with rising and sink processes. First, the tensional fault is formed along the passive continental margin during Mesoproterozoic Era, with basic magmatic rising activities along the fault, so the tensional fault filled with the littoral clastic rocks and carbonate rocks, basic volcano rocks depositing in some places. Then, this ancient fault expanded again and developed into a collision rift zone in the late Paleozoic, filling with paralic coal-bearing strata and terrestrial clastic rocks. The Helan mountain depression basin is expanded by the Alashan block, thrusting to the east after the Triassic period, filled with very thick deposition of lake, lake delta, alluvial fan. Lastly, the basin subsided greatly by compression in late Mesozoic and expanded to the west gradually. Yinchuan basin continued subsided and the west boundary has been extended to the east of Helan mountain during the late Tertiary period [10,11]. Therefore, the Yinchuan basin is complete and the maximum thickness of Quaternary sediments in the study area is 2000 m [12]. Data from drilling and diverse geological studies show that the pore water in sedimentary rocks of the Yinchuan region is divided into two zones: a single phreatic water zone and a multilayered structure area (Figure 3). The phreatic water zone is mainly in the west near Helan Mountain, with soils composed of gravel, sand and clay. The second zone includes all other areas, characterized by three aquifer layers within a depth of 250 m: moving down from the surface they are phreatic aquifer, upper confined aquifer and lower confined aquifer, with nearly continuous, 3–10 m-thick aquitards between them [9].

Hydraulic head measurements indicate a successive variability in the hydrodynamic relationship among the geological units that constitute the groundwater level of phreatic aquifer, namely the leaning pluvial plain, pluvial-alluvial plain, alluvial-lacustrine plain and alluvial flat. As depicted in Figure 2, the groundwater runoff of phreatic aquifer flows from west to east in general, but the runoff directions and runoff conditions possess certain differences in different regions. In the leaning pluvial plain, groundwater flows from west to east and the hydraulic gradient is larger, at 0.07%–0.15%.

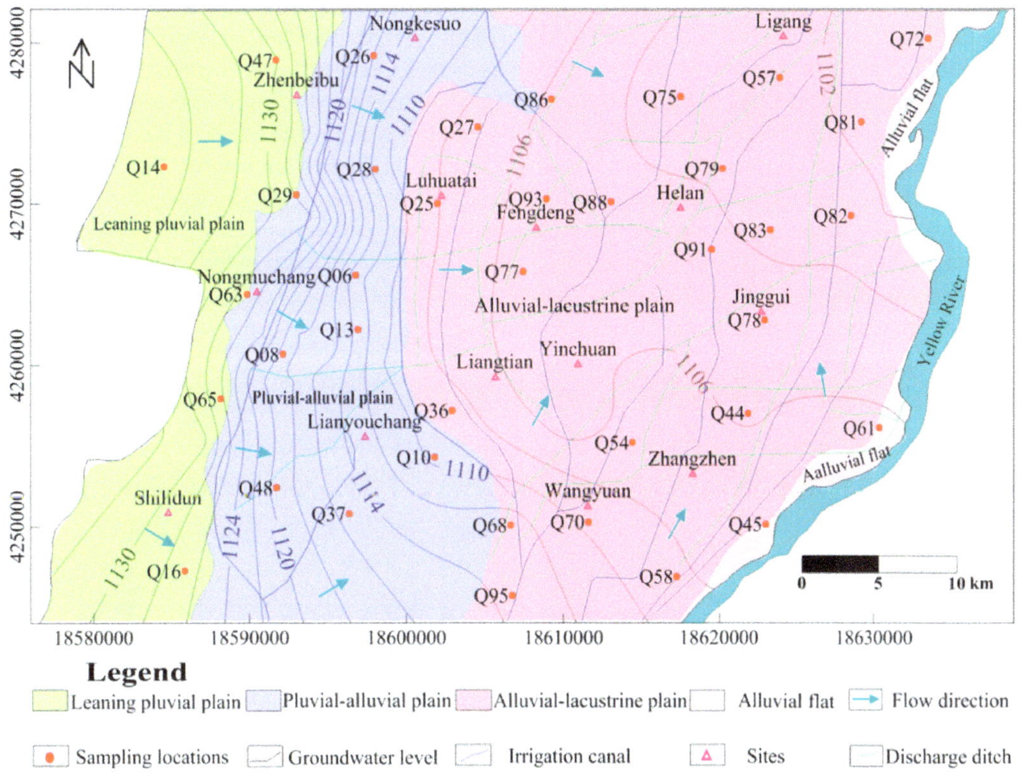

Figure 2. Hydrogeological map of phreatic aquifer and sampling stations in the study area (m).

Along the flow path, because of the existence of fine-grained zone (Figure 3), located between leaning pluvial plain and the pluvial-alluvial plain, groundwater runoff is blocked and hydraulic gradient increases significantly, reached 3%–5%. The direction of groundwater runoff of phreatic aquifer in the alluvial-lacustrine plain diverts into from southwest to northeast and hydraulic gradient decreases significantly, down to 0.07%–0.05%, suggesting that phreatic water flows very slowly in this area (Figure 2). Similar to phreatic aquifer, groundwater runoff of confined water flows mainly from west to east and the hydraulic gradient is relatively large. In order to supply the consumption of industrial and drinking of Yinchuan city, the long-term, concentrated exploitation (212 wells) of groundwater from confined aquifer (8.7×10^7 m^3/a, accounting for 44.03% of the total pumping (1.98×10^8 m^3/a), which were groundwater source areas), has formed a depression cone in the confined aquifer in area between Liangtian and Luhuatai, where groundwater flows down to the center of depression cone and phreatic water recharges confined water for a difference in hydraulic head. However, because hydraulic head of confined aquifer is higher than phreatic aquifer's, confined water will supply phreatic aquifer in other parts of the study area.

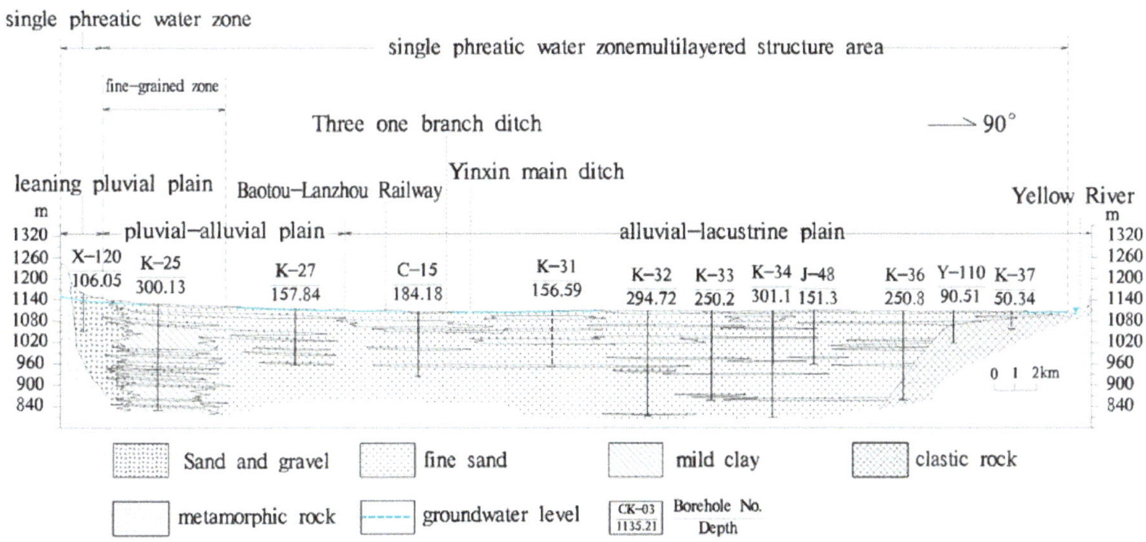

Figure 3. Hydrogeological cross section from A to A'.

Within the region's dense network of agricultural land, canals, and ditches are many scattered lakes, but few rivers. Recharge of phreatic water therefore occurs mainly via seepage from irrigation canals, accounting for more than 80% of the total recharge (about 6.5 × 10^8 m^3/a), and to a lesser extent from precipitation (about 6.2 × 10^7 m^3/a). Summer-autumn irrigation occurs annually from late April to late September, and winter irrigation is from late October to the middle of November (Figure 1). Field studies [9,12] have assessed the depth of phreatic water in the alluvial-lacustrine plain at 1–3 m in spring (before the summer-autumn irrigation), and mostly less than 2 m (Figure 4). Due to the groundwater's shallow depth, slow movement, and the lithology of the aeration zone, evaporation is intensive and accounts for more than 50% of the total discharge (about 6.5 × 10^8 m^3/a) of phreatic water (Figure 4). Another key pathway is via drainage ditches that leads ultimately to the Yellow River [12], accounting for about 16% of the total discharge in the form of farmland irrigation water, industrial waste water, and domestic sewage (Figure 1). These main ditches are perennial and relatively stable, except when their water levels increase during flood periods. They include the Yinxin main ditch, the second drainage ditch and the Yong'er main ditch in the study area.

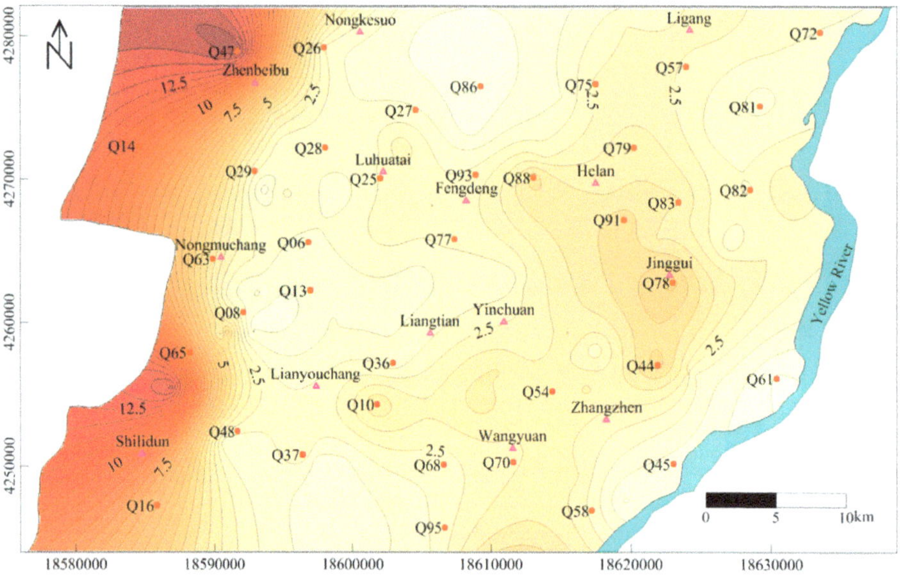

Figure 4. Distribution of groundwater depth (m) (March 2011).

Ecosystems of the Yinchuan region are affected seriously by numerous human activities, such as farming, irrigation, animal husbandry, fisheries, and the extraction of groundwater from aquifers. Water diversion for irrigation from the Yellow River to the Yinchuan Plain has occurred for more than 2000 years [13]. The average amount of water moved from the River to the plain over the last 20 years is 57.46 × 10^8 m^3/a, 21.7 times the water availability from precipitation and inflow of groundwater [13]. Average salinity of the Yellow River water is 0.4 g/L. This long-term practice has increased the water table across the whole plain, and caused more intensive evaporation and hence a higher concentrations of ions in groundwater [9]. Wells on the alluvial-lacustrine plain, especially east of the Tanglai Canal, are less sensitive to rainfall infiltration, but irrigation and extraction are supported mainly by irrigation waters [5]. The main pollution sources of the study area are the point source pollution as industrial factories, non-point source pollution as the agricultural production for using chemical fertilizers and pesticide, and linear pollution sources as drainage ditches accepting industrial and municipal wastewater. NH_4^+ pollution comes mainly from farm planting for using chemical fertilizers and pesticide, livestock breeding and the diverting water from Yellow River.

MATERIALS AND METHODS

Sampling Locations and Analytical Procedures

Samples of phreatic water were collected from 39 wells in 2011, before the summer-autumn irrigation period. They were labeled with a 3-character code, QAB, where Q is phreatic water and AB is the two-digit number of the sampling station (Figure 2). Water temperature and pH were measured at each station. Water was collected in pre-cleaned 1 L plastic polyethylene bottles for physicochemical analysis, after a thorough rinse of the bottle with well water. Handling and preservation of samples followed the Standard Examination Methods for Drinking Water (Ministry of Health of the People's Republic of China, 2006) to maximize quality and consistency [14]. A duplicate sample from each station was acidified to pH 2 with the addition of HNO_3, to assess the constituent cations. In the laboratory of Ningxia Monitoring Station for the Geological Environment, we analyzed sampled by measuring major ions (Na^+, Ca^{2+}, Mg^{2+}, HCO_3^-, SO_4^{2-} and Cl^-), total dissolved solids (TDS), total hardness (TH), K^+, NH_4^+, NO_3^- and fluoride (F^-), using standard procedures recommended by the Chinese Ministry of Water Resources, listed as appendix in Table A1.

We checked those measurements for accuracy by calculating percent charge balance errors (%CBE), with Equation (1).

$$\%CBE = \frac{\sum cations - \sum anions}{\sum cations + \sum anions} \times 100\% \tag{1}$$

where all cations and anions are expressed as milliequivalents per liter. We verified that the TDS measure did not differ much from the sum of all seven major ions. All samples had a small relative error (%CBE) of < ±5%, therefore all 39 samples were used in analyses.

Multivariate Statistical Analysis

Multivariate statistical techniques, such as PCA and CA, are often used as "unbiased methods" to summarize associations between samples and/or variables [15,16]. Such associations based on similar magnitudes and variations in chemical and physical composition may reveal the effects of climate or human activity on water quality. Hierarchical agglomerative CA groups samples by linking their similarities, and thus can be used to illustrate the overall similarity of variables within a dataset. In order to avoid misclassification due to wide differences in the dimensions of data, we standardized measurements with z-scale transformation [17,18]. Standardization tends to increase the influence of factors with small variance, and vice versa. All mathematical and statistical computations were made using the Statistical Package for Social Sciences software (SPSS 16.0) [19].

PCA was used to detect associations among variables (standardized data), thus reducing the number of dimensions in the data table. This was accomplished by diagonalization of the correlation matrix, which transformed the 13 original variables into 13 uncorrelated (orthogonal) ones, i.e., weighted linear combinations of the original variables, or Principal Components (PCs). The eigenvalues of the PCs are a measure of their associated variance, the most participation of the original variables in the PCs is shown by the loadings, and the individual transformed observations are called scores. To reduce the overlap between original variables in each PC, a varimax rotation was conducted [17]. The dataset for PCA included all 13 measures, thereby providing a simultaneous analysis of the whole hydrochemical dataset.

CA is an unsupervised pattern recognition technique that reveals the intrinsic structure or underlying behavior of a dataset, without making a priori assumptions about the data, in order to classify objects into groups or clusters, based on their similarity. There are two types of CA: R-mode and Q-mode [20]. Q-mode CA can be used to assess common water quality data, grouping samples into clusters of distinct quality, based on similarities (i.e., Euclidean distance) in their hydrochemistry. R-mode CA is often used to assess the associations among different variables. In hydrological studies, it has been used to determine the association among water quality parameters, and ultimately the sources and processes that influence them [21,22,23,24]. We chose Ward's method of hierarchical agglomerative CA, a powerful grouping mechanism that uses squared Euclidean distances as a measure of similarity between samples and/or variables (standardized data), has a small space distorting effect, and uses more information about cluster contents than other hierarchical methods [18].

RESULTS AND DISCUSSION

Hydrochemical Characteristics

Analysis of the 13 hydrochemical variables of phreatic water in the study area is summarized inTable 1. The water samples were weakly alkaline, with a pH range of 7.71–8.38. Concentrations of the principal ions varied greatly, with coefficients of variation varying from 0.4 to 1.9, especially Na^+, Mg^{2+}, SO_4^{2-} and Cl^-. For example, the concentration of TDS varied from 272 to 5664 mg/L (mean 1065 mg/L). Measures of all major ions exceeded China's acceptable limits for groundwater quality (Bureau of Quality and Technical Supervision of China 1994). Most samples from the central and eastern parts of the study area had high TDS values above 1000 mg/L, along the Yellow River (Figure 5a), indicating they are unsuitable for drinking. TH measured as $CaCO_3$ varied from 146 to 2845 mg/L, (mean 509 mg/L), and exceeded 450 mg/L in most areas (except the west and small parts of the central study area), suggesting very hard water that is unsuitable for drinking (Figure 5b). Generally, the groundwater from the samples at higher elevation contained mostly Ca^{2+} and HCO_3^-, and the concentrations of Na^+, SO_4^{2-} and Cl^- increased at lower elevations due to mineral dissolution along the paths of groundwater flow [5].

Table 1. Descriptive statistics of hydrochemistry in Yinchuan, China.

Parameters	Phreatic water					Yellow river	National standard
Category	Minimum	Maximum	Mean	SD	CV	/	/
pH	7.71	8.38	8.08	0.11	0.01	8.10	6.5–8.5
TH	146	2845	509	417	0.8	245	450
TDS	272	5664	1065	885	0.8	386	1000
Ca^{2+}	23.0	439.7	83.9	66.3	0.8	53.8	-
Mg^{2+}	23.3	509	85.7	81.4	0.9	27.9	-
K^+	1.5	17.0	4.9	3.4	0.7	3.0	-
Na^+	24.0	860	181	167	0.9	43.0	200
Cl^-	27.6	2592	222	406	1.8	55.2	250
SO_4^{2+}	35.2	986	214	184	0.9	83.6	250
HCO_3^-	220	1106	501	210	0.4	207	-
NO_3^-	0.0	21	2.1	4.4	2.0	-	20
NH_4^+	0.0	1.0	0.1	0.2	4.1	-	0.20
F^-	0.1	1.2	0.3	0.2	0.9	-	1

Note: Units: mg/L (except for pH).

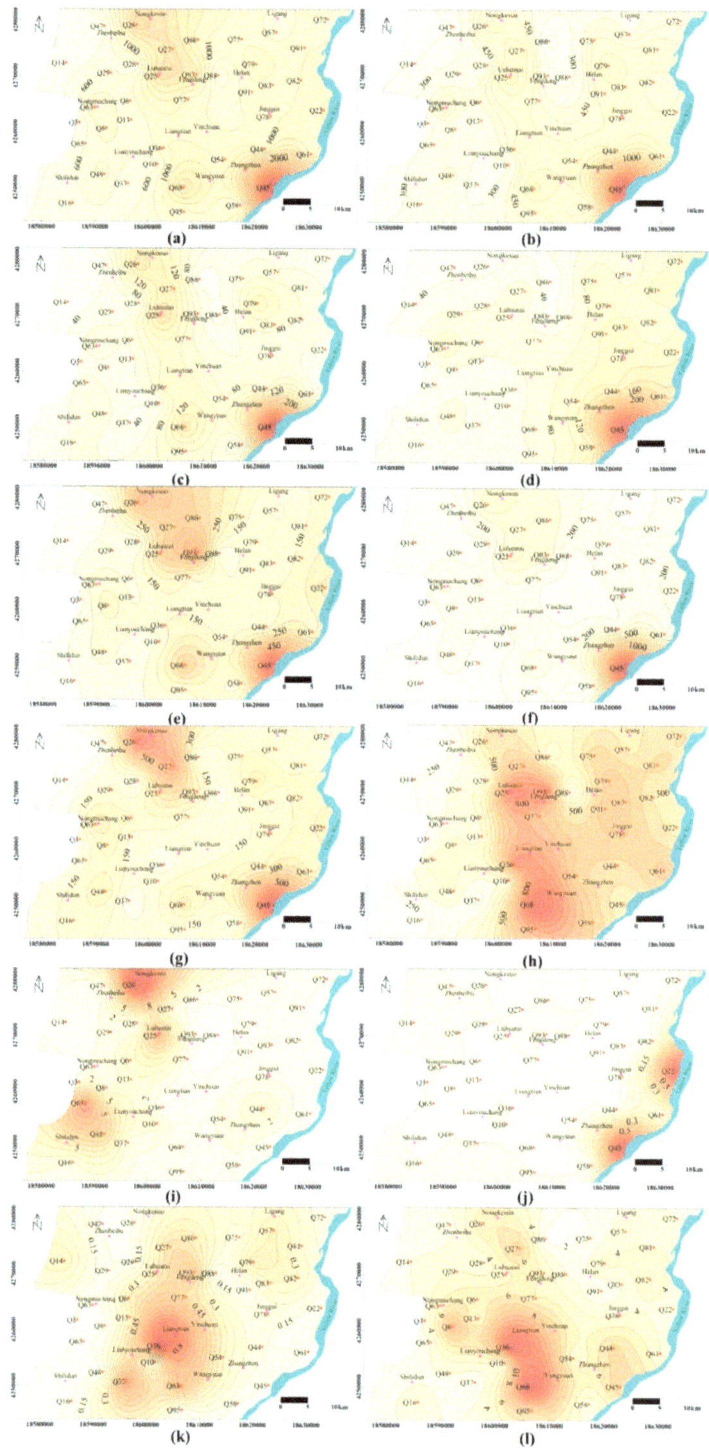

Figure 5. Contour maps of hydrochemical measurements (mg/L) for phreatic water (**a**) TDS; (**b**) TH; (**c**) Mg^{2+}; (**d**) Ca^{2+}; (**e**) Na^{+}; (**f**) Cl^{-}; (**g**) SO_4^{2-}; (**h**) HCO_3^{-}; (**i**) NO_3^{-}; (**j**) NH_4^{+}; (**k**) F^{-}; (**l**) K^{+}.

The spatial distribution of TH and Mg^{2+} were similar (Figure 5b,c), with higher values in the central and eastern areas. The spatial distribution of Ca^{2+} increased from the southwest and northwest parts of the study area to the east (Figure 5d). Na^{+}, SO_4^{2-} and Cl^{-} concentrations were higher (beyond the acceptable limits) in the vicinity of Luhuatai and Nongkesuo in the north, and Zhangzhen in the southeast, and lower in other parts of the study area (Figure 5e,f,g). Unusually high concentrations were found at three sample points: Q26, Q27 and Q45. The distribution of SO_4^{2-} increased from the outer parts of the study area toward the center (Figure 5g). Concentrations of HCO_3^{-} increased gradually from west to east, forming four areas of high concentration near sites Q25, Q36, Q68 and Q93 in the central (Figure 5h). NO_3^{-} concentrations were relatively high in the vicinity of Shilidun and Luhuatai, beyond China's acceptable limits in Nongkesuo (in the north), and lower in other parts of the study area (Figure 5i). In contrast, NH_4^{+}concentrations were markedly higher in the banded areas along the Yellow River beyond the national limits for drinking water (Figure 5j). F^{-} concentrations had a very different distribution, being highest in the center and northeast (Figure 5k). Concentrations of K^{+} increased from the southwest and northwest toward the center, peaking in the northeast (Figure 5l).

Concentrations of nearly all major ions were highest near Q26, Q27 (in the north) and Q45 (in the southeast), which may have been caused by similar hydrogeochemical processes. Evaporation in this area is intense, and the groundwater depth is shallow at less than 2 m, while the extinction evaporation depth is 3 m; the amount of annual evaporation is 8 times that of precipitation [5,9,12]. Furthermore, the decrease in ion concentrations at lower elevations may be attributed to leakage from canals and the seepage of irrigation water, which have very low concentrations of dissolved ions. Local activities such as sewage disposal and agricultural irrigation may further recharge the phreatic water levels, causing the quantity to change.

Piper trilinear diagrams are used commonly to identify hydrochemical patterns in ion data [5,19,25]. In terms of cations, most samples were plotted in the central zone of the right delta-shaped region of our Piper diagram (Figure 6), suggesting that some stations had sodium-type water while most were mixed-type. For anions, most samples were located in the left zone of the lower right delta-shaped region, indicating the dominance of bicarbonate-type water, while some stations had mixed-type.

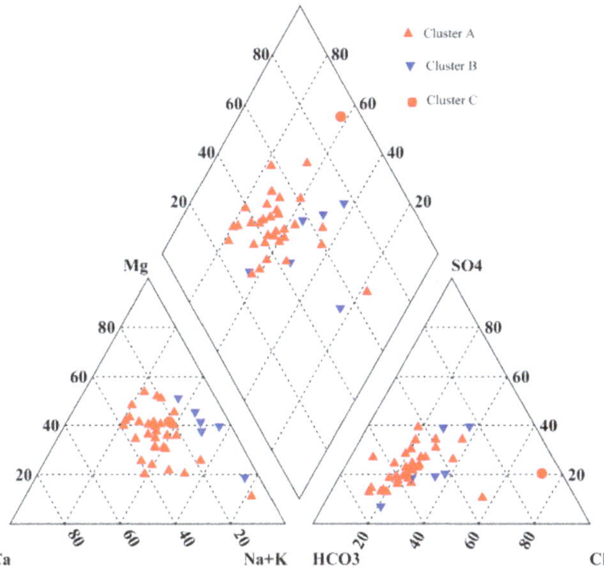

Figure 6. Piper diagram of phreatic water drawn with AquaChem 4.0.

Correlation among Parameters

The correlations among water-quality variables (Table 2) allow us to distinguish several relevant hydrochemical relationships. Measures of the major ions and trace elements, except for K^+, HCO_3^-, NO_3^- and F^-, were significantly and positively correlated with TDS. Concentrations of Cl^-, SO_4^{2-}, Na^+ and Mg^{2+} were related, with correlation coefficients (r) ranging from 0.805 to 0.919, implying that there is possibly a common cause for an increase in these ions. Concentrations of Ca^{2+} and Mg^{2+} were strongly correlated with TH (r = 0.928 and 0.953, respectively), which is intuitive as hardness is an approximate measure of Ca^{2+} and Mg^{2+}. TH was also strongly correlated with SO_4^{2-}, (r = 0.809), and weakly correlated with HCO_3^- (r = 0.245), implying that TH is essentially a permanent hardness. As calcite and dolomite dissolve and precipitate, concentrations of Ca^{2+} and Mg^{2+} increase, and HCO_3^- decreases, resulting in higher TH and lower TDS [26]. The correlation of Ca^{2+} with SO_4^{2-} was not as strong as that of Na^+ and Mg^{2+}, so the dissolution of gypsum cannot be assumed as a main source of Ca^{2+} and SO_4^{2-}. Additional reactions involving Ca^{2+}, such as carbonate dissolution/precipitation and cation exchange, could account for the correlation with SO_4^{2-}.

Table 2. Correlation matrix of the 13 physico-chemical water parameters. The values are the correlations coefficients (r).

Catalog	pH	TH	TDS	Ca^{2+}	Mg^{2+}	K^+	Na^+	Cl^-	SO_4^{2+}	HCO_3^-	NO_3^-	NH_4^+	F^-
pH	1												
TH	−0.50	1											
TDS	−0.57	0.935	1										
Ca^{2+}	−0.41	0.928	0.792	1									
Mg^{2+}	−0.51	0.953	0.952	0.771	1								
K^+	0.04	0.293	0.345	0.106	0.415	1							
Na^+	−0.54	0.742	0.929	0.568	0.805	0.343	1						
Cl^-	−0.62	0.941	0.955	0.846	0.919	0.216	0.839	1					
SO_4^{2-}	−0.58	0.809	0.907	0.641	0.861	0.277	0.851	0.817	1				
HCO_3^-	0.17	0.245	0.347	0.106	0.332	0.557	0.438	0.109	0.227	1			
NO_3^-	−0.27	0.00	0.126	−0.19	0.163	−0.033	0.173	0.044	0.344	−0.02	1		
NH_4^+	−0.41	0.688	0.657	0.719	0.591	0.046	0.556	0.73	0.55	−0.054	−0.11	1	
F^-	0.243	−0.06	0.013	−0.22	0.071	0.551	0.092	−0.06	−0.11	0.443	−0.16	−0.17	1

Notes: **Correlation** = significant at the 0.01 level (2-tailed); *Correlation* = significant at the 0.05 level (2-tailed).

Notably, the chemical reactions within a groundwater system are numerous and highly complex. Although correlation analysis is a useful tool, it can provide only a general insight into phreatic water hydrochemistry [19]. To know more about the exact reactions taking place, more comprehensive analyses of aquifer mineralogy and groundwater flow are needed.

Principal Component Analysis (PCA)

A Bartlett's sphericity test confirmed that the 13 variables were not orthogonal but rather were correlated (Bartlett $\chi^2 = 1354$, 78 df, $p \leq 0.001$). This enables explanation of the variation in these hydrochemical data with a lower number of variables. Additionally, the Kaiser–Meyer–Olkin method (KMO) showed that the measure of sampling adequacy (MSA) was 0.623, indicating that the degree of correlation among the variables and the appropriateness of factor analysis was middling.

PCA considered the hydrochemical variables common to all 39 samples, including pH, TH, TDS, K^+, Na^+, Ca^{2+}, Mg^{2+}, HCO_3^-, SO_4^{2-}, Cl^-, NH_4^+, NO_3^- and F^-. The correlation matrix of the 13 variables (standardized data) revealed several relevant hydrochemical relationships that can help to interpret the primary factors that have influenced current water chemistry in the region [27]. Most variables were strongly correlated with more than one other component (Table 2). In PCA, the correlations among the variables were used to estimate PCs with common associations and associated constituent loadings on PCs [28]. To reduce the overlap between original variables in each PC, a varimax rotation was conducted [17]. The total variance explained by each PC, their loadings, and eigenvalues are shown in Table 3. Based on eigenvalues greater than 1, 4 PCs were extracted that accounted for 87.6% of the total variance in the original dataset (Table 3). The first two PCs explained for most of the variance, at 53.7% and 14.6%, respectively. While PCs 3 and 4 explained 11.2% and 8.1%.

Table 3. Loadings of variables on 4 Principal Components (PCs) of water quality measurements.

Variable	PC1	PC2	PC3	PC4
TH	**0.97**	0.04	−0.04	0.07
Cl^-	**0.97**	0.06	0.03	−0.13
TDS	**0.97**	0.16	0.15	0.07
Mg^{2+}	**0.93**	0.21	0.16	0.08
Ca^{2+}	**0.90**	−0.18	−0.28	0.06
SO_4^{2+}	**0.86**	0.05	0.39	0.04
Na^+	**0.84**	0.25	0.27	0.09
NH_4^+	**0.77**	−0.13	−0.25	−0.20
pH	**−0.59**	0.04	−0.37	0.57
F^-	−0.12	**0.92**	−0.13	0.04
K^+	0.24	**0.79**	0.03	0.26
NO_3^-	0.02	−0.10	**0.94**	−0.02
HCO_3^-	0.21	0.49	0.09	**0.76**
Eigenvalue	6.99	1.89	1.45	1.06
% of Variance explained	53.7	14.6	11.2	8.1
% Cumulative variance	53.7	68.3	79.5	87.6

PC1 was highly related to TH, Cl^-, TDS, Mg^{2+}, Ca^{2+}, SO_4^{2-} and Na^+, indicating that these variables have common patterns. PC1 also had a strongly positive loading for NH_4^+, implying an effect of human activity (e.g., domestic sewage, chemical fertilizers, such as NH_4HCO_3, $CO(NH_2)_2$, NH_4NO_3, $(NH_4)_2SO_4$, $Ca(H_2PO_4)_2$ and KCl) on phreatic water. Further, it had a moderately negative loading for pH, showing that samples with high TDS and much more SO_4^{2-}, so pH will drop. This can be interpreted as a strong evaporation effect due to the drought climate in this area. Major ions, such as Na^+, Mg^{2+}, Ca^{2+}, SO_4^{2-} and Cl^-, are important components of TDS and/or TH.

Due to the fact that the majority of discharge from the phreatic aquifer occurs by evaporation, large amounts of salts remain in the soil and accumulate in phreatic water, lowering its quality. Compounds whose solubility products are small (e.g., the solubility product K_{sp} of $CaCO_3$ is 4.96×10^{-9} at 25 °C) can reach saturation and then precipitate [26], so the main ions that remain dissolved in phreatic water are Mg^{2+}, Ca^{2+}, Na^+, SO_4^{2-}, and Cl^-. The loading of those ions on PC1 ranged from 0.84–0.97. As the concentration of K^+, Na^+, Ca^{2+}, Mg^{2+}, HCO_3^-, SO_4^{2-} and Cl^- in the phreatic water gradually increases with TDS, groundwater will be under saturated with calcite, dolomite and gypsum, with respect to the solubility product. Therefore, we infer that PC1 reflected the strong evaporation effect, shallow groundwater sources, patterns of land use, and intensity of human activities in the Yinchuan region. Indeed, three parts of the region had high values of PC1: Luhuatai and Nongkesuo (in the north), and east of Zhangzheng along the Yellow River (Figure 7a), in which groundwater starts at shallow depth (Figure 4). In contrast, PC1 values were lower in the west, where larger groundwater depths are associated with less evaporation, so the water quality of phreatic water there is hardly affected by the external environment and human activities. The concentrations of ions there were also relatively lower and water type was relatively simpler, for example HCO_3-Mg, HCO_3-Mg.Ca and HCO_3-Mg.Na types.

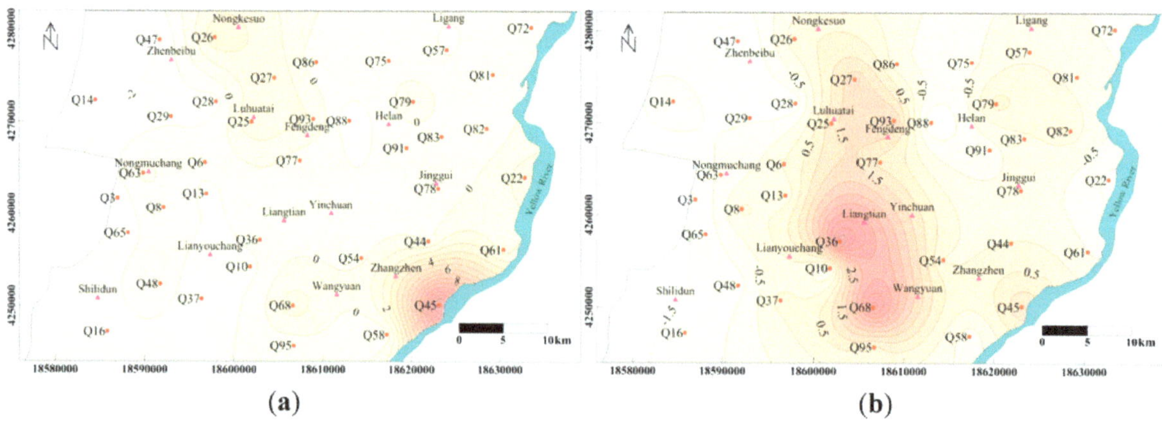

Figure 7. Spatial distribution of sampling stations by (**a**) PC1 scores and (**b**) PC2 scores.

Loadings on PC2 showed that F^- and K^- were most important for that component; in addition, the weak positive correlation of PC2 with HCO_3^- represents dissolution of minerals that contain $F^-$and K^-. Sampling stations in the center of the study area and southeast of Zhangzhen had high PC2 scores, which were otherwise relatively small. This suggests that there is strong dissolution of minerals with F^- and K^- in the central area (Figure 7b). The spatial pattern of PC2 scores was different from that of PC1, i.e., the distribution of high PC2 scores was concentrated in the part of the study area with excessive exploitation of confined water (around Yinchuan City), where groundwater levels fluctuate greatly and the dissolution of mineral is common.

PC3 was greatly influenced by NO_3^-, related to the use of nitrate fertilizers used in irrigation. For example, nitrate measured at sampling station Q26 (near Nongkesuo, in the north) was 21.47 mg/L, the highest NO_3^- concentration that we found, significantly beyond Grade 3 of the water quality standards (Table 1; Figure 5j). The highest positive loading on PC4 was by HCO_3^-, indicating an influence of the dissolution of carbonate minerals.

Cluster Analysis

Q-Mode Cluster Analysis

The classified results of Q-mode CA are presented as dendrograms. For our results, an imaginary horizontal line (phenon line) is drawn across the dendrogram at a linkage distance of about seven reveals three distinct groups, clusters A through C (Figure 8), whose member sampling stations are listed in Table 4. Stiff diagrams for each cluster of water samples show the mean concentrations of their main constituent ions (Figure 8), which combined with the dendrogram and the summary of geochemical and physical data (Table 5) demonstrate that the three clusters are geochemically independent groups.

Each cluster appeared to have a distinct water quality, different from the other two groups. Looking at TDS for example, cluster A includes six subgroups, A1 to A6, where the average TDS concentration is 790 mg/L, and ranged from 272 to 1236 mg/L (characteristic of fresh water). Cluster B includes three subgroups, B1 to B3, with an average TDS of 1765 mg/L (ranging from 985 to 2200 mg/L), which is characteristic of moderately salty water influenced by evaporation and pollution. However, Cluster C includes only one sample, Q45, with TDS of 5664 mg/L, characteristic of salt water. TDS, TH and most major ions (other than Ca^{2+} and HCO_3^-) increased significantly from Cluster A to Cluster C, yet minor ions (K^+, NO_3^- and F^-) were most concentrated in Cluster B. There was no clear trend for Ca^{2+}, probably due to its involvement in so many complex reactions, including carbonate dissolution/precipitation and cation exchange. Similarly, by considering each of the other water quality variables (major ions, minor ions and trace elements), the water quality or chemistry associated with each cluster can be assessed in detail [27].

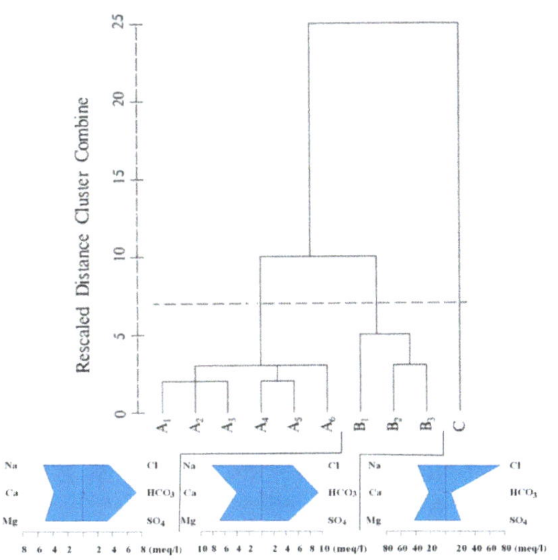

Figure 8. Dendrogram from cluster analysis (Q-mode) of water samples showing the division into three clusters and the stiff diagram of mean concentrations for each cluster.

Table 4. Samples contained in each of the clusters of the dendrogram shown in Figure 8.

Cluster	Samples
A1	Q3,Q8,Q10,Q16, Q28,Q29, Q47,Q63,Q88
A2	Q48, Q65
A3	Q6, Q13, Q37, Q57, Q77, Q86
A4	Q14, Q82
A5	Q44,Q54, Q58,Q61,Q72,Q75,Q78,Q79,Q81,Q83,Q91,Q95
A6	Q22
B1	Q36, Q68,
B2	Q25, Q27, Q93
B3	Q26
C	Q45

Table 5. Mean concentrations of chemical parameters for different clusters of water samples.

Index	Cluster A	Cluster B	Cluster C
pH	8.10	8.06	7.71
TH	417	610	2845
TDS	790	1765	5664
Ca^{2+}	78.0	56.0	439.7
Mg^{2+}	60.9	147.4	508.7
K^+	3.9	9.6	8.0
Na^+	122	381	860
Cl^-	123	351	2592
SO_4^{2+}	159	378	986
HCO_3^-	444	815	440
NO_3^-	1.53	5.76	0
NH_4^+	0.03	0.00	1.01
F^-	0.21	0.55	0.10

Note: Units: mg/L (except for pH).

The water samples of Cluster A were dominated by magnesium, sodium and bicarbonate, however, calcium, chloride and sulfate were also present, and in high concentration in some areas. Therefore, the dominant hydrochemical facies are HCO_3-Mg, HCO_3-Mg.Na and HCO_3-Na.Ca types. In terms of chemical composition, this water had the lowest mean concentrations of some of the main ions and K^+, compared to other waters of the study area, with the exception of Ca^{2+} and HCO_3^-. However, its pH values were the highest. Most samples in this group were obtained in the west of the study area, near the leaning pluvial plain of Helan Mountain, and in the northeast. These phreatic water samples were found in stations where groundwater flows quickly and the depth down to groundwater is relatively large, resulting in relatively little influence of pollution discharge and evaporation, so water quality was better in this part of the study area (Figure 4 andFigure 6).

The water samples of Cluster B were dominated by sodium, magnesium, bicarbonate and chloride, and they also had relatively high concentrations of sulfate and calcium. So the dominant hydrochemical facies are HCO_3-Na.Mg, HCO_3·Cl-Na.Mg and HCO_3·Cl-Na types. As for chemical elements, this water type had the highest mean concentrations of HCO_3^-, K^+, NO_3^- and F^-, while Ca^{2+} was the lowest among the clusters. Cluster B stations were found in the center of the study area (Figure 4) and had higher ion concentrations than Cluster A (except Ca^{2+} and NH_4^+), which may indicate that they are more influenced by evaporation than group A stations, due to run off underground. These samples were collected mainly in the vicinity of Luhuatai and Nongkesuo (in the north) and Wangyuan town, where depths down to groundwater are shallow (1 to 2.5 m), making water there more susceptible to evaporation and pollution discharge.

Finally, the water sample of Cluster C was dominated by sodium, magnesium and chloride, and it also had relatively high concentrations of sulfate, bicarbonate and calcium, so the hydrochemical facies is Cl-Na.Mg type. This water had the highest mean concentrations of main ions and NH_4^+, while pH, HCO_3^-, NO_3^- and F^- were the lowest. The single sample station of this cluster, Q45, is located east of Zhangzheng in the alluvial flat near the Yellow River. The depth to groundwater level there is less than 1.5m (Figure 4), so evaporation is high.

R-Mode Cluster Analysis

Based on this cluster analysis of our standardized, z-transformed water quality data [17,18], there were two distinct groups or clusters of sampling stations (Figure 9). Cluster 1 represented measures of potassium (K^+), bicarbonate (HCO_3^-), fluoride (F^-), nitrate (NO_3^-) and pH. Cluster 2 represented measures of total dissolved solid (TDS), total hardness (TH), magnesium (Mg^{2+}), calcium (Ca^{2+}), sodium (Na^+), chloride (Cl^-), sulfate (SO_4^{2-}), and ammonium (NH_4^+).

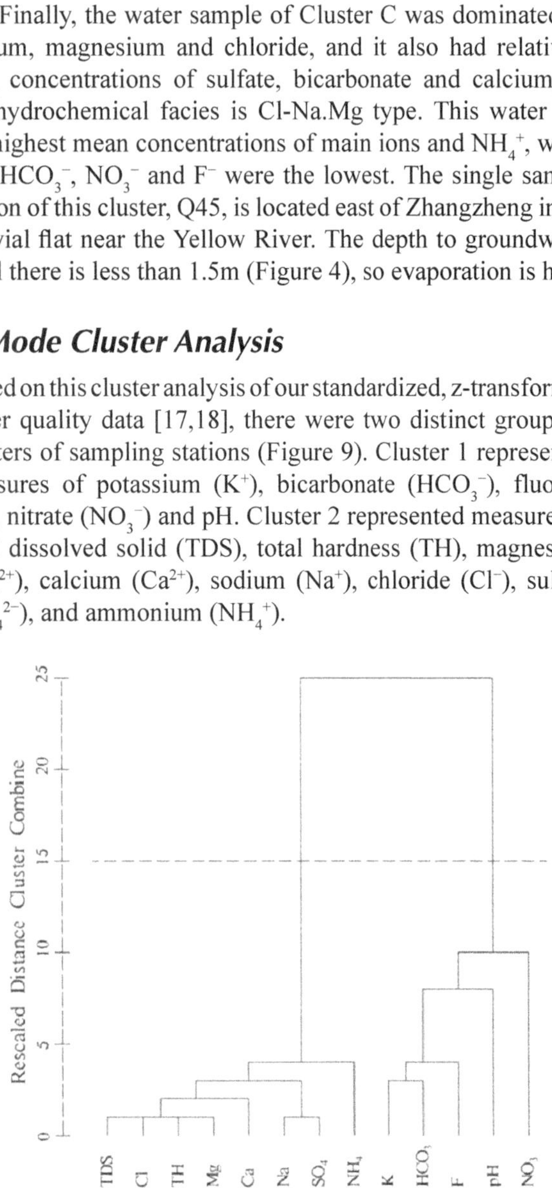

Figure 9. Dendrogram from cluster analysis (R-mode) of 13 variables for water samples.

The presence of two distinct clusters suggests that there may be two distinct sets of influences that are affecting water samples in our study area, i.e., a combination of natural factors, pollution, and other human activities, known as "poles" [30]. To characterize the two poles revealed by clustering, we inferred that the influence on Cluster 1 is most likely the effect of mineral dissolution, because the concentrations of these elements increased with flow down the hydraulic gradient, from west to east. The pole for Cluster 2, however, cannot be attributed to a single factor; there may be many processes that influencing the water quality of this pole simultaneously, including natural processes such as evaporation, and anthropogenic sources such as agricultural practices. Natural processes, such as strong evaporation, as the concentrations of these main ions are not easy to precipitate with other ions during flowing underground and these ions increased greatly. The concentration of TDS for phreatic water ranged between 272 mg/L and 5664 mg/L, with only the precipitation of carbonate will occur. In addition, anthropogenic sources, such as the application of pesticides, insecticides and fertilizers in irrigation farming and domestic sewage, also exert significant impact.

CONCLUSIONS

Phreatic water is an invaluable resource that supports numerous activities in the Yinchuan region, including agricultural irrigation, industry and drinking, and has done so for centuries. However, it is being heavily contaminated by human activities. This paper investigated the hydrochemistry of major ions in phreatic water samples from across the region, using methods such as correlation analysis, PCA, CA and tri-linear diagrams to assess a multivariate data set. Our comprehensive analysis of the Yinchuan region's hydrogeological system can be summarized as follows:

1. The quality of phreatic water varied greatly across the Yinchuan region and its composition changed greatly in alluvial-lacustrine plain, the east of study area. Changes of TDS and concentrations of major ions generally increased along the hydraulic flow path, from west to east. Samples with TDS < 1 g/L were mainly from the west part of the plain, while samples in the central part of the region usually had TDS of 1 to 1.5 g/L. Moderate salt water with TDS > 1.5 g/L was located at three single areas: west of Wangyuan Town, near Luhuatai and Nongkesuo in the north, and east of Zhangzheng, along the Yellow River (Figure 5a).

2. The quality of phreatic water in study area is mainly controlled by the strong evaporation effect caused by the dry climate, dissolution of carbonate minerals and those containing F− and K−, and human activities including the treatment of industrial and municipal wastewater, the discharge of domestic sewage, and utilization of chemical fertilizer. This is confirmed by the results of correlation analysis and principal component analysis. The main hydrogeochemical processes include the strong evaporation effect caused by the dry climate, the dissolution/precipitation of limestone, dolomite and calcite. Cation exchange and dissolution of other minerals, such as fluorite, feldspar and mica, also took place. Mixing with local irrigation water can also explain some of the observed variation of ions.

3. Two types of clustering analysis helped to further characterize the water quality. CA in Q-mode identified three clusters, representing a significant gradient from A to C of increasing hydrochemical measures (except Ca^{2+} and HCO_3^-), from fresh water to almost saline water, which is consisted with the general flow

direction (from west to east). R-mode CA revealed two distinct clusters of variables, related to two distinct sets of factors that are likely influencing Yinchuan region's phreatic water, which may include natural processes pollution, and/or other human impacts.

4. In order to improve and protect the quality of phreatic water, greater importance should be attached to water resources management and planning in the study area. In leaning pluvial plain, as TDS < 1 g/L and the concentrations of other ions are small, groundwater depth is much larger than 3 m (the extinction evaporation depth), so the evaporation effect is weak and the quality is good for drinking, agricultural irrigation and industry. In addition, the area is the recharge zone of groundwater for the whole plain, so water resources management shall be strictly performed in this area for the pollution from industrial enterprises and livestock breeding. However, in other parts, the water qualities of phreatic water have been polluted more or less due to the lessening groundwater depth and extensive influences of human activities. Therefore, the best solution is reducing the amount of evaporation by increasing pumping groundwater and decreasing canal irrigation water in irrigation period in order to reduce groundwater level into a more proper level.

ACKNOWLEDGMENTS

This research was supported by the Doctor Postgraduate Technical Project of Chang'an University (2013G5290003) and the National Natural Science Foundation of China (41172212). The authors would like to acknowledge the anonymous reviewers for their critical review of the manuscript and suggestions for further improvements.

Author Contributions

The work presented here was carried out in collaboration between all authors. Zhang Xuedi and Qian Hui defined the research theme, designed methods and experiments. Chen Jie and Qiao Liang carried out the laboratory experiments. Zhang Xuedi analyzed the data, interpreted the results and wrote the paper. All authors have contributed to, seen and approved the manuscript.

APPENDIX

Table A1. General information of samples.

Samples	pH	TH	TDS	Ca^{2+}	Mg^{2+}	K^+	Na^+	Cl^-	SO_4^{2-}	HCO_3^-	NO_3^-	NH_4^+	F^-
Q3	8.0	254.5	368	49.9	34.9	3.00	31.0	51.8	58.2	245.1	1.24	0.00	0.10
Q6	8.1	405.2	727	71.0	64.0	6.00	89.2	89.8	209.6	352.0	3.62	0.00	0.36
Q8	8.0	317.2	933	55.7	50.1	10.00	197.0	158.8	235.4	377.1	1.47	0.00	0.08
Q10	8.1	260.3	528	51.8	34.9	3.00	86.0	62.1	127.1	289.1	2.49	0.00	0.16
Q13	8.1	334.7	621	55.7	55.9	4.50	86.0	103.6	129.6	333.1	2.71	0.00	0.30
Q14	8.4	197.8	272	40.3	25.6	3.00	24.0	27.6	36.4	251.4	0.57	0.03	0.36
Q16	8.1	314.1	517	44.2	60.5	3.50	44.0	58.7	172.9	232.6	1.70	0.00	0.10
Q22	8.1	513.2	1236	115.2	55.9	3.00	268.0	281.8	272.1	465.1	0.34	0.93	0.10
Q25	8.0	789.3	1905	71.0	192.1	4.50	356.8	466.0	337.0	892.5	11.30	0.00	0.36
Q26	7.9	635.7	2200	32.6	179.3	6.00	480.0	490.2	701.9	534.2	21.47	0.00	0.10
Q27	8.1	661.1	1850	74.9	145.5	8.00	370.0	314.1	592.4	647.4	1.81	0.00	0.44
Q28	8.0	305.5	439	57.6	44.2	3.00	40.0	58.7	53.1	333.1	0.00	0.00	0.10
Q29	8.1	349.2	631	61.4	54.7	4.00	83.0	79.4	164.7	333.1	1.70	0.00	0.11
Q36	8.2	515.2	985	51.8	119.9	17.00	149.0	145.0	62.6	842.2	0.00	0.00	1.20
Q37	8.0	213.0	359	38.4	32.6	2.00	44.0	62.1	53.0	220.0	1.47	0.00	0.76
Q44	8.1	600.2	951	126.7	73.3	7.50	107.0	141.5	205.2	534.2	3.39	0.00	0.16
Q45	7.7	2845.0	5664	439.7	508.7	8.00	860.0	2592.5	986.3	440.0	0.00	1.01	0.10
Q47	8.1	201.5	290	34.4	32.6	2.50	24.0	34.5	35.2	226.3	0.36	0.00	0.08
Q48	8.0	436.4	896	76.8	68.7	3.00	138.0	148.4	256.6	358.3	8.50	0.00	0.10

Q54	8.1	571.5	969	111.4	79.2	4.00	128.0	155.3	187.6	565.7	1.47	0.00	0.30
Q57	8.1	380.9	673	65.3	61.7	3.50	86.0	34.5	159.1	471.4	0.23	0.00	0.36
Q58	8.1	474.6	810	121.0	37.3	3.00	130.4	96.7	100.5	603.4	0.90	0.00	0.10
Q61	8.1	634.4	1157	149.8	61.7	4.00	185.0	189.9	231.6	628.5	0.00	0.08	0.10
Q63	8.0	547.5	1025	99.8	82.6	5.00	138.0	231.3	295.8	308.0	1.36	0.00	0.08
Q65	8.1	332.8	600	42.2	68.7	3.00	72.0	65.6	130.3	383.4	13.56	0.00	0.10
Q68	8.2	753.2	1886	73.0	178.1	17.00	388.0	310.7	290.1	1106.2	0.00	0.00	0.64
Q72	8.1	438.0	915	101.8	44.2	3.00	174.0	145.0	146.4	553.1	0.00	0.00	0.16
Q75	8.1	511.3	988	101.8	68.7	2.00	164.0	138.1	199.5	590.8	0.11	0.00	0.16
Q77	8.1	462.3	934	69.1	85.0	5.00	153.0	110.5	156.3	653.6	0.00	0.00	0.44
Q78	8.2	555.1	1002	115.2	69.8	4.00	153.0	151.9	181.3	609.7	0.68	0.00	0.16
Q79	8.0	706.0	1217	92.2	143.2	4.50	138.0	200.2	294.9	641.1	0.57	0.00	0.40
Q81	8.1	558.6	988	115.2	71.0	4.00	153.0	145.0	188.2	565.7	0.00	0.00	0.24
Q82	8.4	467.9	1003	76.8	79.2	5.00	172.0	169.2	233.6	490.2	0.14	0.00	0.30
Q83	8.1	577.5	894	107.5	85.0	3.00	98.4	131.2	132.4	615.9	0.00	0.00	0.30
Q86	8.0	145.9	1048	23.0	25.6	1.50	344.0	359.0	95.9	364.5	0.23	0.00	0.36
Q88	8.0	242.6	397	57.6	23.3	3.00	57.0	55.2	65.2	238.8	0.00	0.00	0.08
Q91	8.2	547.3	976	105.6	76.8	2.00	144.0	127.7	168.6	647.4	0.00	0.00	0.11
Q93	8.0	307.4	1765	32.6	69.8	5.00	540.0	379.7	282.6	867.3	0.00	0.00	0.56
Q95	8.1	477.7	908	61.4	97.8	7.00	145.0	82.9	122.3	735.4	0.00	0.00	0.10

Conflicts of Interest

The authors declare no conflict of interest.

REFERENCES

1. Brunke, M.; Gonser, T. The ecological significance of exchange processes between rivers and groundwater. Freshw. Biol. **1997**, 37, 1–33.
2. Dillon, P.J.; Kirchner, W.B. The effects of geology and land use on the export of phosphorous from watersheds. Water Res. **1975**, 9, 125–148.
3. Sophocleous, M. Interactions between groundwater and surface water: The state of the science. Hydrogeology **2002**, 10, 52–67.
4. Vega, M.; Pardo, R.; Barrado, E.; Deban, L. Assessment of seasonal and polluting effects on the quality of river water by exploratory data analysis. Water Res. **1998**, 32, 3581–3592.
5. Qian, H.; Li, P.Y. Hydrochemical characteristics of groundwater in Yinchuan plain and their control factors. Asian J. Chem. **2011**, 23, 2927–2938.
6. Jin, X.; Wan, L.; Zhang, Y.; Xue, Z.; Yin, Y. A study of the relationship between vegetation growth and groundwater in the Yinchuan Plain. Earth Sci. Front. **2007**, 14, 197–203.
7. Andradea, E.M.; Palacio, H.A.Q.; Souzab, I.H.; Leaoa, R.A.O.; Guerreiroc, M.J. Land use effects in groundwater composition of an alluvial aquifer (Trussu River, Brazil) by multivariate techniques. Environ. Res. **2008**, 106, 170–177.
8. Reisenhofer, E.; Adami, G.; Barbieri, P. Using chemical and physical parameters to define the quality of karstic freshwaters (Timavo River, Northeastern Italy): A chemometric approach.Wat. Res. **1998**, 32, 1193–1203.
9. Qian, H.; Wu, J.; Zhou, Y.H.; Li, P.Y. Stable oxygen and hydrogen isotopes as indicators of lake water recharge and evaporation in the lakes of the Yinchuan Plain. Hydrol. Process. **2013**.
10. Liu, P.G.; Fan, S.X.; Li, X.J. The geochemical element characteristics and paleo-sedimentary environment of the Quaternary deposits in Yinchuan Basin. J. Geomech. **2000**, 6, 43–50.
11. Zhang, L.; Wang, L. Groundwater Resources in Ningxia; Ningxia People Publishing House: Yinchuan, China, 2003.
12. Wu, X.H.; Qian, H.; Yu, D.M. Investigation and Assessment of Rational Allocation of Groundwater Resources in the Yinchuan Plain; Geology Publishing House: Beijing, China, 2008.
13. Lu, D.M. History of Diverting Yellow River Water for Irrigation in Ningxia; China Water Power Press: Beijing, China, 1987.
14. Ministry of Health of the People's Republic of China. Standards for drinking water quality; GB 5749-2006; Standards Press of China: Beijing, China, 2007.
15. Kaufman, L. Finding Groups in Data: An Introduction to Cluster Analysis; Wiley: New York, NY, USA, 1990.
16. Tang, Q.Y. DPS© Data Processing System: Experimental Design, Statistical Analysis and Data Mining, 2nd ed.; Science Press: Beijing, China, 2010. (In Chinese)
17. Helena, B.; Pardo, R.; Vega, M.; Barrado, E.; Fernandez, J.M.; Fernandez, L. Temporal evolution of groundwater composition in an alluvial aquifer (Pisuerga River, Spain) by principal component analysis. Water Res. **2000**, 34, 807–816.
18. Willet, P. Similarity and Clustering in Chemical Information Systems; Research Studies Press, Wiley: New York, NY, USA, 1987.

19. Li, P.Y.; Qian, H.; Wu, J.H.; Zhang, Y.Q.; Zhang, H.B. Major ion chemistry of shallow groundwater in the Dongsheng Coalfield, Ordos Basin, China. Mine Water Environ. **2013**, 32, 195–206.

20. Brown, C. Applied Multivariate Statistics in Geohydrology and Related Science, 1st ed.; Springer: Berlin, Germany, 1998; pp. 1–248.

21. Hopke, P.K.; Gladney, E.S.; Gordon, G.E.; Zoller, W.H.; Jones, A.G. The use of multivariate analysis to identify sources of selected elements in the Boston urban aerosol. Atmos. Environ.**1976**, 10, 1015–1025.

22. Voudouris, K.; Panagopolous, A.; Koumanatakis, J. Multivariate statistical analysis in the assessment of hydrochemistry of the Northern Korinthia Prefecture alluvial system (Peloponnese, Greece). Nat. Resour. Res. **2000**, 9, 135–146.

23. Hussein, M.T. Hydrochemical evaluation of groundwater in the Blue Nile Basin, eastern Sudan, using conventional and multivariate techniques. Hydrogeol. J. **2004**, 12, 12–144.

24. Mahlknecht, J.; Steinich, B.; Navarro de Leon, I. Groundwater chemistry and mass transfers in the Independence aquifer, central Mexico, by using multivariate statistics and mass-balance models. Environ. Geol. **2004**, 45, 781–795.

25. Hussain, M.; Ahmed, S.M.; Abderrahman, W. Cluster analysis and quality assessment of logged water at an irrigation project, eastern Saudi Arabia. J. Environ. Manag. **2008**, 86, 297–307.

26. Qian, H.; Zhang, X.D.; Li, P.Y. Calculation of $CaCO_3$ solubility (precipitability) in natural waters. Asian J. Chem. **2012**, 24, 668–672.

27. Li, P.Y.; Qian, H.; Wu, J.H. Hydrochemical characteristics and evolution laws of drinking groundwater in Pengyang County, Ningxia, Northwest China. E J. Chem. **2011**, 8, 565–575.

28. Lucas, L.; Jauzein, M. Use of principal component analysis to profile temporal and spatial variations of chlorinated solvent concentration in groundwater. Environ. Pollut. **2008**, 151, 205–212.

29. Wu, J.H.; Li, P.Y.; Qian, H.; Duan, Z.; Zhang, X.D. Using correlation and multivariate statistical analysis to identify hydrogeochemical processes affecting the major ion chemistry of waters: a case study in Laoheba phosphorite mine in Sichuan, China. Arab. J. Geosci. **2013**.

30. Rapin, F. Anthropogenic Effects on Sediment from Bay of Nice and Villefranche-sur-Mer on the French Coast. Ph.D. Dissertation, University of Geneva, Geneva, Switzerland, 1980; p. 221.

CHAPTER 13 UNCERTAINTY ANALYSIS OF MULTIPLE HYDROLOGIC MODELS USING THE BAYESIAN MODEL AVERAGING METHOD

Leihua Dong, Lihua Xiong, and Kun-xia Yu

INTRODUCTION

To date, various hydrological models have been put forward and widely used in flood forecasting, planning, and water resources management [1, 2]. Since different models have strengths in capturing different aspects of the real world processes, combining the results from diverse models by weighting procedures can present a better performance than any individual model [3–5]. The early model combination researches in hydrologic forecasting employed such tools as neural network [6] and fuzzy system [7]. Recently, Bayesian Model Averaging (BMA), a method for averaging over different competing models, has been introduced to ensemble hydrologic predictions.

Bayesian Model Averaging came to prominence in statistics in the mid-1990s, and Madigan and Raftery [8] were the first to propose this method for combining predictions. Subsequently, Raftery [9] and Draper [10] gave more detailed discussion about BMA. It has been applied in diverse fields such as economics [11], biology [12], ecology [13], public health [14], toxicology [15], meteorology [16], and management science [17]. In many case studies, BMA produces accurate and reliable predictions and was shown to be a better scheme than other model-combining methods [18–20]. In recent years, hydrologists have also applied BMA to hydrologic modeling, such as groundwater [21] and rainfall-runoff modeling [22–24].

A prediction from a single model has been recognized to be associated with a certain degree of uncertainty, and so is the prediction from combining a number of different single models. Thus, uncertainty analysis is an indispensable element for any hydrologic modeling study. The uncertainty usually arises from errors during the calibration of parameters, the design of model structure, and measurements of input and output data [25, 26]. To account for these uncertainties, many uncertainty analysis techniques have been developed and applied to diverse catchments, such as Generalized Likelihood Uncertainty Estimation (GLUE), Parameter Solution (ParaSol), and Bayesian inference based on Markov chain Monte Carlo (MCMC) [27, 28]. Each of those techniques has its own advantage in uncertainty analysis. In the uncertainty analysis of BMA scheme, the composition of Monte Carlo method [29] is used to generate BMA probabilistic ensemble predictions, and then the 90% uncertainty intervals can be derived within the range of the 5% and 95% quantiles.

Previous studies of BMA in hydrology mostly focused on the comparison of the BMA mean prediction with each individual model's prediction, to prove the better performance of the prediction after weighted averaging. As BMA also has the ability to provide a statistical distribution of the quantity to be forecasted, the research focus in this study is shifted onto the comparison of the prediction uncertainty interval generated by the BMA with that of each individual model, in order to see if BMA can also improve the prediction reliability. The technical route of the research in this paper is described in Figure 1. Another purpose of this paper is that by calibrating different hydrological models under different objective functions, each of which has distinctive advantages in better modeling certain flow ranges, we can construct different sets of ensemble members for combination in order to fully explore the superiority of BMA. Therefore, two kinds of BMA combination schemes are designed, analyzed, and compared. In the first BMA scheme, we calibrate each of the three models under the same Nash-Sutcliffe efficiency objective function, thus providing three-member predictions ensemble for the

Citation: Leihua Dong, Lihua Xiong, and Kun-xia Yu, "Uncertainty Analysis of Multiple Hydrologic Models Using the Bayesian Model Averaging Method," Journal of Applied Mathematics, vol. 2013, Article ID 346045, 11 pages, 2013. doi:10.1155/2013/346045.

BMA combination. In the second BMA scheme, three different objective functions other than Nash-Sutcliffe efficiency are adopted, each of which is supposed to have some advantage of better simulating a certain range of flows (low flow, medium flow, and high flow). All three models are, respectively, calibrated for each of three objective functions to obtain the optimized parameter sets.

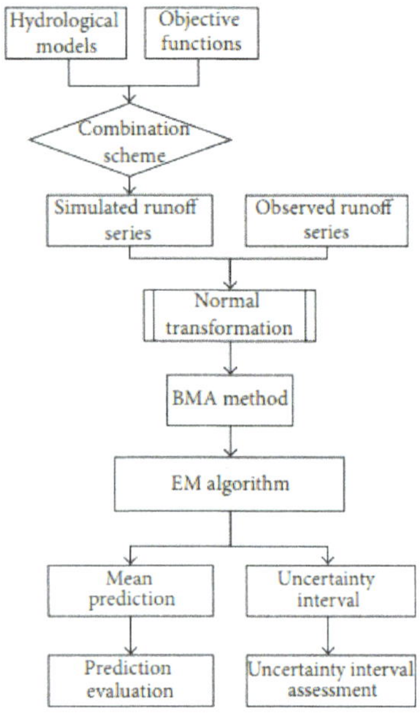

Figure 1: Flowchart of using BMA scheme for hydrological ensemble prediction as well as for prediction uncertainty analysis.

METHODS

Bayesian Model Averaging

Bayesian Model Averaging (BMA) is a statistical technique designed to infer a prediction by weighted averaging over many different competing models. This method is not only a scheme for model combination but also a coherent approach for accounting for between-model and within-model uncertainty [22]. Below is a brief description of the basic ideas of this method.

Let us consider a quantity Q to be predicted on the basis of input data $D = [X, Y]$ (X denotes the input forcing data, and Y stands for the observational flow data). $f = [f_1, f_2, \ldots, _K]$is the ensemble of the K-member predictions. The probabilistic prediction of BMA is given by

$$p(Q \mid D) = \sum_{k=1}^{K} p(f_k \mid D) \cdot p_k(Q \mid f_k, D). \quad (1)$$

The terms in (1) are explained as follows. $(f_k \mid D)$ is the posterior probability of the prediction f_k given the input data D and reflects how well model f_k fits Y. Actually $p(f_k \mid \mathrm{D})$ is just the BMA weight w_k, and better performing predictions receive higher weights than the worse performing ones; all weights are positive and should add up to 1. $p_k(Q \mid f_k, D)$ is the conditional probability density function (PDF) of the predictand Q conditional onf_k and D. For computation convenience, $p(Q \mid f_k, D)$ is always assumed to be a normal PDF and is represented as$g(Q \mid f_k, \sigma_k^2) \sim N(f_k, \sigma_k^2)$, where σ_k^2 is the variance associated with model prediction f_k and observations Y. In order to make this assumption valid, some techniques such as Box-Cox transformation are needed to make the data approximately normally distributed and to narrow the data range. The BMA mean prediction is a weighted average of the individual model's predictions, with their posterior probabilities being the weights. In the case that the observations and individual model predictions are all normally distributed, the BMA mean prediction can be expressed as

$$E[Q \mid D] = \sum_{k=1}^{K} p(f_k \mid D) \cdot E\left[g\left(Q \mid f_k, \sigma_k^2\right)\right] = \sum_{k=1}^{K} w_k f_k. \quad (2)$$

EM Algorithm for BMA Parameter Estimation

To estimate BMA weight w_k and model prediction variance σ_k^2, the Expectation-Maximization (EM) algorithm, which has proved to be an efficient technique for BMA calculation based on the assumption that K-member predictions are normally distributed, is described in this section [23].

Firstly, if we denote the set of BMA parameters to be estimated by $\theta = \{w_k, \sigma_k^2, k = 1, 2, \ldots, K\}$, the log form of likelihood function can be represented as

$$l(\theta) = \log(p(Q \mid D)) = \log\left(\sum_{k=1}^{K} w_k \cdot g\left(Q \mid f_k, \sigma_k^2\right)\right). \quad (3)$$

It is difficult to maximize the function (3) by analytical method. The EM algorithm is a method for finding the maximum likelihood by alternating between two steps, the expectation step and maximization step. The two steps are iterated to convergence when there is no significant change between two consecutive iterative log-likelihood estimations. In EM algorithm, a latent variable (unobserved quantity) z_k^t is used as an assistant for estimating BMA weight w_k. The procedure of EM algorithm for BMA scheme is described as follows.

(1) Initialization.

Set Iter = 0. Initialize

$$w_k^{(0)} = \frac{1}{K},$$

$$\sigma_k^{2(0)} = \frac{\sum_{k=1}^{K} \sum_{t=1}^{T} \left(Y^t - f_k^t\right)^2}{K \cdot T}, \quad (4)$$

where Iter is the number of iteration and T is the number of data in the calibration period. Y^t and f_k^t are denoted as the observation and the corresponding prediction by the kth model for the time t.

(2) Calculate the Initial Likelihood:

$$l(\theta)^{(0)} = \sum_{t=1}^{T} \log\left(\sum_{k=1}^{K} \left(w_k^{(0)} \cdot g\left(Q \mid f_k^t, \sigma_k^{2(0)}\right)\right)\right). \quad (5)$$

(3) Compute the Latent Variable. Set Iter = Iter + 1, then calculate

$$z_k^{t\,(\text{Iter})} = \frac{g\left(Q \mid f_k^t, \sigma_k^{2\,(\text{Iter}-1)}\right)}{\sum_{k=1}^{K} g\left(Q \mid f_k^t, \sigma_k^{2\,(\text{Iter}-1)}\right)}. \tag{6}$$

(4) Update the Weight:

$$w_k^{(\text{Iter})} = \frac{1}{T}\left(\sum_{t=1}^{T} z_k^{t\,(\text{Iter})}\right). \tag{7}$$

(5) Update the Variance:

$$\sigma_k^{2\,(\text{Iter})} = \frac{\sum_{t=1}^{T} z_k^{t\,(\text{Iter})} \cdot \left(Y^t - f_k^t\right)^2}{\sum_{t=1}^{T} z_k^{t\,(\text{Iter})}}. \tag{8}$$

(6) Update the Likelihood:

$$l(\theta)^{(\text{Iter})} = \sum_{t=1}^{T} \log\left(\sum_{k=1}^{K}\left(w_k^{(\text{Iter})} \cdot g\left(Q \mid f_k^t, \sigma_k^{2\,(\text{Iter})}\right)\right)\right). \tag{9}$$

(7) Check for Convergence. If $(\theta)^{(\text{Iter})} - l(\theta)^{(\text{Iter}-1)}$ is less than a prespecified tolerance level, stop the whole estimation procedure; else go back to Step (3).

Estimation of Prediction Uncertainty Interval

After BMA weight w_k and prediction variance σ_k^2 being estimated, we use the composition of Monte Carlo method to generate BMA probabilistic predictions for any time t [29]. The procedures are described as follows.

(1) Generate an integer value of k from $[1, 2, \ldots, K]$ with probability $[w_1, w_2, \ldots, w_K]$. A specific procedure is described as follows.

- (1a) Set the cumulative weight $w_0' = 0$ and compute

  $$w_k' = w_{k-1}' + w_k \text{ for } k = 1, 2, \ldots, K.$$

- (1b) Generate a random number u between 0 and 1.
- (1c) If $w_{k-1}' \le u < w_k'$, it indicates that we choose the$k$th member of the ensemble predictions.

(2) Generate a value of Qt from the PDF of $g(Q_t \mid f_k^t, \sigma_k^2)$. Here, $g(Q_t \mid f_k^t, \sigma_k^2)$ represents the normal distribution with mean f_k^t and variance σ_k^2.

(3) Repeat the above steps (1) and (2) for M times. M is the probabilistic ensemble size. In this paper, we set $M = 100$.

After generating the BMA probabilistic ensemble predictions, sort them in the ascending order. Then the 90% uncertainty intervals can be derived within the range of the 5% and 95% quantiles. For each individual model in the BMA scheme, the prediction uncertainty interval can also be constructed, with the Monte Carlo sampling method still being used to approximate the assumed PDF of $g(Q_t \mid f_k^t, \sigma_k^2)$.

MATERIALS

Study Area and Data

The study area is Mumahe catchment, a branch of Han River. It is located in Shanxi Province of China and the total area is 1224 km^2. The basin has a subtropical climate, and the area is humid with fairly high precipitation. The mean annual rainfall for the period of 1980–1987 is 1070 mm, and the mean annual runoff is 687 mm, or roughly 64% of the annual rainfall.The hydrological data include daily runoff, rainfall, and evaporation. There are 2992 data points in total, and 1825 (the period of 1980.1.1– 1985.12.31) of them are used for calibration, while the rest 1167 data points (the period of 1986.1.1–1987.12.31) are used for validation.

Hydrological Models and Optimization Algorithm

In this study, three conceptual hydrological models are employed for testing the capability of BMA: the Xinanjiang Rainfall-Runoff Model (XAJ), the Soil Moisture Accounting and Routing Model (SMAR), and SIMHYD Rainfall-Runoff Model. Xinanjiang Rainfall-Runoff Model was developed in 1970s. It is a conceptual hydrologic model, which has been widely used in humid and semihumid regions of China. And all the 15 parameters of this model have strong physical meanings. SMAR model is a lumped conceptual model with soil moisture as a central theme. The model consists of two components in sequence: a water balance component with 5 water balance parameters and a routing component with 4 routing parameters. SIMHYD model is a daily conceptual model that estimates daily stream flow from daily rainfall and areal potential evapotranspiration data and it contains 7 parameters [30]. For calibrating these hydrological models, Shuffled Complex Evolution (SCE-UA) method is employed here for parameter optimization [31].

Objective Functions

The selection of objective function (OF) is of great importance since it will have great influence on the values of calibrated parameters and thus on simulation results of the rainfall-runoff model. Different objective functions can be adopted for different kinds of practical issues. For example, the objective function of squared model errors of squared transformed flow can be applied in high flow studies, and the objective function of squared model errors of logarithmic transformed flow can be applied in low flow studies [32]. In this study, four objective functions have been used for the parameter calibration.

(1) OF1: The Nash-Sutcliffe Coefficient of Efficiency (R^2):

$$R^2 = 1.0 - \frac{\sum_{t=1}^{T}\left(Q_{\text{obs}}^t - Q_{\text{sim}}^t\right)^2}{\sum_{t=1}^{T}\left(Q_{\text{obs}}^t - \overline{Q}_{\text{obs}}\right)^2}, \tag{10}$$

where Q_{obs}^t and Q_{sim}^t are observed and simulated data at time

t and $\overline{Q_{obs}}$ is the average of observed data in the calibration period.

(2) OF2: Mean Squared Error of Squared Transformed (*MSEST*):

$$\text{MSEST} = \frac{\sum_{t=1}^{T} \left({Q_{obs}^{t}}^2 - {Q_{sim}^{t}}^2 \right)^2}{T}. \tag{11}$$

Transforming the observed data in squared form puts great emphasis on fitting peak values.

(3) OF3: Mean Squared Error of Squared Root Transformed (*MSESRT*):

$$\text{MSESRT} = \frac{\sum_{t=1}^{T} \left(\sqrt{Q_{obs}^{t}} - \sqrt{Q_{sim}^{t}} \right)^2}{T}. \tag{12}$$

MSESRT can be employed in the medium flow simulation.

(4) OF4: Mean Squared Error of Logarithmic Transformed (*MSELT*):

$$\text{MSELT} = \frac{\sum_{t=1}^{T} \left(\ln Q_{obs}^{t} - \ln Q_{sim}^{t} \right)^2}{T}. \tag{13}$$

This transformation helps model parameterization to better fit the low flow values.

Construction of BMA(3) and BMA(9) Schemes

When the prediction data are highly non-Gaussian, we should firstly transform the data to be normally distributed by Box-Cox transformation before using EM algorithm. OF1 is the most widely used objective function for parameter optimization and is used in calibrating each of three hydrological models mentioned above to generate three different predictions.

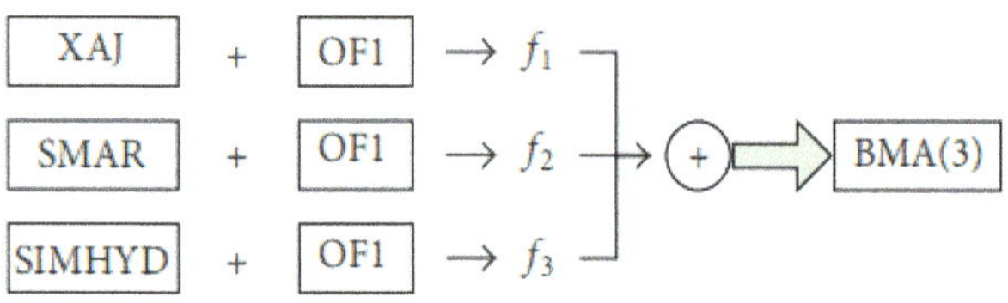

Figure 2: Diagram of BMA(3) combination scheme.

We combine these three different predictions by BMA to construct a three-member predictions ensemble; thus, we denote the first BMA scheme as BMA(3). Figure 2 shows the procedure of BMA(3) combination scheme. The other three objective functions, that is, OF2, OF3, and OF4 are, respectively, fit for high, medium, and low flow simulation. All three hydrological models are, respectively, calibrated for each of these three objective functions to obtain the optimized parameter sets. As the same model with different parameter sets will give rise to different outcomes, nine different predictions are generated. We can use BMA method to combine these nine different predictions to construct a nine-member predictions ensemble, which is just the second BMA scheme denoted as BMA(9). The procedure of BMA(9) combination scheme is described in Figure 3.

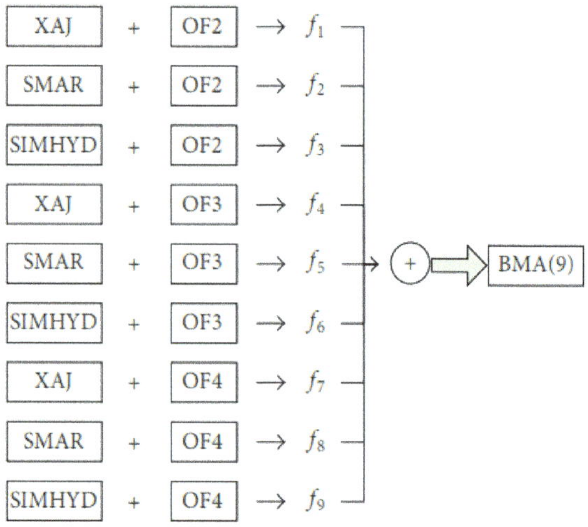

Figure 3: Diagram of BMA(9) combination scheme.

Let E denote the uncertainty of the forecast, and it can be written as $E = [E_h, E_m, E_l]$, including three components, that is, the high flow simulation uncertainty E_h, the medium flow simulation uncertainty E_m, and the low flow simulation uncertainty E_l. In BMA(9), the forecasts which are generated under OF2 have relatively small E_h, so they can get higher weights than other forecasts in high flow simulation. Similarly, the forecasts generated under OF3 have relatively high weights in medium flow simulation, while the ones generated under OF4 have higher weights than others in low flow simulation. By averaging the forecasts from a set of different combinations of hydrological model and objective function, the advantage of BMA(9) is its ability to reduce the simulation errors by giving weights to each of the nine-member forecasts according to their performance in different flow ranges.

Performance Criteria for Evaluating the Mean Prediction

There are three indices for evaluating the mean prediction.

- The Nash-Sutcliffe Coefficient of Efficiency (R^2). The definition of R^2 has expressed in (10). R^2 is not only an objective function but also a widely used performance criterion. It ranges from minus infinity to 1.0, with higher values indicating better agreement. It is difficult to evaluate the performance of the model with R^2 in all flow ranges, since the value of R^2 is always negative in the medium flow range.
- Daily Root Mean Square Error (*DRMS*):

$$\text{DRMS} = \sqrt{\frac{\sum_{t=1}^{T} \left(Q_{obs}^{t} - Q_{sim}^{t} \right)^2}{T}}, \tag{14}$$

where Q_{obs}^{t} and Q_{sim}^{t} are observed and simulated data at time t. *DRMS* is sensitive to the differences between the observations and simulations. The lower the DRMS value is, the better the prediction performance is.

- Relative Error of Total Runoff (*RE*):

$$\mathrm{RE} = 1.0 - \frac{\sum_{t=1}^{T} Q_{\mathrm{sim}}^{t}}{\sum_{t=1}^{T} Q_{\mathrm{obs}}^{t}}. \tag{15}$$

It reflects the performance in the simulation of the total runoff amount. Lower values of RE indicate better agreement of total surface runoff.

Performance Criteria for Assessing the Prediction Uncertainty Interval

Xiong et al. [33] have presented a set of indices for assessing the prediction uncertainty intervals generated by the uncertainty analysis methods. Three main indices are selected here to assess the prediction uncertainty intervals produced by BMA schemes as well as from each individual hydrological model.

(1) Containing Ratio (*CR*). The containing ratio is used for assessing the goodness of the uncertainty interval. It is defined as the percentage of observed data points that are covered in the prediction bounds.

(2) Average Band-Width (*B*). Consider

$$B = \frac{1}{T}\sum_{t=1}^{T}\left(q_u^t - q_l^t\right), \tag{16}$$

where q_u^t and q_l^t are denoted as upper and lower prediction bounds at time t. The average band-width B is also an index for measuring the performance of estimated uncertainty interval.

(3) Average Deviation Amplitude (*D*).The average deviation amplitude *D* is an index to quantify the average deflection of the curve of the middle points of the prediction bounds from the observed streamflow hydrograph.

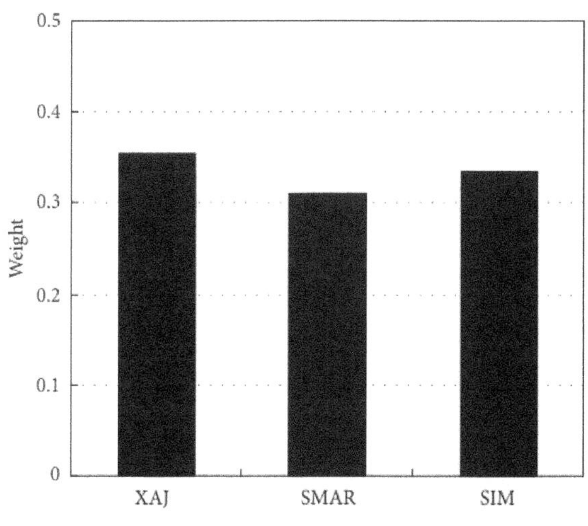

Figure 4: Histogram of weights of individual model predictions in BMA(3) scheme.

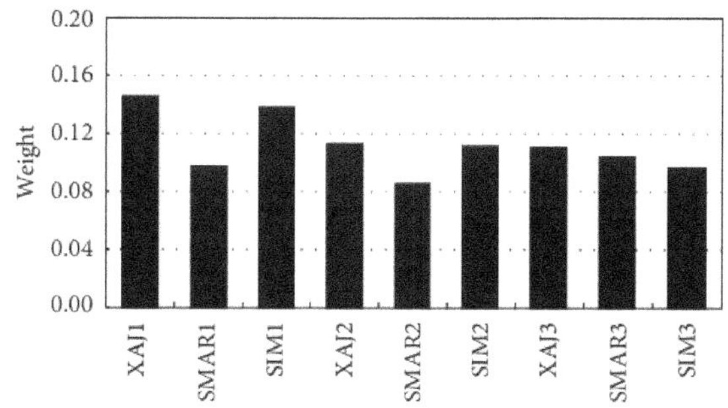

Figure 5: Histogram of weights of the individual model predictions in BMA(9) scheme.

It is defined as

$$D = \frac{1}{T}\sum_{t=1}^{T}\left|\frac{1}{2}\left(q_u^t + q_l^t\right) - Q_{\mathrm{obs}}^{t}\right|, \tag{17}$$

where Q_{obs}^{t} is the observed discharge at time t.

RESULTS AND DISCUSSION

The weights of individual models in BMA(3) scheme are displayed in Figure 4, while the weights in BMA(9) are showed in Figure 5. Moreover, in order to compare the performance of two BMA schemes in different flow ranges, according to the characteristics of the streamflow values of Mumahe catchment, data are broken into three flow ranges: high flow (top 10%), medium flow (middle 50%), and low flow (bottom 40%).

BMA(3) Results

We check the mean prediction of BMA(3) using three criteria illustrated in Section 4.1. Results of BMA(3) and its 3 individual models in the mean prediction for the whole flow series are presented in Table 1.

In terms of R^2 , the mean prediction of BMA(3) can achieve 90.68% in calibration period and 86.98% in validation period, which is better than its best individual model prediction (XAJ). However, in terms of RE, the mean prediction of BMA(3) performs much worse than its best individual model prediction. Three indices illustrated in Section 4.2 are used for assessing the prediction uncertainty intervals of both BMA(3) and its three individual models. The results for the whole flow series are also showed in Table 1. It is clear that BMA(3) uncertainty interval has the largest values of CR and *B*, and almost the smallest *D*, in both calibration and validation periods. In other words, BMA(3) uncertainty interval has better properties than any individual model's uncertainty interval in terms of CR and *D*, but worse in terms of *B*.Then we compare the differences between BMA(3) and its individual model in uncertainty interval by the graph. Figure 6 displays the mean prediction and 90% uncertainty interval of both BMA(3) and its 3 individual models for Mumahe catchment in the year of 1983 during the calibration period. The observations of 1983 are shown as dots, and the BMA(3) mean prediction and its individual models' predictions are represented by solid curve. As the statistical results showed in Table 1, the uncertainty intervals of the individual models have low containing ratio and large deviation amplitude. But the uncertainty interval of BMA(3) is much broader than that of any of its individuals. It can be found from Figure 7 that the results of validation period are similar to that of the calibration period. In general, the uncertainty interval of BMA(3) has better performance than its individual models for the whole flow series.

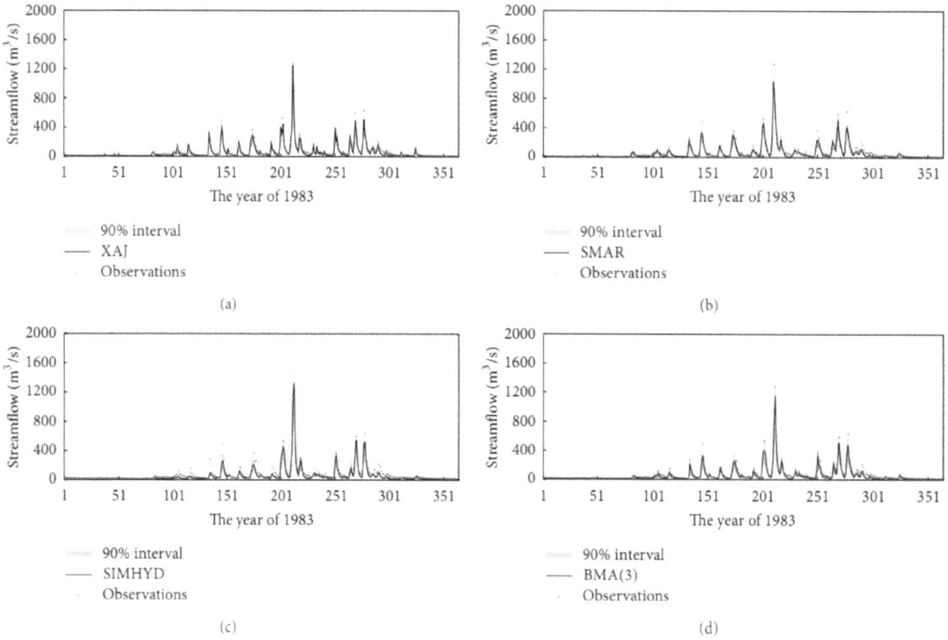

Figure 6: The mean prediction and 90% uncertainty interval of both BMA(3) and 3 individual models for the Mumahe catchment in 1983 during the calibration period.

BMA(9) Results

Table 2 lists the results of BMA(9) and its 9 individual models in the mean prediction for the whole flow series. And from it we can easily find that in calibration period, the mean prediction of BMA(9) performs better than its best individual prediction according to the value of *R*2 and DRMS, though the mean prediction of BMA(9) does not have any advantage in comparison to its individual model predictions in terms of RE. The results of the uncertainty intervals of BMA(9) and its 9 individual models are also listed in Table 2.

The containing ratio of BMA(9) uncertainty interval reaches 91.11% in calibration period and 90.23% in validation period, which are much higher than those of the uncertainty intervals of any individual model. The average deviation amplitude of the BMA(9) uncertainty interval is smaller than that of the uncertainty intervals of most of its nine individual models. From Figures 8 and 9, the similar conclusion can be concluded both in calibration and validation periods.

Comparison of BMA(3) and BMA(9)

The results of both BMA(3) and BMA(9) in terms of the mean prediction and 90% uncertainty interval for the whole flow series are listed in Table 3 for comparison.

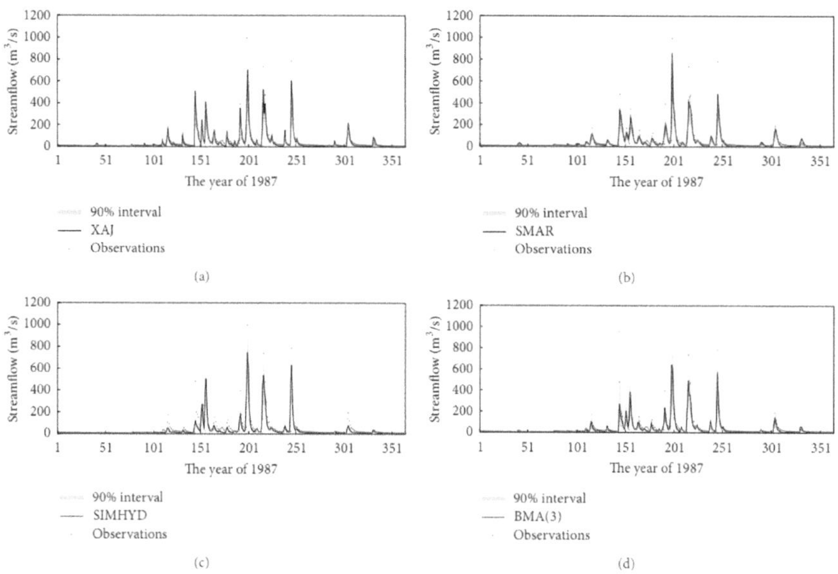

Figure 7: The mean prediction and 90% confidence interval of both BMA(3) and 3 individual models for the Mumahe catchment in 1987 during the validation period.

Table 1: Results of BMA(3) and its 3 individual models in the mean prediction as well as 90% uncertainty interval for the whole flow series.

Models	Mean prediction			90% uncertainty interval		
	R^2 (%)	DRMS	RE (%)	CR (%)	B (m^3/s)	D (m^3/s)
			Calibration period:			
XAJ	88.69	30.77	21.04	24.83	31.41	16.69
SMAR	87.69	32.11	**16.21**	32.83	32.80	17.21
SIM	80.73	40.17	31.51	14.83	**27.38**	22.33
BMA(3)	**90.68**	**27.92**	27.87	**40.72**	43.76	**16.06**
			Validation period:			
XAJ	85.77	29.22	17.79	24.28	24.66	14.09
SMAR	85.30	29.70	**14.19**	31.91	25.52	14.56
SIM	69.81	42.56	39.48	14.33	**18.40**	20.07
BMA(3)	**86.98**	**27.95**	30.72	**40.65**	36.71	14.13

Note: bolded values represent the best results.

BMA(3) mean prediction has slightly better performance than BMA(9) mean prediction in terms of $R2$ and DRMS in both calibration and validation periods, while BMA(3) mean prediction is slightly worse than BMA(9) mean prediction in terms of RE. For the uncertainty intervals, some findings are listed as follows: (1) in terms of CR, BMA(9) uncertainty interval is much higher than BMA(3) uncertainty interval in both calibration and validation periods; (2) in terms of B, BMA(9) uncertainty interval is obviously larger than BMA(3) uncertainty interval in both calibration and validation periods; (3) in terms of D, BMA(9) uncertainty interval performs slightly better than BMA(3) uncertainty interval in both calibration and validation periods. Further, we compare the BMA(3) and BMA(9) mean predictions with respect to three flow ranges in Table 4. According to the values of three indices for mean prediction, BMA(3) mean prediction has better performance than BMA(9) mean prediction in high flow range, but has worse performance in medium and low flow ranges, during both calibration and validation periods.

Table 2: Results of BMA(9) and its 9 individual models in the mean prediction and 90% uncertainty interval for the whole flow series.

Objective function	Models	Mean prediction			90% uncertainty interval		
		R^2 (%)	DRMS	RE (%)	CR (%)	B (m^3/s)	D (m^3/s)
			Calibration period				
OF2 (MSEST)	XAJ	85.45	34.89	30.24	17.89	29.43	21.46
	SMAR	84.61	35.89	6.96	31.67	36.51	19.30
	SIM	80.73	40.17	31.51	15.39	**28.47**	22.67
OF3 (MSESRT)	XAJ	89.78	29.25	10.44	68.06	33.37	**11.75**
	SMAR	80.25	40.66	10.13	44.17	35.37	17.39
	SIM	72.42	48.05	**−5.82**	47.72	42.57	21.26
OF4 (MSELT)	XAJ	79.99	40.93	12.39	63.94	33.92	14.75
	SMAR	58.01	59.29	−9.22	42.28	43.45	28.32
	SIM	52.71	62.92	−41.07	38.89	55.51	26.93
BMA(9)		**90.49**	**28.22**	21.40	**91.11**	70.98	14.54
			Validation period				
OF2 (MSEST)	XAJ	82.70	32.21	31.92	14.79	**21.56**	18.20
	SMAR	80.05	34.59	**0.66**	30.23	29.52	16.64
	SIM	69.81	42.56	39.48	20.84	24.43	22.32
OF3 (MSESRT)	XAJ	**88.52**	**26.25**	4.54	68.56	26.95	**9.62**
	SMAR	78.26	36.11	7.48	44.56	27.59	14.53
	SIM	71.09	41.64	8.98	53.86	27.69	16.47
OF4 (MSELT)	XAJ	77.25	36.94	8.74	63.07	26.68	11.85
	SMAR	43.43	58.25	−18.79	35.53	35.76	27.36
	SIM	72.27	40.79	−21.69	34.05	36.22	18.96
BMA(9)		84.54	30.46	25.42	**90.23**	55.91	13.20

Note: bolded values represent the best results.

Then we compare the uncertainty intervals of BMA(3) and BMA(9) in three different flow ranges and have some findings as follows: (1) the CR value of BMA(9) uncertainty interval has absolute predominance in comparison with that of BMA(3) uncertainty interval for each of three flow ranges in both calibration and validation periods; (2) the B value of BMA(9) uncertainty interval is larger than that of BMA(3) uncertainty interval for all three flow ranges in both calibration and validation periods; (3) the D value of BMA(9) uncertainty interval is slightly larger than that of BMA(3) in high flow range but smaller in medium and low flow ranges in both calibration and validation periods.

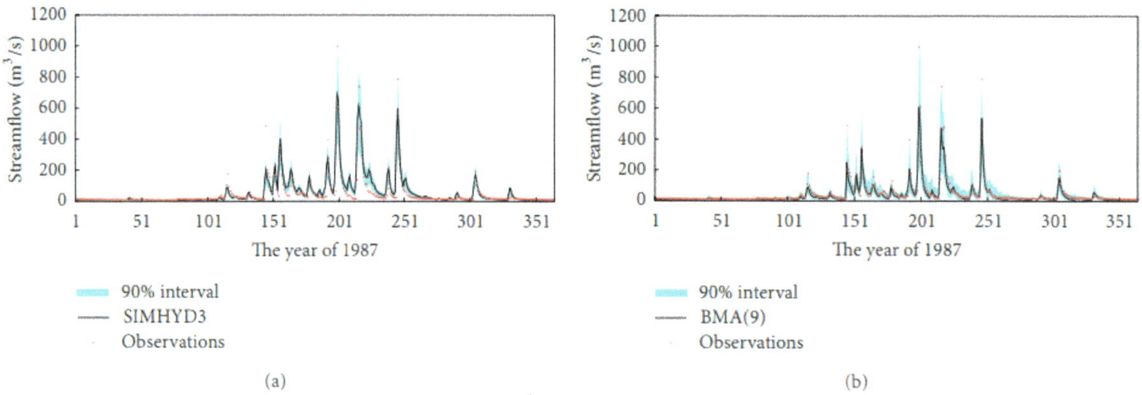

Figure 9: The mean prediction and 90% confidence interval of BMA(9) and SIMHYD3 model (the SIMHYD with the objective function OF3) for the Mumahe catchment in 1987 during the validation period.

Table 3: The comparison of BMA(3) and BMA(9) in the mean prediction and 90% uncertainty interval for the whole flow series.

Indices	Calibration		Validation	
	BMA(3)	BMA(9)	BMA(3)	BMA(9)
Mean Prediction				
R^2 (%)	90.68	90.49	86.98	84.54
DRMS (m^3/s)	27.92	28.22	27.95	30.46
RE (%)	27.87	21.40	30.72	25.42
90% uncertainty interval				
CR (%)	40.72	91.11	40.65	90.23
B (m^3/s)	43.76	70.98	36.71	55.91
D (m^3/s)	16.06	14.54	14.13	13.20

Table 4: The comparison of BMA(3) and BMA(9) in the mean prediction and 90% uncertainty interval for three flow ranges.

Indices	High flow		Medium flow		Low flow	
	BMA(3)	BMA(9)	BMA(3)	BMA(9)	BMA(3)	BMA(9)
Calibration period						
Mean prediction						
R^2 (%)	93.01	91.74	32.28	52.76	95.83	96.39
DRMS (m^3/s)	78.15	84.90	23.24	19.41	7.81	7.27
RE (%)	15.48	17.44	35.66	21.51	69.29	46.73
90% uncertainty interval						
CR (%)	88.74	92.05	45.91	91.32	27.40	90.75
B (m^3/s)	273.17	342.61	40.34	74.97	6.39	19.23
D (m^3/s)	59.78	63.66	18.33	15.44	6.21	5.02
Validation period						
Mean prediction						
R^2 (%)	89.00	85.47	22.03	41.82	93.66	94.94
DRMS (m^3/s)	92.51	106.35	19.01	16.42	6.87	6.14
RE (%)	22.49	27.68	31.35	17.66	67.48	45.11
90% uncertainty interval						
CR (%)	85.33	88.00	46.76	90.81	28.60	90.02
B (m^3/s)	252.88	282.17	34.97	61.22	7.19	18.45
D (m^3/s)	65.67	66.12	14.90	14.03	5.99	4.82

CONCLUSIONS

In this paper, the Bayesian Model Averaging (BMA) method is employed to construct a three-member predictions ensemble, denoted by BMA(3), and a nine-member predictions ensemble, denoted by BMA(9), for ensemble prediction as well as for prediction uncertainty analysis. There are three kinds of comparisons made in terms of both mean prediction and prediction uncertainty interval in this study: BMA(3) with its three individual models, BMA(9) with its nine individual models, and BMA(3) with BMA(9). In particular, we break observational flows into three different ranges for detailed comparison and analysis. The performance of two BMA schemes can be summarized as follows.

(1) In terms of mean predictions, BMA(3) performs generally better than any of its individual models. And BMA(9) mean prediction has generally higher accuracy than each of its individual model predictions. The comparison between BMA(3) and BMA(9) in mean predictions indicates that BMA(9) does not have any advantage compared to BMA(3) as far as the entire flow series is concerned. The performance of BMA(9) mean prediction is better than that of BMA(3) in both medium and low flow ranges, however, worse in the high flow range.

(2) In terms of the containing ratio for assessing the uncertainty intervals, the BMA(3) has a larger CR value than any of its individual models. And the containing ratio of BMA(9) uncertainty interval is also markedly larger than that of all its individual models when the CR value is calculated for the whole flow series. When the CR value is compared for different flow ranges, BMA(9) uncertainty interval performs better than its individual models in high, medium, and low flow ranges. In comparison with BMA(3), BMA(9) uncertainty interval also has absolute predominance in terms of CR.

(3) The average band-width B of BMA(3) uncertainty interval is larger than that of all its individuals. And the average band-width of BMA(9) uncertainty interval is even larger than that of BMA(3). It is found that, for uncertainty intervals, the increase of containing ratio is accompanied by the increase of band-width, which has already been pointed out by Xiong et al. [33].

(4) The average deviation amplitude D of BMA(3) uncertainty interval is generally smaller than the best individual in the ensemble. In terms of D, BMA(9) uncertainty interval also has a better performance than the best individual among its nine-member ensemble, especially in high flow range. Moreover, in terms of D, BMA(9) uncertainty interval performs better than BMA(3) uncertainty interval in medium and low flow ranges, but worse in the high flow range.

Based on this study, it is found that BMA is a particularly useful method for dealing with two issues. Firstly, when there are two or more competing models or methods available for the same problem, BMA can assess the relative performances of all models by assigning weights to each model or method and then produce more accurate mean prediction by weighted averaging of all predictions from those models or methods. Secondly, BMA can be used when there is uncertainty over control variables. The uncertainty intervals for both individual predictions and the BMA prediction can be derived when the distribution of the data is known or assumed.

Two issues from this study of BMA also need to be pointed out. The first is about the data transformation process. It is obvious that the daily flow data do not strictly obey the normal distribution even after the Box-Cox transformation. In fact, it is impossible to make every prediction from every model be normally distributed by using only a uniform transformation coefficient. Another problem is about the quality of the hydrological models chosen for combination. In this paper, the models employed here are all conceptual hydrological models. If better models are chosen as the ensemble members, then it is expected that the better results will come out of the BMA combination.

ACKNOWLEDGMENTS

This research is supported by the National Natural Science Foundation of China (Grant nos. 51190094, 51079098), which is greatly appreciated. The comments and suggestions from the editor and the reviewers are very helpful in the improvement of the paper and are greatly appreciated.

REFERENCES

1. WMO, "Intercomparison of conceptual models used in hydrological forecasting," Operational Hydrology Report 7, WMO, Geneva, Switzerland, 1975.
2. V. P. Singh, Computer Models of Watershed Hydrology, Water Resources Publications, 1995.
3. D. J. Reid, "Combing three estimates of gross domestic products," Economica, vol. 35, pp. 431–444, 1968.
4. J. M. Bates and C. W. J. Granger, "The combination of forecasts," Operational Research Quarterly, vol. 20, pp. 451–468, 1969.
5. J. P. Dickinson, "Some statistical results in the combination of forecasts," Operational Research Quarterly, vol. 24, no. 2, pp. 253–260, 1973. ·
6. Y. Shamseldin, K. M. O'Connor, and G. C. Liang, "Methods for combining the outputs of different rainfall-runoff models," Journal of Hydrology, vol. 197, no. 1–4, pp. 203–229, 1997. · ·
7. L. Xiong, A. Y. Shamseldin, and K. M. O'Connor, "A non-linear combination of the forecasts of rainfall-runoff models by the first-order Takagi-Sugeno fuzzy system," Journal of Hydrology, vol. 245, no. 1–4, pp. 196–217, 2001. · ·
8. D. Madigan and A. E. Raftery, "Model selection and accounting for model uncertainty in graphical models using Occam's window," Journal of the American Statistical Association, vol. 89, pp. 1535–1546, 1994.
9. E. Raftery, "Bayesian model selection in social research," Sociological Methodology, vol. 25, pp. 111–163, 1995.
10. D. Draper, "Assessment and propagation of model uncertainty," Journal of the Royal Statistical Society B, vol. 57, no. 1, pp. 45–97, 1995. · ·
11. C. Fernández, E. Ley, and M. Steel, "Benchmark priors for Bayesian model averaging," Journal of Econometrics, vol. 100, no. 2, pp. 381–427, 2001. · · ·
12. K. Y. Yeung, R. E. Bumgarner, and A. E. Raftery, "Bayesian model averaging: development of an improved multi-class, gene selection and classification tool for microarray data," Bioinformatics, vol. 21, no. 10, pp. 2394–2402, 2005. · ·
13. B. A. Wintle, M. A. McCarthy, C. T. Volinsky, and R. P. Kavanagh, "The use of bayesian model averaging to better represent uncertainty in ecological models," Conservation Biology, vol. 17, no. 12, pp. 1579–1590, 2003. · ·

14. K. H. Morales, J. G. Ibrahim, C.-J. Chen, and L. M. Ryan, "Bayesian model averaging with applications to benchmark dose estimation for arsenic in drinking water," Journal of the American Statistical Association, vol. 101, no. 473, pp. 9–17, 2006. · · ·
15. G. Koop and L. Tole, "Measuring the health effects of air pollution: to what extent can we really say that people are dying from bad air?" Journal of Environmental Economics and Management, vol. 47, no. 1, pp. 30–54, 2004. · ·
16. E. Raftery, F. Balabdaoui, T. Gneiting, and M. Polakowski, "Using bayesian model averaging to calibrate forecast ensembles," Technical Report 440, Department of Statistics, University of Washington, 2003.
17. V. Viallefont, A. E. Raftery, and S. Richardson, "Variable selection and Bayesian model averaging in case-control studies," Statistics in Medicine, vol. 20, no. 21, pp. 3215–3230, 2001. · ·
18. J. A. Hoeting, D. Madigan, A. E. Raftery, and C. T. Volinsky, "Bayesian model averaging: a tutorial," Statistical Science, vol. 14, no. 4, pp. 382–417, 1999. · · ·
19. M. A. Clyde, "Bayesian model averaging and model search strategies," in Bayesian Statistics, J. M. Bernardo, A. P. Dawid, J. O. Berger, and A. F. M. Smith, Eds., vol. 6, pp. 157–185, Oxford University Press, 1999.
20. E. Raftery and Y. Zheng, "Discussion: performance of bayesian model averaging," Journal of the American Statistical Association, vol. 98, no. 464, pp. 931–938, 2003. ·
21. S. P. Neuman, "Maximum likelihood Bayesian averaging of uncertain model predictions," Stochastic Environmental Research and Risk Assessment, vol. 17, no. 5, pp. 291–305, 2003. · ·
22. N. K. Ajami, Q. Duan, and S. Sorooshian, "An integrated hydrologic Bayesian multimodel combination framework: confronting input, parameter, and model structural uncertainty in hydrologic prediction," Water Resources Research, vol. 43, no. 1, Article ID W01403, 2007. · ·
23. Q. Duan, N. K. Ajami, X. Gao, and S. Sorooshian, "Multi-model ensemble hydrologic prediction using Bayesian model averaging," Advances in Water Resources, vol. 30, no. 5, pp. 1371–1386, 2007. · ·
24. X. Zhang, R. Srinivasan, and D. Bosch, "Calibration and uncertainty analysis of the SWAT model using genetic algorithms and Bayesian model averaging," Journal of Hydrology, vol. 374, no. 3-4, pp. 307–317, 2009. · ·
25. K. Beven and A. Binley, "The future of distributed models: model calibration and uncertainty prediction," Hydrological Processes, vol. 6, no. 3, pp. 279–298, 1992. ·
26. H. V. Gupta, K. J. Beven, and T. Wagener, "Calibration and uncertainty estimation," in Encyclopedia of Hydrological Sciences, John Wiley and Sons, Chichester, UK, 2003.
27. L. Marshall, D. Nott, and A. Sharma, "A comparative study of Markov chain Monte Carlo methods for conceptual rainfall-runoff modeling," Water Resources Research, vol. 40, no. 2, Article ID W02501, 2004. · ·
28. J. A. Vrugt, H. V. Gupta, W. Bouten, and S. Sorooshian, "A Sshuffled complex evolution metropolis algorithm for optimization and uncertainty assessment of hydrologic model parameters," Water Resources Research, vol. 39, no. 8, 2003. ·
29. J. M. Hammersley and D. C. Handscomb, Monte Carlo Methods, Methuen, London, UK, 1975.
30. F. H. S. Chiew, M. C. Peel, and A. W. Western, "Application and testing of the simple rainfall runoff model SIMHYD," in Mathematical Models of Small Watershed Hydrology and Applications, V. P. Singh and D. Frevert, Eds., pp. 335–366, Water Resources Publications, 2002.
31. Q. Duan, S. Sorooshian, and V. Gupta, "Effective and efficient global optimization for conceptual rainfall-runoff models," Water Resources Research, vol. 28, no. 4, pp. 1015–1031, 1992. · ·
32. L. Oudin, V. Andréassian, T. Mathevet, C. Perrin, and C. Michel, "Dynamic averaging of rainfall-runoff model simulations from complementary model parameterizations," Water Resources Research, vol. 42, no. 7, Article ID W07410, 2006. · ·
33. L. Xiong, M. Wan, X. Wei, and K. M. O'Connor, "Indices for assessing the prediction bounds of hydrological models and application by generalised likelihood uncertainty estimation," Hydrological Sciences Journal, vol. 54, no. 5, pp. 852–871, 2009.

CHAPTER 14

MODELING THE PROJECTED CHANGES OF RIVER FLOW IN CENTRAL VIETNAM UNDER DIFFERENT CLIMATE CHANGE SCENARIOS

Tuan B. Le and Hatim O. Sharif

Keywords: Asian Precipitation–Highly Resolved Observational Data Integration Towards Evaluation of water resources (APHRODITE), climate change, Global Climate Models (GCMs), Long Ashton Research Station-Weather Generator(LARS-WG), hydrologic modeling, Soil and Water Assessment Tool (SWAT), Huong River, flood control

INTRODUCTION

Numerous studies have shown that climate change significantly affects the hydrologic cycle and alters runoff regimes and water availability in different regions across the Globe [1,2,3,4,5]. In its fifth assessment report [6], the Intergovernmental Panel on Climate Change (IPCC) affirms that human development activities have caused large increases in greenhouse gases concentration in the atmosphere since 1850, mostly from burning fossil fuels and deforestation. These elevated concentrations cause temperature to rise on a global scale. The global increases in temperature appear to directly and significantly alter the hydrologic cycle through changing evapotranspiration (ET) and accumulation and melting of snow and ice [6]. In addition, climate change, as indicated by changes of atmospheric circulation, also significantly modifies precipitation patterns, intensities, and amounts at global scales [7,8]. It could consequently disturb the balance of temporal and spatial distribution of water resources and cause large scale floods or long-term droughts. At the regional scale, climate change causes changes in discharge of rivers and streams that potentially reduce or increase the needed water availability for human activities [9,10]. Future climate changes associated with specific emission scenarios can be projected using Global Climate Models (GCMs). GCMs' output data can be used regionally, after disaggregation, to force hydrological models to simulate projected runoff and river discharge. An approach that is broadly used in many previous studies is to generate a baseline simulation based on observed meteorological data in the past and use the baseline meteorological data to force a hydrologic model to generate baseline hydrologic conditions [3,4,8,11,12,13,14]. The evaluation of future changes in river discharge is based on the comparison of projected results with the baseline and their differences. This method may eliminate or substantially reduce the internal bias or error of the models. A study by Wu et al. [14] indicated that projected climate change will have great impacts on snow dominant streams, leading to lower summer streamflows and warmer summer stream temperature with smaller effects on transient and rain dominant regimes in the northwestern region of the United States. Hamlet [15] studied the projected impacts of climate change on seasonal rainfall in the Pacific Northwest region of North America and performed hydrologic simulations to support water resources management, flood control, hydropower development and other water related environmental services in the region. Thodsen [16] combined climate model generated precipitation and potential evapotranspiration and a hydrologic model to study the impact of climate change on six major Danish catchments and reported potential future increases in the 100-year and 1000-year precipitation leading to increases in discharge that are approximately twice as much.

River discharge in central Vietnam is highly fluctuating. A report of the Institute of Strategy and Policy On Natural

Citation: Le, T.B.; Sharif, H.O. Modeling the Projected Changes of River Flow in Central Vietnam under Different Climate Change Scenarios. Water 2015, 7, 3579-3598. doi:10.3390/w7073579.

Resources and Environment of Vietnam (ISPONRE) supported by the United Nations Environment Programme (UNEP) identified Vietnam as one of the most adversely effected countries by climate change during the last 50 years [17]. Recent studies on future global water availability and major continental rivers' future discharges also projected considerable increases in runoff in South East Asia region, and Vietnam in particular, in the twenty-first century [13,18,19]. Such projected increases in runoff vary from 10%–20% to 40%–50% among different studies depending on which GCMs and emission scenarios were selected for simulation. Weiland et al. [13] and Nohara et al. [19] analyzed seasonal water availability in Vietnam and neighboring countries and both studies reported that peak river discharges typically occurred in the summer and fall with river discharges significantly lower in the winter and spring. In the twenty-first century, Nohara et al. [19] projected an increase in discharge mostly in the summer and fall for the period from 2006 to 2100 and Weiland et al. [13] confirmed these conclusions. However, the projection of river discharge from these studies is limited for only IPCC Special Report on Emissions Scenarios (IPCC-SRES) scenario A1B and was not based on hydrologic modeling.

The rainy season in Central Vietnam spans the summer and fall which are characterized by heavy precipitation and flood events [20]. The region has minimal precipitation from December to April and precipitation increases significantly from May to November with high intensity and density precipitation events in July and August. Seasonal precipitation, especially in the summer and fall, has increased in the second half of the twentieth century and grown more obviously in the beginning of the twenty-first century [21]. This most likely has increased runoff and annual discharges of rivers in the region; however, it has not been analyzed and demonstrated in any previous study.

On the other hand, flood control policy of the local governments requires quantitative evaluation of river flow changes in the past as well as in the future. In fact, water resources related policies and laws have been recently established by the government of Vietnam as well as local governments in response to water resources degradation and negative impacts of global climate change on water resources. One of the latest decisions approved by the Prime Minister of Vietnam in September, 2013 regarding the national strategy for sustainably exploitation and usage of natural resources until 2020 with the vision to 2030 states that it is necessary to practice water saving nationwide and focus on solving the water shortage problem during the dry season in the major river basins [22]. Due to the fact that water resources are unevenly distributed both spatially and temporally, especially in central Vietnam where precipitation could range from 400 to 4000 mm/year among different places and about 70%–85% annual total water budget occur in the rainy season [23], it is suggested that reservoirs and dams be efficiently operated to store water in the rainy months and supply in the dry months. Dam and reservoir development is also an option due to the fact that Vietnam is facing seasonal water shortage and cannot ensure water safety (at the standard of 4000 m^3/person/year) because about 63% of the national water budget comes from outside the country and might not be used for domestic supply [23]. The Huong River basin itself had three major dam projects built on its three main tributaries: Binh Dien dam on the Bo tributary, Huong Dien dam on the Huu Trach tributary and Ta Trach dam on the Ta Trach tributary. These dams are currently playing an important role in flood control besides generating hydro-power for the area. Both Binh Dien dam and Huong Dien dam are currently operating at their maximum flood control capacities, 180 million m^3 and 400 million m^3respectively; while Ta Trach dam is designed with about 400 million m^3 capacity for flood control purposes [24,25]. Flood control within the Huong River basin could be improved to reduce the damage caused by floods if flow is forecasted ahead of time. Ngo & Nguyen [24] suggested that doubling the flood control capacities of Binh Dien and Huong Dien dam would reduce the inundation time with water level above three meters downstream of the Huong River during an extreme flood event to only three hours, instead of several days under the current capacities. In many years before 2010 when the hydroelectric dams had not been built, the Huong River's water was under serious salinization far from the coast into mainland reaches [26].

With all the issues mentioned above, this study aims to answer three major questions: How has river discharge in Central Vietnam changed in the past 60 years? How will seasonal total discharges vary in the 21st century because of climate change under the consideration of different IPCC-SRES scenarios? And what are the differences among projections produced by different GCMs? The study is focused on one mid-size river basin, the Huong River basin. Future climate data used in this study are based on three IPCC-SRES scenarios: A2, A1B and B1 [27]. The choice of the GCMs that simulate the scenarios strongly affects the projected changes of meteorological parameters [2]. In this study, three GCMs including ECHAM5-OM (Germany), HadCM3 (UK) and GFDL-CM2.1 (USA) are selected to simulate future daily meteorological data which serve as input to the hydrological model SWAT. River discharge simulations of SWAT are performed separately for each input dataset produced by a pair of GCM and IPCC-SRES. This will create nine different projections of river discharge.

DATA AND METHODS

Study Area and Hydrologic Model

The study area in this paper is the Huong River basin that is located in Thua Thien Hue province in Central Vietnam. The basin has a drainage area of 2830 km^2 with three major tributaries that originate in the southern and western mountainous areas: the Ta Trach and Huu Trach tributaries, which unite and form the main stem of the Huong River, and the Bo tributary, which merges into the Huong River downstream (Figure 1). This basin is also the study area of Le et al. [28] and its topographic and meteorological characteristics are fully described therein.

Similarly to Le et al. [28], the Soil and Water Assessment Tool (SWAT) hydrologic model was used in this study with only discharge simulation components were applied. Le et al. [28] also provides description on how discharge is simulated in SWAT.

Historical Meteorological Data

Daily precipitation observations for the period from 1951 to 2007 were extracted from APHRODITE gage-based precipitation data at a spatial resolution of 0.25° × 0.25° while for the period from 2008 to 2012, the daily precipitation values were collected from local rain gages within the basin. The centers of the eight 0.25° × 0.25° grids of APHRODITE are shown in Figure 1. Precipitation data are re-grouped into seasonal precipitation; spring (March through May), summer (June through August), fall (September through November) and winter (December through February). Temperature, humidity, and wind speed data were obtained for the period from 1951 to 2012 from the National Climatic Data Center (NCDC) of the US National Oceanic and Atmospheric Administration (NOAA) for station in Hue city (Figure 1). The satellite-based solar radiation data were obtained from the Atmospheric Science Data Center of the US National Aeronautics and Space Administration (NASA). These daily data have low spatial resolution of 1° × 1°. For that reason, the study area falls into only one cell whose boundaries are 17° N, 16° S, 107° W and 108° E. The location of the center of this grid is shown as the Solar Gage in Figure 1.

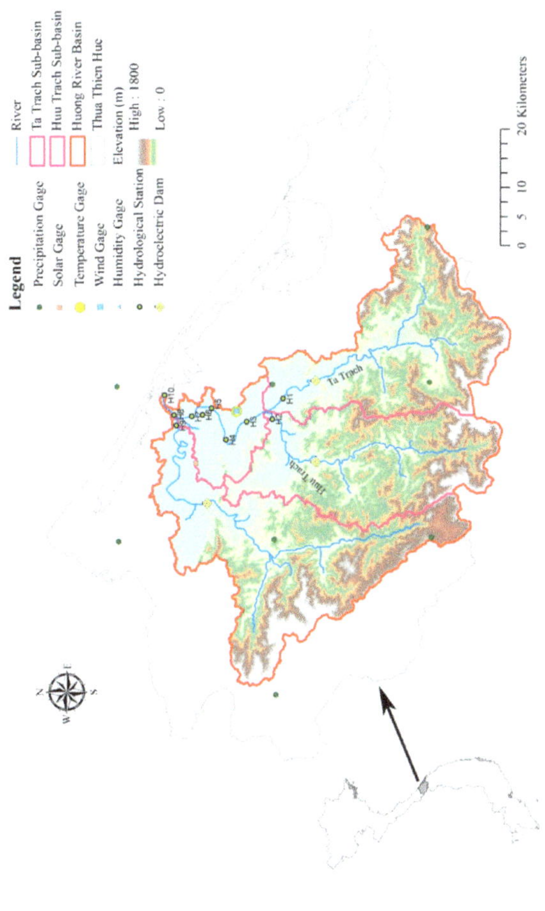

Figure 1. The Huong River basin in Central Vietnam [28]. Following SWAT model description of input data, what is described as "Gage" in the figure is actually the center of a grid for gridded data.

Modeled Future Climatic Changes

Projections from three GMCs included in the IPCC Fourth Assessment Report [27] are used in this study. Future climate simulations by each model were based on three future IPCC-SRES emission scenarios: A2 (high emission), A1B (medium emission) and B1 (low emission). The Long Ashton Research Station Weather Generator (LARS-WG) stochastic weather generator model was used to disaggregate GCM model output. Observed daily meteorological data for a site are required to generate the probability distributions of each meteorological variable and their correlations by using semi-empirical distribution [29]. The simulation of precipitation in LARS-WG is based on alternating wet (nonzero precipitation) and dry periods with the length of each period being chosen arbitrarily from the appropriate distributions for a given month. On a wet day, the precipitation magnitude is estimated by using the semi-empirical distribution of the precipitation for the month regardless of precipitation value of the previous days [30]. Daily maximum and minimum temperature are generated by using their statistical distributions and correlation with wet and dry conditions [31]. In general, daily temperature is lower for a wet day than a dry one. Similarly, solar radiation is generated using its semi-empirical distribution for wet and dry periods separately [30]. Details on the weather generation by LARS-WG were described by Racsko et al. [31], Semenov & Stratonovitch [29] and Semenov et al. [30]. LARS-WG version 5.5 was downloaded from the LARS-WG website [32].

There are three meteorological variables projected by the GCMs, precipitation, temperature and solar radiation. GCMs' outputs were disaggregated by LARS-WG into daily meteorological data at single sites (grid centers) based on the statistical properties of observation. Observed daily precipitation data from 1951 to 2012 were used to compute the probability distribution of wet and dry series as well as other LARS-WG parameters. Maximum and minimum daily temperature and solar radiation data were similarly used to generate probability distributions for the same period. The model then generated future meteorological data from 2013 to 2100 for precipitation, temperature, and solar radiation for each site based on statistics of historical data and future climate scenarios. Eight time series for eight precipitation gages (grid centers) were stochastically generated by LARS-WG. The ECHAM5-OM, HadCM3 and GFDL-CM2.1 models were used to simulate future climate conditions based on A2, A1B, and B1 scenarios. The A2 scenario describes a heterogeneous world with continuous increase in population and regionally oriented economic development that produce high emission of greenhouse gases and aerosols. The A1B scenario describes a rapid economic growth world with global population increases in the first half of the twenty-first century and declines in the second half of the century. The A1B scenario assumes a balance on fossil and non-fossil energy consumption of the world that produces medium emission of greenhouse gases and aerosols. The B1 scenario describes a world with social and environmental sustainability; greenhouse gases and aerosols emission is very low in this scenario [27]. These GCMs are widely used models that are regularly updated and improved by comparing their results against observed data [33].

Observed River Discharge Data

The Institute of Resources, Environment, and Biotechnology of Hue University, Vietnam (IREB-Hue) perform field measurements of flow discharge of the Huong River and its tributaries on a monthly basis. Measurements are taken on an arbitrary day of the month depending on weather conditions and the availability of employees. These discharge measurements are only available from 2007 to 2010. There are 10 measurement locations in the basin: three at the outlets of each tributary of Huong River (site H1, H2 and H9), six along Huong River from the confluence of Ta Trach River and Huu

Trach River to the confluence of Huong River and Bo River (site H3, H4, H5, H6, H7 and H8), and one at the outlet of Huong River (site H10) before it pours into the lagoon (Figure 1). These monthly data are available in the annual reports of IREB-Hue.

Discharge measurements of the Huong River and its tributaries are calculated from measured flow velocities by multiplying them with the equivalent cross-sectional surface areas. Flow velocities are measured manually at the depth of 1.5 m below the water surface at specific locations across the channel. The equipment used to measure flow velocities is the Global Digital Water Flow Probe [34]. At each channel cross-section, there are five to seven flow velocity measurements depending on the width of the river and the values are then averaged for that cross-section. Cross-sectional geometries and areas are constructed and calculated based on multiple depth measurements between the two riversides. These depth measurements of the riverbed are conducted by a Cole-Parmer Portable Water Depth Sounder Gage [34]. The measured discharges at the outlets of Ta Trach and Huu Trach tributaries as well as at the outlet of Huong River are used for calibration and validation of the SWAT model.

Projected Future Changes in River Discharge

Daily meteorological data including precipitation, maximum and minimum air temperature, and solar radiation at eight locations that correspond to the centers of the eight precipitation grids (Figure 1) are generated for the baseline period as well as for 2011–2100 periods by three GCMs ECHAM5-OM, HadCM3 and GFDL-CM2.1 for the three IPCC-SRES scenarios A2, A1B and B1. The simulated outputs include one dataset for the baseline and nine datasets for the future which are corresponding to nine pairs of GCM and emission scenario selected. These daily datasets then serve as meteorological input data for the calibrated SWAT model to simulate the Huong River discharge of corresponding periods. Average seasonal values were also calculated to identify the impact of climate change on seasonal discharge of the Huong River.

Outputs from the GCMs, disaggregated by LARS-WG to generate input to SWAT, were used to estimate projected river discharge. The average total seasonal discharge of each 30 year period was calculated and compared to the baseline discharge to see whether there are any significant changes in river discharge induced by projected climate change. The baseline used in this study is the 1961–1990 period. Climate data for the other 30-year periods: the 1951–1980, 1971–2000, 1981–2010 and 1991–2020 periods were also calculated. Projected values for the rest of the 21st century were calculated for the 2011–2040, 2021–2050, 2031–2060, 2041–2070, 2051–2080, 2061–2090 and 2071–2100 periods. The 30 year period used in this analysis is appropriate to represent the impact of climate change in the central decade due to the fact that the variability of water availability within a decade is not only impacted by climate change within that decade, but also the decade before [12].

The analysis of projected changes in the Huong River discharge includes the biases of the climate models that are contained in the models' outputs. Since climate models' biases do not only exist in the projected data; but also in the baseline; the comparison of projected future discharge to the baseline is still meaningful in evaluating the impact of climate change on changing the river discharge. In other words; the analysis of projected changes in the Huong River discharge is based on an assumption that climate models' biases are systematic and comparable in all models' outputs.

RESULTS AND DISCUSSION

Calibration and Validation of River Discharge Simulations

SWAT calibration was performed over the catchment of Ta Trach tributary (site H1) for the period of 2007–2010, where the hydroelectric dam has no impact. The calibrated model parameters for the Ta Trach sub-basin were used to validate the model over the Huu Trach sub-basin at site H2 and the Huong River basin at site H10 (Figure 1). The processes for SWAT model setup, simulation, calibration and validation for this region are described in detail in Le et al. [28].

The ArcSWAT for ArcGIS 10 interface of SWAT 2009 is used in this study to set up the model. Calibration was performed manually with ArcSWAT by adjusting the value of one model parameter at a time to attain better agreement between simulated and observed values. Veith et al.[35] reported that CH_N2 (Manning's N value for stream channels), CN2 (SCS curve number), SOL_K (soil hydraulic conductivity) and SOL_AWC (available water capacity of the soil) were the most significant SWAT model parameters that affect discharge simulation outputs. They also noted that the effects of ALPHA_BF (base flow alpha factor), ESCO (soil evaporation compensation factor) and GW_REVAP (groundwater "revap" coefficient) on model simulation could be considerable.

The model performance was evaluated through two standard performance measures: the Nash-Sutcliffe Efficiency (NSE) and the Pearson product–moment correlation coefficient (r). The NSE describes the predictive capability of the model and takes values that are smaller than or equal to one. Larger values of NSE that are closer to one indicate the model's ability to reproduce observed data. Pearson's r is a popular statistical measure that is used here to evaluate the degree of linear correlation between the simulated and observed variable. These measures are defined as follow:

$$\mathrm{NSE} = 1 - \frac{\sum_{t=1}^{N} (Qo(t) - Qs(t))^2}{\sum_{t=1}^{N} \left(Qo(t) - \overline{Qo}\right)^2} \tag{1}$$

$$r = \frac{\sum_{t=1}^{N} \left(Qs(t) - \overline{Qs}\right)\left(Qo(t) - \overline{Qo}\right)}{\sqrt{\sum_{t=1}^{N} \left(Qs(t) - \overline{Qs}\right)^2 \sum_{t=1}^{N} \left(Qo(t) - \overline{Qo}\right)^2}} \tag{2}$$

where Qo(t) and Qs(t) are the observed and simulated discharge at day t, respectively. Qo and Qs are the average observed and simulated discharge, respectively. N is the total number of observation/simulation days.

The discharge calibration process started with the adjustment of CH_N2 parameter. The NSE and r were calculated for every simulation and the value of adjusted model parameter that generated higher NSE and r than previous simulation was saved. The calibration process is continued by adjusting CN2, SOL_K, SOL_AWC, ALPHA_BF, ESCO and GW_REVAP, respectively, following the same procedure. The calibration process demonstrated that Huong River discharge was most sensitive to CH_N2 (Manning's N value for stream channels), CN2 (SCS curve number), and SOL_K (soil hydraulic conductivity). It was moderately sensitive to SOL_AWC (available water capacity of the soil) and ALPHA_BF (base flow alpha factor) and has low sensitivity to changes in ESCO (soil evaporation compensation factor) and GW_REVAP (groundwater "revap" coefficient).

Table 1 lists the values of the calibrated model parameters for discharge simulations. The values represent the calibrated parameters of the Huong River basin such as channel roughness, and hydraulic conductivity of soils that fall within their acceptable ranges in previous SWAT studies. The SCS curve number (CN) values, which are a function of land use, soil permeability, and antecedent soil moisture range between a lowest value of 42 to a highest of 97. The lower CN values are associated with the areas where forests dominate and sandy soils while the higher CN values apply to urban areas and wetlands. The hydraulic conductivity of each soil group slightly changed from their initial values after calibration. Dystric Gleysols at wetland areas had the lowest calibrated hydraulic conductivity of 0.318 mm/h, Luvic Arenosols with mostly sand at the coastal areas had the highest calibrated hydraulic conductivity of 690.0 mm/h.

Table 1. Model parameters used to calibrate discharge at H1.

Calibrated Model Parameters		Calibrated Values
Parameters	*Definitions (Unit)*	
CH_N2	Manning's N value for stream channels	0.035
CN2	SCS curve number	42–97
SOL_K	Soil hydraulic conductivity (mm/h)	0.318–690.0
SOL_AWC	Available water capacity of the soil (mm/mm)	0.075–0.315
ALPHA_BF	Base flow alpha factor (days)	0.25
GW_REVAP	Groundwater "revap" coefficient	0.02
ESCO	Soil evaporation compensation factor	1.0

The calibrated discharge of the Ta Trach sub-basin (H1) for the 2007–2010 period is shown inFigure 2. Due to the frequency of observed discharge, the comparison of simulated and observed discharge had to be performed at a daily time scale but with just one observed value for each month. The discharge hydrographs are shown in logarithmic scale because of the large differences between high and low flows (more than two orders of magnitudes). As can be seen from Figure 2, simulated discharge follows the fluctuations of observed discharge. The base flows during the dry seasons are relatively well simulated. However, the peak flows in the rainy seasons of 2007 and 2008 are slightly underestimated while overestimated in 2009 and 2010. Comparison of simulated and observed data points results in a Nash-Sutcliffe efficiency of 0.88 and a Pearson correlation coefficient of 0.94 (Table 2). In fact, errors in rainfall input and discharge measurement affect discharge simulation and calibration. APHRODITE rainfall products suffer from its very coarse resolution and cannot adequately capture the spatial and temporal distributions of extreme rainfall events. Discharge measurements were performed manually from a boat and have high potential of errors and bias. Nonetheless, the results are satisfactory based on model evaluation guidelines as well as compared to published SWAT simulations of daily discharge [36,37,38]. The calibration results were also improved compared to the previous study Le et al. [28] despite the same model parameters being used during calibration. This improvement is caused by the usage of input precipitation products. The higher accuracy of precipitation estimates brought by rain gage-based APHRODITE products than satellite-based TRMM products [39] used in Le et al. [28] contributed to better Nash-Sutcliffe efficiency and Pearson correlation coefficient of discharge calibration.

SWAT was validated over the Huu Trach sub-basin (H2) and the Huong River basin (H10). Discharge simulation at H2 (Figure 3) during the rainy season of 2007 has the same problem of underestimation observed at H1. Examining the APHRODITE precipitation for this period shows that it agrees with the simulated discharge pattern and does not match discharge observations. This indicates that the apparent underestimation is most probably a result of errors in either precipitation estimates or observed flow and is not caused by SWAT model errors. In general, observed discharges of the Huu Trach sub-basin (H2) and Huong River (H10) basin are reasonably estimated by SWAT. The fluctuations of simulated discharge match observed ones except for the deviation in 2007 rainy season at H2 (Figure 3 and Figure 4). Simulated and observed discharge comparisons at H2 and H10 from January 2007 to December 2009 have Nash-Sutcliffe efficiencies of 0.72 and 0.73 and Pearson correlation coefficients of 0.85 and 0.91, respectively (Table 2).

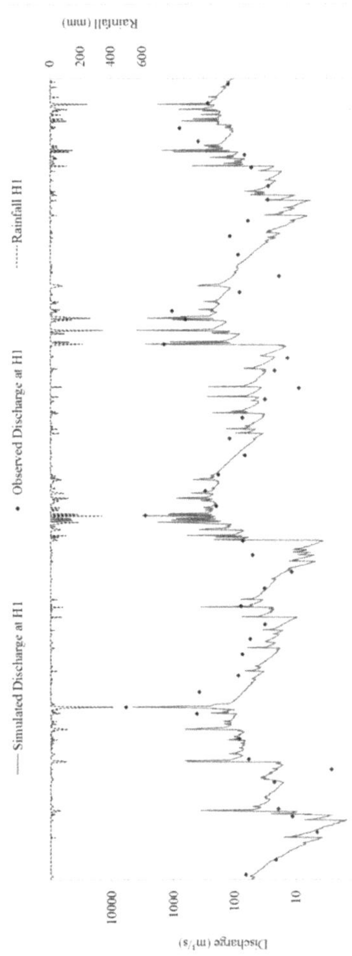

Figure 2. Discharge calibration at H1.

Table 2. Performance evaluation statistics of discharge simulation.

Station	Discharge Simulation	
	NSE (≤1)	*r (0–1)*
H1 (*Calibration*)	0.88	0.94
H2 (*Validation*)	0.72	0.85
H10 (*Validation*)	0.73	0.91

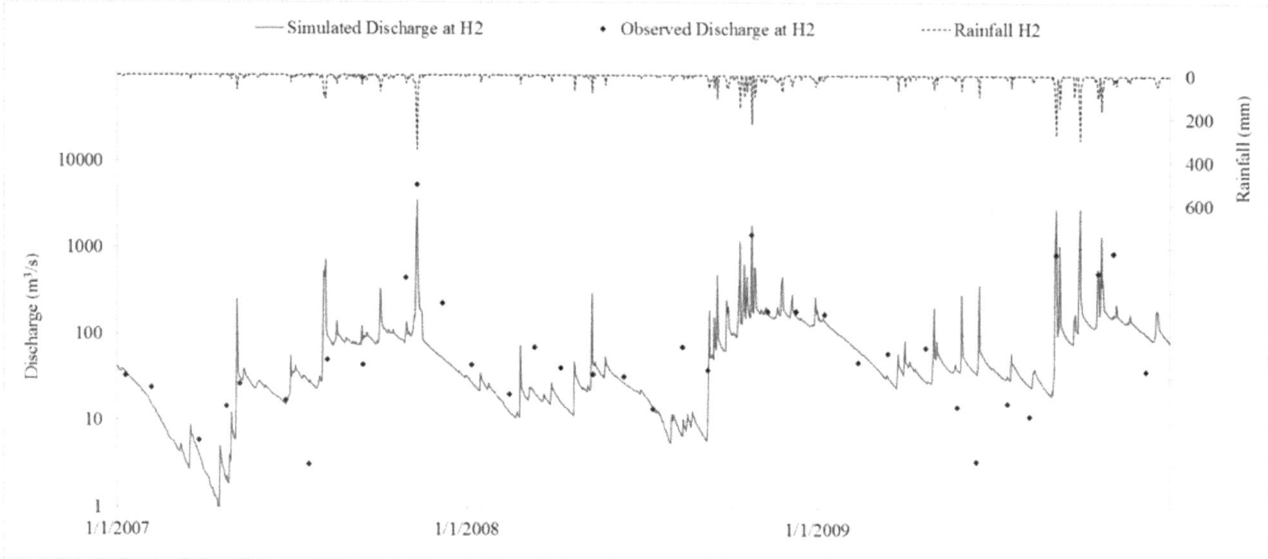

Figure 3. Discharge validation at H2.

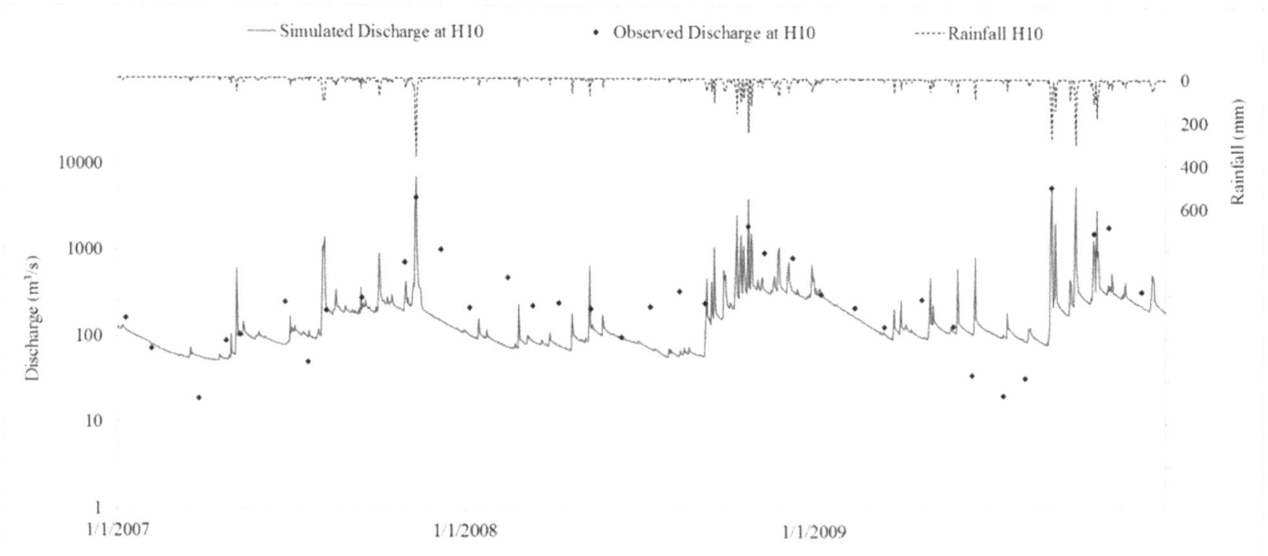

Figure 4. Discharge validation at H10.

Historical and Future Climate Data

Figure 5 illustrates the total seasonal precipitation and its standard deviation in the Huong River basin. Spatial distribution of the seasonal precipitation during the 1951–2012 period shows highest precipitation during the summer season: precipitation in excess of 1000 mm over the mountainous area of the basin with lower precipitation of about 800–950 mm in the delta area. The spatial pattern of fall precipitation is different, higher values of around 550–600 mm over the delta area with lower values of 450–500 mm in the mountain region. Precipitation in the spring is average, about 400–450 mm, and is relatively homogeneous over the basin. Winter is known as the driest season with low precipitation, below 100 mm, falling uniformly over the whole basin.

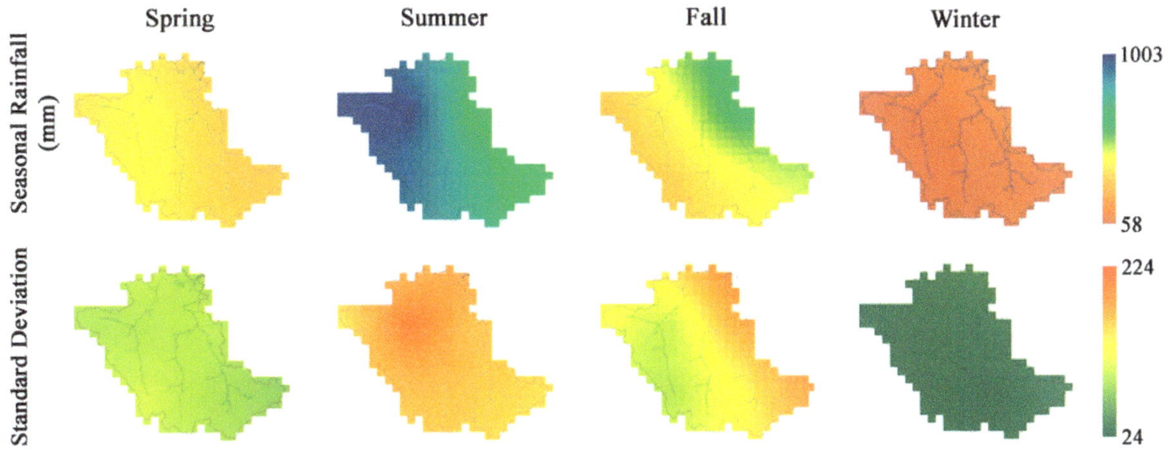

Figure 5. Seasonal precipitation distribution and their standard deviation.

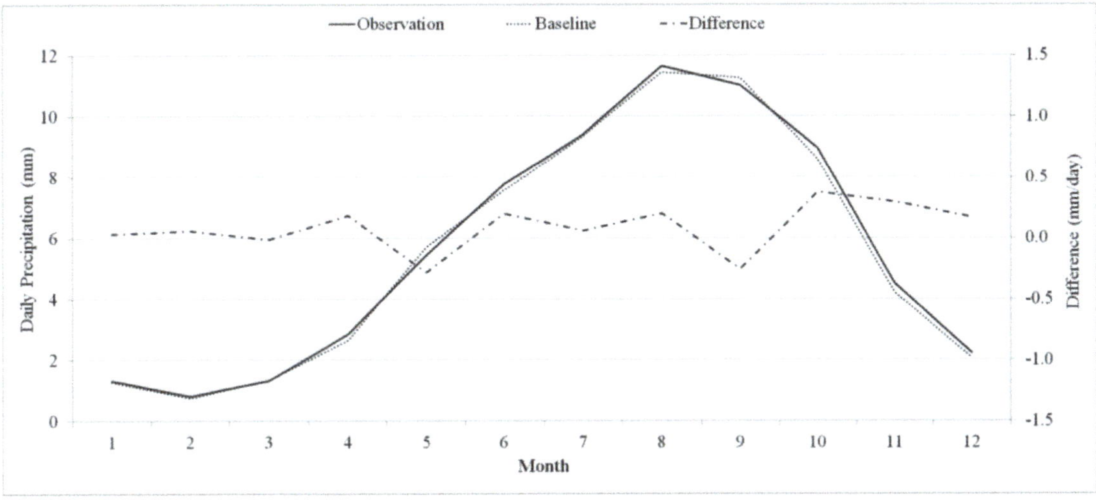

Figure 6. Average daily precipitation in 12 months from observed data (solid line–left scale) and baseline data (dotted line–left scale). The dashed line represents the differences between the two datasets (right scale).

Maximum daily temperature in the summer is always above 35 °C and usually around 37–39 °C; it reaches above 40 °C on rare occasions. In the winter, maximum daily temperature is usually around 13–15 °C but can reach 17–18 °C during warmer years. Summer minimum daily temperature fluctuates between 27 and 30 °C; in warmer summers the minimum daily temperature reaches above 32 °C. The minimum daily temperature fluctuates between 12 and 13 °C in the in some years fell below 10 °C. Humidity within the Huong River basin area is always very high at 95%–100% during the rainy season. Humidity significantly decreases when temperature increases in the summer to about 45%–60%.

Historical meteorological data are input into LARS-WG to reproduce the baseline of the 1961–1990 period. Reproduced data were compared with observed data to evaluate the performance of the model. The main purpose of this step is to identify internal bias and errors of the model before simulating the future climate projections. Figure 6 illustrates the average daily precipitation graph of each month based on observed and LARS-WG-simulated values. The Nash-Sutcliffe efficiency and Pearson correlation coefficient of these two sets of data are 0.997 and 0.999 respectively; which indicates that baseline values are very well reproduced and that internal model bias and error are minimal. Calculated differences in daily precipitation are lower than 0.5 mm.

The spatial distribution of the changes in seasonal precipitation projected by the GCMs for the 2011–2040, 2021–2050, 2031–2060, 2041–2070, 2051–2080, 2061–2090 and 2071–2100 periods are shown in Figure 7. On average, the GCMs' output data show an increase in precipitation from the present to the 2011–2040 period followed by a slight decrease until the second half of the century then a gradual increase until the end of the century. The spring season is projected to witness the most change in precipitation, a decrease of about 9% in the southern mountainous region during the 2011–2040 period. Summer and fall will stay as the wettest seasons. A considerable drop in precipitation is also projected for the winter season with the highest drop of about 8.5% in the 2071–2100 period in the south of the basin. In general, the GCMs project an increase in maximum and minimum temperature for all three IPCC-SRES scenarios in the twenty-first century. This will raise the amount of ET within the basin and reduce the total runoff that can reach the Huong River.

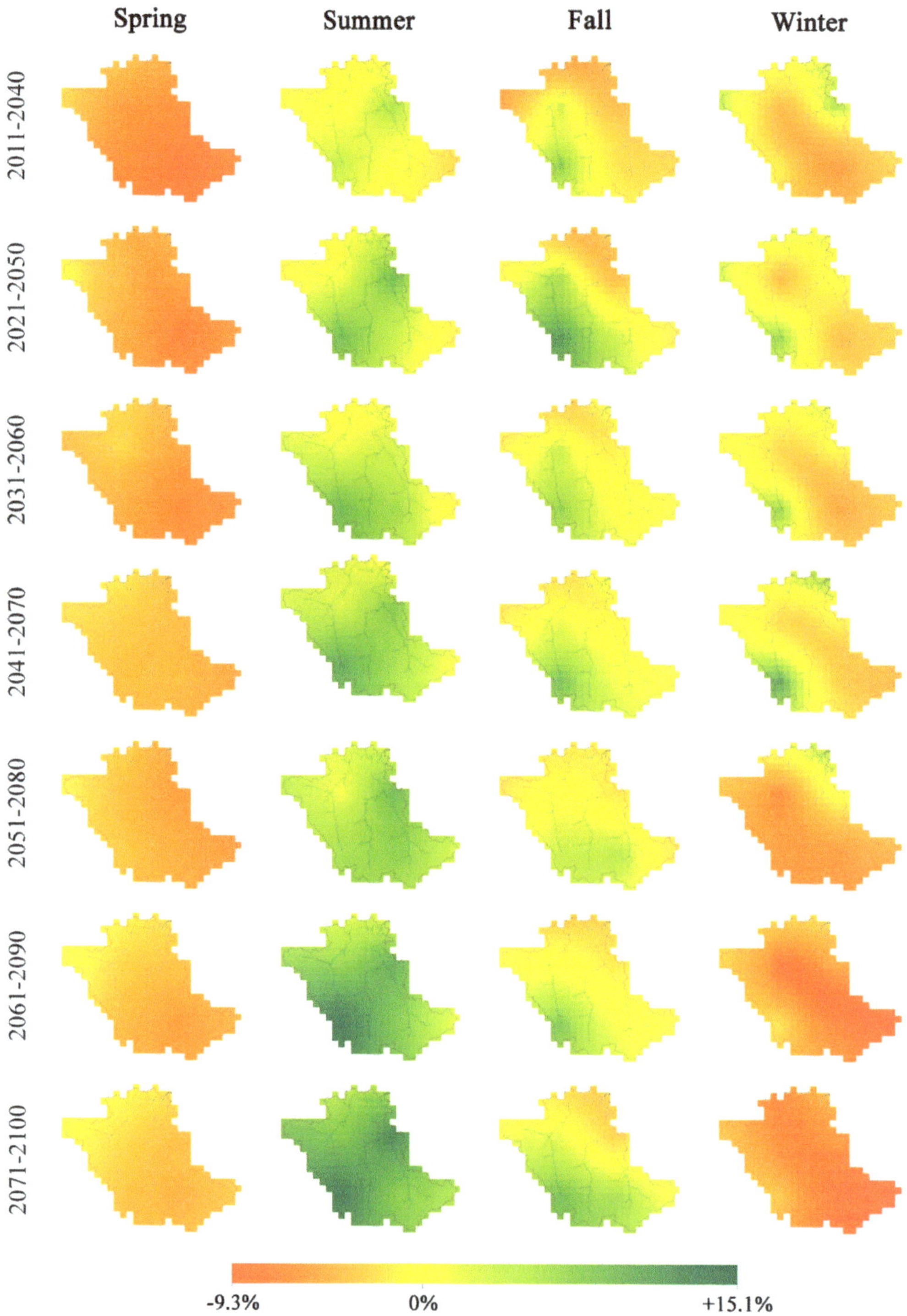

Figure 7. Change in seasonal precipitation in the 2011–2040, 2021–2050, 2031–2060, 2041–2070, 2051–2080, 2061–2090 and 2071–2100 periods compared to the baseline.

Changes in River Discharge during 1951–2012 Period

Model simulations of the Huong River discharge for the 1951–2012 period (and observations) show an increase in annual discharge. On average, the increase in annual discharge is relatively small until the end of the 20th century; however, extreme peak flows occurred more often during the last decade. Daily peak flows that exceed 2000 m^3/s occurred only few times during 1951–2007 period (in 1974, 2000 and 2005); however peak flows of over 2000 m^3/s have occurred in every rainy season from 2008 through 2012. Since 2000, the Huong River discharge in the rainy season increased considerably, especially from September to November, but decreased during the dry season. These changes are in response to the unusual extreme precipitation events that were observed in the first decade of the twenty-first century. Seasonal river discharge has also risen during the 1951–2012 period in general (Figure 8). Decadal average discharge indicates an increase of seasonal discharge in the fall from 1971–2000 period to 1991–2020 period. Summer discharge also increased during this period. The increase in spring and winter discharge was much more modest.

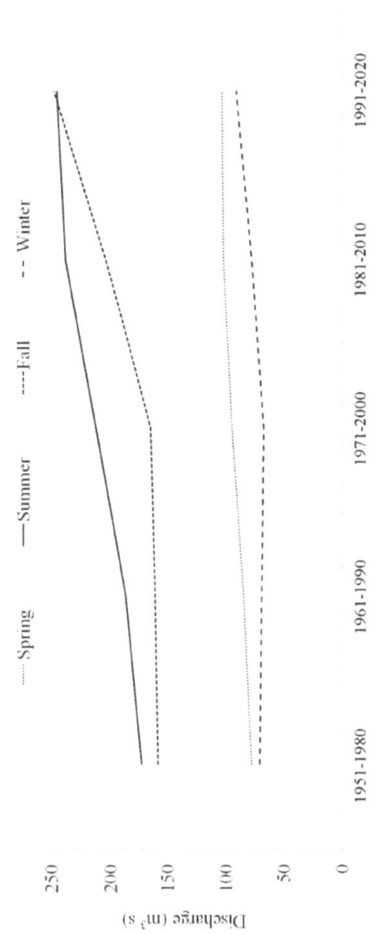

Figure 8. Average seasonal discharge from 1951 to 1980 period to 1991–2020 period.

Impact of Climate Change on River Discharge

The potential impacts of global and local climate change on water resources of the region were estimated by running the SWAT model driven by projected meteorological variables. Like other tropical rivers, the Huong River's water budget is primarily contributed mainly by direct rainfall; for that reason, any change in precipitation would alter the amount of discharge directly. Future discharge of the Huong River was simulated by forcing SWAT model with GCM outputs. Results are presented for the three preset emission scenarios for seven 30-year periods (Figure 9). Each projected discharge result is corresponding to a combination of one GCM and one emission scenario. The authors are not aware of any major land use changes in the Huong River basin and have no information on projected land use changes in the region. For these reasons, the existing land use patterns are used in the SWAT simulations. In general, the Huong River discharge is projected to continue its increase for the rest of the 21st century. Modeled ET calculations by SWAT agree with the effect of average temperature rise, especially in the winters of the 2071–2100 period and springs of the 2011–2040 period. The warming of local climate in the future that is simulated by the GCMs causes a decrease of river discharge for some periods even though precipitation is projected to increase.

In the scenario of high emission A2, there are similarities between ECHAM5-OM and HadCM3 projections. Summer is the season with the most amount of discharge, followed by fall. Spring and winter continue to have much lower discharge and river discharge is slightly higher in the spring than in the winter. The Huong River discharge is projected to increase in the summer and the fall for the rest of this century. However, HadCM3 model projects a drop in fall discharge from the 2031–2060 period until the 2051–2080 period and then a gradual increase to the level of the 2011–2040 period. Slight decreases in seasonal discharge in the winter and the spring are also projected by ECHAM5-OM and HadCM3 models. On the contrary, GFDL-CM2.1 model projects a small decrease in discharge in the summer, especially from the 2011–2040 period to the 2051–2080 period. This model also projects winter discharge of the Huong River will increase and become higher than the spring discharge which was rarely observed in the past. The increase of fall discharge in A2 scenario is once more reinforced by GFDL-CM2.1 model.

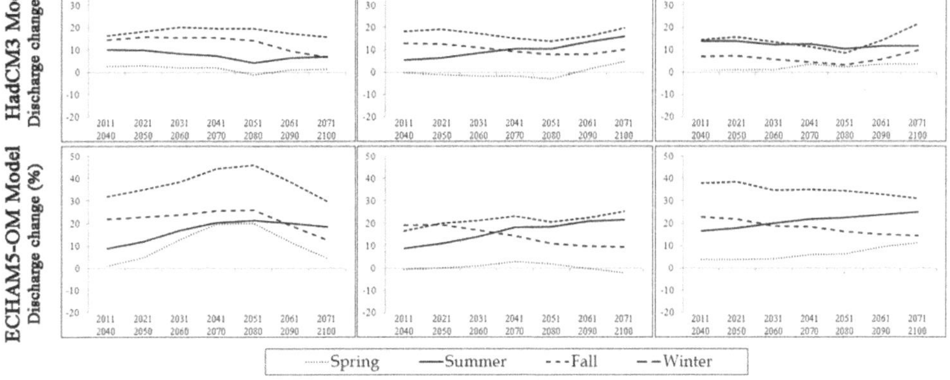

Figure 9. Projected change in total seasonal discharge in the rest of the twenty-first century produced by different GCMs in different IPCC-SRES scenarios.

Projections in the scenario of medium emission A1B show disagreements among the three climate models. Both HadCM3 and GFDL-CM2.1 models project a continuous decrease of river discharge until the 2051–2080 period and then an increase until the end of the 21st century. The ECHAM5-OM model projects that it will increase until the 2051–2080 period and then decrease for the rest of the century. HadCM3 and GFDL-CM2.1 models have similar projections for spring discharge. GFDL-CM2.1 once more projects a winter discharge that stays higher than spring discharge while HadCM3 and ECHAM5-OM models still produce normal winter discharge values that are lower than those of the spring. The Huong River discharge in the fall tends to decrease slightly in the middle of the century and increase at the end of the century as for GFDL-CM2.1 model results. HadCM3 model projects this discharge will be rather stable except a small increase in the 2051–2080 period. ECHAM5-OM model projects similar scenarios of discharge change for the four seasons at varied levels. This model shows that in any season of the year, river discharge will keep increasing until the 2051–2080 period and then decreases until the end of the century. Seasonal discharge in the fall will sometimes become higher than summer discharge, especially in the 2011–2040, 2021–2050 and 2041–2070 periods.

Discharge projections under low emission scenario B1 also show disagreements among the three climate models, especially in the summer and the fall seasons. The GFDL-CM2.1 model projects the river discharge to increase significantly in the fall from the 2011–2040 period to the 2041–2070 period and will maintain that level to the end of the century. Summer discharge is also projected to increase in the twenty-first century. The generated winter discharge values in B1 scenario from this model are higher than the spring values and the gap between these two lines seems to widen over time. The HadCM3 model has an opposite projection that summer discharge will decrease through the twenty-first century and fall discharge will decrease similarly just until the 2051–2080 period and will climb up afterwards. Both HadCM3 and ECHAM5-OM models project minimal change in winter and spring discharge within the whole twenty-first century. However, ECHAM5-OM model projects summer discharge to increase and fall discharge to decrease.

The disagreement among three GCMs' projections is caused by different numerical equations used in the GCMs. However, all three GCMs agree that the Huong River discharge in the spring will not change significantly in the twenty-first century under any emission scenario. Winter discharge is also projected to undergo quantitatively insignificant changes, except the divergence between GFDL-CM2.1 model and the other two models due to model performance. Summer and fall discharge will vary with different possibilities of increasing and decreasing and those variations are significant. Figure 10 provides details on annual projection of seasonal discharge of the Huong River. Previous studies generated projections for South East Asia indicate an increase in runoff from 10%–20% to 40%–50% in the 21st century depending on different GCMs and emission scenarios (e.g., [13,19]). This also agrees with the projections that we obtained in this study.

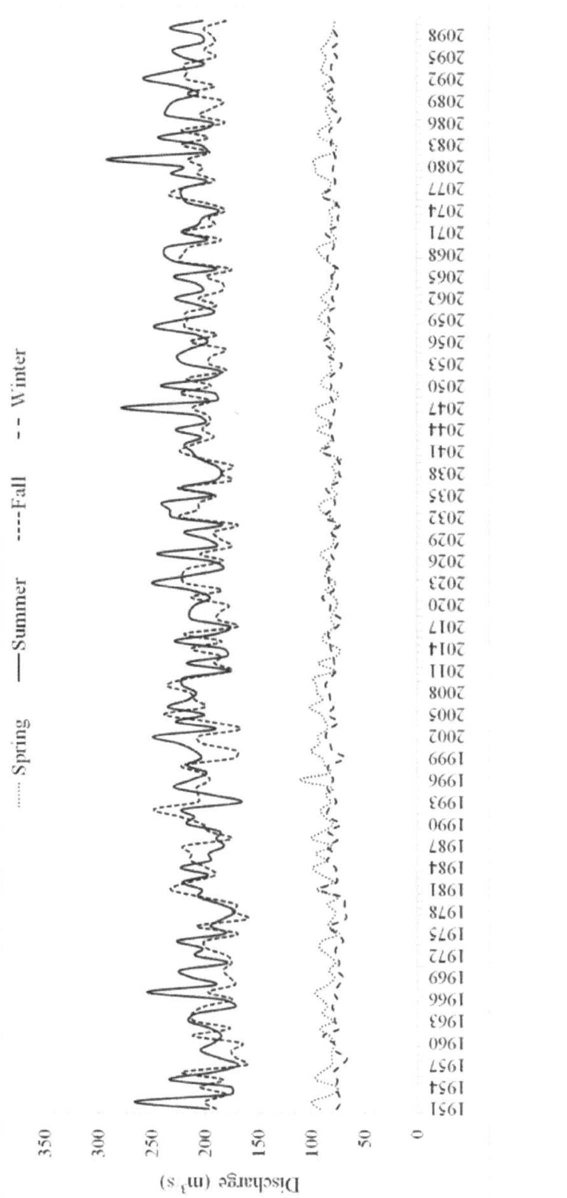

Figure 10. Historical and projected seasonal discharge from 1951 to 2100 (average of all GCMs and IPCC-SRES scenarios).

SUMMARY AND CONCLUSIONS

This paper used the stochastic weather generator LARS-WG and the hydrologic model SWAT to study projected future water availability in the Huong River basin in Central Vietnam. Future meteorological parameters are projected by three popular GCMs, ECHAM5-OM, HadCM3 and GFDL-CM2.1, for three IPCC-SRES scenarios A2, A1B and B1. Future projections of climate and discharge data were compared to baseline period from 1951 to 2012. The main source of the precipitation data for the baseline period was the gage-based APHRODITE precipitation product, which is complemented by observation in the later years. Other historical meteorological data were based on in-situ measurements and satellite observations. LARS-WG and SWAT were first calibrated and validated over the study area. LARS-WG was used to simulate meteorological conditions during the historical period and its results compared very well against observations. SWAT model was also calibrated and validated over the Huong River basin against monthly observed discharge data. Results of SWAT model calibration and validation were adequate with a Nash-Sutcliff efficiency coefficient of 0.88 and a correlation coefficient of 0.94.

Over the historical period, the Huong River discharge has been increasing since the 1951–1980 period. In particular, the fall discharge has increased significantly since the 1971–2000 period while the summer discharge has increased most between the 1970's and 1981–2010 period. According to models' projections for seasonal Huong River discharge in the twenty-first century, this increase in the summer and fall will continue with A2 and B1 emission scenarios modeled by the three GCMs. Changes in winter and spring discharge appear insignificant. In most of the cases, it is projected to slightly decrease during the 21st century. However, detailed analyses indicate differences in possible changes of seasonal river discharge in the future among different GCMs and climate scenarios. Although precipitation is the most important agent in changing in discharge, the projection of the Huong River discharge is not always proportional to the projected change in precipitation under the same IPCC-SRES scenario and using the same GCM. An obvious explanation is that ET will rise significantly due to the increase of temperature and that reduces the total amount of runoff.

The intent of this paper is to generate data that may aid decision makers in making decisions that will ensure sustainability of water resources in the region. The Huong River is witnessing intense on-going water resources management activities, the details of which are not completely available to the authors, e.g., dam construction and operation policies. Thus, simulations of future discharge do not include the impacts of management and flow regulation. Nonetheless, the study presents results that may help guide water resources planning and management in the region by quantitatively providing potential changes in river flow in the 21st century.

ACKNOWLEDGMENTS

Support of the Environmental Science and Engineering program, University of Texas at San Antonio to the first author is cordially acknowledged.

Author Contributions

This paper presents part of Tuan Le study research. Tuan Le's study was advised by Hatim O. Sharif who guided his research and contributed significantly to preparing the manuscript for publication.

Conflicts of Interest

The authors declare no conflicts of interest.

REFERENCES

1. Arnell, N.W. Climate change and global water resources. Glob. Environ. Chang. **1999**, 9, 831–849.
2. Arnell, N.W. Effects of IPCC SRES emissions scenarios on river runoff: A global perspective.Hydrol. Earth Syst. Sci. **2003**, 7, 619–641.
3. Arnell, N.W. Climate change and global water resources: SRES emissions and socioeconomic scenarios. Glob. Environ. Chang. **2004**, 14, 31–52.
4. Vorosmarty, C.J.; Green, P.; Salisbury, J.; Lammers, R.B. Global water resources: Vulnerability from climate change and population growth. Science **2000**, 289, 284–288.
5. Oki, T.; Kanae, S. Global hydrological cycles and world water resources. Science **2006**, 313, 1068–1072.
6. Intergovernmental Panel on Climate Change (IPCC). Climate Change 2013: The Physical Science Basis. Contribution of Working Group I to the Fifth Assessment Report of the Intergovermental Panel on Climate Change; Cambridge University Press: Cambridge, UK, 2013.
7. Felzer, B.; Heard, P. Precipitation Differences amongst GCMs Used for the U.S. National Assessment. J. Am. Water Resour. Assoc. **1999**, 35, 1327–1340.
8. Nijseen, B.; O'Donnell, G.M.; Hamlet, A.F.; Lettenmaier, D.P. Hydrologic sensitivity of global rivers to climate change. Clim. Chang. **2001**, 50, 143–175.
9. Arreguin-Cortes, F.I.; Lopez-Perez, M. Impacts of climate change on the hydrological cycle in Mexico. Int. J. Water Resour. Dev. **2013**, 29, 172–183.
10. Rasilla, D.F.; Garmendia, C.; Carlos, J. Climate change projections of streamflow in the Iberian Peninsula. Int. J. Water Resour. Dev. **2013**, 29, 184–200.
11. Hashmi, M.Z.; Shamseldin, A.Y.; Melville, B.W. Comparison of SDSM and LARS-WG for simulation and downscaling of extreme precipitation events in a watershed. Stoch. Environ. Res. Risk Assess. **2011**, 25, 475–484.
12. Sanderson, M.G.; Wiltshire, A.J.; Betts, R.A. Projected changes in water availability in the United Kingdom. Water Resour. Res. **2012**, 48, W08512.
13. Weiland, F.C.S.; van Beek, L.P.H.; Kwadijk, J.C.J.; Bierkens, M.F.P. Global patterns of change in discharge regimes for 2100. Hydrol. Earth Syst. Sci. **2012**, 16, 1047–1062.
14. Wu, H.; Kimball, J.S.; Elsner, M.M.; Mantua, N.; Adler, R.F.; Stanford, J. Projected climate change impacts on the hydrology and temperature of Pacific Northwest rivers. Water Resour. Res. **2012**, 48, W11530.
15. Hamlet, A.F. Assessing water resources adaptive capacity to climate change impacts in the Pacific Northwest Region of North America. Hydrol. Earth Syst. Sci. **2011**, 15, 1427–1443.
16. Thodsen, H. The influence of climate change on stream flow in Danish rivers. J. Hydrol. **2007**,333, 226–238.
17. Institute of Strategy and Policy On Natural Resources and Environment of Vietnam (ISPONRE). Vietnam Assessment Report on Climate Change; Vanhoa-Thongtin Publishing House: Hanoi, Vietnam, 2009.
18. Milly, P.C.D.; Dunne, K.A.; Vecchia, A.V. Global pattern of trends in streamflow and water availability in a changing climate. Nature **2005**, 438, 347–350.
19. Nohara, D.; Kitoh, A.; Hosaka, M.; Oki, T. Impact of climate change on river discharge projected by multimodel ensemble. J. Hydrometeorol. **2006**, 7, 1076–1089.
20. Chen, T.C.; Tsay, J.D.; Yen, M.C.; Matsumoto, J.

Interannual variation of the late rainfall in central Vietnam. J. Clim. **2012**, 25, 392–413.

21. Yen, M.C.; Chen, T.C.; Hu, H.L.; Tzeng, R.Y.; Dinh, D.T.; Nguyen, T.T.T.; Wong, C.J. Interannual variation of the Fall rainfall in Central Vietnam. J. Meteorol. Soc. Jpn. **2011**, 89A, 259–270.
22. Prime Minister of Vietnam (PM-Vietnam). Decision No 1570/QD-TTg: Approved Strategy for Sustainably Exploitation and Usage of Natural Resources and Marine Environment Protection until 2020, Vision to 2030; PM-Vietnam: Hanoi, Vietnam, 2013.
23. Ngo, D.T. Effective Usage and Development of Water Resources; Biodiversity and Climate Change Conference: Connection to Poverty and Sustainable Development: Hanoi, Vietnam, 2007.
24. Applications of the MIKE 11 Model for Evaluating the Effectiveness in Flood Control of the Reservoirs in Upstream of the Huong River. Available online: http://lib.wru.edu.vn/index.php?option=com_content&view=category&layout=blog&id=15&Itemid=196&limitstart=8(accessed on 3 February 2014).
25. Hydropower Development on Huong River Basin, Limitations and the Measures to Improve the Benefits. Available online: http://lib.wru.edu.vn/index.php?option=com_content&view=category&layout=blog&id=15&Itemid=196&limitstart=5 (accessed on 3 February 2014).
26. Nguyen, T.; Nguyen, H.S. Impact of climate change in the watershed of Huong River, Thua Thien Hue Province. J. Sci. Hue Univ. **2010**, 58, 107–119.
27. Intergovernmental Panel on Climate Change (IPCC). Climate Change 2007: The Physical Science Basis. Contribution of Working Group I to the Fourth Assessment Report of the Intergovermental Panel on Climate Change; Cambridge University Press: Cambridge, UK, 2007.
28. Le, T.B.; Al-Juaidi, F.H.; Sharif, H. Hydrologic Simulations Driven by Satellite Rainfall to Study the Hydroelectric Development Impacts on River Flow. Water **2014**, 6, 3631–3651.
29. Semenov, M.A.; Stratonovitch, P. Use of multi-model ensembles from global climate models for assessment of climate change impacts. Clim. Res. **2010**, 41, 1–14.
30. Semenov, M.A.; Brooks, R.J.; Barrow, E.M.; Richardson, C.W. Comparison of the WGEN and LARS-WG stochastic weather generators for diverse climates. Clim. Res. **1998**, 10, 95–107.
31. Racsko, P.; Szeidl, L.; Semenov, M.A. A serial approach to local stochastic weather models.Ecol. Model. **1991**, 57, 27–41.
32. LARS-WG Stochastic Weather Generator. Available online: http://www.rothamsted.ac.uk/mas-models/larswg.php (accessed on 1 June 2015).
33. Covey, C.; AchutaRao, K.M.; Cubasch, U.; Jones, P.; Lambert, S.J.; Mann, M.E.; Phillips, T.J.; Taylor, K.E. An overview of results from the Coupled Model Intercomparison Project. Glob. Planet. Chang. **2003**, 37, 103–133.
34. Institute of Resources, Environment, and Biotechnology of Hue University, Vietnam (IREB-Hue). Annual Reports on Water Quality Monitoring Results of Huong River; IREB-Hue: Hue, Vietnam, 2003–2010.
35. Veith, T.L.; van Liew, M.W.; Bosch, D.D.; Arnold, J.G. Parameter sensitivity and uncertainty in SWAT: A comparison across five USDA-ARS watersheds. Am. Soc. Agric. Biol. Eng. **2009**,53, 1477–1486.
36. Santhi, C.; Arnold, J.G.; Williams, J.R.; Dugas, W.A.; Srinivasan, R.; Hauck, L.M. Validation of the SWAT model on a large river basin with point and nonpoint sources. J. Am. Water Resour. Assoc. **2001**, 37, 1169–1188.
37. Van Liew, M.W.; Garbrecht, J. Hydrologic simulation of the Little Washita River experimental watershed using SWAT. J. Am. Water Resour. Assoc. **2003**, 39, 413–426.
38. Moriasi, D.N.; Arnold, J.G.; van Liew, M.W.; Bingner, R.L.; Harmel, R.D.; Veith, T.L. Model evaluation guidelines for systematic quantification of accuracy in watershed simulations.Trans. ASABE **2007**, 50, 885–900.
39. Yatagai, A.; Krishnamurti, T.N.; Humar, V.; Mishra, A.K.; Simon, A. Use of APHRODITE rain gauge-based precipitation and TRMM 3B43 products for improving Asian monsoon seasonal precipitation forecasts by the superensemble method. J. Clim. **2014**, 27, 1062–1069.

CHAPTER 15 MODELLING HYDROLOGY OF A SINGLE BIORETENTION SYSTEM WITH HYDRUS-1D

Yingying Meng, Huixiao Wang, Jiangang Chen, and Shuhan Zhang

INTRODUCTION

Rapid urbanization in watershed, with the increasing impervious area, implies both larger stormwater runoff volumes and peak flows and consequently reduces other components of the hydrologic cycle, for example, infiltration and evapotranspiration. Moreover, stormwater directly transports harmful substances from urban surfaces to downstream water systems, thus degrading the water quality. The negative impacts of urban stormwater have received widespread recognition [1], and maintaining stormwater quantity (e.g., flood peak and total volume) and quality (e.g., pollution) as close as the predevelopment levels has become increasingly popular. Bioretention, also known as rain garden, biofilter, or biofiltration, is a terrestrial-based water quantity and quality control practice that can be designed to mimic predevelopment hydrology (PGCo, 2007). It is thus commonly used as a source control technique to manage stormwater runoff in areas under urbanization and a retrofit technique in already developed areas [2]. Bioretention has also played an important role in the implementation of best management practice (BMP) and low impact development (LID) in America, water sensitive urban design (WSUD) in Australia, and sustainable urban drainage system (SUDS) in England.

There are many factors influencing the performance of bioretention systems, such as type of vegetation, depth of the filter media, size of the system relative to its catchment, and type of soil. Sizing, vegetation, construction technique, and soil mixture were all reported to have an important influence on the hydraulic conductivity of bioretention [3, 4]. The sizing of biofilters was also emphasized by Brown and Hunt III [5] who presented better reductions in runoff volume with deeper media depth. Furthermore, the hydraulic conductivity of the underlying soil and the internal water storage zone depth were also considered as primary factors influencing water reduction [6]. Overall, this research work on factors influencing the performance of bioretention was mainly based on column studies in laboratories or field studies [4, 5]. Because experimental observations were easily restricted by test conditions, unexpected results were sometimes reached. In the field study of Brown and Hunt III [5], for example, the surface storage volume of two bioretention cells was undersized because of design and construction errors, having substantial negative impacts on cell performance. Therefore, there is an increasing need to predict the hydrologic and water quality performance of bioretention systems using hydrologic model, which could be conveniently used in design, evaluation, or other purposes.

Initial model studies about bioretention did not include underdrains; for example, Heasom et al. [7] have attempted to predict the overflow volume in a bioinfiltration cell using one-dimensional hydrological model HEC-HMS. Considering an underdrain, He and Davis [8] developed a two-dimensional model simulating the subsurface flow. However, both models were based on individual rainfalls and were unable to perform continuous simulations and therefore they could not account for the changes in soil moisture conditions from previous rainfall events. The RECARGA model [9], widely used in the design and performance assessment of bioretention systems [10], allows for both continuous modelling and single-event modelling, but its minimum hourly rainfall interval makes it unable to conduct simulations for very short periods. Moreover, some parameters such as the number of underdrains and their depths and types of filter media could not be specified by the user, limiting the model's applications in some situations. As the water movement process in bioretention cells installed

Citation: Yingying Meng, Huixiao Wang, Jiangang Chen, and Shuhan Zhang, "Modelling Hydrology of a Single Bioretention System with HYDRUS-1D," The Scientific World Journal, vol. 2014, Article ID 521047, 10 pages, 2014. doi:10.1155/2014/521047.

with underdrains is very similar to agricultural drainage pipes, modelling hydrologic performance in bioretention systems with DRAINMOD, an agricultural drainage model, has been common in recent years [11]. But DRANIMOD is unsuitable for conducting short-term simulations with a minimum calculation time of 1 month. Other models used in bioretention simulations involve SWMM (USEPA, 2010), SUSTAIN (USEPA, 2013), or MUSIC (eWater, 2013), but because of scale problem they are not appropriate for a single facility simulation.

One potential solution to reduce the frequent urban waterlogging disasters in Beijing in recent years is to control urban runoff at source as much as possible. As a source control technique, bioretention systems have advantages in ultraurban areas such as Beijing where land is unavailable for large control practices such as retention ponds, grassed swales, and constructed wetlands. The main objective of this study is to evaluate the hydrologic performance of bioretention facilities and provide instructive guidance for their design and application in Beijing. We developed a model tool to predict the hydrologic performance of a single bioretention facility and discuss the influence of different design parameters. Overall, the above-mentioned models have their own shortcomings in terms of scale, calculation time, configuration design, and other aspects. Comparatively, the HYDRUS-1D model [12] is a more appropriate model, with flexible water flow boundary conditions, a minimum calculation interval of 1 s, and unlimited simulation time. Hilten et al. [13] and Ladu et al. [14] both simulated the stormwater performance of a green roof using HYDRUS-1D. As a heterogeneous multilayer soil medium such as the green roof system, a bioretention system has the potential to be simulated by this model, while no data have yet been reported in the literature on the model's ability to model bioretention performance. In our study, the HYDRUS-1D model was devised based on input variables measured at two bioretention cells constructed in Beijing. The hydrology processes were measured at these facilities for nearly 4 years, and the collected data were used to calibrate and validate the model. The factors affecting the bioretention performance are discussed based on the simulation results for different design storms, filter media depths and types, and surface areas.

METHODS

Principle of the HYDRUS-1D Model

The hydrologic processes in bioretention systems consist of evapotranspiration, infiltration, and runoff generation. The water balance equation is given as follows:where ET is the evapotranspiration, P is the precipitation, I is the infiltration, R is the runoff, and DSW is the change in soil water content. The total runoff is given as follows:where RS is the surface runoff and RB is the bottom runoff from the drainage layer.

Field Study Site

Two parallel bioretention cells (Cells A and B) were constructed in Mentougou district in Beijing in 2010 (Figure 1). Each cell was 3 m * 2 m on the top surface, 2.2 m * 1.1 m on the bottom surface, and 1.1 m deep. Being designed to capture and treat stormwater runoff from a 60 m^2 impervious roof, each cell covers 10% of the catchment area. The cells were built with the following composition (Figure 2):(i)a drainage layer at the base, containing a 110 mm diameter slotted PVC pipe (connected to the observation well for flow measurement) surrounded by 30 mm gravel with the diameter of 5–10 mm;(ii) sixty centimeters of filter media: conventional media in Cell A, 97% of sand and soil, 3% of peat (both by volume); two layers in Cell B, one-third of sand and soil and two-thirds of blast furnace slag with the diameter of 5 mm on the top layer of 25 cm (both by volume), vermiculite on the bottom layer of 35 cm with the diameter of 0.5–5 mm to increase soil porosity [4];(iii)five centimeters of mulch: shredded pine bark;(iv)vegetation cover, with native plants of Ophiopogon japonicas and Iris tectorum for Cells A and B. Before choosing vegetation to conduct experiments on, we defined some criteria to ensure the vegetation suitable for bioretention construction. The criteria were defined to fit the climate in Beijing, specifically, drought, flood, pollution, salt, shade, and cold tolerant, with ornamental value locally selected, low cost, and low maintenance. Based on these requirements, we finally chose Ophiopogon japonicas andIris tectorum;(v)an overflow drain connected to the PVC pipe allowing a maximum ponding depth of 15 cm;(vi)an impervious geotextile on the sides and bottom to minimize migration of water into or out of the system.

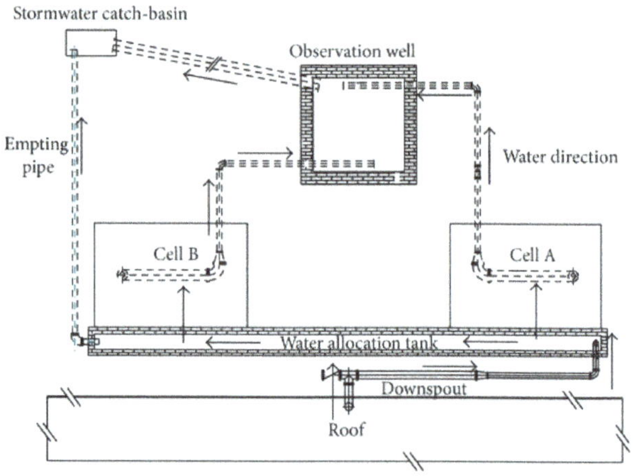

Figure 1: Plan of the bioretention study area (not to scale).

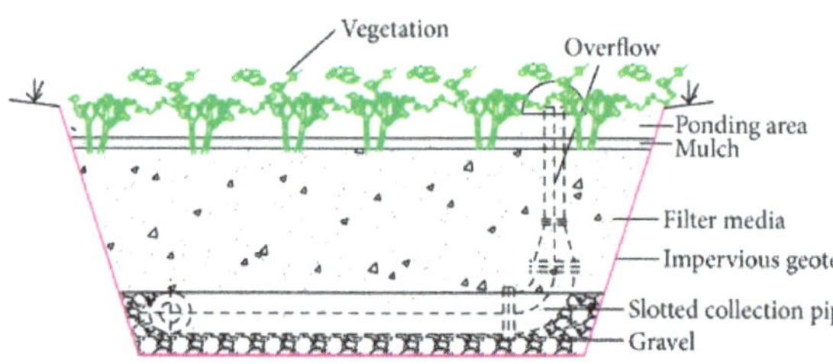

Figure 2: Schematic view of the bioretention cell.

Field Measurements

The infiltration and overflow volume which contributed to the bottom runoff both discharged through the PVC pipe, and its flow rate was measured using a 5 L measuring cup and a stopwatch every 5 min after the PVC pipe started to drain off water in the observation well, until there is no pipe flow. If ponding occurred, the water level in the ponding area was measured by a meter ruler every 5 min until ponding disappeared. These in situ data were collected during June to September in 2010, 2011, 2012, and 2013.

In the absence of natural rainfall, simulated stormwater was sometimes used for the experiments, using the reference method provided in Hsieh and Davis [15] to prepare the simulated stormwater. In artificial rainfall, the simulated stormwater with the theoretical volume from each cell's catchment area was mixed well and pumped into the cells evenly over an hour. The bottom runoff and ponding processes were also monitored.

Thirty-eight rainfall events were monitored, with thirty-three artificial events and five natural events. The artificial rainfall test results were mainly used for parameter calibration and validation of the HYDRUS-1D model, and the natural rainfall results were all used for model validation.

Modelling with HYDRUS-1D

Input requirements for HYDRUS-1D include geometry and time information, soil hydraulic and vegetation properties, initial and boundary conditions, and meteorological information, whose values are listed in Table 1. The soil hydraulic parameters used in the van Genuchten model were measured using a high-speed centrifuge method (Table 2). Daily measurements of meteorological variables, including air temperature and humidity, atmospheric pressure, precipitation, wind speed and direction, and incoming shortwave and longwave radiation, were collected from the meteorological station in Beijing (number 54511) near the experimental site. The initial conditions are given in terms of water content, which is linearly distributed in the soil profile. The other parameters were specified according to the default values.

Table 1: The input information in the model for Cells A and B.

Input information	Parameters	Values	
		Cell A	Cell B
Initial condition	Minimum water content	0.36	0.34
	Maximum water content	0.38	0.36
	Maximum height at soil (cm)	15	15
Geometry information	Number of layers	1	2
	Depth of the soil (cm)	60	25, 35
Time information	Time duration (min)	300	120
	Time step (min)	0.00001~5	0.001~0.25
Water flow-soil hydraulic property model	Model	van Genuchten-Mualem	
Water flow boundary conditions	Upper boundary condition	Atmospheric BC with surface layer	
	Lower boundary condition	Seepage face	
Vegetation properties	Water uptake reduction model	Feddes	
	Crop height (cm)	30	60
	Root depth (cm)	20	30

Table 2: The soil hydraulic parameters for the van Genuchten model in different structure layers.

Structural layer	Residual water content (cm^3/cm^3)	Saturation water content (cm^3/cm^3)	α (min^{-1})	n (—)	Saturated hydraulic conductivity (cm/min)	l (—)
Soil media in Cell A	0.065	0.410	0.025	1.69	0.083	0.5
Top media in Cell B	0.071	0.490	0.460	1.30	0.252	0.5
Bottom vermiculite in Cell B	0	0.448	0.002	1.44	0.518	0.5

In this study, the model was validated based on the measured bottom runoff and water level. Because the artificial and natural rainfalls were both individual events, we chose an artificial rainfall and a natural rainfall, respectively, as the rainfall events for validation. Statistics of the root-mean-square error (RMSE), mean relative error (MRE), and correlation coefficient (R^2) were then used to assess the accuracy of the simulation. The smaller the RMSE and MRE are, the more closely the R^2 approximated to zero and the better the model performed.

Model Application

Upon verifying the accuracy of the HYDRUS-1D model, simulations were run using different input variables to optimize the design of the bioretention systems. Four general scenarios with different rainfall return periods, filter media depths and types, and surface areas were investigated, and, through scenario analysis using the model, the influence of the different design parameters on the hydrologic performance of bioretention systems was studied. In the above simulations, storms were simulated as independent events. As HYDRUS-1D had no limitation on simulation time, the long-term hydrologic performance of bioretention systems was also assessed by inputting the annual meteorological data in 2012.

The hydrologic performance of bioretention systems could be described in terms of the hydrologic effectiveness and the water detention and retention effects. Hydrologic effectiveness (R_{hydro}) denotes the total rainfall runoff treated by the bioretention in whatever form, specified using (3). Water detention means water temporarily reserved by the system, which was demonstrated by bottom runoff delay (Δt_{dd}), bottom runoff peak flow delay (Δt_{pd}), and bottom runoff peak flow reduction (R_{pr}), given in (4)–(6). Comparatively, water retention (R_{reten}) refers to the water completely reserved by the system, given by (7). The ponding duration (t_{pond}) can also reflect the hydrologic performance to some extent. Consider

$$R_{hydro} = \frac{V_{inflow}}{V_{runoff}} \times 100\%, \tag{3}$$

where V_{inflow} is the inflow volume of the bioretention cell and V_{runoff} is the runoff volume from the catchment area. Consider

$$\Delta t_{dd} = t_{drain} - t_{inflow}; \tag{4}$$

where t_{drain} is the time the bottom runoff appears and $t$inflow is the time the inflow enters the bioretention cell. Consider

$$\Delta t_{pd} = t_{pdrain} - t_{pinflow} \tag{5}$$

where t_{pdrain} is the time the bottom runoff peak appears and tpinflow is the time the inflow peak appears. Consider

$$R_{pr} = \frac{q_{pinflow} - q_{pdrain}}{q_{pinflow}} \times 100\%, \tag{6}$$

where $q_{pinflow}$ is the inflow peak and q_{pdrain} is the bottom runoff peak. Consider

$$R_{reten} = \frac{V_{inflow} - V_{drain}}{V_{runoff}} \times 100\%, \tag{7}$$

where V_{drain} is the bottom runoff volume.

RESULTS AND DISCUSSION

Model Validation.

In our experiments, overflow never occurred; thus, the bottom runoff volume was equal to the infiltration volume. Dividing the infiltration volume by the surface area of the bioretention cell, we obtain the infiltration rate. The comparison of the modelled infiltration rate by HYDRUS-1D and the observation values in Cells A and B are shown in Figure 3, with all of the simulated values in good agreement with observed ones. Because the filter media of Cell B had added blast furnace slag and vermiculite whose particle sizes were larger than soil particles, it was reasonable that the infiltration rate was higher in Cell B in both the artificial and the natural rainfalls, and the bottom runoff delay was obviously later in Cell A in the artificial rainfall.

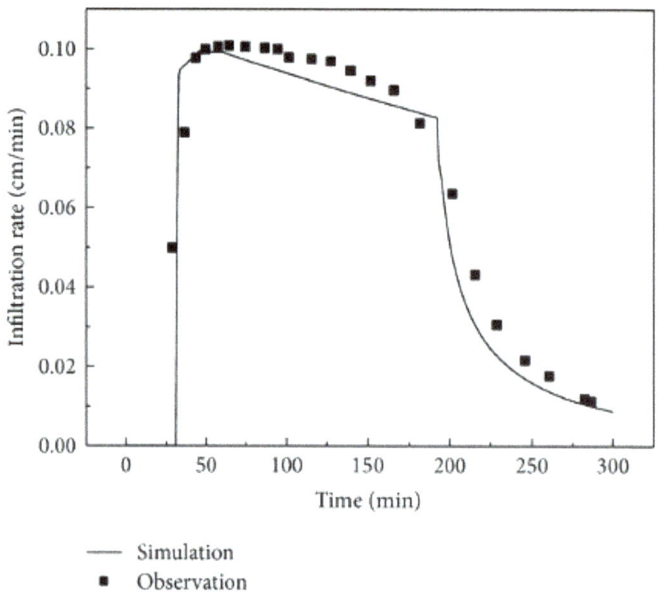

(a)

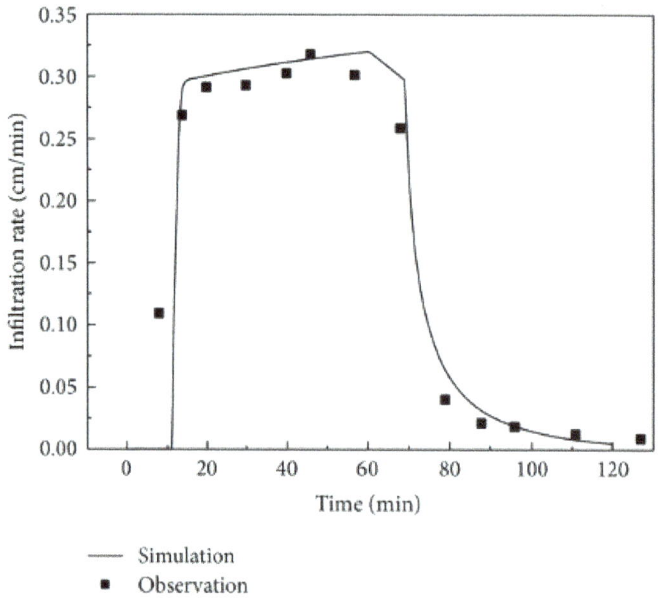

(b)

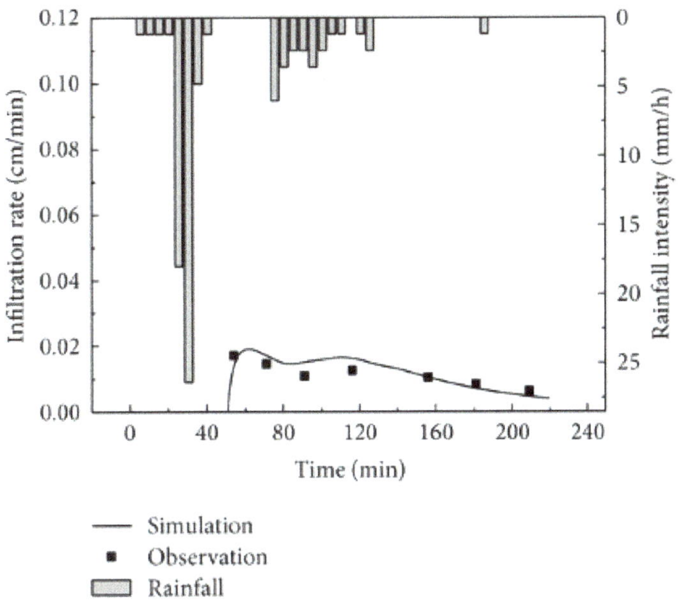

(c)

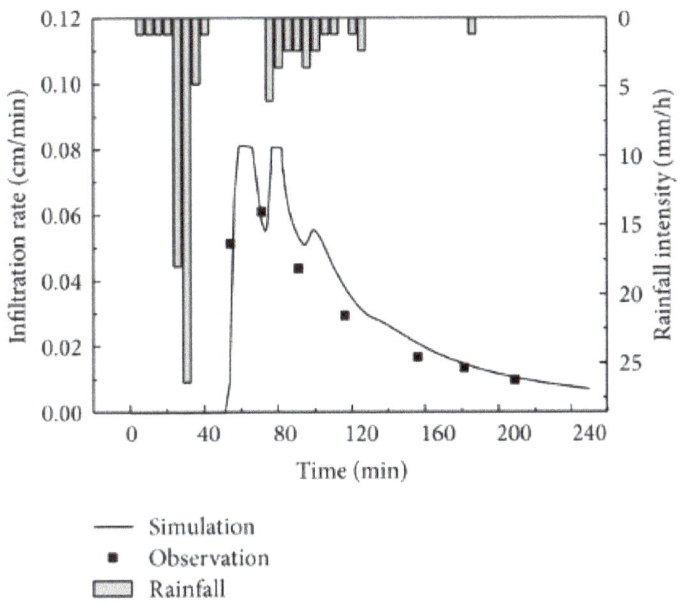

(d)

Figure 3: Observed and simulated infiltration rate in the artificial rainfall for validation in (a) Cell A and (b) Cell B and in the natural rainfall for validation in (c) Cell A and (d) Cell B.

The statistics for the RMSE, MRE, and R^2 between the simulated and observed values are shown in Table 3. The evaluation results were acceptable with the RMSEs all approximating zero, the MRE almost at 0.20, a higher R^2 (>0.9) in the artificial rainfall, and a relatively smaller R^2 (>0.6) in the natural rainfall, which might be the result of fewer observed data.

Table 3: Statistics for simulation accuracy assessment.

Cell ID	RMSE		MRE		R^2	
	Artificial rainfall	Natural rainfall	Artificial rainfall	Natural rainfall	Artificial rainfall	Natural rainfall
A	0.014	0.003	0.16	0.22	0.95	0.76
B	0.036	0.017	0.23	0.25	0.97	0.61

For the ponding process, the simulated and measured water levels in Cell A in the artificial rainfall for validation are presented in Figure 4. The observed and calculated values were well matched, as the RMSE, MRE, and R^2 were 0.34, 0.059, and

0.99, respectively. Flooding never occurred in Cell B in the experiments, while a thin layer of water (less than 2 cm) appeared in the simulation results, probably because the water retention of mulch is not taken into account by the model.

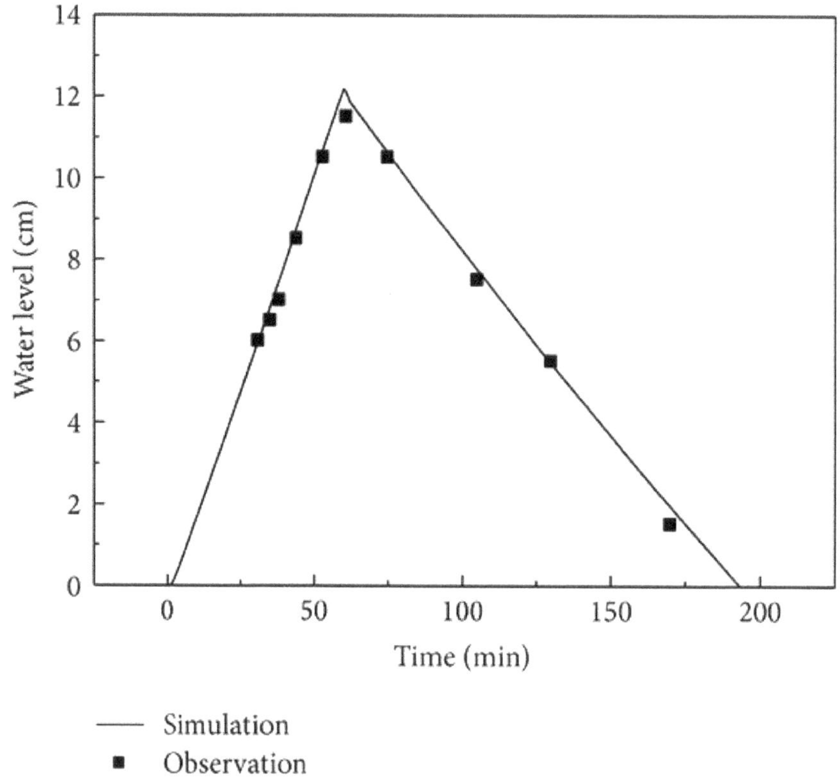

Figure 4: Observed and simulated water level in the artificial rainfall for validation in Cell A.

Overall, the HYDRUS-1D model was capable of capturing the hydrologic processes in bioretention systems with reasonable accuracy and can thus be used to assess bioretention performance under different circumstances.

Design Parameters Optimization

Rainfall Return Periods

Twenty-four-hour design storms with different return periods in the central area of Beijing were calculated according to the Hydrologic Manual of Beijing [16]. With the other input requirements the same as those in the Cell A simulation, the simulated results for different design storms are presented in Figure 5. From Figure 5, it is clear that the hydrologic effectiveness (R_{hydro}) decreased with the rainfall return period because of the increased total runoff volume from the catchment area. Thus, in rainfalls with return periods greater than 5 years, the peak flow reduction (R_{pr}) was higher because R_{hydro} was lower, and only in rainfalls with return periods smaller than 2 years did *R*pr really increase because of the higher R_{hydr}o. Therefore, greater hydrologic effectiveness is the premise for better water treatment. The water detention effect can be further evaluated using the bottom runoff delay (Δt_{dd}) and bottom runoff peak flow delay (Δt_{pd}) besides R_{pr}, and Δt_{dd} gradually decreased when the rainfall return period was greater than 5 years, while an obvious decrease was observed with a rainfall return period smaller than 5 years. However, Δt_{pd} changed dramatically with the rainfall return period, with a sharp increase for a return period smaller than 5 years, a significant decrease for a return period between 5 and 10 years, and a gentle decrease for a return period greater than 10 years. This is because of the two peaks of the 24-hour design storm in Beijing, and the bioretention could only abate the first peak with a rainfall return period greater than 10 years, while, in rainfalls with a smaller return period, it successfully eliminated the first peak, resulting in the delayed bottom runoff peak caused by the second rainfall peak. Furthermore, the variation in ponding duration (t_{pond}) was very similar to Δt_{dd}, with a gradual increase in rainfalls for a return period greater than 5 years, while there was a significant increase in rainfalls with a return period smaller than 5 years. It should be noted that to prevent mosquito breeding and maintain vegetation growth ponding duration is required to be shorter than 48 h (PGCo, 2007), and the simulated *t*pond between 4 and 20 h under different design storms all met the requirement. In summary, the water detention effect in bioretention systems is much better in rainfalls with smaller return periods. Referring to the total water retention effect, the decrease in water retention (R_{reten}) with the rainfall return period was not obvious, and R_{reten} was smaller than 10% even with the smallest storm return period of 1 year, showing that bioretention measures could store very limited stormwater, which might be because of the completely impervious sides and bottom surface in our study site.

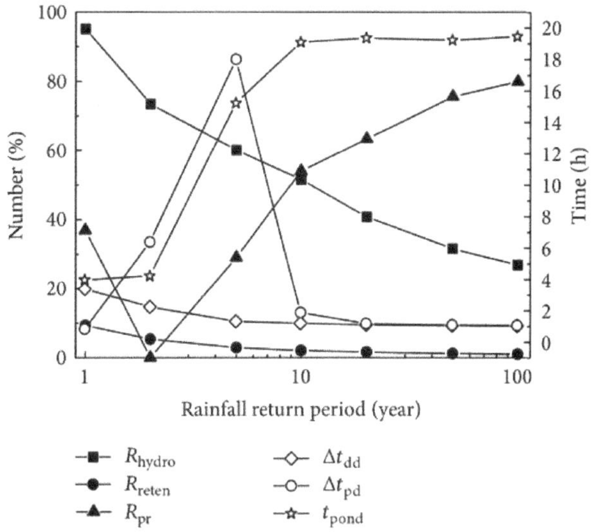

Figure 5: Hydrologic performance of bioretention measures under rainfalls with different return periods.

Conclusively, with impervious surroundings, bioretention facilities only regulate the inflow runoff discharge process, their total water retention effect is very limited, and their detention effect is significantly better in small rainfalls with return periods smaller than 2 years. This agrees with the research results of Davis [17] and Li et al. [18], who indicated that the greatest impact was noted for the smaller events. Thus, a 2-year storm or less is recommended for use as the design storm for bioretention measures. Therefore, the following discussion on the design parameters is based on a 1-year storm.

Filter Media Depths

Taking into account the demand for water quality improvement [19] and the underdrains successfully connected with the municipal storm sewer, the input filter media depths were from 30 to 90 cm. The variation in hydrologic performance is presented in Figure 6. The bottom runoff peak flow reduction (R_{pr}) and bottom runoff delay (Δt_{dd}) increased significantly with the filter media depth, while the total water treated (R_{hydro}), water retention (R_{reten}), bottom runoff peak flow delay (Δt_{pd}), and ponding duration (t_{pond}) showed almost no change.

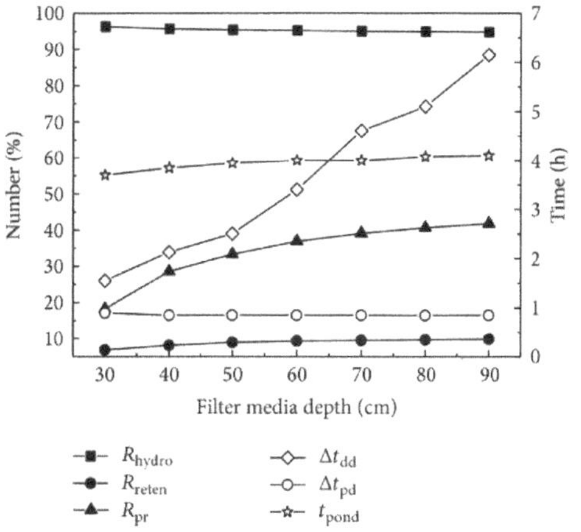

Figure 6: Hydrologic performance of bioretention measures under different filter media depths.

As one of the main costs in constructing bioretention cells, the filter media depth should not be too great; thus, a cost-effective depth should be used in bioretention design. This could also refer to the media depth requirements in North Carolina, where vegetation is the determined factor and the minimum media depth is 0.6 m for cells vegetated with grass or shallow rooted plants and 0.9 m for cells vegetated with shrubs or trees (NCDENR, 2009).

This indicated that the filter media depth only has a beneficial impact on water detention, with better water detention in deeper filter media, while it has little or no influence on hydrologic effectiveness, water retention, and ponding control, which may also be attributed to the impervious surroundings. As in the study of Brown and Hunt III [5], the exfiltration volume in bioretention systems without the impermeable membrane is much higher in the deeper media cells because of greater storage volume in the media and more exposure to the side walls, leading to the better reductions in runoff volume.

Types of Filter Media

Nine soils, from sand to clay with declining saturated hydraulic conductivity, included in the soil catalog of the HYDRUS-1D model, were used as input filter media, and in each case the hydrologic performance was simulated with default soil hydraulic parameters in the model (Figure 7). It was obvious that filter soil played an important role in hydrologic effectiveness and the total water treated by the system (R_{hydro}) decreased with declining hydraulic conductivity from sand to clay. This agreed with the research results of Le Coustumer et al. [2] who even proposed that the initial specified hydraulic conductivity of filter media was the critical determinant of the long-term hydraulic behavior of a biofilter. In this study, with a high infiltration rate, the sandy texture soils treated all the rainfall runoff without ponding, while with low infiltration rate the clay soil only treated 23.6% of the total rainfall runoff while it overflowed the other runoff and was always ponding.

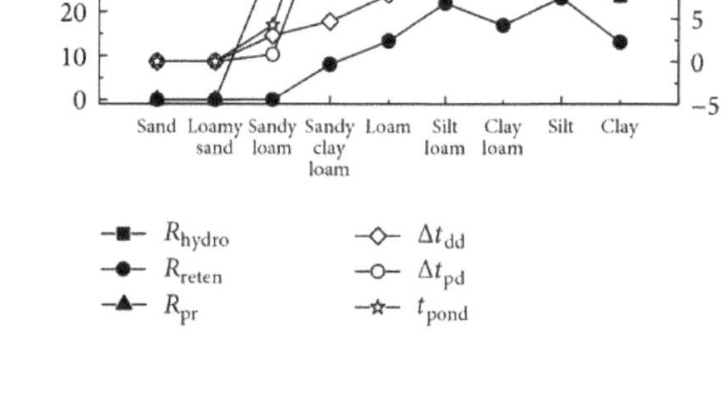

Figure 7: Hydrologic performance of bioretention measures under different filter soils.

Without higher hydrologic effectiveness, the obvious increase in the water detention parameters of bottom runoff peak reduction (R_{pr}), bottom runoff delay (Δt_{dd}), and bottom runoff peak flow delay (Δt_{pd}) with the declining infiltration rate makes no sense. Moreover, the slight increase in *R*reten with declining hydraulic conductivity was also attributed to the poor retention effect of the whole bioretention system, which has already been mentioned above.

Recommendations for hydraulic conductivity of soil media vary from one country to another [2], with at least 12.5 mm/h in New Zealand and America, between 36 and 360 mm/h in Austria and between 50 and 200 mm/h in Australia. The saturated hydraulic conductivity of soils included in the soil catalog of the HYDRUS-1D model, taken from Carsel and Parrish [20], is shown in Table 4. Considering the conductivity requirement in different countries and its potential reduction with time, it is recommended that soils between loamy sand and sandy loam are used as the filter media in bioretention design enabling high hydrologic effectiveness (R_{hydro}) and medium detention and retention effects (R_{pr}, Δt_{dd}, Δtpd, and R_{reten}). Moreover, the ponding duration (*t*pond) in the simulations of loamy sand and sandy loam was shorter than 5 h, also maintaining better vegetation growth. In USEPA (1999), the sandy loam soil has already been recommended to be used in bioretention systems. In our experiment site, the saturated hydraulic conductivity of soil media in Cells A and B was, respectively, 49.6 and 151 mm/h (0.083 and 0.252 cm/min in Table 2), very close to the infiltration performance of sandy loam and loamy sand; thus, the two facilities both perform well in terms of water infiltration.

Table 4: The saturated hydraulic conductivity of soils included in the soil catalog of the HYDRUS-1D model.

Soil type	Sand	Loamy sand	Sandy loam	Sandy clay loam	Loam	Silt loam	Clay loam	Silt	Clay
Saturated hydraulic conductivity (mm/h)	297	146	44	13.1	10.4	4.5	2.6	2.5	2

Surface Areas

General design guidelines suggest that the bioretention basin is approximately 5–7% of the effective upslope drainage area contributing to runoff (USEPA, 1999). With a surface area of 1–100% of the catchment area, the hydrologic performance of the bioretention is presented in Figure 8. As the surface area increased, the hydrologic effectiveness (R_{hydro}) and water retention effect (R_{reten}) both increased; meanwhile, the ponding duration (t_{pond}) decreased. This agreed with the research results of Jones and Hunt [21] who suggested that large bioretentionareas could reduce surface ponding times. However, the three hydrologic parameters for the water detention effect, bottom runoff delay (Δt_{dd}), bottom runoff peak flow delay (Δt_{pd}), and bottom runoff peak flow reduction (R_{pr}), behaved differently with Δt_dd increasing with the surface area, while Δt_{pd} and R_{pr} changed drastically. This may be a function of the variation in hydrologic effectiveness, and, with a quick increase in R_{hydro} between area ratios of 1% and 10%, both Δt_{pd} and R_pr changed irregularly. When R_{hydro} approached 100% with an area ratio larger than 10%, Δt_{pd} increased gradually, but R_{pr} still had irregular variations, probably because of the combined effects of different inflow volumes and the constant hydraulic capacity of the filter media. However, it was evident that larger surface areas achieved better hydrologic performance, which was also confirmed by Le Coustumer et al. [2] who found that a larger surface area compensated for low conductivity by providing a greater filter area and ponding volume. From another perspective, Dussaillant et al. [22] showed that bioretention with an area of 10–20% of the contributing impervious area maximized groundwater recharge. However, in reality, as the bioretention area increases, the value of land increases especially in current Chinese cities and the facility becomes more costly. Considering the cost of land, a costeffective surface area is recommended in bioretention design. From Figure 8, when the bioretention covers more than 10% of the catchment area, the total hydrologic effectiveness tends towards stability, and the water detention and retention effects change for the better; moreover, a ponding duration shorter than 4 h is also acceptable, and thus the surface area of 10% of the catchment area may be a reasonable compromise.

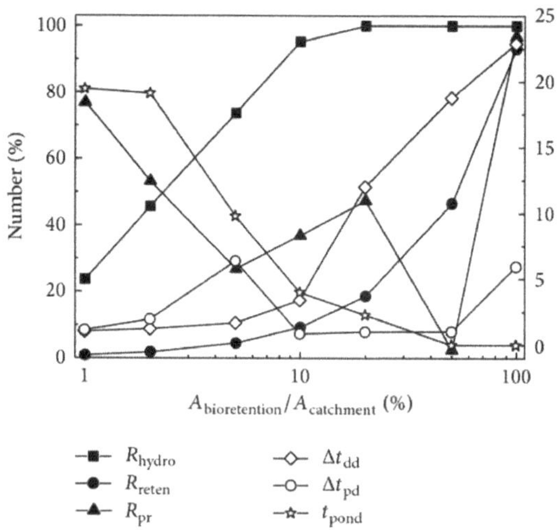

Figure 8: Hydrologic performance of bioretention measures under different bioretention areas of the catchment area.

Long-Term Hydrologic Performance

Using a medium filter depth of 60 cm, the filter type used in Cell A, which approximated the recommended sandy loam soil, and the recommended surface area of 10% of the catchment area, the long-term hydrologic performance of bioretention systems was assessed by inputting the annual meteorological data in 2012. The water volumes variation results are given in Figure 9. Because of the impervious sides and bottom surface of the bioretention system, it could be expected that the infiltration volume would take a large share of the total rainfall and the soil retention water may occupy a much lower percentage, which is proved in Figure 9 with the infiltration volume increasing rapidly with time, while the soil retention volume always fluctuates in the year. Meanwhile, the vegetation transpiration volume also increases evidently with time, which showed that, in the long run, evapotranspiration played an important role in the hydrology efficiency of bioretention systems. This was also confirmed by Dussaillant et al. [22], who reported that plant evapotranspiration during interstorm periods provided a greater available soil water storage capacity for the next rainfall event.

It could be seen in Figure 9 that, after a year of operation, the infiltration, evaporation, transpiration, soil retention, and overflow volumes in the bioretention system were 560 mm, 6.3

mm, 146 mm, 1.4 mm, and 20 mm, respectively, contributing to 75.7%, 0.9%, 19.7%, 0.2%, and 2.7% of the total rainfall in 2012.

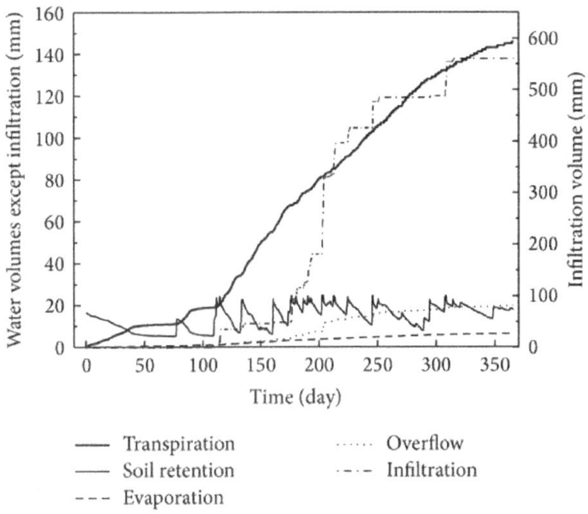

Figure 9: Accumulative water volumes variation with time in the long-term simulation using the annual meteorological data in 2012.

Furthermore, some researchers provided that plants improved filter performance; for example, Archer et al. [23] reported that root growth increased hydraulic conductivity as a result of macropores created by root dieback, and Le Coustumer et al. [4] showed that plants with thick roots maintained system permeability over time compared with plants with finer roots. However, plants only influence evapotranspiration through their growth characteristics of height and root depth in HYDRUS-1D, and, as shown in Figure 10, evapotranspiration volume increased evidently with plant height. Presently, the model is unable to simulate the effect of plants on the permeability of the system.

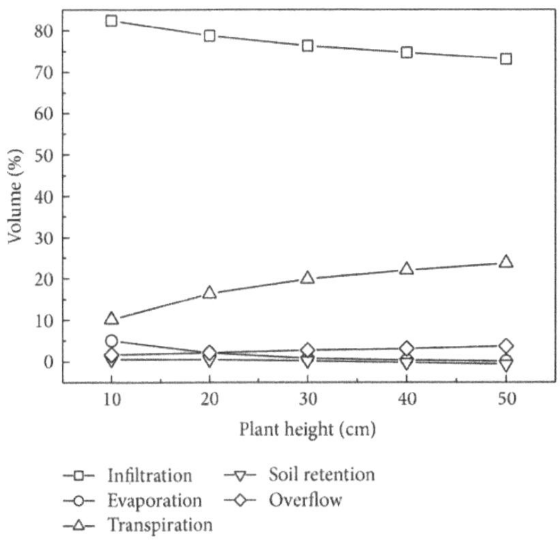

Figure 10: Water volume variation in bioretention systems with different plant heights.

CONCLUSIONS

Because stormwater management in urbanized areas has become ubiquitous, bioretention systems have been introduced as an effective source control technique to reduce runoff from impervious surfaces.

In this study, the hydrologic performance of two bioretention cells is modelled using HYDRUS-1D, with simulation results verified by field data. In the study, HYDRUS-1D accurately predicted infiltration and ponding processes in the bioretention cells.

The influence of different design parameters on the rainfall return period, media depth and type, and surface area to hydrologic performance was evaluated using the calibrated HYDRUS-1D model. It was shown that bioretention systems with underdrains and impervious boundaries have only some detention effect on bottom runoff delay, bottom runoff peak flow delay, and bottom runoff peak flow reduction, and their total water retention effect was very limited. Better detention effect was noted for smaller rainfall events, and a 2-year or less design storm was consequently recommended.

Filter media depth also had a significant impact on water detention but little or no effect on the total water treated. Better water detention appeared in deeper filter media, while, considering the filter cost, a cost-effective depth was recommended in bioretention design.

Both the hydraulic conductivity of filter media and surface area size influenced hydrologic effectiveness greatly, and better hydrologic effectiveness was reached with higher hydraulic conductivity and surface area ratio of the catchment area. Filter media with conductivity between loamy sand and sandy loam was recommended in bioretention design, enabling some conductive and retention effect as well as vegetation growth. Considering the cost of land, the cost-effective surface area was recommended in bioretention design, and the surface area of 10% of the catchment area may be a reasonable compromise.

Using the optimized design parameters for the rainfall return period, filter media depth and type, and surface area size, the long-term hydrologic performance of bioretention systems was further evaluated. As expected, the runoff inflow into the bioretention cell was mainly attenuated via infiltration, while at the same time evapotranspiration played an important role in the long run, contributing to 20.6% of the total rainfall in 2012.

Filter media play a very important role in hydrologic performance of bioretention measures, as conductivity and water retention capacity directly affect the infiltration, storage, and pollutant removal of inflow runoff. The pollutant transport process through the bioretention was not included in this study; thus, simulations of water quality improvement performance could be tried using the solute transport function of the HYDRUS-1D model, and, on this basis, the potential design parameters for better pollutant removal could be discussed, providing more references for the promotion and application of bioretention measures.

Conflict of Interests

The authors declare that there is no conflict of interests regarding the publication of this paper.

Authors' Contribution

Huixiao Wang, Jiangang Chen, and Shuhan Zhang contributed equally to this work.

ACKNOWLEDGMENTS

This work is supported by the National Science Foundation of China under Grant no. NSFC-51179009 and National Major Science and Technology Programs under Grant no. 2013ZX07304-001.

REFERENCES

1. B. E. Hatt, N. Siriwardene, A. Deletic, and T. D. Fletcher, "Filter media for stormwater treatment and recycling: the influence of hydraulic properties of flow on pollutant removal," Water Science and Technology, vol. 54, no. 6-7, pp. 263–271, 2006. · ·
2. S. le Coustumer, T. D. Fletcher, A. Deletic, S. Barraud, and J. F. Lewis, "Hydraulic performance of biofilter systems for stormwater management: Influences of design and operation," Journal of Hydrology, vol. 376, no. 1-2, pp. 16–23, 2009. · ·
3. D. D. Carpenter and L. Hallam, "Influence of planting soil mix characteristics on bioretention cell design and performance," Journal of Hydrologic Engineering, vol. 15, no. 6, pp. 404–416, 2010. · ·
4. S. Le Coustumer, T. D. Fletcher, A. Deletic, S. Barraud, and P. Poelsma, "The influence of design parameters on clogging of stormwater biofilters: a large-scale column study," Water Research, vol. 46, no. 20, pp. 6743–6752, 2012. · ·
5. R. A. Brown and W. F. Hunt, "Impacts of media depth on effluent water quality and hydrologic performance of undersized bioretention cells," Journal of Irrigation and Drainage Engineering, vol. 137, no. 3, pp. 132–143, 2011. · ·
6. R. A. Brown and W. F. Hunt, "Underdrain configuration to enhance bioretention exfiltration to reduce pollutant loads," Journal of Environmental Engineering, vol. 137, no. 11, pp. 1082–1091, 2011. · ·
7. W. Heasom, R. G. Traver, and A. Welker, "Hydrologic modeling of a bioinfiltration best management practice," Journal of the American Water Resources Association, vol. 42, no. 5, pp. 1329–1347, 2006. · ·
8. Z. He and A. P. Davis, "Process modeling of storm-water flow in a bioretention cell," Journal of Irrigation and Drainage Engineering, vol. 137, no. 3, pp. 121–131, 2011. · ·
9. A. R. Dussaillant, A. Cuevas, and K. W. Potter, "Raingardens for stormwater infiltration and focused groundwater recharge: simulations for different world climates," Water Science and Technology: Water Supply, vol. 5, no. 3-4, pp. 173–179, 2005. ·
10. T. M. Muthanna, M. Viklander, and T. T. Thorolfsson, "An evaluation of applying existing bioretention sizing methods to cold climates with snow storage conditions," Water Science and Technology, vol. 56, no. 10, pp. 73–81, 2007. · ·
11. R. A. Brown, W. F. Hunt, and R. W. Skaggs, "Modeling bioretention hydrology with DRAINMOD," Low Impact Development, pp. 441–450, 2010. ·
12. J. Šimůnek, M. Šejna, H. Saito, M. Sakai, and M. T. Van Genuchten, "The HYDRUS-1D software package for simulating the movement of water, heat, and multiple solutes in variably saturated media, Version 4.0," HYDRUS software series, 3: 315, 2008.
13. R. N. Hilten, T. M. Lawrence, and E. W. Tollner, "Modeling stormwater runoff from green roofs with HYDRUS-1D," Journal of Hydrology, vol. 358, no. 3-4, pp. 288–293, 2008. · ·
14. J. L. C. Ladu, P. L. Demetry, T. O. Henry, and X. Lijun, Modeling Storm Water Runoff from Green Roofs with HYDRUS-1D, 2010.
15. C. Hsieh and A. P. Davis, "Evaluation and optimization of bioretention media for treatment of urban storm water runoff," Journal of Environmental Engineering, vol. 131, no. 11, pp. 1521–1531, 2005. · ·
16. Water Bureau of Beijing, Hydrologic Manual of Beijing, Water Bureau of Beijing, Beijing, China, 1999.
17. A. P. Davis, "Field performance of bioretention: hydrology impacts," Journal of Hydrologic Engineering, vol. 13, no. 2, pp. 90–95, 2008. · ·
18. H. Li, L. J. Sharkey, W. F. Hunt, and A. P. Davis, "Mitigation of impervious surface hydrology using bioretention in North Carolina and Maryland," Journal of Hydrologic Engineering, vol. 14, no. 4, pp. 407–415, 2009. · ·
19. H. Li and A. P. Davis, "Urban particle capture in bioretention media. I: laboratory and field studies," Journal of Environmental Engineering, vol. 134, no. 6, pp. 409–418, 2008. · ·
20. R. F. Carsel and R. S. Parrish, "Developing joint probability distributions of soil water retention characteristics," Water Resources Research, vol. 24, no. 5, pp. 755–769, 1988. ·
21. M. Jones and W. F. Hunt, "Effect of bioretention on runoff temperature in trout sensitive regions," inProceedings of the World Environmental and Water Resources Congress, pp. 1582–1588, May 2009. · ·
22. A. Dussaillant, K. Cozzetto, K. Brander, and K. Potter, "Green-Ampt model of a rain garden and comparison to Richards equation model," Sustainable World, vol. 6, pp. 891–900, 2003. ·
23. N. A. L. Archer, J. N. Quinton, and T. M. Hess, "Below-ground relationships of soil texture, roots and hydraulic conductivity in two-phase mosaic vegetation in South-East Spain," Journal of Arid Environments, vol. 52, no. 4, pp. 535–553, 2002. · ·

CHAPTER 16 SALTWATER INTRUSION IN COASTAL AQUIFERS: A PRIMARY CASE STUDY ALONG THE ADRIATIC COAST INVESTIGATED WITHIN A PROBABILISTIC FRAMEWORK

Giada Felisa, Valentina Ciriello and Vittorio Di Federico

Keywords: Saltwater intrusion, coastal vegetation, stochastic approach, pumping well, risk analysis

INTRODUCTION

Contamination of freshwater bodies caused by saltwater intrusion (SI) is a global issue, affecting water quality, vegetation, and soil conditions along coastal lines. Deterioration of this freshwater resource threatens the sustainability of the water supply of coastal communities and their economic development [1,2]. Freshwater stored in coastal aquifers is particularly vulnerable to degradation due to (i) its close proximity to seawater, and (ii) the significant water demand associated with coastal areas whereby groundwater is often the main source of drinking water [3]. Continual degradation of groundwater quality in coastal aquifers has become an increasing concern in heavily urbanized regions across the world. In many countries, including Cyprus, Mexico, Oman, and Israel, hundreds of wells along the coastline have had to be shut down because of SI [4]; in California, saltwater from the Pacific Ocean has seeped into the Los Angeles Basin coastal aquifer, replacing freshwater [5]. Extensive research is currently being carried out worldwide in order to understand the mechanisms of SI and develop strategies to control the process. For example, costly technological solutions, such as desalination, barriers or injection wells have been implemented to deal with the reduction in freshwater availability [6].

Citation: Felisa, G.; Ciriello, V.; Di Federico, V. Saltwater Intrusion in Coastal Aquifers: A Primary Case Study along the Adriatic Coast Investigated within a Probabilistic Framework. Water 2013, 5, 1830-1847. doi:10.3390/w5041830.

Over the last several decades SI has become a crucial issue in Italy as denoted by many studies conducted on different coastal areas [7,8]. SI represents an even more serious threat along the coast of Emilia-Romagna (Figure 1), as it is combined with a strong anthropogenic subsidence, only partially limited in recent years. Though the subsidence in this area has a natural component, with an average of 0.5 mm/year [9], it is exacerbated by gas extraction [10], drainage to reclaim farmland, and groundwater over-exploitation, limited only since the late 1980s [11,12]. The presence of vegetation and complex ecosystems in such areas (i.e., lagoons, marshes, and pine forests), also increases potential adverse impacts from an ecological point of view: understanding the mechanisms whereby these sensitive environments/ecosystems can be affected by SI represents a critical research challenge, as evidenced by recent literature findings [13]. During the past twenty years, the coastal pine forest of Ravenna, along the Adriatic coast, has been affected by a progressive lowering in vegetation density, uprooting of tall plants, drying of pine plants, and changes in the richness of plant species. These impacts have been attributed to groundwater salinity, shallow water tables, drought, and air pollution [14,15]. The trend in groundwater salinization underneath the coastal pine forest is a process that is progressing in time, as recent studies have demonstrated [16].

In recent years, numerical approaches have been increasingly adopted to represent the detailed and complex processes involved in SI [3]. The characterization of the mixing zone as well as the evolution of intrusion temporally represent some applications in which the adoption of numerical models is relevant [17,18]. Similarly, analytical formulations can provide a prompt insight into relevant physical processes involved and, for this reason, a significant number of studies have been conducted in this realm [19,20,21,22]. In particular, the theory developed by Strack [23], based on the sharp interface approximation, has been widely adopted to predict the extension of SI in steady-state flow under the Dupuit-Forchheimer assumption [3,20,24,25]. Several models describing situations of potential engineering interest have been developed and serve as a suitable reference to be re-examined from a risk analysis (RA) point of view [26,27].

A rigorous application of RA in environmental problems has been promoted by several environmental agencies and institutions [28,29]. In the context of climate change, taking into account the uncertainty affecting hydrological parameters is even more important to make accurate predictions. In general, uncertainties are linked to (i) the incomplete knowledge of system dynamics (epistemic uncertainty), and (ii) the randomness which is inherent to natural phenomena (aleatory uncertainty) [30,31]. This uncertainty, propagating towards the state variables of interest through a selected analytical or numerical model, has to be properly considered in a probabilistic framework. In this work we expand upon this approach by providing a general methodology applicable to different contexts and model complexities. In particular, using examples and characteristics of the coastal aquifers found in the area of Ravenna in Emilia-Romagna, we propose a basic scheme representing the key features of steady-state SI in a probabilistic framework. First, an analytical model based on the theory of Strack [25] whereby groundwater withdrawal from a well within a phreatic aquifer is considered. Subsequently, the effects of well pumping on SI and coastal vegetation are investigated where the problem parameters exhibit associated uncertainty. This problem is addressed utilizing model reduction techniques [32,33], which provide an alternative to onerous and time consuming traditional simulation methods (e.g., Monte Carlo simulations). This methodology may be easily applied to complex case studies, assisting in the definition of proper management solutions for coastal aquifers. Regarding this point, in the exemplificative application presented here, we derive the dependence of the risk for the vegetation on the parameters characterizing the well field, quantifying the primary role of the pumping discharge.

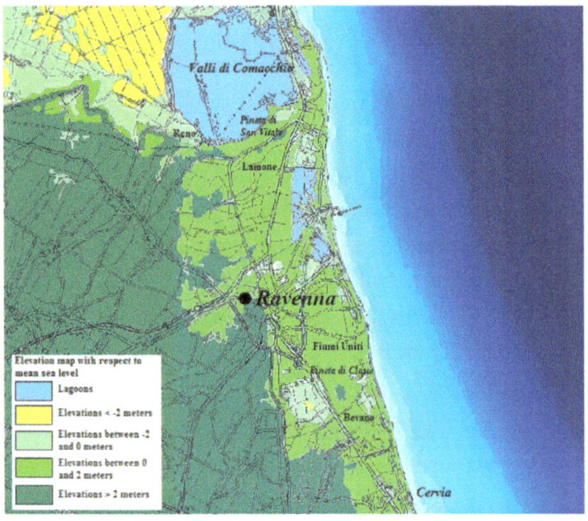

Figure 1. Map of the Adriatic coast around Ravenna [34].

CHARACTERIZATION OF THE AREA OF INTEREST

The Adriatic Sea is a shallow basin whose depth rises southwards and is characterized by a modest tidal excursion. The Adriatic coastline south of the Po River was originally part of its delta and alluvial plain. The province of Ravenna, bounded to the north by the Comacchio Valleys and to the south by the city of Cervia, is characterized by a high population, about 350,000 inhabitants whose density more than triples during the summer. In this area, many unpermitted and poorly-constructed wells have been drilled by bathing establishments and farms, resulting in environmental damage [35]. Anthropogenic activities have altered the original ecosystem and now the once sandy coastline is characterized by river and canal mouths, lagoons, pine forests (e.g., San Vitale and Classe), heavily urbanized areas (e.g., the cities of Marina Romea and Cervia), industrial facilities, and reclaimed agricultural lands [36]. The entire coast is threatened by SI in the phreatic aquifer and seawater encroachment inland along the rivers (e.g., Reno, Lamone, Fiumi Uniti, and Bevano rivers), natural and anthropogenic land subsidence, direct contamination from water bodies open to the sea, destruction of coastal dunes with consequent reduction of their barrier effect, insufficient aquifer recharge and sea level rise. In particular, over the last few decades the land surface has dropped to 4 m below mean sea level in some areas [36]. This in turn has modified the rivers and regular groundwater flow regimes. These inland areas require protection from river flooding and marine intrusion with dikes and levees [36]. A partially artificial drainage network was built in the coastal area to lower the phreatic level and keep the agricultural land dry. Unfortunately, this management process is complex and the presence of these modified drainage networks has not been able to mitigate SI and keep the pine forests healthy [36].

The coastal phreatic aquifer of Ravenna, the focus of this study, is primarily located within the littoral sands, and locally in the shallow marine wedge deposits. The groundwater basin is unconfined in the east, but 3–4 km from the coast it becomes overlain and confined by the most recent alluvial fine-grained continental deposits. The basin thickness varies from a maximum of 30 m in the area of San Vitale Forest to a minimum of about 15 m in the vicinity of Cervia [36]. From east to west, the major sedimentary units consist of: (i) a wedge of fine-grained (fine sand to silty clays) sediments representative of a shallow marine environment; (ii) littoral sands representative of the foreshore, deep-shore, and the sand dune environments adjacent and inland of the beach. The westernmost portion of the study area is characterized by the presence of clay back-shore lagoon deposits [37,38]. Some continental-type alluvial deposits (clay and silt) overlay the littoral sands and lagoon deposits in the west. Based on geologic data from slug tests and grain size analyses [36,39], the hydraulic conductivity of the aquifer is relatively high near the San Vitale Forest (1.1–24.8 m/day) and Cervia (11–21 m/day), and is generally lower in the Classe Forest area (0.2–7.7 m/day). Along the entire coast, the hydraulic head measured in wells was almost always above mean sea level, with values ranging from 0.5 to 4 m in winter and from −0.5 to 3.5 m in summer [36]. It was observed that water salinity (groundwater at the water table) varied seasonally. For instance, salinity values were higher (22–44 g/L) during the summer periods (e.g., near Comacchio

Valleys) compared to winter months where values generally ranged between 3 and 22 g/L. Moreover, encroachment of sea water along rivers and canals, combined with subsidence and drainage from agricultural lands, contributes to enhance the salinization process [36].

Along the coast of Ravenna, pine forests are situated mostly on sandy dunes and were planted in the last century by farmers to protect their land from sea-salt spray. Plant ecologists have reported that the distribution of plant species in the coastal area varies depending on elevation, soil salinity, water salinity, and the duration of various stress periods by which the vegetation is affected [40]. Based on soil and water salinity, plants along the coast can be divided into categories including very tolerant to salt (halophyte plants), tolerant/fairly, and tolerant/not tolerant to water/soil/sea-spray salinity [41]. Increases in salinity within a coastal habitat will have direct and indirect effects on vegetation [15]. In the study area of interest, new sand is added to the beaches every few years from dredging the offshore Adriatic Sea in order to counteract the continuous erosion along the coast. The pine forests (i.e., 50–150 m from the coast) and the adjacent farmlands are drained in winter to keep the groundwater level low. However, this increases the extent of SI in the phreatic aquifer, which is further enhanced by the transpiration of the pine trees [14]. Natural pine seedling germination does not occur and new small pine trees are planted each year to replace dead plants. The most dramatic decrease in the diversity of plant species occurs where water salinity exceeds the 3 g/L threshold, above which most trees species become stressed, with the exception of a few,i.e., Pinus pinaster (William Aiton, 1789) and Pinus pinea (Carl von Linné, 1753) [15].

MATERIALS AND METHODS

Theoretical Framework

In this study, the theory developed by Strack [23] was used to predict the extension of SI into a coastal phreatic aquifer for the case of steady state flow under the Dupuit-Forchheimer assumption [20,22,24,25]. The sharp interface approximation and the homogeneity assumption allow for the derivation of a closed form solution that describes the main processes involved in SI. Consider a semi-infinite domain bounded by a straight coastline (Figure 2a), and divided into two zones, with freshwater in zone 1 and freshwater and saltwater in zone 2. The aquifer thickness b [L] is:

$$b = h_f \qquad \text{zone 1} \tag{1}$$

$$b = h_f - d + \xi \qquad \text{zone 2} \tag{2}$$

where d [L] is the mean sea level (MSL) with respect to the bottom of the aquifer, ξ [L] the interface depth with respect to the MSL, h_f [L] the hydraulic head and q [L^2T^{-1}] the uniform discharge to the sea per unit width of aquifer [42]. In zone 2, the Ghyben-Herzberg assumption $h_f - d = ((\gamma_s - \gamma_f)/\gamma_f)\xi$ is imposed to express the relationship between hydraulic head and interface depth (with respect to MSL).

Following the approach of Strack for an unconfined aquifer [23], the potential φ [L^2] is defined as:

$$\phi = \frac{1}{2}\left[h_f^2 - \frac{\gamma_s}{\gamma_f}d^2\right] \qquad \text{zone 1} \tag{3}$$

$$\phi = \frac{1}{2}\frac{\gamma_s}{\gamma_s - \gamma_f}\left(h_f - d\right)^2 \qquad \text{zone 2} \tag{4}$$

these functions and their first derivatives are continuous across the multiple-zone aquifer and satisfy the Laplace equation, $\nabla^2\phi = 0$, in the x-z plane. At the interface, substituting in Equations (3) or (4) the condition $\xi = d$, hence $h_f = (\gamma_s/\gamma_f)\,\xi$, the potential is obtained as:

$$\phi_{toe} = \frac{1}{2}\frac{\gamma_s(\gamma_s - \gamma_f)}{\gamma_f^2}d^2 \tag{5}$$

Now consider then the presence of a pumping/recharging well with discharge Q_w [L^3T^{-1}], located at a distance x_w from the coast (Figure 2b,c). The solution of the problem, in terms of potential, can be found by resorting to the superposition principle and the method of images as [23]:

$$\phi = \frac{q}{K}x + \frac{Q_w}{4\pi K}\ln\left[\frac{(x - x_w)^2 + y^2}{(x + x_w)^2 + y^2}\right] \tag{6}$$

where K [LT^{-1}] represents the hydraulic conductivity of the aquifer. The location of the leading edge, or "toe", of the saltwater wedge can be computed from Equations (5) and (6) as:

$$\phi_{toe} = \frac{q}{K}x_{toe} + \frac{Q_w}{4\pi K}\ln\left[\frac{(x_{toe} - x_w)^2 + y^2}{(x_{toe} + x_w)^2 + y^2}\right] \tag{7}$$

Finally, combining expressions (3) and (4) with (6) allows for the determination of the elevation of the hydraulic head above the MSL:

$$h_f - d = \sqrt{\frac{2q}{K}x + \frac{Q_w}{2\pi K}\ln\left[\frac{(x - x_w)^2 + y^2}{(x + x_w)^2 + y^2}\right] + \frac{\gamma_s}{\gamma_f}d^2} - d \qquad \text{zone 1} \tag{8}$$

$$h_f - d = \sqrt{\frac{\gamma_s - \gamma_f}{\gamma_s}\left[\frac{2q}{K}x + \frac{Q_w}{2\pi K}\ln\left[\frac{(x - x_w)^2 + y^2}{(x + x_w)^2 + y^2}\right]\right]} \qquad \text{zone 2} \tag{9}$$

utilizing the Ghyben-Herzberg assumption, the interface depth is $\xi = (\gamma_f/(\gamma_s - \gamma_f))(h_f - d)$ for zone 2, and $\xi = d$ for zone 1, whereby no saltwater is present.

For multiple pumping wells, Equation (6) can be extended to:

$$\phi = \frac{q}{K}x + \sum_{i=1}^{n}\frac{Q_i}{4\pi K}\ln\left[\frac{(x - x_i)^2 + (y - y_i)^2}{(x + x_i)^2 + (y - y_i)^2}\right] \tag{10}$$

where (x_i, y_i) are the coordinates and Q_i is the pumping rate of well i, which can be taken as positive or negative to represent pumping and recharging mode, respectively. With these assumptions, Equations (8) and (9) become for zone 1 and 2, respectively:

$$h_f - d = \sqrt{\frac{2q}{K}x + \sum_{i=1}^{n}\frac{Q_i}{2\pi K}\ln\left[\frac{(x-x_i)^2+(y-y_i)^2}{(x+x_i)^2+(y-y_i)^2}\right] + \frac{\gamma_s}{\gamma_f}d^2} - d \quad (11)$$

$$h_f - d = \sqrt{\frac{\gamma_s-\gamma_f}{\gamma_s}\left[\frac{2q}{K}x + \sum_{i=1}^{n}\frac{Q_i}{2\pi K}\ln\left[\frac{(x-x_i)^2+(y-y_i)^2}{(x+x_i)^2+(y-y_i)^2}\right]\right]} \quad (12)$$

Figure 2 represents the effects on the freshwater-saltwater interface due to the presence of one or more wells. While under natural undisturbed conditions the interface maintains a state of equilibrium (Figure 2a), water extraction lowers the water table with respect to its initial position (Figure 2b, blue solid line) allowing the interface to move inland. As the pumping rate increases, the interface advances towards the well until an unstable critical situation is reached (Figure 2c, red solid line). Any further increase in the pumping rate causes a jump of the toe position landward of the well, which thereby becomes contaminated by saltwater.

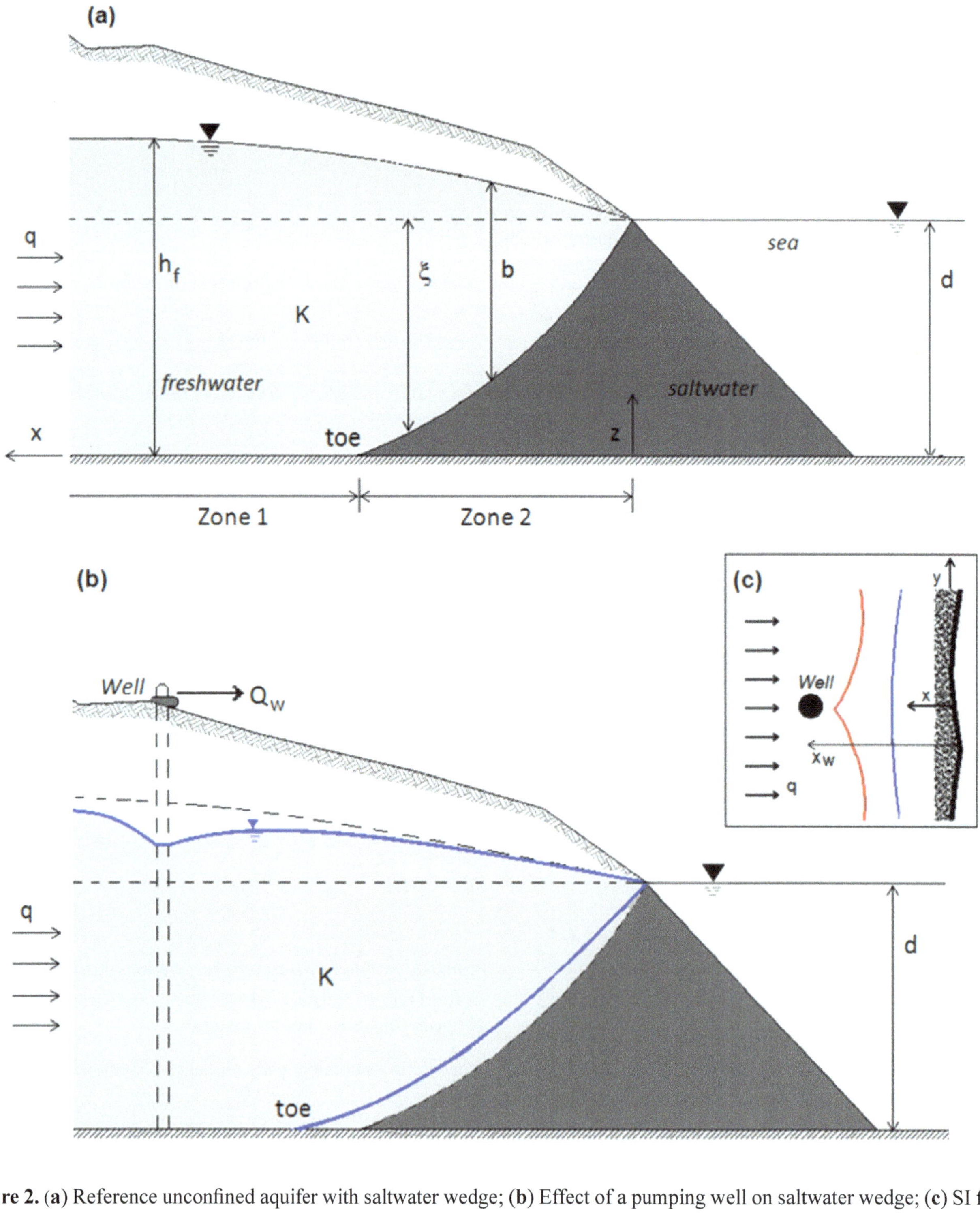

Figure 2. (**a**) Reference unconfined aquifer with saltwater wedge; (**b**) Effect of a pumping well on saltwater wedge; (**c**) SI fronts near the pumping well (blue and red lines), in the x-yplane, for two different values of pumping rate.

Problem Formulation

In order to evaluate the effects of SI on coastal vegetation due to both natural and anthropogenic causes, a simplified conceptualization representing specific areas of the Emilia-Romagna coast was configured (Figure 3). An 'environmentally sensitive area' representing a pinewood forest (an infinitely long strip) was located at a minimum distance $x^* = 50$ m from the coastline. The lower limit of the plants capture zone (the volume of soil affected by water uptake by roots) is defined such that the sensitive zone surrounding the plant is not impacted by the intruding saltwater and any adverse effects on the plants are prevented. The extension of this zone depends on both the length of roots and the width of the transpiration zone; its minimum elevation from the bottom of the aquifer is set to $z_{lim} = 12$ m in the present case. A well field established for agricultural purposes was located inland of the pinewood, as depicted in Figure 3. The well field was comprised of three wells, each located in the middle of an area of dimensions b·e, distant a from the coast; between every pumping area there are non-overlapping zones of width equal to c, to reduce mutual interactions among the wells. The total discharge Q was equally divided among the wells. Values assumed for the design parameters of the well field are summarized in Table 1.

Following a probabilistic approach, we model the hydrogeological parameters K, q and d, as independent random variables. Based on previous studies [42], all parameters were assumed to follow a normal distribution, characterized by the average values and the coefficients of variation reported in Table 2. Note that the coefficient of variation of d was selected according to the prediction of annual sea level rise [12] and subsidence for the area of interest.

Once the uncertain parameters are defined, the extension x_{toe} of SI in the x-y plane is evaluated through Equations (5) and (10). The computed values of x_{toe} are affected by the uncertainty propagated from the input parameters through the selected model equations. Since the domain considered is symmetric, the maximum value of x_{toe} occurs along the middle, at $y = y_{crit}$. The extent of intrusion can be determined in the x-z plane corresponding to y_{crit}, by computing the elevation of the interface z_i with respect to the bottom of the aquifer via Equation (12). This is also a random variable. The risk for the vegetation, here considered as the probability that the saltwater interface reaches the sensitive zone below the plants roots, is then derived following two steps. First, the probability that plant roots occur in the zone impacted by SI (i.e., zone 2 in Figure 4) is computed as:

$$P_1 = P(x^* < x_{toe}) = 1 - F_{x_{toe}}(x^*) \quad (13)$$

where $F_{x_{toe}}(x^*)$ is the cumulative distribution function of the random variables x_{toe}. According to the data results $P_1 \approx 1$, i.e., the vegetation falls into zone 2 almost certainly within the selected case study. Secondly, the probability that the saltwater interface reaches the sensitive zone below the plants roots is evaluated in correspondence of the maximum intrusion, i.e., $y = y_{crit}$. This probability is conditional on the previous one as follows:

$$P_2 = P\left(z_i(x^*; y_{crit}) > z_{\lim} \middle| x^* < x_{toe}\right) \quad (14)$$

In this selected case study, as $P_1 \approx 1$, Equation (14) reduces to $P_2 = P(z_i(x^*; y_{crit}) > z_{\lim}) = 1 - F_{z_i}(z_{\lim})$, where F_{z_i} is the cumulative distribution function of the random variable z_i. Consequently, in the following section the simplified formulation for the computation of the risk, i.e., $R = P_2$, was considered.

The proposed methodology is suitable to predict the risk to which coastal vegetation is subjected, considering the uncertainty of the parameters involved. For this purpose, the evaluation of the mean position of the toe (considering the schematic scenario described above) and its variance, applying global sensitivity analysis [32], must be conducted in order to evaluate the influence (sensitivity) of each uncertain parameter. Subsequently, the risk affecting the vegetation is computed for different scenarios defined through the parameters that characterize the well field and modifying their values with respect to those reported in Table 1. In particular, variations in (i) the overall pumping discharge, Q; (ii) the distance of wells from the coast, a; and (iii) the width of non-overlapping zones, c, were considered.

Table 1. Design parameters of the well field represented in Figure 3.

Parameters	Values
a (m)	650
b (m)	300
c (m)	100
e (m)	300
x_1 (m)	800
y_1 (m)	950
x_2 (m)	800
y_2 (m)	550
x_3 (m)	800
y_3 (m)	150
$Q_1, Q_2\ Q_3$ (m^3/day)	200
Q (m^3/day)	600

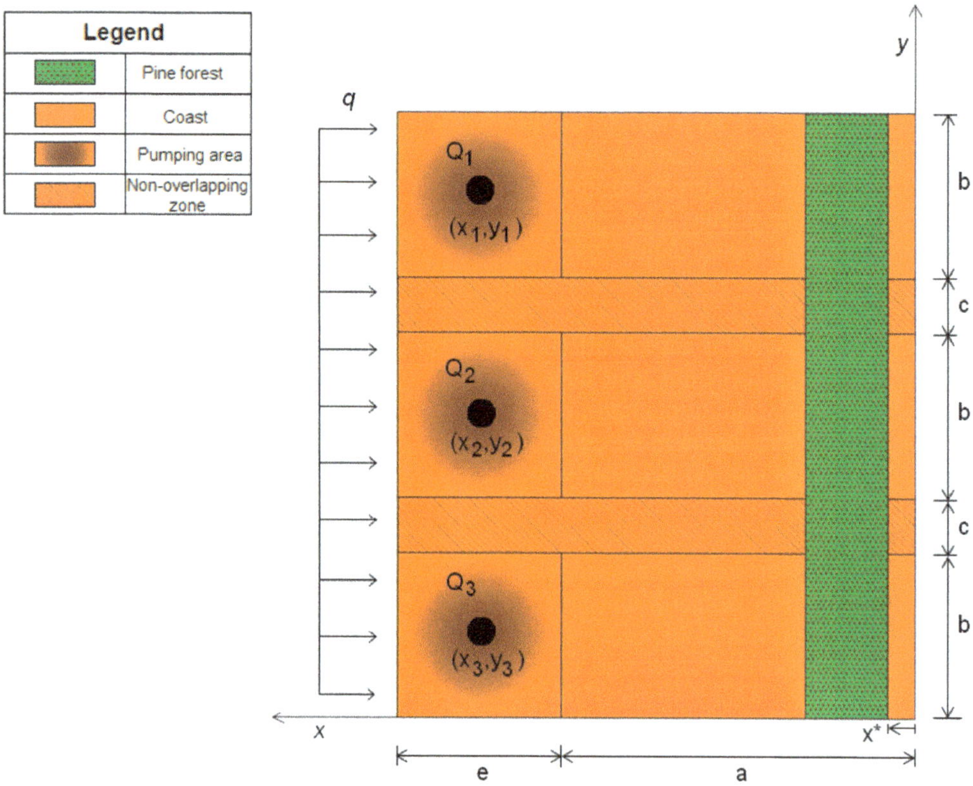

Figure 3. Well field and location of pine forest in x-y plane.

Table 2. Main characteristic parameters of the model.

Parameters	Distribution
K (m/day)	$N\ (\mu_K = 40;\ cv_K = 0.1)$
q (m^2/day)	$N\ (\mu_q = 0.9;\ cv_q = 0.1)$
d (m)	$N\ (\mu_d = 20;\ cv_d = 0.001)$

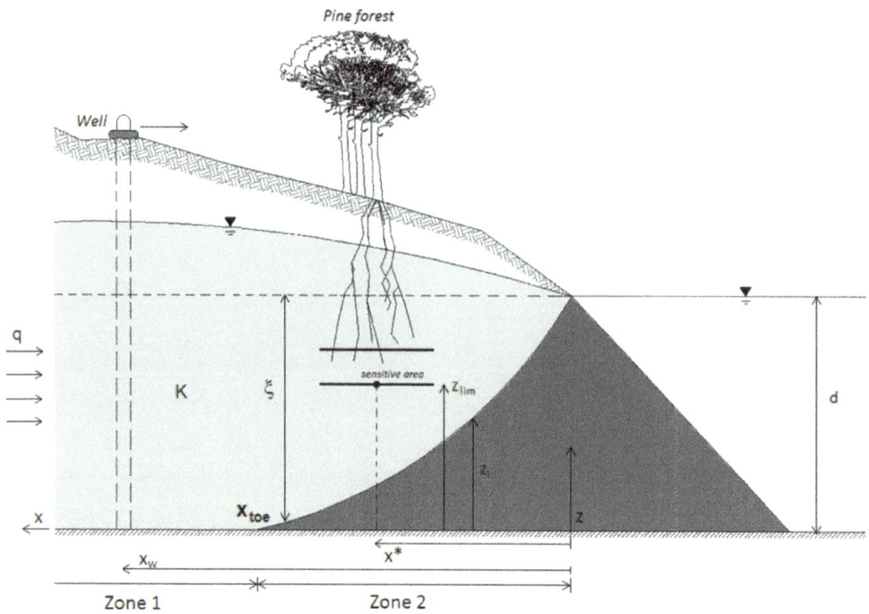

Figure 4. Simplified conceptualization of the problem, in x-zplane, for $y = y_{crit}$.

Model Reduction Approach for Risk Analysis

In order to perform the stochastic analysis described above, time intensive simulation methods based on the Monte Carlo (MC) approach are typically applied. For this analysis, a different, more efficient strategy was implemented that relied on model reduction via Polynomial Chaos Expansion (PCE) [43]. The basic idea of model reduction methods consists in modifying the form of the selected model without introducing any simplifying assumptions in the description of the physical processes involved. For the PCE approach, a surrogate model was derived in a simple polynomial form depending only on the uncertain input parameters. The computation of the PCE approximation requires a very limited number of MC simulations performed on the original model.

Let the model response of interest, y, belong to the Hilbert space of second-order random variables. Under this assumption, y can be approximated through the PCE technique [43,44] as follows:

$$\tilde{y} = \tilde{f}(\mathbf{x}, t, \zeta(\mathbf{p})) = \sum_{j=0}^{P-1} a_j(\mathbf{x}, t)\Psi_j(\zeta(\mathbf{p})) \qquad (15)$$

The resulting formulation constitutes a metamodel, $\tilde{y}$, of y. This metamodel is a simple polynomial function which is expressed in terms of a set of independent random variables, collected in vector ζ. Here, $P = (M+q)!/(M!q!)$, is the number of terms of the expansion, and qis the maximum degree considered; a_j represent unknown deterministic coefficients, while $\Psi_j$denote the suitable multivariate polynomial basis in the Hilbert space containing the response.

A non-intrusive regression-based approach can be employed to calculate the coefficients a_j [32]. As mentioned above, this method requires a number N of simulations performed on the original model proportional to P.

Once the PCE is available, simple analytical post-processing allows derivation of (i) the first two moments of the approximated response, and (ii) the global sensitivity indices of Sobol, that represent dimensionless ratios among the partial variance of the response, due to a generic subset of uncertain model parameters, and the total variance of the response [32]. Specifically the mean of the model response coincides with the coefficient of the zero-order term, a_0 in the Equation (15), and the total variance of the response and the generic Sobol index are given, respectively, by:

$$V^{(\tilde{y})} = Var\left[\sum_{j=0}^{P-1} a_j\Psi_j(\zeta)\right] = \sum_{j=1}^{P-1} a_j^2 E\left[\Psi_j^2(\zeta)\right] \qquad (16)$$

$$S_{i_1,\ldots i_s}^{(\tilde{y})} S = \frac{\sum_{\alpha\in\varphi_{i_1\ldots i_s}} a_\alpha^2 E\left[\Psi_\alpha^2\right]}{V^{(\tilde{y})}} \qquad (17)$$

for the computation of $E\left[\Psi_\alpha^2\right]$ see e.g., [45].

Furthermore, the PCE metamodel represents a suitable basis to perform Monte Carlo simulations, allowing for a more efficient means to perform risk analysis or to solve inverse problems with reduced computational time/cost. This methodology has demonstrated excellent promise for use in many other applications related to flow and transport processes in porous media (e.g., [46,47]).

RESULTS AND DISCUSSION

Applying the methodology outlined in Section 3.2, the extension of SI was first evaluated by using the values (parameters) collected in Table 1 to represent the well field. The uncertainty in the hydrogeological parameters was modeled using probability distributions as reported in Table 2. Under this framework, the first two moments associated with x_{toe} and the global sensitivity indices were evaluated as described in Section 3.3.

The mean value of x_{toe} as a function of y is depicted in Figure 5a, showing the intervals of width ±σ and ±2σ surrounding it. The maximum value for the mean of x_{toe} (i.e., the leading edge of the intrusion) is reached at $y_{crit} = 3b/2 + c$, since the domain is symmetric. The maximum intrusion extent was equal to 303.6 m for the selected case study; this value is consistent with those typically observed for the area of Ravenna, and with results reported in numerical studies of the effects of subsidence on SI [35]. The uncertainty in the hydrogeological parameters resulted in a standard deviation of x_{toe} with a maximum of ~50 m (i.e., more than the 15% of the correspondent mean value).

The standard deviation σ of x_{toe} is illustrated in Figure 5b as a function of y, showing the individual contributions due to each random parameter value (K, q and d) in order to investigate how these different values influence SI. The results indicate that the primary influence on SI was due to the variability in q, whereby, in contrast, the mean sea level, d, was relatively inconsequential. However, the impact of d was shown to produce significant effects on SI in a long-term period of time due to climatic changes and the combined increase in natural/anthropogenic subsidence. As expected, the effect of domain heterogeneity, exemplified through the variability ofK, was also significant. These results highlight the importance of proper site characterization in order to obtain accurate predictions for SI. Other studies (e.g., [22]) have relied on the theory presented by Strack [23], whereby a sensitivity-based approach was adopted in order to define vulnerability indicators for SI. However, these global sensitivity analysis models do not account for the simultaneous variability of all the parameters of interest when evaluating effects on SI.

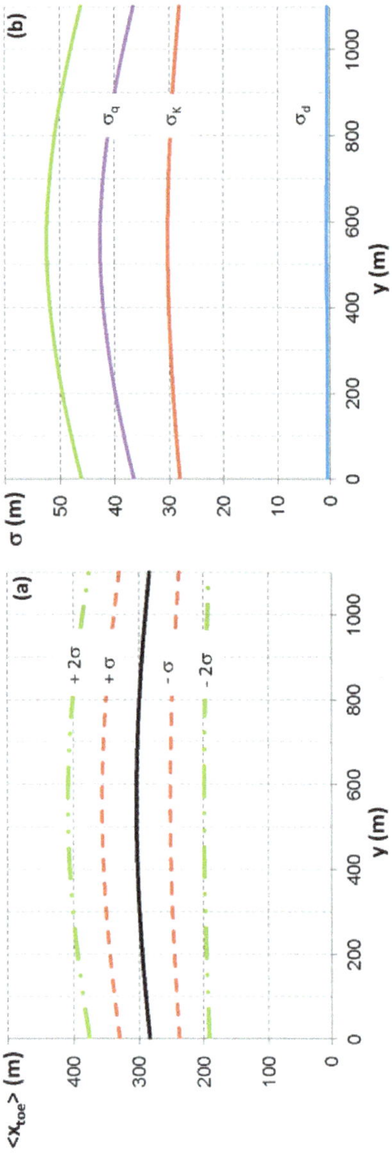

Figure 5. (**a**) Mean interface location x_{toe} (solid line) $\pm\sigma$ and $\pm 2\sigma$(dashed line), versus y; (**b**) Standard deviation σ of x_{toe}, and partial contribution of each individual parameter, i.e., σ_q, $\sigma_K$and σ_d.

This study also evaluated he influence on SI on the risk for vegetation health due to variations of the deterministic parameters Q (overall pumping discharge), a (distance of the well field from the coast), and c (width of non-overlapping zones).

The mean and the standard deviation of x_{toe} versus y for different values of total pumping rate Qare depicted in Figure 6a and Figure 6b respectively. In particular, it is shown that the mean of x_{toe}reaches its maximum at $y = y_{crit}$, ranging between 280 and 365 m. This revealed a relatively high degree of variability for different values of Q_i; specifically, the largest extent of SI was reached for the maximum pumping rate considered (i.e., $Q_i = Q/3 = 300$ m^3/day) (Figure 6a). The standard deviation σ of x_{toe} followed the same trend (Figure 6b), whereby values ranged between 45.7 and 74.8 m, primarily as a function of the variability of domain properties, q and K, as described above.

A similar analysis is illustrated in Figure 7a,b, with the exception that different values of a were considered to evaluate effects on SI. As previously noted, the results of this study revealed that the distance of the well field from the coast, within the range considered, seemed to be inconsequential, as the two moments of x_{toe} did not exhibit significant changes in the extent of SI. In particular, the maximum value of the mean of x_{toe} ranged between 294 and 308 m (i.e., trend increasing with a) (Figure 7a), and the associated standard deviation varied between 49.7 and 54 m (Figure 7b).

Finally, the influence of c (the width of the non-overlapping zones among the wells) on the first two moments of x_{toe} is depicted in Figure 8a,b. By varying the width of the non-overlapping zones, the size of the entire domain is modified. For example, maximum values of mean x_{toe} and standard deviations were obtained for smaller values of c, whereby $y_{crit} = 3b/2 + c$ tended to decrease. For c = 50 m, y_{crit} = 500 m, with a mean value of x_{toe} equal to 307 m and an associated standard deviation of 53.6 m; for a value of c = 200 m, y_{crit} becomes 660 m, whereby the mean value of x_{toe} was reduced to 297 m with a standard deviation of 50.2 m.

For this analysis the design parameters of the well field demonstrated that SI was extremely sensitive to the pumping discharge Q, in contrast to the other parameters considered (i.e., a and c) which seem to have only a minimal influence on the optimization of groundwater extraction and effect on SI.

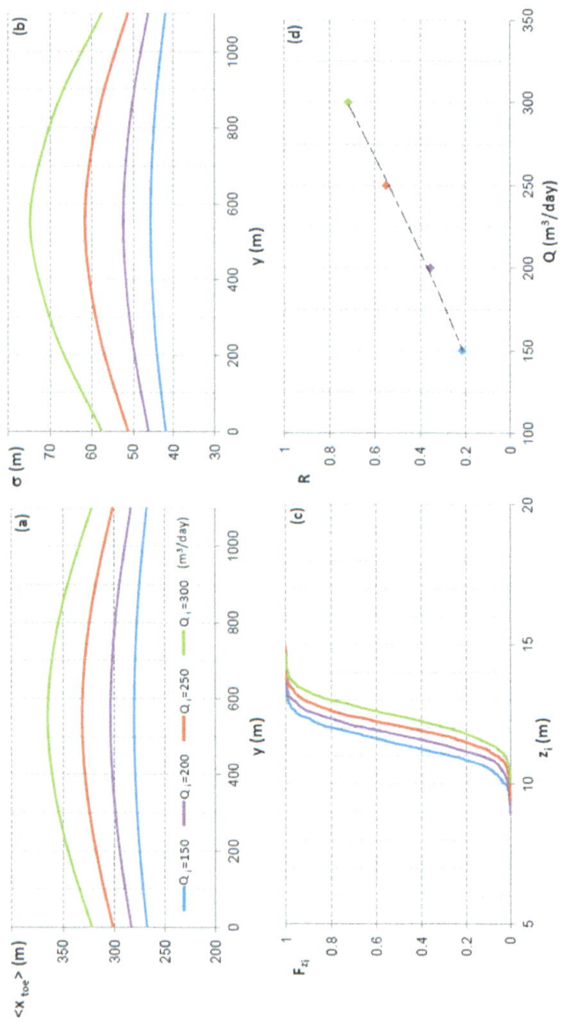

Figure 6. (**a**) Mean value of x_{toe}; (**b**) Standard deviation σ of x_{toe}; (**c**) Cumulative distribution function F_{z_i} of the

surface elevation z_i computed in $z_{\lim}$ (Figure 4); and (**d**) Computed risk [see Equation (14)] resulting from different values of single pumping discharge Q_i.

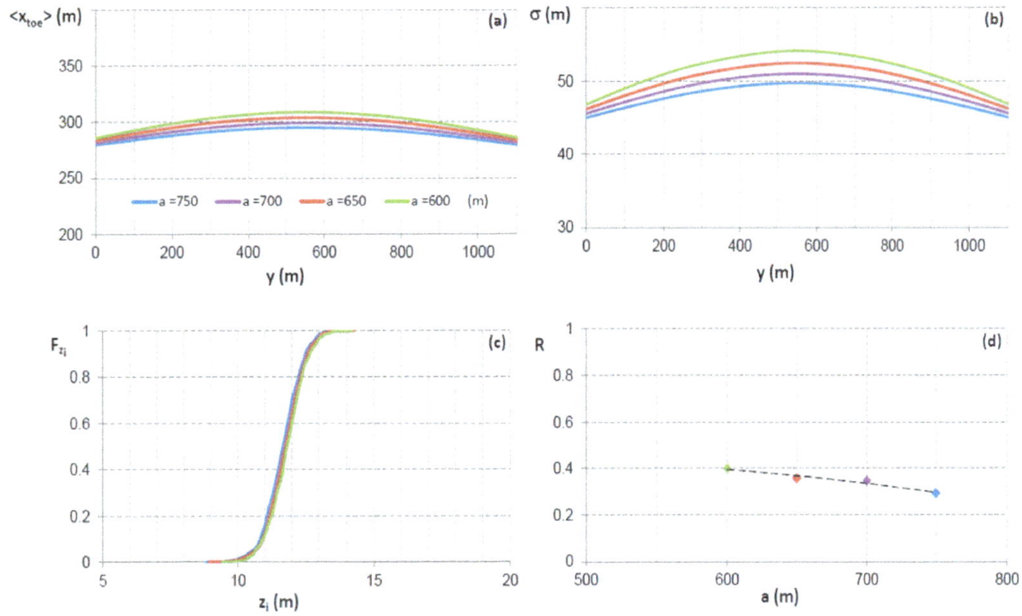

Figure 7. (**a–d**) As Figure 6 but for different values of distance from the coast a.

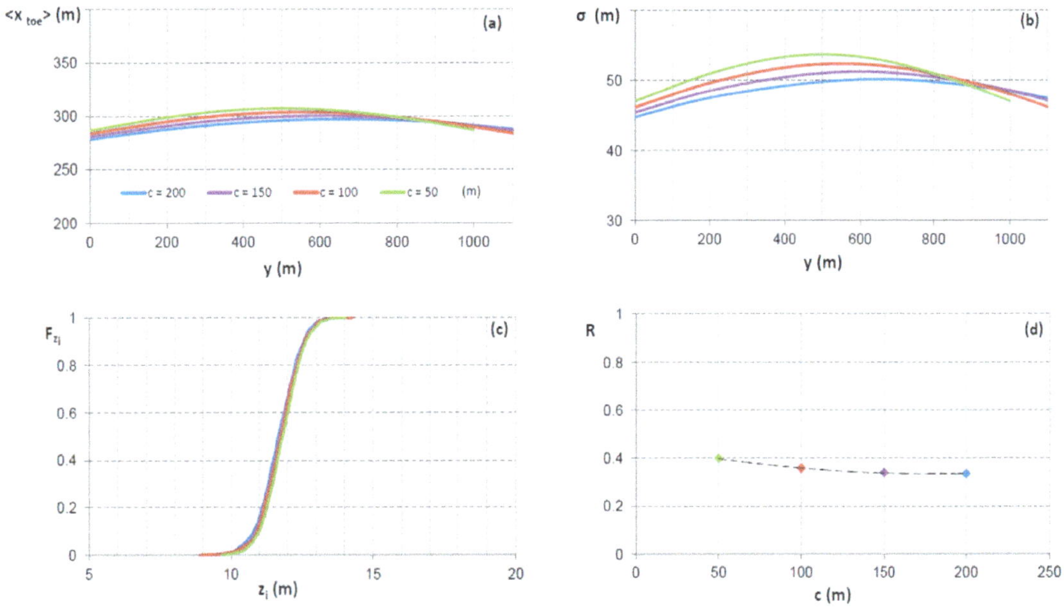

Figure 8. (**a–d**) As Figure 6 but for different values of width of non-overlapping zones c.

In order to complete the analysis, the effects of the variability in the parameters characterizing the well field on the risk of contamination were evaluated. The results obtained increasing Q_i are presented in Figure 6c. It can be observed that the cumulative distribution functions of the interface elevation computed in $z_{\lim}$, $F_{z_i}\left(z_{\lim}\right)$, tend to decrease with higher values of Q_i and consequently the risk increases quasi-linearly (Figure 6d). When considering the value of $z_{\lim}$ = 12 m as a fixed variable herein, $F_{z_i}\left(z_{\lim}\right)$ ranged from 0.78 to 0.28, as Q_i varied between 150 and 300 m³/day. In Figure 7c, obtained for different values of a, a different trend is revealed, whereby $F_{z_i}\left(z_{\lim}\right)$ increases and risk decreases, as expected for higher values of a (Figure 7d). In this case, $F_{z_i}(z_{lim})$ increases from 0.60 to 0.71 while varying a between 600–750 m. Finally, in the last simulation whereby different values of c were utilized, a similar trend to the one described above was observed, albeit less pronounced (Figure 8c,d). For example, a value of z_{lim} = 12 m, produced a narrow range of values of $F_{z_i}(z_{lim})$ from 0.61 to 0.67 when varying c between 50 and 200 m.

Overall, when comparing all of the different simulation results obtained for the risk of SI (Figure 6d, Figure 7d, Figure 8d), the increase in Q seems to produce the most detrimental effects in terms of extent of SI and risk to the preservation of vegetation health in coastal areas. For example, with values of z_{lim} = 12 m, the results show that risk ranged between 0.2 and 0.7 when varying parameter Q (Figure 6d), while values of risk lower than 0.4 resulted when parameters a and c were varied (Figure 7d and Figure 8d).

It should be noted that the parameters of the well field can also be modeled as random variables and investigated via a global sensitivity analysis to appropriately understand their influence on the risk of contamination. This could be particularly useful for problems that require optimization solutions [e.g., 20]. Although the analyses herein primarily focused on a risk assessment for vegetation in coastal areas influenced by SI, this study also revealed that both the pumping rate and domain heterogeneity (i.e., permeability distribution) significantly influenced the extent of SI, consistent with findings reported by other studies [20,48].

CONCLUSIONS

This study presents a computationally efficient technique to evaluate risk to ecosystems (i.e., vegetation health) along coastal regions that are impacted by saltwater intrusion processes. The methods used were applied to the coastline of Emilia-Romagna, Italy, a region that has been and continues to be impacted by saltwater intrusion. A stochastic approach to the problem was adopted to represent the random nature of the hydrogeological parameters. Computational efficiency was further increased by incorporating model reduction techniques via Polynomial Chaos Expansion theory for modeling saltwater intrusion effects and associated risk analyses.

A simplified analytical formulation was adopted to allow for the evaluation of the probability that the vegetation capture zone overlaps with the intruding saltwater wedge. Anthropogenic influence on the risk to vegetation due to saltwater intrusion was modeled through the presence of a well field, specifically accounted for through adjustment of (i) the pumping discharge; (ii) the wells distance from the coast; and (iii) the mutual distance between wells. For the simplified scenario representative of the Ravenna coast, the comprehensive results of the modeling analyses revealed the parameters most responsible for influencing the extent of saltwater intrusion. For example, increases in the overall pumping discharge of the well field produced the most significant increase in the probability that saltwater intrusion reaches the vegetation capture zone.

The computational methodology used herein can be applied to more complex systems whereby a closed-form solution is not available and would require more time-consuming numerical approaches (i.e., to resolve the shape and size of the intruding saltwater wedge). In such cases, the model reduction technique adopted for these analyses allowed for both excellent representation of the physical system and an increase of computational efficiency. Moreover, the current approach can assist in the development of appropriate management strategies for coastal aquifers. These techniques allow the parameters describing the anthropogenic influences to be considered as random variables, which provide an improved means of discriminating the parameters most responsible for saltwater intrusion and for solving optimization problems that require a specific set of environmental constraints.

ACKNOWLEDGMENTS

Marco Marcaccio and Andrea Chahoud of ARPA Emilia-Romagna are gratefully acknowledged for helpful discussions.

Conflicts of Interest

The authors declare no conflict of interest.

REFERENCES

1. Cassardo, C.; Jones, J.A.J. Managing water in a changing world. Water **2011**, 3, 618–628.
2. Gain, A.K.; Giupponi, C.; Renaud, F.G. Climate change adaptation and vulnerability assessment of water resources systems in developing countries: A generalized framework and a feasibility study in Bangladesh. Water **2012**, 4, 345–366.
3. Werner, A.D.; Bakker, M.; Post, V.E.A.; Vandenbohede, A.; Lu, C.; Ataie-Ashtiani, B.; Simmons, C.T.; Barry, D.A. Seawater intrusion processes, investigation and management: Recent advances and future challenges. Adv. Water Resour. **2013**, 51, 3–26.
4. Bolster, D.T.; Tartakovsky, D.M.; Dentz, M. Analytical models of contaminant transport in coastal aquifers. Adv. Water Resour. **2007**, 30, 1962–1972.
5. Edwards, B.D.; Evans, K.R. Saltwater intrusion in Los Angeles area coastal aquifers—The marine connection; USGS Fact Sheet 030–02. Available online: http://pubs.usgs.gov/fs/2002/fs030-02/ (accessed on 6 June 2013).
6. Bray, B.; Tsai, F.T.C.; Sim, Y.; Yeh, W.W.G. Model development and calibration of a saltwater intrusion model in southern California. J. Am. Water Resour. Assoc. **2007**, 43, 1329–1343.
7. Capaccioni, B.; Didero, M.; Paletta, C.; Didero, L. Saline intrusion and refreshening in a multilayer coastal aquifer in the Catania Plain (Sicily, southern Italy): Dynamics of degradation processes according to the hydrochemical characteristics of groundwater. J. Hydrol. **2005**, 307, 1–16.
8. De Luca, A.; Preziosi, E.; Giuliano, G.; Mastroianni, D.; Falconi, F. First evaluation of the saltwater intrusion in the Tiber delta area (Rome, central Italy). In Proceedings of 18th Salt Water Intrusion Meeting, Cartagena, Spain,

31 May–3 June 2004; p. 34.

9. Carbognin, L.; Tosi, L.; Teatini, P. Analysis of actual land subsidence in Venice and its hinterland (Italy). In Land Subsidence; Barends, F.B.J., Brouwer, F.J.J., Schröder, F.H., Eds.; Balkema: Rotterdam, The Netherlands, 1995; pp. 129–137.
10. Bau, D.; Gambolati, G.; Teatini, P. Residual land settlement near abandoned gas fields raises concern over Northern Adriatic coastland. Eos Trans. Am. Geophys. Union **2000**, 81, 245–249.
11. Teatini, P.; Ferronato, M.; Gambolati, G.; Gonella, M. Groundwater pumping and land subsidence in the Emilia-Romagna coastland, Italy: Modeling the past occurence and the future trend. Water Resour. Res. **2006**, 42, 1–19.
12. Preti, M.; De Nigris, N.; Morelli, M.; Monti, M.; Bonsignore, F.; Aguzzi, M. Stato del litorale emiliano-romagnolo all'anno 2007 e piano decennale di gestione; Agenzia regionale per la prevenzione e l'ambiente (ARPA): Emilia-Romagna, Italy, 2009; p. 270.
13. Manfreda, S.; Caylor, K.K. On the vulnerability of water limited ecosystems to climate change. Water **2013**, 5, 819–833.
14. Antonellini, M.; Mollema, P.N. Impact of groundwater salinity on vegetation species richness in the coastal pine forests and wetlands of Ravenna, Italy. Ecol. Eng. **2010**, 36, 1201–1211.
15. Giambastiani, B.M.S.; Mollema, P.N.; Antonellini, M. Groundwater management in the Northern Adriatic Coast (Ravenna, Italy): New strategies to protect the coastal aquifer from saltwater intrusion. In Groundwater. Modeling, Management, and Contamination; König, L.F., Weiss, J.L., Eds.; Nova Science Publishers: New York, NY, USA, 2009; pp. 203–230.
16. Marconi, V.; Dinelli, E.; Antonellini, M.; Capaccioni, B.; Balugani, E.; Gabbianelli, G. Hydrogeochemical characterization of the phreatic system of the coastal wetland located between Fiumi Uniti and Bevano rivers in the southern Po plain (Northern Italy). In Proceedings of European Geosciences Union General Assembly, Vienna, Austria, 19–24 April 2009; p. 9771.
17. Henry, H.R. Salt intrusion into fresh-water aquifers. J. Geophys. Res. **1959**, 64, 1911–1919.
18. Servan-Camas, B.; Tsai, F.T.-C. Two-relaxation-time lattice Boltzmann method for the anisotropic dispersive Henry problem. Water Resour. Res. **2010**, 46.
19. Cheng, A.H.-D.; Halhal, D.; Naji, A.; Ouazar, D. Pumping optimization in saltwater-intruded coastal aquifers. Water Resour. Res. **2000**, 36, 2155–2165.
20. Mantoglou, A. Pumping management of coastal aquifers using analytical models of saltwater intrusion. Water Resour. Res. **2003**, 39.
21. Mantoglou, A.; Papantoniou, M. Optimal design of pumping networks in coastal aquifers using sharp interface models. J. Hydrol. **2008**, 361, 52–63.
22. Werner, A.D.; Ward, J.D.; Morgan, L.K.; Simmons, C.T.; Robinson, N.I.; Teubner, M.D. Vulnerability indicators of sea water intrusion. Ground Water **2012**, 50, 48–58.
23. Strack, O.D.L. A single-potential solution for regional interface problems in coastal aquifers.Water Resour. Res. **1976**, 12, 1165–1174.
24. Naji, A.; Cheng, A.H.-D.; Ouazar, D. Analytical stochastic solutions of saltwater/freshwater interface in coastal aquifers. Stoch. Hydrol. Hydraul. **1998**, 12, 413–430.
25. Lu, C.; Chen, Y.; Luo, J. Boundary condition effects on maximum groundwater withdrawal in coastal aquifers. Ground Water **2012**, 50, 386–393.
26. De Barros, F.P.J.; Bolster, D.; Sanchez-Vila, X.; Nowak, W. A divide and conquer approach to cope with uncertainty, human health risk, and decision making in contaminant hydrology.Water Resour. Res. **2011**, 47.
27. Tartakovsky, D.M. Assessment and management of risk in subsurface hydrology: A review and perspective. Adv. Water Resour. **2012**, 51, 247–260.
28. US National Research Council. Review of Recommendations for Probabilistic Seismic Hazard Analysis: Guidance on Uncertainty and Use of Experts; National Academy Press: Washington, DC, USA, 1997; p. 84.
29. European Communities (EC). Common Implementation Strategy for the Water Framework Directive (2000/60/EC); Guidance Document No. 11; Planning Processes, Working Group 2.9.; Office for the Official Publications of the European Communities: Luxembourg, 2003.
30. Tartakovsky, D.M. Probabilistic risk analysis in subsurface hydrology. Geophys. Res. Lett. **2007**,34.
31. Rubin, Y. Applied Stochastic Hydrogeology, 1st ed.; Oxford University Press: New York, NY, USA, 2003; p. 416.
32. Sudret, B. Global sensitivity analysis using polynomial chaos expansions. Reliab. Eng. Syst. Safety **2008**, 93, 964–979.
33. Villa-Vialaneix, N.; Follador, M.; Ratto, M.; Leip, A. A comparison of eight metamodeling techniques for the simulation of N_2O fluxes and N leaching from corn crops. Environ. Model. Softw. **2012**, 34, 51–66.
34. E-R Ambiente Home Page. Available online: http://ambiente.regione.emilia-romagna.it/geologia/cartografia/webgis-banchedati (accessed on 3 October 2013).
35. Giambastiani, B.M.S.; Antonellini, M.; Oude Essink, G.H.P.; Stuurman, R.J. Saltwater intrusion in the unconfined coastal aquifer of Ravenna (Italy): A numerical model. J. Hydrol.**2007**, 340, 91–104.
36. Antonellini, M.; Mollema, P.; Giambastiani, B.; Bishop, K.; Caruso, L.; Minchio, A.; Pellegrini, L.; Sabia, M.; Ulazzi, E.; Gabbianelli, G. Saltwater intrusion in the coastal aquifer of the southern Po Plain, Italy. Hydrogeol. J. **2008**, 16, 1541–1556.
37. Bondesan, M.; Favero, V.; Viñals, M.J. New evidence on the evolution of the Po-delta coastal plain during the Holocene. Quat. Int. **1995**, 29–30, 105–110.
38. Amorosi, A.; Centineo, M.C.; Dinelli, E.; Lucchini, F.; Tateo, F. Geochemical and mineralogical variations as indicators of provenance changes in late Quaternary deposits of SE Po Plain. Sediment. Geol. **2002**, 151,

273–292.

39. Ulazzi, E.; Antonellini, A.; Gabbianelli, G. Coastal phreatic aquifer characterization of the Cervia Area. In Stratigraphy and Sedimentology of Aquifers; Vallone, R., Ed.; APAT: Parma, Italy, 2007.
40. Maas, E.V. Salt tolerance of plants. In Handbook of Plant Science in Agriculture; CRC Press: Boca Raton, FL, USA, 1984; pp. 57–75.
41. Hutchinson, I. Salinity Tolerances of Plants of Estuarine Wetlands and Associated Uplands; Washington State Department of Ecology Report; Washington State Shorelands and Coastal Zone Management Program, Wetlands Section: Washington, DC, USA, 1991.
42. Cheng, A.H.-D.; Ouazar, D. Analytical solutions. In Seawater Intrusion in Coastal Aquifers—Concepts, Methods and Practices, 1st ed.; Bear, J., Cheng, A.H.-D., Sorek, S., Ouazar, D., Herrera, I., Eds.; Springer: Dordrecht, the Netherlands, 1999; Volume 14, pp. 163–191.
43. Wiener, N. The homogeneous chaos. Am. J. Math. **1938**, 60, 897–936.
44. Ghanem, R.G.; Spanos, P.D. Stochastic Finite Elements: A Spectral Approach; Springer: Berlin, Germany, 1991.
45. Abramowitz, M.; Stegun, I.A. Handbook of Mathematical Functions; Dover Publications: New York, NY, USA, 1970.
46. Ciriello, V.; Di Federico, V.; Riva, M.; Cadini, F.; De Sanctis, J.; Zio, E.; Guadagnini, A. Polynomial chaos expansion for global sensitivity analysis applied to a model of radionuclide migration in a randomly heterogeneous aquifer. Stoch. Environ. Res. Risk Assess.**2013**, 27, 945–954.
47. Ciriello, V.; Di Federico, V. Analysis of a benchmark solution for non-Newtonian radial displacement in porous media. Int. J. Non Linear Mech. **2013**, 52, 46–57.
48. Park, C.H.; Aral, M.M. Multi-objective optimization of pumping rates and well placement in coastal aquifers. J. Hydrol. **2004**, 290, 80–99.

CHAPTER 17

FORECASTING DAILY PRECIPITATION USING HYBRID MODEL OF WAVELET-ARTIFICIAL NEURAL NETWORK AND COMPARISON WITH ADAPTIVE NEUROFUZZY INFERENCE SYSTEM

Abazar Solgi, Vahid Nourani, and Amir Pourhaghi

INTRODUCTION

Estimation and forecasting of precipitation and its runoff have played effective and critical role in the watershed management and proper utilization of watershed, dams, and reservoirs and finally minimizing the damage caused by floods and drought. Therefore, this subject is the hydrologist's interest. Predicting any event forms the basis of crisis management, and when this goal can be achieved, the predicting model could be accessed. Several methods are used for predicting hydrological events such as precipitation. Using each of these methods is always with some error in results. Accurate prediction of hydrological signals such as precipitation can provide useful information to predict amount of precipitation for water resources and soil management in a basin. In addition, correct prediction of hydrological signals plays an important role in reducing the effects of drought on water resources systems. Hydrological systems are affected by many factors such as climate, land cover, soil infiltration rates, evapotranspiration which is dependent on stochastic components, multitemporal scales, and above-mentioned nonlinear characteristics. Despite nonlinear relationships, uncertainty, and high lack of precision and variables temporal and spatial characteristics in water circulation system, none of the statistical and conceptual models which are proposed for accurate precipitation and runoff modeling were able to become a superior and capable model. Today, nonlinear networks are widely used as one intelligent system in predicting such a complex phenomenon. One of the methods that have been considered in recent years in the field of hydrology is the application of wavelet transform as new and effective method in signals and time series analysis. Monthly modeling and predicting of river flow D4 structure was used with application of wavelet neural network method and monthly flow data from two stations (Gerdelli Station on Canakdere River and Isakoy Station on Goksudere River) in the Eastern Black Sea, Turkey. The neurowavelet model improved by two methods: combining discrete wavelet transform and multilayer perceptron (MLP) to predict monthly flow, compared with multilayer perceptron models of multilinear regression (MLR) and autocorrelated model (AR). The comparison of the results revealed that the suggested model could increase the forecast accuracy and perform better than MLP, MLR, and AR models [1]. Improvement in the performance of artificial neural networks (ANN) to predict seasonal time series was reviewed. So several structures of proposed artificial neural network are presented to predict seasonal time series. The model for four full time series was tested. The results of proposed neural network had been compared with the results of current statistical models and other structures of neural network. This comparison showed that the proposed model of neural network has less prediction error than other methods [2]. A combined neural-wavelet model is provided for prediction of Ligvan Chai (Tabriz, Iran) watershed precipitation. For this purpose, the original time series by wavelet transform converted into several subseries, which these converted series were used as an input to the Neural network for Monthly precipitation prediction. The results showed that the proposed model can predict both short-

Citation: Abazar Solgi, Vahid Nourani, and Amir Pourhaghi, "Forecasting Daily Precipitation Using Hybrid Model of Wavelet-Artificial Neural Network and Comparison with Adaptive Neurofuzzy Inference System," Advances in Civil Engineering, vol. 2014, Article ID 279368, 12 pages, 2014. doi:10.1155/2014/279368.

and long-term precipitation events because of using multiscale time series as the ANN input layer [3]. The approach with surface water modeling and wavelet analysis is performed for understanding the base flow components of rivers discharge in Karst environments. This investigation examined the advantages of wavelet analysis in this regard [4]. Approach improvement based on the precipitation-runoff modeling using a combination of artificial neural network-wavelet is performed, which shows the model that precipitation and discharge data entered as an input outperformed than the model just precipitation was entered as an input [5]. Precipitation-runoff model using a combination of wavelet-neural network model is presented. According to the fitted coefficients ($R2$), Root Mean Square Error (RMSE) concluded that the hybrid model of waveletneural network is more efficient than the neural network and regression [6]. A method based on transform discrete wavelet and artificial neural networks to predict applied flow in seasonal river in semiarid watershed in Cyprus was presented. Wavelet coefficients as input Levenberg Marquardt (LM) artificial neural network models were used to predict the flow. The relative performance of the wavelet-neural network (WA-ANN) and artificial neural network (ANN) models was compared to lead times of 1 and 3 days flow forecasting for two different rivers. In both cases, neural network-wavelet model for flow predictions was more accurate than artificial neural network. The results indicate that wavelet-neural network models are a promising new method of short-term flow forecasting in nonperennial rivers in semiarid watersheds such as those found in Cyprus [7]. Two combination methods of artificial intelligence for modeling precipitation-runoff for two watersheds in Azerbaijan, Iran, are presented. Two hybrid AI-based models which are reliable in capturing the periodicity features of the process are introduced for watershed precipitation-runoff modeling. In the first model, the SARIMAX1 -ANA (seasonal autoregressive integrated moving average with exogenous input) model is used and and in the second model the wavelet-ANFIS model is used. The results showed that the proposed models can predict both short- and long-term runoff discharges, considering seasonality effects, but the second model is relatively more appropriate because it used the multiscale time series of precipitation and runoff data in the ANFIS input layer [8]. A new combination of neural networks for modeling precipitation-runoff in the basin Aq Chay, Iran, is presented. The model was combined with data processing methods, genetic algorithms, and Levenberg Marquardt algorithm for training the neural network input. Results showed that this method has more accurately predicted runoff from artificial neural networks and adaptive neuralfuzzy inference system [9]. Feature extraction method based on the self-organizing map and the combined wavelet-neural network method was combined and presented for modeling the precipitation-runoff. Two-stage procedure to model the precipitation-runoff process of the Delaney Creek and Payne Creek Basins, Florida, USA, was presented. The two-stage procedure includes data preprocessing and model building. The results proved that the proposed model leads to better outcome especially in terms of determination coefficient for detecting peak points (DC peak) [10]. For modeling daily precipitation-runoff in the Eel River Basin, California, at several stations, the artificial intelligence (AI) techniques have been used. In this way, different ANFIS structures were trained, using Sugeno algorithm in order to estimate daily discharge values at different stations. In order to improve the model efficiency, the input data were then classified into some clusters by the means of fuzzy C-means (FCMs) method. The goodness-of-fit measures support the gainful use of the IGANFIS and FCM methods in spatiotemporal modeling of hydrological processes [11]. Because predicting the amount and timing of precipitation can be an important aid to managers, planners, and farmers in conjunction with the planning and optimal use of water resources, different methods are used to predict precipitation. One of these methods that has recently received considerable attention is the use of wavelet theory. Due to widespread use of artificial neural networks and wavelet transform in various disciplines, especially science related to water, and according to rarely use of wavelet transform in Iran, forecast precipitation with a hybrid model of neural network-wavelet and adaptive fuzzy wavelet neural network model is examined in this research.

MATERIALS AND METHODS

Study Area

Verayneh Rain gauge station in the Nahavand city is in geographical position 48 degrees, 24 minutes, and 15 seconds East longitude and 34 degrees, 04 minutes,

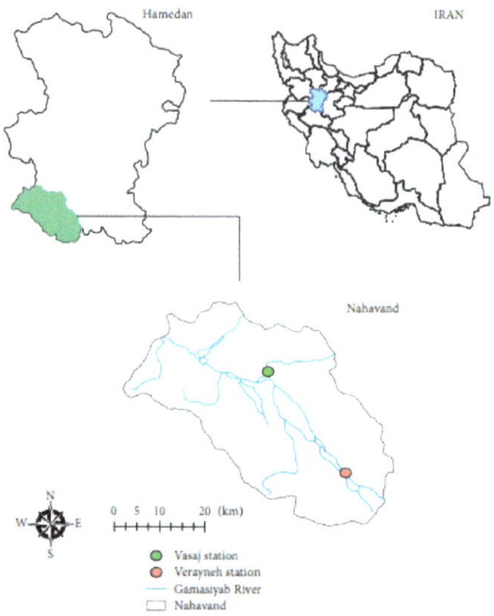

Figure 1: Location of the Nahavand in Hamedan and Iran.

and 32 seconds North latitude. Figure 1 shows Location of the Nahavand in Hamedan, Iran. The station was established in 1348 and has a height of 1795 meters above sea level; code stations are 21–534 with long-term average annual precipitation of 521 mm. Precipitation data is presented as average of 1.43, maximum of 96, minimum of 0, standard deviation of 5.31, and variance of 28.29.

Wavelet Transform

The wavelet transform has increased in usage and popularity in recent years since its inception in the early 1980s, yet still does not enjoy the wide spread usage of the Fourier transform. Fourier analysis has a serious drawback. In transforming to the frequency domain, time information is lost. When looking at a Fourier transform of a signal, it is impossible to tell when a particular event took place but wavelet analysis allows the use of long time intervals where we want more precise low-frequency information and shorter regions where we want high-frequency information. Figure 2 compares Fourier transform and wavelet transform [3]. In the field of earth sciences, Grossmann and Morlet [12], who worked especially on geophysical seismic signals, introduced the wavelet transform application. A comprehensive literature survey of wavelet in geosciences can be found in Fofola-Georgiou and Kumar [13] and the most recent contributions are cited by Labat [14]. As there are many good books and articles introducing the wavelet transform, this paper will not delve into the theory behind wavelets and only the main concepts of the transform are briefly presented; recommended literature for the wavelet novice includes Mallat [15] or Labat et al. [16].The time-scale wavelet transform of a continuous time signal, $x(t)$, is defined as follows:

$$T(a,b) = \frac{1}{\sqrt{a}} \int_{-\infty}^{+\infty} g^{*}\left(\frac{t-b}{a}\right) x(t)\,dt, \tag{1}$$

where g^{*} corresponds to the complex conjugate and $g(t)$ is called wavelet function or mother wavelet. The parameter a acts as a dilation factor, while b corresponds to a temporal translation of the function (t), which allows for the study of the signal around b [17]. The main property of wavelet transform is to provide a time-scale localization of processes, which derives from

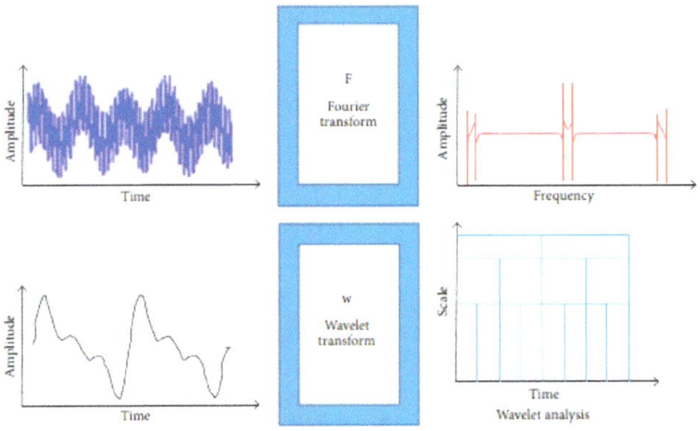

Figure 2: Comparison of Fourier and wavelet transforms.

the compact support of its basic function. This is opposed to the classical trigonometric function of Fourier analysis. The wavelet transform searches for correlations between the signal and wavelet function. This calculation is done at different scales of a and locally around the time of b. The result is a wavelet coefficient $((a, b))$ contour map known as a scalogram. In order to be classified as a wavelet, a function must have finite energy, and it must satisfy the following "admissibility conditions" [15]:

$$\int_{-\infty}^{+\infty} g(t)\,dt = 0, \qquad C_g = \int_{-\infty}^{+\infty} \frac{|\hat{g}(\omega)|^2}{|\omega|} d\omega < \infty, \tag{2}$$

where $g(\omega)^{\wedge}$ is Fourier transform of $g(t)$; that is, the wavelet must have no zero frequency component. In order to obtain a reconstruction formula for the studied signal, it is necessary to add "regularity conditions" to the previous ones. Consider

$$\int_{-\infty}^{+\infty} t^{k} g(t)\,dt = 0, \quad \text{where } k = 1, 2, \ldots, n-1. \tag{3}$$

So the original signal may be reconstructed using the inverse wavelet transform as follows:

$$x(t) = \frac{1}{C_g} \int_{-\infty}^{+\infty} \int_{0}^{\infty} \frac{1}{\sqrt{a}} g\left(\frac{t-b}{a}\right) T(a,b) \frac{da \cdot db}{a^2}. \tag{4}$$

For practical applications, the hydrologist does not have at his or her disposal a continuous-time signal process but rather a discrete-time signal. A discretization of (2) based on the trapezoidal rule maybe is the simplest discretization of the continuous wavelet transform. This transform produces $N2$ coefficients from a data set of length N; hence redundant information is locked up within the coefficients, which may or may not be a desirable property [17]

To overcome the mentioned redundancy, logarithmic uniform spacing can be used for the a scale discretization with correspondingly coarser resolution of the b locations, which allows for N transform coefficients to completely describe a signal of length N. Such a discrete wavelet has the form [15]

$$g_{m\cdot n}(t) = \frac{1}{\sqrt{{a_0}^{m}}} g\left(\frac{t - nb_0{a_0}^{m}}{{a_0}^{m}}\right), \tag{5}$$

where m and n are integers that control the wavelet dilation and translation, respectively, a_0 is a specified fined dilation step greater than 1, and b_0 is the location parameter and must be greater than zero. The most common and simplest choices for parameters are $a_0 = 2$ and $b_0 = 1$. This power-of-two logarithmic scaling of the translation and dilation is known as the dyadic grid arrangement. The dyadic wavelet can be written in more compact notation as [15]

$$g_{m,n} = 2^{-m/2} g\left(2^{-m} i - n\right). \tag{6}$$

Discrete dyadic wavelets of this form are commonly chosen to be orthonormal; that is, [15]

$$\int_{-\infty}^{+\infty} g_{m,n}(t)\, g_{m',n'}(t)\,dt = \delta_{m,m'} \delta_{n,n'}, \tag{7}$$

where δ is Kronecker delta. This allows for the complete regeneration of the original signal as an expansion of a linear combination of translates and dilates of orthonormal wavelets. For a discrete-time series, x_i, the dyadic wavelet transform becomes

$$T_{m,n} = 2^{-m/2} \sum_{i=0}^{N-1} g\left(2^{-m} i - n\right) x_i, \tag{8}$$

where $T_{m,n}$ is wavelet coefficient for the discrete wavelet of scale $a=2^m$ and location $b=2^m n$. Equation (9) considers a finite time series x_i, $i = 0, 1, 2, \ldots, N-1$; N is an integer power of 2: $N=2M$. This gives the ranges of m and n as, respectively, $0 < n < 2^{M-m} - 1$ and $1 < m < M$. At the largest wavelet scale (i.e., $2m$, where $m=M$) only one wavelet is required to cover the time interval, and only one coefficient is produced.

At the next scale (2^{m-1}), two wavelets cover the time interval; hence, two coefficients are produced and so on down to m=1. At m=1, the a scale is 21; that is, 2^{M-1} or $N/2$ coefficients are required to describe the signal at this scale. The total number of wavelet coefficients for a discrete-time series of length $N=2^M$ is then 1+2+4+8+⋯+2M−1 =N−1. In addition to this, a signal smoothed component $\overline{T}$ is left, which is the signal mean. Thus, a time series of length N is broken into N components, that is, with zero redundancy. The inverse discrete transform is given by [15].

$$x_i = \overline{T} + \sum_{m=1}^{M} \sum_{n=0}^{2^{M-n}-1} T_{m,n} 2^{-m/2} g\left(2^{-m} i - n\right) \tag{9}$$

or in a simple format as [15]

$$x_i = \overline{T}(t) + \sum_{m=1}^{M} W_m(t), \tag{10}$$

where $\overline{T}(t)$ is called approximation subsignal at level M and $Wm(t)$ are details subsignals, at levels m = 1, 2, . . . , M. The wavelet coefficients, $W(t)$ (m = 1, 2, . . . , M), provide the details signals, which can capture small features of interpretational value in the data; the residual term, $\overline{T}(t)$, represents the background information of data. Because of simplicity of $W_1(t)$, $W2(t)$, . . . , $W_M(t)$, $\overline{T}(t)$, some interesting characteristics, such as period, hidden period, dependence, and jump, can be diagnosed easily through wavelet components [8].

Hybrid Wavelet-Artificial Neural Network (WANN) Model

In this study, daily precipitation in a 43-year period of Verayneh Station in the city of Nahavand is used, to check the homogeneity of the data Vasaj Station as an auxiliary stations and the double mass curve used, which shows the homogeneity of data. Then the data were analyzed using wavelet transform. Thus, using MATLAB program and considering different levels and mother wavelet coded in MATLAB, wavelet transform got on the input parameter. For the data set considered in the present study, the input variables as well as the target variables are first normalized linearly in the range of 0 and 1. This range is selected because of the use of the logistic function (which is bounded between 0.0 and 1.0) as the activation function for the output layer [18], that is, (11). The normalization is done using the following equation:

$$y = 0.5 + \left(0.5 \times \left(\frac{x - \overline{x}}{x_{max} - x_{min}}\right)\right).$$

In this equation, x is the data, x is the average data, $\overline{x}$max is the maximum data, $x$min is the minimal data, and y data is normalized. Then, 70% of data as training data, 15% as cross validation data, and 15% as test data were considered. Using artificial neural network and building different models that in each model the type of input is different and by comparing different training methods such as Multilayers Perceptron, GFF, and so forth. Best Education Act, number of neurons and hidden layers and the number of run, number of iterations of each epoch was determined. Then the best model using genetic algorithm and neural network was optimized to obtain the best estimates. Adaptive neurofuzzy system is investigated for two models. In model one, the previous day's precipitation data and tomorrow precipitation were used to predict rain tomorrow. And, on the second model, day precipitation data were used to predict rain tomorrow. Twothirds of the data are as training data, and the remaining onethird are as the test data (test) network model systems by adaptive fuzzy introduced. The coefficient of efficiency (CE) was used to compare the goodness of fit between the measured flow and the simulated flow. CE which is also known as the Nash-Sutcliffe coefficient is defined as [19]

$$\mathrm{CE} = 1 - \frac{\sum \left(p_{\mathrm{obs}} - p_{\mathrm{Pre}}\right)^2}{\sum \left(p_{\mathrm{obs}} - \overline{p}\right)^2},$$

$$\mathrm{RMSE} = \sqrt{\frac{\sum \left(p_{\mathrm{obs}} - p_{\mathrm{pre}}\right)^2}{n}},$$

$$R^2 = 1 - \frac{\sum_{i=1}^{N} \left(p_{\mathrm{obs}} - p_{\mathrm{Pre}}\right)^2}{\sum_{i=1}^{N} \left(p_{\mathrm{Pre}} - \overline{p}\right)^2}, \tag{12}$$

where p_{obs} is the observed precipitation rate, p_{pre} is the simulated flow rate, and p is the average value of the observed flow rate [3, 8]. RMSE was used to measure prediction accuracy which produces a positive value by squaring the errors. The RMSE increases from zero for perfect predictions through large positive values as the discrepancies between predictions and observations become increasingly large. Obviously, high value for CE (up to one) and small value for RMSE indicate high efficiency of the model. Although Legates and McCabe (1999) indicated that a hydrological model can be sufficiently evaluated by DC and RMSE, the Akaike Information Criterion (AIC), which is based on the parsimony, penalizes the RMSE with the number of trained parameters (Npar) which was also utilized as [11]

$$\mathrm{AIC} = m \times \ln(\mathrm{RMSE}) + 2(N\mathrm{par}), \tag{13}$$

where m is number of inputs and smaller value of AIC indicates a better model. The following formula determines the maximum decomposition level [17]:

$$L = \mathrm{int}\left[\log(N)\right], \tag{14}$$

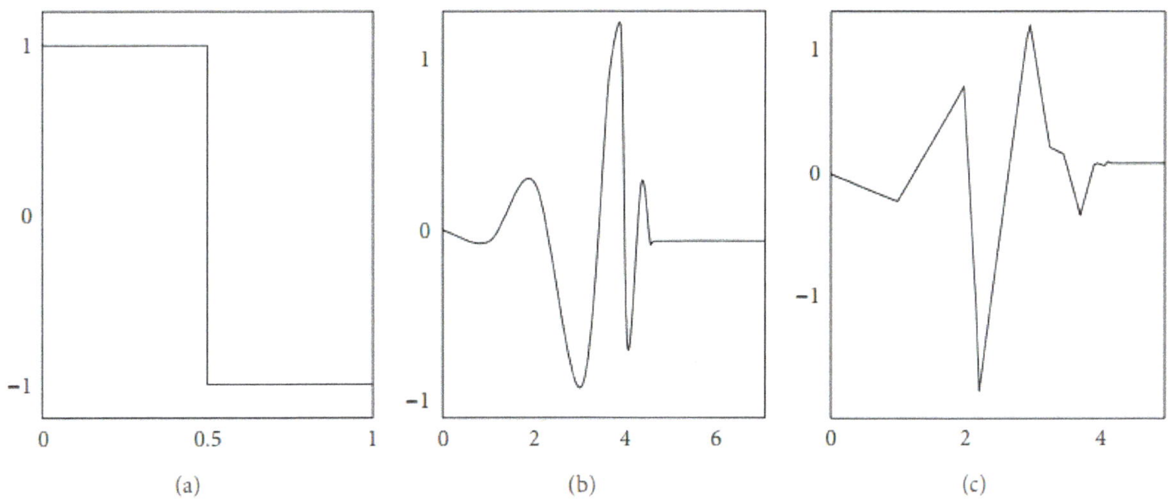

Figure 3: (a) Harr wavlet, (b) db2 wavlet, and (c) sym3 wavelet.

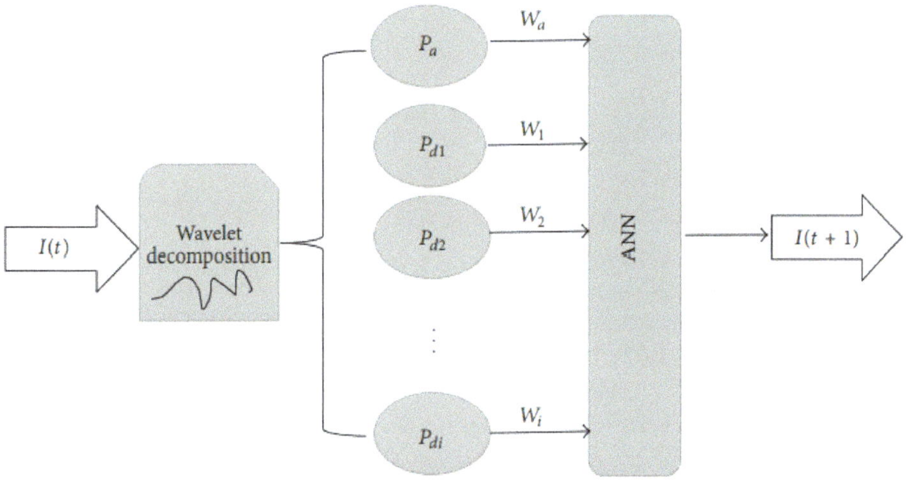

Figure 4: Schematic presentation of WANN model.

where L and N are decomposition level (i and j) and number of time series data, respectively. For the study at hand $N = 15705$, so L=4. In this study it was tried to investigate the effects of the used mother wavelet type as well as decomposition level on the model efficiency. Hence, time series were decomposed to two, three, four, five, six, and seven levels by four different kinds of wavelet transforms, that is, 1— Haar wavelet, a simple wavelet, 4—Daubechies wavelet (db4), a most popular wavelet, and 3—sym3 wavelet. These wavelets are shown in Figure 3. Figure 4 shows schematic structures of WANN model.

Adaptive Neurofuzzy Inference System (ANFIS). Of course the leading theory in quantifying uncertainty in scientific models from the late nineteenth century until the late twentieth century had been the probability theory. However, the gradual evolution of the expression of uncertainty using probability theory was challenged first in 1937 by Max Black with his studies in vagueness and then with the introduction of fuzzy sets by Zadeh 1965. Zadeh's paper had a profound influence on thinking about uncertainty because it challenged not only probability theory as the sole representation for uncertainty but also the very foundations upon which probability theory was based: classical binary (two-valued) logic [20]. Each fuzzy system contains three main parts: fuzzifier, fuzzy data base, and defuzzifier. Fuzzy data base contains two main parts: fuzzy rule base and inference engine. In fuzzy rule base, rules related to fuzzy propositions are described [21]. Thereafter, analysis operation is applied by fuzzy inference engine. There are several fuzzy inference engines which can be employed for this goal, with Sugeno and Mamdani being the two of well-known ones [22]. Neurofuzzy simulation refers to the algorithm of applying different learning techniques produced in the neural network literature to fuzzy modeling or a fuzzy inference system (FIS) [23]. This is done by fuzzification of the input through membership functions (MFs), where a curved relationship maps the input value within the interval of [01]. The parameters associated with input as well as output membership functions are trained using a technique like backpropagation and/or least squares. Therefore, unlike the multilayer perceptron (MLP), where weights are tuned, in ANFIS, fuzzy language rules or conditional (if-then) statements are determined in order to

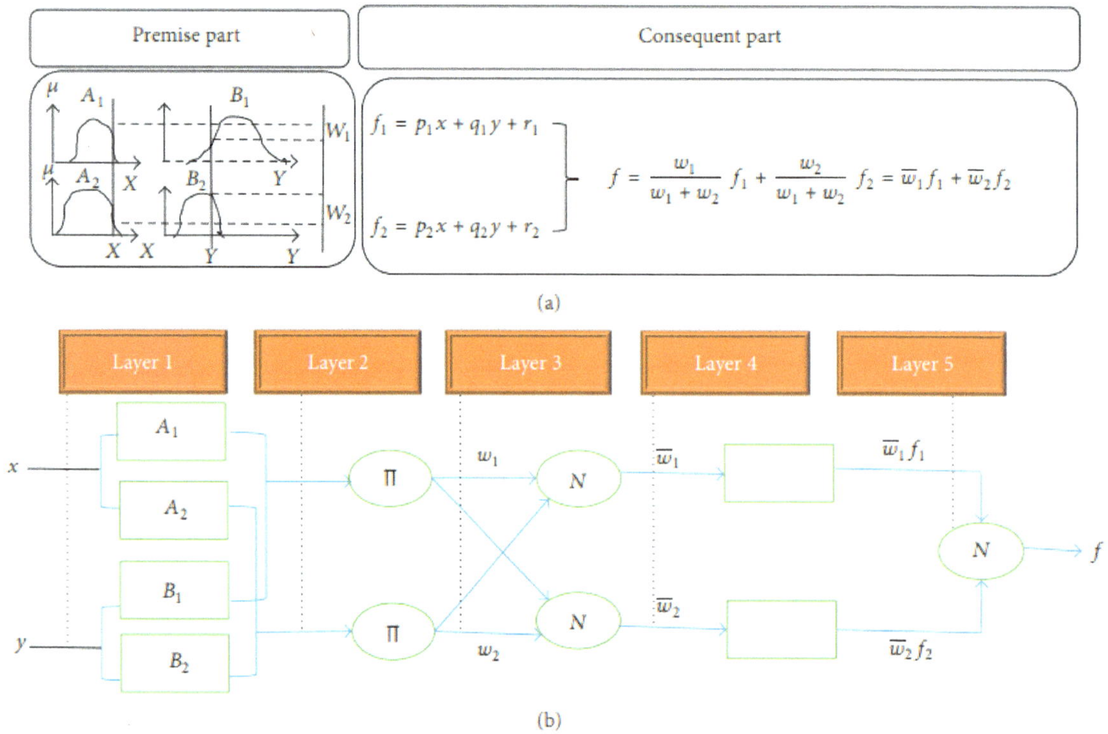

Figure 5: (a) Sugeno's fuzzy if-then rule and fuzzy reasoning mechanism and (b) equivalent ANFIS structure.

train the model [24]. The ANFIS is a universal approximator and as such is capable of approximating any real continuous function on a compact set to any degree of accuracy. The ANFIS is functionally equivalent to fuzzy inference systems [21]. Specifically the ANFIS system of interest here is functionally equivalent to the Sugeno first-order fuzzy model [21]. The general construction of the ANFIS is presented in Figure 5. Figure 5(a) shows the fuzzy reasoning mechanism for the Sugeno model to derive an output function f from a given input vector $[x, y]$. The corresponding equivalent ANFIS construction is shown in Figure 5(b). According to this figure, it is assumed that the FIS has two inputs x and y and one output f. For the first-order Sugeno fuzzy model, atypical rule set with two fuzzy if-then rules can be expressed as [25]

rule (1): if $\mu(x)$ is A_1 and $\mu(y)$ is B_1, then $f_1 = P_1x + q_1y + r_1$;

rule (2): if $\mu(x)$ is A_2 and $\mu(y)$ is B_2, then $f_2 = P_2x + q_2y + r_2$,

where A_1, A_2 and B_1, B_2 are the MFs for inputs x and y, respectively; P_1, q_1, and r_1 and P_2, q_2, and r_2 are the parameters of the output function. The functioning of the ANFIS is as follows. Layer 1. Each node in this layer produces membership grades of an input variable. The output of ith node inlayer k is denoted by $Q_i^{\ k}$ .Assuming a generalized bell function (gbellmf) as the membership function (MF), the output $Q^i_{\ 1}$ can be computed as [26]

$$Q_i^{\ 1} = \mu_{Ai}(x) = \frac{1}{1 + ((x - c_i)/a_i)^{2b_i}}, \quad (15)$$

where $\{a_i, b_i, c_i\}a$ are adaptable variable s known as premise parameters.

Layer 2. Every node in this layer multiplies the incoming signals. Consider

$$Q_i^{\ 2} = w_i = \mu_{Ai}(x) \cdot \mu_{Ai}(x) \qquad i = 1, 2. \quad (16)$$

Layer 3. The ith node of this layer calculates the normalized firing strengths as

$$Q_i^{\ 3} = \overline{w}_i = \frac{w_i}{w_1 + w_2} \qquad i = 1, 2. \quad (17)$$

Layer 4. Node i in this layer calculates the contribution of the ith rule towards the model output, with the following node function [26]:

$$Q_i^{\ 4} = \overline{w}_i(p_ix + q_iy + r_i) = \overline{w}_if_i, \quad (18)$$

where w is the output of layer 3 and $\{P_1, q_1, r_1\}$ is the parameter set. Layer 5. The single node in this layer calculates the overall output of the ANFIS as [26]

$$Q_i^{\ 5} = \sum_i \overline{w}_if_i = \frac{\sum_i w_if_i}{\sum_i w_i}. \quad (19)$$

Table 1: Details and structures of WANN model

Structure	Model	Learning rule	Transfer function	Momentum	Maximum epochs	Runs	Processing elements	Hidden layers
D1	MLP	momentum	Tan Axon	0.7	1000	3	50	1
D2	MLP	momentum	Tan Axon	0.7	1000	3	50	1
D3	MLP	momentum	Tan Axon	0.8	1000	3	50	1
D4	MLP	momentum	Tan Axon	0.6	1000	3	50	1
D5	MLP	momentum	Tan Axon	0.7	1000	3	50	1
D6	MLP	momentum	Tan Axon	0.7	1000	3	50	1
D7	MLP	momentum	Tan Axon	0.8	1000	3	50	1
D8	MLP	momentum	Tan Axon	0.7	1000	3	50	1
D9	MLP	momentum	Tan Axon	0.7	1000	3	50	1
D10	MLP	momentum	Tan Axon	0.9	1000	3	50	1
D11	MLP	momentum	Tan Axon	0.7	1000	3	50	1
D12	MLP	momentum	Tan Axon	0.7	1000	3	50	1
D13	MLP	momentum	Tan Axon	0.7	1000	3	50	1
D14	MLP	momentum	Tan Axon	0.6	1000	3	50	1
D15	MLP	momentum	Tan Axon	0.7	1000	3	50	1
D16	MLP	momentum	Tan Axon	0.7	1000	3	50	1
D17	MLP	momentum	Tan Axon	0.7	1000	3	50	1
D18	MLP	momentum	Tan Axon	0.7	1000	3	50	1

Table 2: Results and structures of WANN model

Structure	Mother wavelet	Levels	MAE	MSE	*R*
D1	Haar	2	0.0037	0.00003	0.8863
D2	Haar	3	0.0034	0.00026	0.9392
D3	Haar	4	0.0034	0.00036	0.9321
D4	Haar	5	0.0031	0.00002	0.9502
D5	Haar	6	0.0035	0.00067	0.9001
D6	Haar	7	0.0032	0.00032	0.9404
D7	Db4	2	0.0032	0.00022	0.9482
D8	Db4	3	0.0033	0.00026	0.9343
D9	Db4	4	0.0035	0.00029	0.9016
D10	Db4	5	0.0033	0.00031	0.9333
D11	Db4	6	0.0036	0.00037	0.8915
D12	Db4	7	0.0031	0.00049	0.9413
D13	Sym3	2	0.0039	0.00041	0.8704
D14	Sym3	3	0.0043	0.00045	0.8516
D15	Sym3	4	0.0039	0.00036	0.8711
D16	Sym3	5	0.0035	0.0003	0.8993
D17	Sym3	6	0.0035	0.0004	0.9031
D18	Sym3	7	0.0036	0.00034	0.8918

Table 3: Evaluation of model performance (structure D4)

Performance	*P*
MSE	0.0002
NMSE	0.032
MAE	0.003
Min Abs Error	0.001
Max Abs Error	0.037
r	0.950

Table 4: Sensitivity analysis of structure D4.

Sensitivity	P
A	0.0157
d_1	0.0917
d_2	0.0546
d_3	0.0263
d_4	0.0232
d_5	0.0114

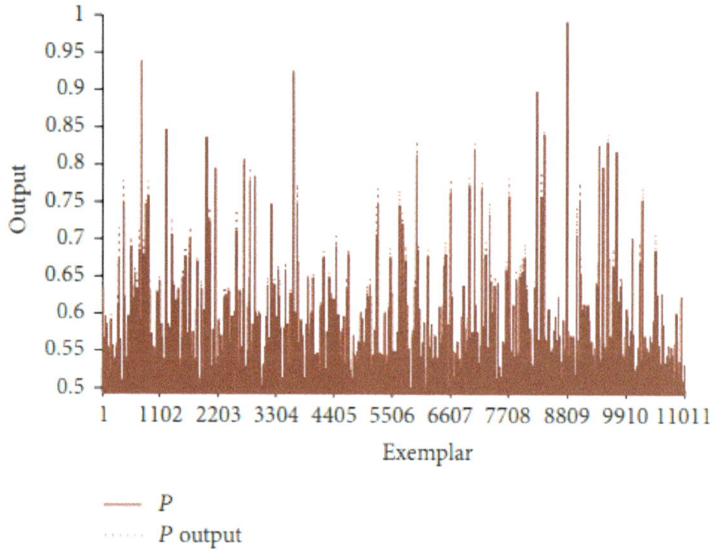

Figure 6: Computed and observed precipitation time series (structure D4).

The learning algorithm for ANFIS is a hybrid algorithm, which is a combination of the gradient descent and leastsquares method [25]. The parameters for optimization are

Table 5: Results and structures of ANFIS model

Model	Structure	Membership function	EPOCH	RMSE	Test error	R
1	D19	Trimf	20	0.0271	0.0264	0.6464
	D20	Trapmf	20	0.0271	0.0265	0.6326
	D21	Gbellmf	50	0.0268	0.0256	0.7835
	D22	Gaussmf	40	0.0269	0.0257	0.7757
	D23	Gauss2mf	40	0.0289	0.0265	0.6265
	D24	Pimf	40	0.0269	0.0257	0.7757
	D25	Dsigmf	40	0.0270	0.0259	0.7559
	D26	Psigmf	40	0.0270	0.0259	0.7559
2	D27	Trimf	20	0.0271	0.0263	0.6563
	D28	Trapmf	20	0.0272	0.0264	0.6453
	D29	Gbellmf	30	0.0269	0.0256	0.7806
	D30	Gaussmf	30	0.0269	0.0256	0.7826
	D31	Gauss2mf	40	0.0269	0.0257	0.7757
	D32	Pimf	40	0.0269	0.0256	0.7786
	D33	Dsigmf	40	0.0270	0.0256	0.7786
	D34	Psigmf	30	0.0270	0.0259	0.7759

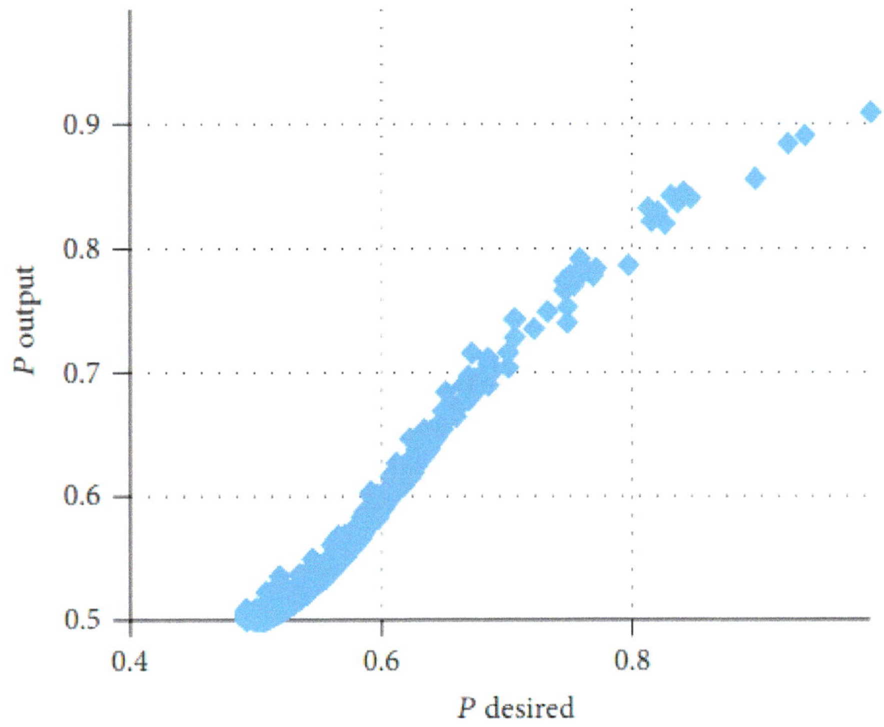

Figure 7: Output precipitation against observational precipitation

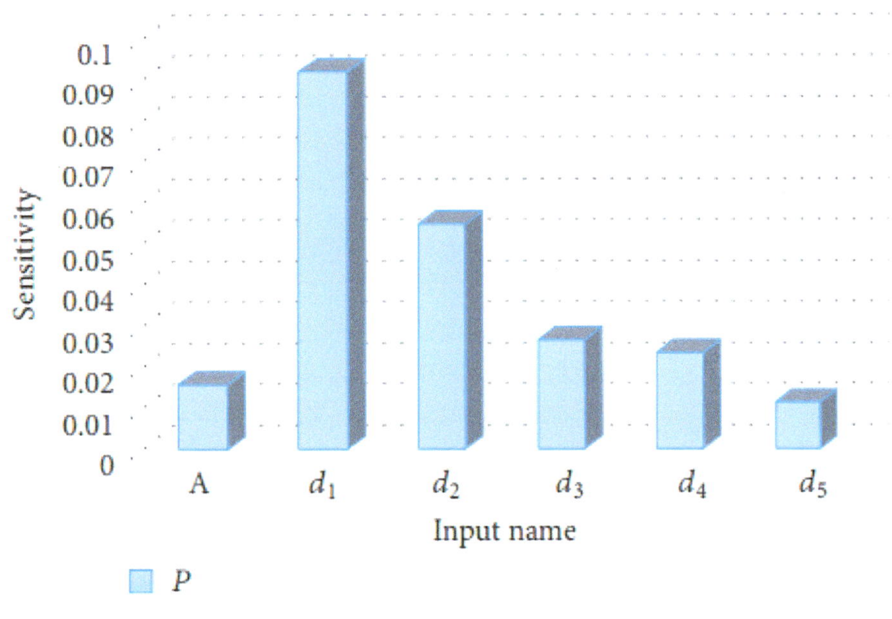

Figure 8: Sensitivity analysis of structure D4.

Table 6: Computed precipitation of ANFIS and WANN models.

Model	Validation step				Calibration step			
	RMSE	R^2	CE	AIC	RMSE	R^2	CE	AIC
WANN	0.021	0.903	0.743	18.22	0.028	0.774	0.525	20.17
ANFIS	0.027	0.613	0.562	19.84	0.033	0.599	0.632	20.62

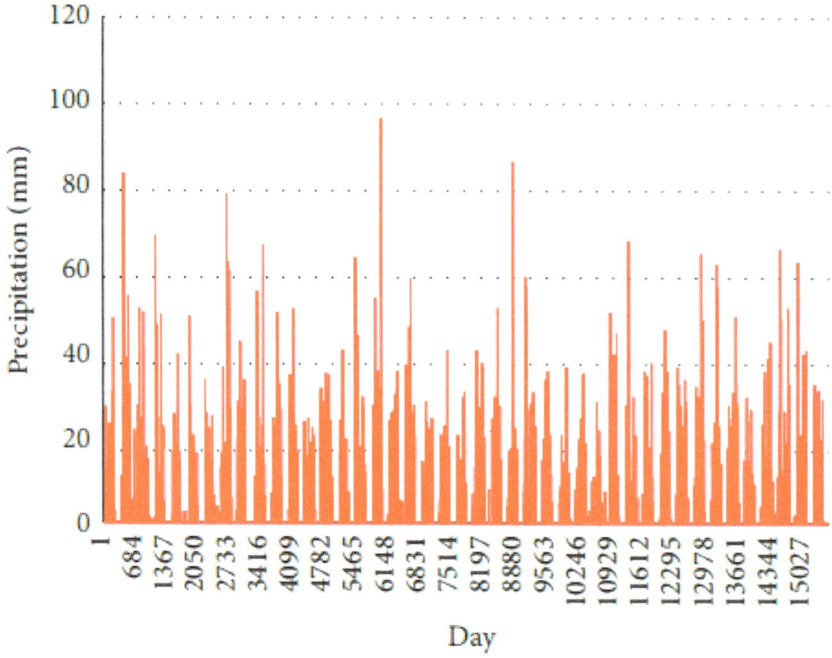

Figure 9: Precipitation time series.

the premise parameters $\{a_i, b_i, c_i\}$ and the consequent parameters $\{P_1, q_1, r_1\}$. In the forward pass of the hybrid learning approach, node outputs go forward until layer (4) and the consequent parameters are identified by the least-squares technique. In the backward pass, the error signals propagate backward and the premise parameters are updated by gradient descent. More information for ANFIS can be found in related literatures [21, 26].

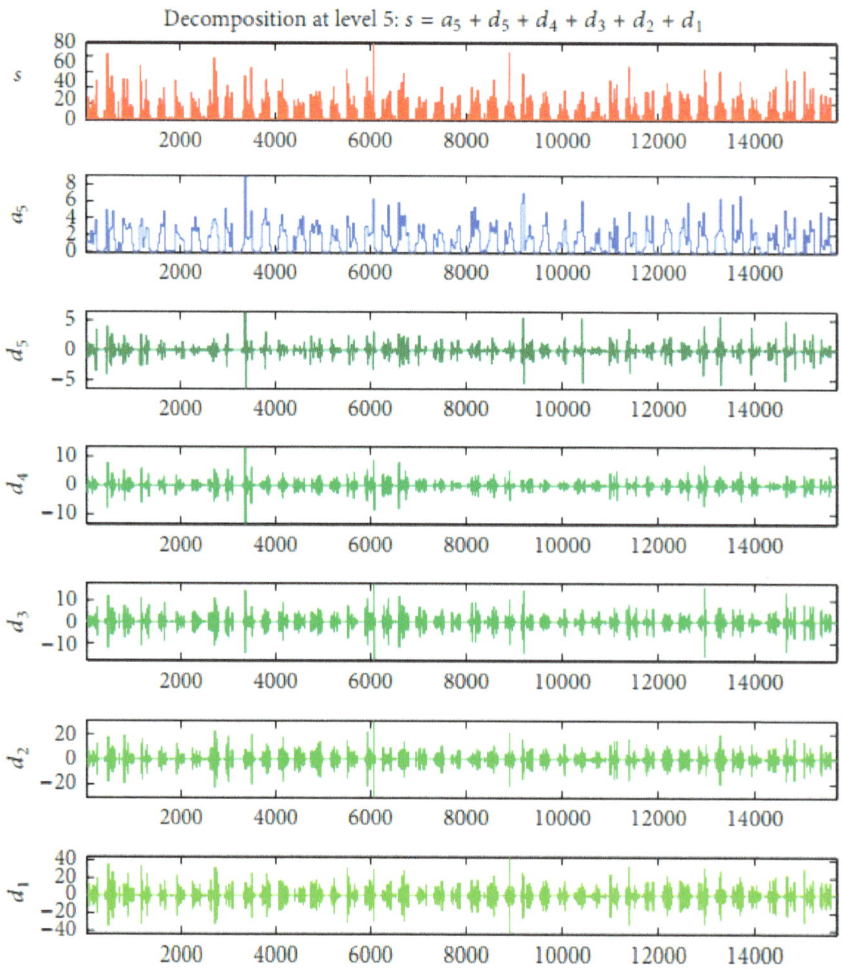

Figure 10: Approximation and details subsignals of precipitation time series by Haar wavelet (level 5)

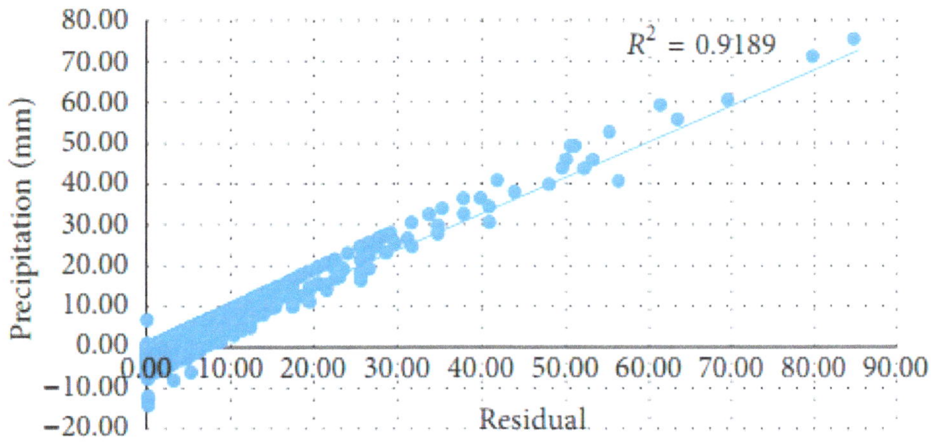

Figure 11: Actual fluctuations precipitation against the residuals.

ANFIS uses system of learning algorithms, neural network and fuzzy logic in order to design a nonlinear mapping between the input and output. Also due to capability in combining of linguistic power fuzzy systems with a numerical strength of a neural network, the modeling of processes such as hydrology reservoir management and estimating suspended sediment load is very powerful [27, 28]. Adaptive neurofuzzy is based on changes in the amount and range of functions belonging to different iterations to achieve the appropriate network based on the minimum error functions. Takagi Sugeno inference method is used in the ANFIS model. The number and type of inputs and the membership functions shape are affected by neurofuzzy model [21].

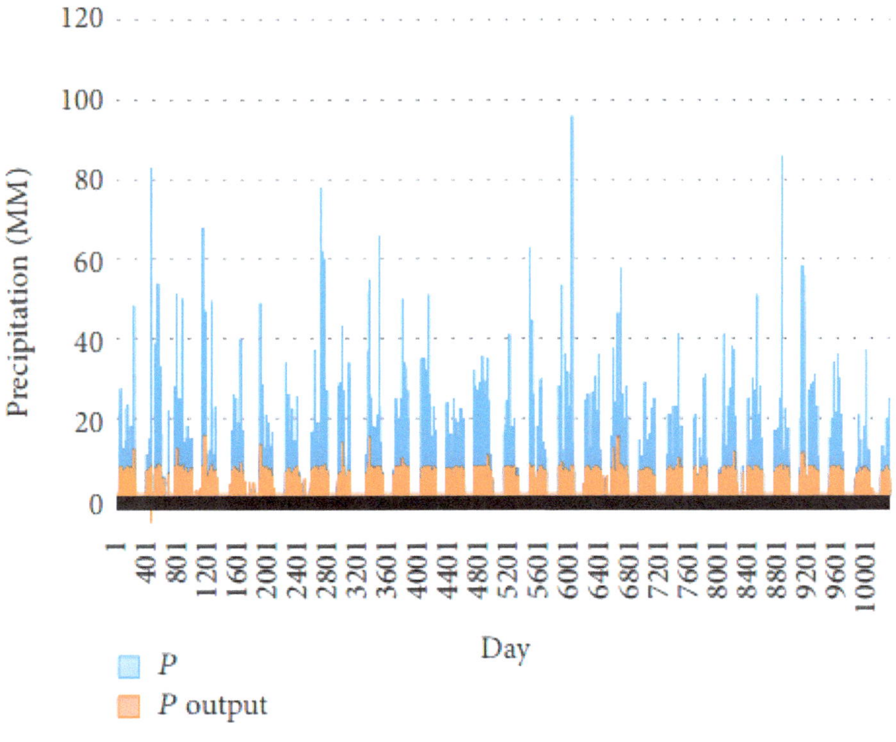

Figure 12: Computed and observed precipitation time series (structure D4).

Results and Discussion

The different models were examined. Thus, using different wavelets, subseries at different levels were studied for different mother's wavelet. Tables 1 and 2 show the results. To do this, the momentum learning rule and Tan Axon transfer function

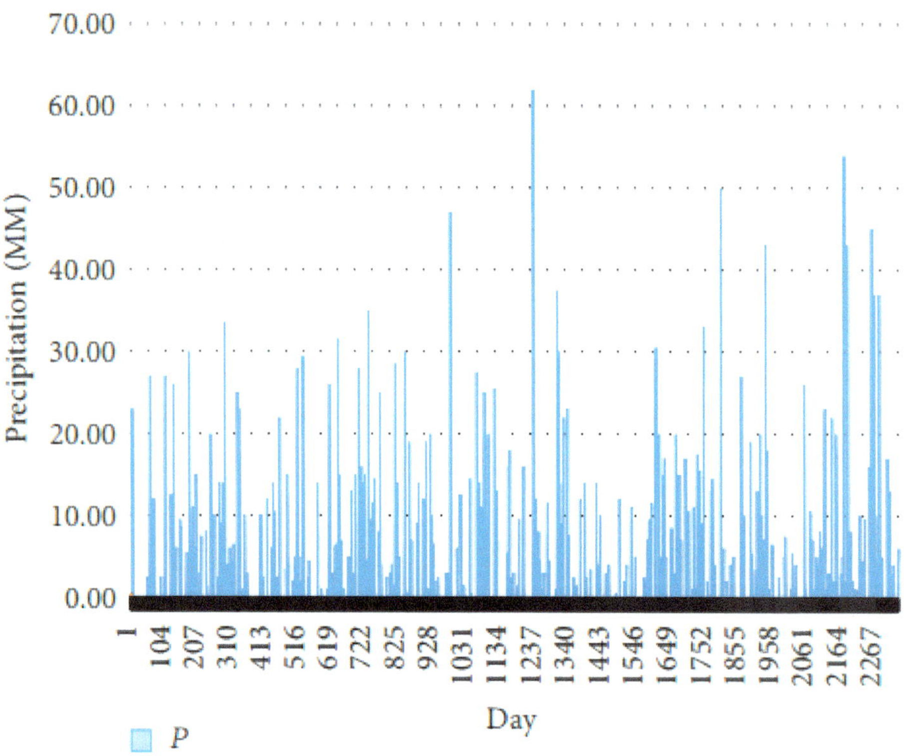

Figure 13: Computed and observed precipitation time series (structure D21).

with 1000 epoch and three runs of network in different hidden layers are used. In Table 2, the mother wavelet is shown using various levels for each structure in a hybrid model of waveletneural network. According to Table 2, the best model is wavelet-neural network with D4 structure with level 5 and the wavelet mother Haar. Figure 6 shows computed precipitation time series and observed precipitation. In Figure 7, precipitation output against observational precipitation was showed. In Table 3, evaluation of D4 structure performance has been examined. Sensitivity analysis showed that the subseries 1 ($d1$) has highest sensitivity for predicting precipitation than the other subseries (Figure 8). Table 4 shows Sensitivity analysis of structure D4. Figure 9 showed precipitation time series and Figure 10 showed wavelet analysis. Different structures in adaptive neurofuzzy inference system, comparing with different membership functions and epochs, were examined. To find the best model, parameters were determined in accordance with Table 6. And the best model was the D21 structure. Membership function of Psigmf and epochs 20 is selected. The results of this study are presented in Table 5. Figure 11 shows the residual values against the observed precipitation data. The results are consistent with observed changes in precipitation and residual values are very high. WANN model was compared with ANFIS model by parameters which are available in Table 6. Based on this comparison WANN model has less error than ANFIS model (i.e., 0.021), correlation coefficient is higher than ANFIS (i.e., 0.95), Nash-Sutcliffe coefficient is higher than ANFIS (i.e., 0.743), and AIC coefficient is smaller than ANFIS (i.e., 18.22). Thus, WANN model is belter.

CONCLUSIONS

In this study, the wavelet transform, which can capture the multiscale features of signals, was used to decompose the Verayneh precipitation time series. Then the subsignals were used as input to the ANN model to predict the precipitation. Furthermore, the effect of wavelet transform type on the model performance was investigated using three different kinds of wavelet transforms. The model results show the high merit of Haar wavelet in comparison with the others (i.e., db4 and Sym3). As a final result, it was deduced that although increasing of decomposition level can progress the model ability an optimum level can be chosen on the basis of the signal length. As the reviews of the structures in this study were collected, it can be concluded that the accuracy of the hybrid model of wavelet-neural network is higher than the accuracy of the adaptive fuzzy neural network model. Also Figures 12 and 13 are confirmed above-mentioned. Results showed that the proposed model can predict precipitation in both short and long terms, because of the use of multiscale time series as entering data to neural network. The obtained results show that the proposed model can predict both short- and long-term precipitation events because of using multiscale time series as the ANN input layer.

CONFLICT OF INTERESTS

The authors declare that there is no conflict of interests regarding the publication of this paper.

REFERENCES

1. O. Kis¨¸i, "Stream flow forecasting using neuro-wavelet technique," Hydrological Processes, vol. 22, no. 20, pp. 4142–4152, 2008.
2. C. Hamzac¸ebi, "Improving artificial neural networks' performance in seasonal time series forecasting," Information Sciences, vol. 178, no. 23, pp. 4550–4559,

2008.

3. V. Nourani, M. T. Alami, and M. H. Aminfar, "A combined neural-wavelet model for prediction of Ligvanchai watershed precipitation," Engineering Applications of Artificial Intelligence, vol. 22, no. 3, pp. 466–472, 2009.
4. F. Salerno and G. Tartari, "A coupled approach of surface hydrological modelling and Wavelet Analysis for understanding the baseflow components of river discharge in karst environments," Journal of Hydrology, vol. 376, no. 1-2, pp. 295–306, 2009.
5. L. H. C. Chua and T. S. W. Wong, "Improving event-based rainfall-runoff modeling using a combined artificial neural network-kinematic wave approach," Journal of Hydrology, vol. 390, no. 1-2, pp. 92–107, 2010.
6. M. Komasi, Modelling rainfall—runoff model using a combination of wavelet—ANN, Tabriz University, Tabriz, Iran, 2007.
7. J. Adamowski and K. Sun, "Development of a coupled wavelet transform and neural network method for flow forecasting of non-perennial rivers in semi-arid watersheds," Journal of Hydrology, vol. 390, no. 1-2, pp. 85–91, 2010.
8. V. Nourani, O. Kisi, and M. Komasi, "Two hybrid artificial ¨ intelligence approaches for modeling rainfall-runoff process," Journal of Hydrology, vol. 402, no. 1-2, pp. 41–59, 2011.
9. S. Asadi, J. Shahrabi, P. Abbaszadeh, and S. Tabanmehr, "A new hybrid artificial neural networks for rainfall-runoff process modeling," Neurocomputing, vol. 121, pp. 470–480, 2013.
10. V. Nourani and M. Parhizkar, "Conjunction of SOM-based feature extraction method and hybrid wavelet-ANN approach for rainfall-runoff modeling," Journal of Hydroinformatics, vol. 15, no. 3, pp. 829–848, 2013.
11. V. Nourani and M. Komasi, "A geomorphology-based ANFIS model for multi-station modeling of rainfall-runoff process," Journal of Hydrology, vol. 490, pp. 41–55, 2013.
12. Grossmann and J. Morlet, "Decomposition of hardy function into square integrable wavelets of constant shape," SIAM Journal on Mathematical Analysis, vol. 5, pp. 723–736, 1984.
13. E. Fofola-Georgiou and P. Kumar, Eds., Wavelet in Geophysiscs, Academic Press, New York, NY, USA, 1995.
14. D. Labat, "Recent advances in wavelet analyses," Journal of Hydrology, vol. 314, no. 1–4, pp. 275–288, 2005.
15. S. G. Mallat, A Wavelet Tour of Signal Processing, Academic Press, San Diego, Calif, USA, 2nd edition, 1998.
16. D. Labat, R. Ababou, and A. Mangin, "Rainfall-runoff relations for karstic springs," Journal of Hydrology, vol. 238, no. 3-4, pp. 149–178, 2000.
17. V. Nourani, M. Komasi, and A. Mano, "A multivariate ANNwavelet approach for rainfall-runoff modeling," Water Resources Management, vol. 23, no. 14, pp. 2877–2894, 2009.
18. S. Riad, J. Mania, L. Bouchaou, and Y. Najjar, "Rainfall-runoff model usingan artificial neural network approach," Mathematical and Computer Modelling, vol. 40, no. 7-8, pp. 839–846, 2004.
19. Talei, L. H. C. Chua, and T. S. W. Wong, "Evaluation of rainfall and discharge inputs used by Adaptive Network-based Fuzzy Inference Systems (ANFIS) in rainfall-runoff modeling," Journal of Hydrology, vol. 391, no. 3-4, pp. 248–262, 2010.
20. T. J. Ross, Fuzzy Logic with Engineering Application, McGraw Hill, New York, NY, USA, 3rd edition, 1995.
21. J. S. R. Jang, C. T. Sun, and E. Mizutani, Neuro-Fuzzy and Soft Computing: A Computational Approach to Learning and Machine Intelligence, Prentice Hall, Englewood Cliffs, NJ, USA, 1997.
22. C. T. Cheng, J.-Y. Lin, Y.-G. Sun, and K. Chau, "Long-term prediction of discharges in manwan hydropower using adaptivenetwork-based fuzzy inference systems models," in Advances in Natural Computation, vol. 3612 of Lecture Notes in Computer Science, pp. 1152–1161, Springer, Berlin, Germany, 2005.
23. [23] M. Brown and C. Harris, Neuro-Fuzzy Adaptive Modeling and Control, Prentice-Hall, New York, NY, USA, 1994.
24. T. Rajaee, S. A. Mirbagheri, M. Zounemat-Kermani, and V. Nourani, "Daily suspended sediment concentration simulation using ANN and neuro-fuzzy models," Science of the Total Environment, vol. 407, no. 17, pp. 4916–4927, 2009.
25. M. Aqil, I. Kita, A. Yano, and S. Nishiyama, "A comparative study of artificial neural networks and neuro-fuzzy in continuous modeling of the daily and hourly behaviour of runoff," Journal of Hydrology, vol. 337, no. 1-2, pp. 22–34, 2007.
26. J. S. R. Jang, C. T. Sun, and E. Mizutani, Neurofuzzy and Soft Computing: A Computational Approach to Learning and Machine Intelligence, Prentice Hall, Englewood Cliffs, NJ, USA, 1995.
27. O. Kis ¨ ̧i, "Evolutionary fuzzy models for river suspended sediment concentration estimation," Journal of Hydrology, vol. 372, no. 1–4, pp. 68–79, 2009.
28. P. C. Nayak, K. P. Sudheer, D. M. Rangan, and K. S. Ramasastri, "A neuro-fuzzy computing technique for modeling hydrological time series," Journal of Hydrology, vol. 291, no. 1-2, pp. 52–66, 2004.

CHAPTER 18

THE EFFECT OF TRAINING IMAGE AND SECONDARY DATA INTEGRATION WITH MULTIPLE-POINT GEOSTATISTICS IN GROUNDWATER MODELLING

X. L. He, T. O. Sonnenborg, F. Jørgensen, and K. H. Jensen

INTRODUCTION

Aquifer heterogeneity is one of the severe challenges in groundwater flow simulation and with limited number of observations it is always difficult to depict the complete subsurface geology. Hence, statistical methods are often used to estimate geological heterogeneity. Various geostatistical methods have been developed, including variogrambased techniques (Delhomme, 1979; Deutsch and Journel, 1992; Wingle and Poeter, 1993; Johnson, 1995; Klise et al., 2009), the transition probability-based method (Carle and Fogg, 1996), object-based modelling (Deutsch and Wang, 1996) and the multiple-point geostatistical approach (MPS) (Journel, 1993; Guardiano and Srivastava, 1993; Strebelle, 2002). Most of these methods require observations to draw correlation, and some even have the ability to integrate multiple sources of observations. With the development in geophysical technology, high-resolution geophysical mapping techniques are now available, such as magnetic resonance sounding (MRS) (Lubczynski and Roy, 2003), airborne electromagnetic system SkyTEM (Sørensen and Auken, 2004), satellite remote sensing (Hoffmann, 2005), and ground penetrating radar (GPR) (Clement and Ward, 2008). Bourges et al. (2012) illustrated different ways of applying gravity data, refraction seismic data and borehole data with geostatistical methods. However, the proper method to incorporate these data into a geostatistical simulation is still a subject of active research. The theory of multiple-point geostatistics has been developed over the last two decades. An important development was the pixel-based single normal equation simulation algorithm (SNESIM) proposed by Strebelle (2002), which allowed for simulations with reasonable computational power. The primary advantage of MPS is its capability to capture multiple-point-based structure information instead of using two-point-based statistics (variogram) (Journel, 2005). The database from which the structural information is retrieved is referred to as a training image (TI). Comunian et al. (2011) pointed out that a 3-D (three-dimensional) TI is necessary for 3-D MPS simulation, but it is not trivial to generate a 3-D TI since geological observations generally only provide 2-D information. Hence, 3-D applications are one of the most important challenges for MPS (Huysmans and Dassargures, 2009). Okabe and Blunt (2005) as well as Okabe and Blunt (2007) generated 3-D TI realizations by applying information from lateral 2-D images on orthogonal directions. Coz et al. (2011) built a 3-D TI by successively replicating a single 2-D TI. These "copy and paste" methods are simple but certain assumptions need to be invoked. Other methods include complicated statistical simulations such as Comunian et al. (2011) and Comunian et al. (2012). They used probability aggregation approaches to retrieve statistical information from 2-D TIs which subsequently were applied to simulate a 3-D TI. Maharaja (2008) proposed a simple object-based algorithm, TiGenerator, to generate parametric images. However, an image purely generated by stochastic methods lacks evidence from geological observations, and its application in MPS can be questioned. Huysmans and Dassargures (2009) concluded that the sensitivity of the model predictions to the training image is

Citation: X. L. He, T. O. Sonnenborg, F. Jørgensen, and K. H. Jensen, The effect of training image and secondary data integration with multiple-point geostatistics in groundwater modelling, doi:10.5194/hess-18-2943-2014.

an interesting topic for further research. Another advantage of MPS is the ability to incorporate multiple sources of data (Liu et al., 2005; Strebelle, 2006; Hu and Chugunova, 2008). With the flourishing of new measurement techniques, geological observations with relatively high resolution and accuracy are available, and integrating them into stochastic simulations is appealing. Liu et al. (2004) demonstrated how the integration of seismic data reduces the uncertainty in geofacies simulation. Other studies (Strebelle et al., 2002; Huysmans and Dassargues, 2012) also applied soft data conditioning in MPS, but the specific effect of soft data conditioning on the groundwater flow regime has rarely been studied. The objective of this study is to evaluate the sensitivity of different training images and soft data in MPS and to explore the extent to which this sensitivity propagates in groundwater flow simulations. Four scenarios of MPS simulations are designed with different combination of TI and soft data to provide suggestions regarding the applicability of MPS. The simulated geological models are incorporated into a steady state groundwater model and each model is calibrated against field observations. The results are analysed based on estimated hydraulic conductivity, simulated hydraulic head and stochastic capture zone.

Study area and data

The study area covers a 14.5 km by 13.9 km region located near the town of Ølgod in western Denmark (Fig. 1). This region is dominated by arable land with inland marsh around seven streams. The land surface elevation in this area reaches to about 64 m above mean sea level (a.s.l.) in the northwestern part, and decreases to around 17 m a.s.l. in the southeastern part. The climate is characterized by mean temperatures ranging from 1.4 °C in January to 16.5 °C in August with an annual average around 8.2 °C. Precipitation is highest in autumn and winter while spring is relatively dry, the average annual precipitation is approximately 1050 mm year^{-1} (Stisen et al., 2011). According to the National Water Resources Model (DK-model; Henriksen et al., 2003), the annual average groundwater recharge is 611 mm year^{-1} .Water consumption primarily relies on groundwater abstraction. In the Danish national geological JUPITER database (http: //www.geus.dk/jupiter/index-dk.htm), there are 165 pumping wells in this area with a total mean abstraction of 3.2 × 10^{6} m^{3} year^{-1} in the period from 2000 to 2010

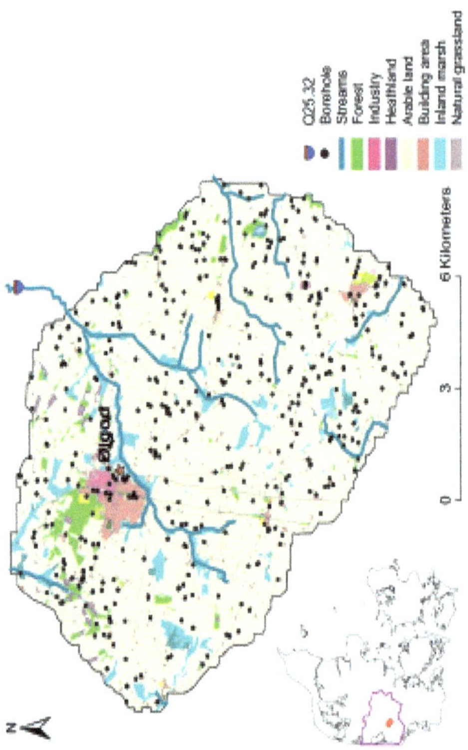

Figure 1. Location and land use of the study area. Black points denote the 525 boreholes whose logs were used as hard data in the geostatistical simulations.

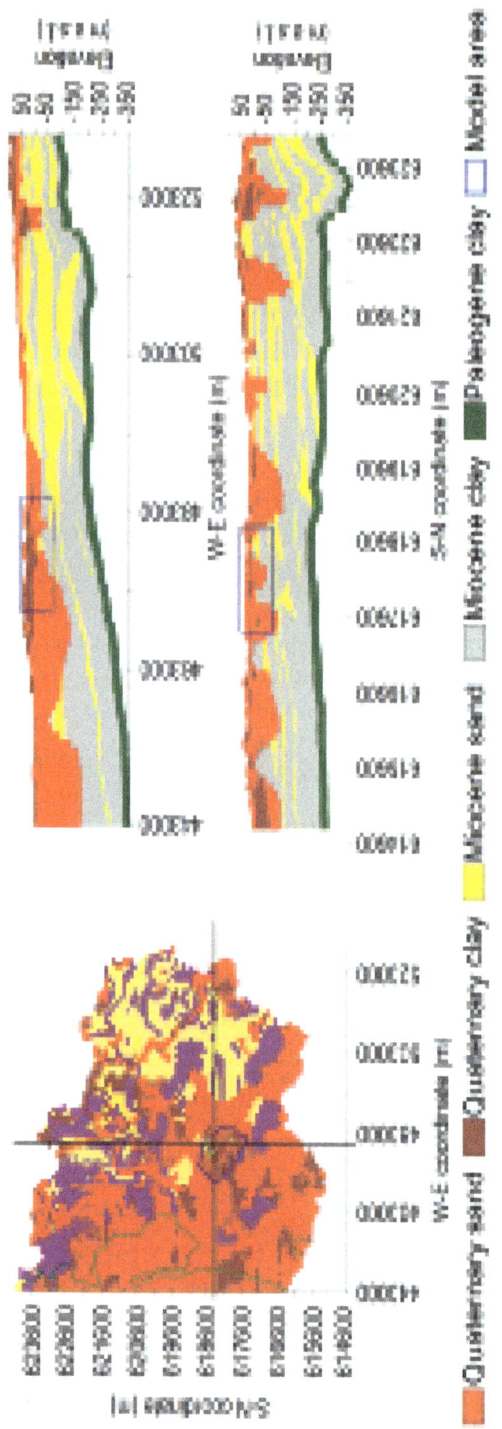

Figure 2. Regional geological structure, model area in this study is marked with a blue frame. The location is marked by the purple polygon in Fig. 1, and the horizontal plane is taken at 10 m a.s.l. (modified after Scharling et al., 2009).

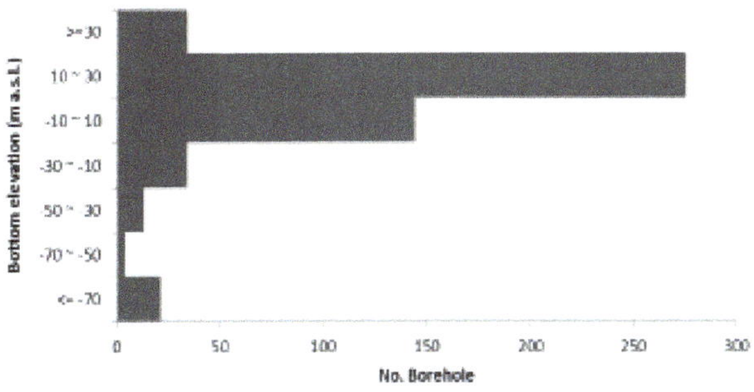

Figure 3. Bottom elevation of boreholes in the model area. In total there are 525 boreholes, but only 22 boreholes are deeper than −70 m a.s.l.

Figure 2 shows the regional geology of the study area as described by Scharling et al. (2009) based on hydrostratigraphic modelling. This regional-scale model was primarily intended to establish the Miocene successions in western Denmark, while less effort was devoted to the modelling of the Quaternary sediments. This model indicates that below elevations of −70 m a.s.l., the geology in the study area is dominated by Miocene sediments. Intensive geological mapping in the study area including borehole logs, seismic surveys and airborne transient electromagnetic (SkyTEM) investigations show that the geology at elevations above −70 m a.s.l. is dominated by highly heterogeneous Quaternary sediments with variable thickness (Høyer et al., 2011). Below, Miocene deposits are located with a thickness of up to about 150 m followed by a Palaeogene clay at the bottom (Rasmussen et al., 2010). Accordingly, the geological settings have been conceptualized into five units: Quaternary sand, Quaternary clay, Miocene sand, Miocene clay and Palaeogene clay. The JUPITER database holds lithology logs of 525 boreholes in the study area (Fig. 1), but most of the boreholes are relatively shallow and only 22 boreholes reach deeper than −70 m a.s.l. (Fig. 3). Therefore, the geological analysis and modelling were performed on Quaternary sediments from the surface down to −70 m a.s.l. Another important source of information is SkyTEM data (Høyer et al., 2011). The subsurface electrical resistivity data were collected with line spacing from 125 to 270 m, and the soundings penetrated more than 200 m down. The resistivity data have been filtered and inverted using a 19-layer "smooth model", resulting in a 3-D resistivity data set with cell size of 100 m × 100 m × 5 m (Høyer et al., 2011). With its high resolution, SkyTEM data are ideal as soft probability data or training image for MPS simulation.

METHODOLOGY

Multiple-point geostatistics

MPS was first presented as a direct algorithm in stochastic simulation by Guardiano and Srivastava (1993). Later, Strebelle (2002) introduced the SNESIM algorithm which combines the flexibility of the pixel-based algorithm and the ability to reproduce crisp shapes of the object-based algorithm.

The critical information of sequential simulation is the conditional probability distribution function (cpdf), and in the SNESIM algorithm the cpdf of a random variable S (U) is solved by the following equation:

$$P(S(U) = S_k|d_n) \cong \frac{c_k(d_n)}{c(d_n)} k = 1, \ldots, K \tag{1}$$

where U denotes an arbitrary location, and k denotes a certain class of lithology, which can have K categories in total. dn denotes the data event with n data points included and is obtained by scanning the TI with a search template consisting of n+1nodes centred at location U. c (d_n) denotes the number of replicates of a certain conditioning data event d_n, and ck (dn) denotes the number of replicates in c (d_n) where the central node U has the random variable S (U) = Sk. Therefore, Eq. (1) implies that the probability of state S_k to occur at location U with conditioning data event d_n equals to the proportion c_k (d_n)/c (d_n) from a training image. The solution is achieved by scanning a training image. A training image is a conceptual 2-D or 3-D map which depicts the expected structure and pattern of facies (Strebelle, 2002).

In the practice of stochastic simulation, the cpdf is integrated with the sequential simulation paradigm (Goovaerts, 1997, p. 376). The hard data is first assigned to the closest grid nodes, and all the unsampled nodes are visited once and only once in a random path. At each unsampled node U, the recorded cpdf corresponding to an actually present hard conditioning data event is retrieved and used to draw the simulated value S (U). The software package SGeMS (Stanford Geostatistical Modelling Software) (Remy et al., 2009) was applied in this study. The MPS simulation was carried out using the simulation package snesim_std provided by SGeMS. The simulation included two categories, Quaternary sand and Quaternary clay, with target proportions set to 0.67 and 0.33, respectively, as indicated by borehole data. As suggested by Liu (2006), the dimension parameter of the search template was based on the primary understanding of the geological structure, which can be indicated by TI1 (Table 1). The ellipsoid search template has 240 nodes, and the geometry was set to 2000, 2000 and 60 m for maximum, mean and minimum separately, without any rotation or affinity.

The Software Package

SGeMS (Stanford Geostatistical Modelling Software) (Remy et al., 2009) was applied in this study. The MPS simulation was carried out using the simulation package snesim_std provided by SGeMS. The simulation included two categories, Quaternary sand and Quaternary clay, with target proportions set to 0.67 and 0.33, respectively, as indicated by borehole data. As suggested by Liu (2006), the dimension parameter of the search template was based on the primary understanding of the geological structure, which can be indicated by TI1 (Table 1). The ellipsoid search template has 240 nodes, and the geometry was set to 2000, 2000 and 60 m for maximum, mean and minimum separately, without any rotation or affinity.

Soft data conditioning Soft data or secondary data provide indirect information on the distribution of geological facies. Typical soft data include geophysical data such as seismic data, spaceborne geodetic observations, and airborne electromagnetic data. To be integrated in SNESIM, soft data first have to be converted into facies probability data (Strebelle, 2006). The cpdf based on both hard and soft data, P (A|B,C), where "B" indicates the hard data set and "C" indicates the soft data set, is solved by deriving a Bayesian-based model of integrating P (A|B) and P (A|C) (Journel, 2002). In Bayesian statistical terms, P (A) is the facies global proportion or prior probability. And in this case, the notation $P(S(U) = S_k|d_n)$ in Eq. (1) can be simplified as P (A|B). P (A|C) denotes the cpdf derived from soft data. The cpdf integrating both hard and soft data is updated as:

$$P(A|B, C) = \frac{a^{\tau}}{a^{\tau} + b * c^{\tau}} \in [0, 1] \quad (2)$$

$$a = \frac{1 - P(A)}{P(A)} \quad (3)$$

$$b = \frac{1 - P(A|B)}{P(A|B)} \quad (4)$$

$$c = \frac{1 - P(A|C)}{P(A|C)}. \quad (5)$$

The parameter τ is used to adjust the contribution from soft information "C". $\tau = 1$ indicates that the two data sets "B" and "C" are independent. For $\tau = 0$ soft information is ignored, while for $\tau > 1$ the influence of soft data "C" is increased, and it is decreased for $\tau < 1$ (Journel, 2002). The sensitivity of parameter τ in SNESIM was analysed by Liu (2006), and the choice of parameter values for τ in this study is based on their results.

SkyTEM data to soft probability

A supervised technique (Liu et al., 2005) was applied to retrieve probabilistic information from SkyTEM data. The resistivity data was converted to facies probability data by correlating the facies occurrence in the borehole lithological logs with SkyTEM data. The borehole logs were subdivided into 5 m intervals in correspondence with the discretization of the SkyTEM data. The lithological logs were categorized as sand or clay and the corresponding resistivity value was recorded. Subsequently, for each resistivity bin (size $1\Omega\cdot m$) the sand probability based on borehole data was computed and plotted against the corresponding resistivity (Fig. 4). The data points were fitted to a truncated polynomial non-linear regression function of resistivity (green line).

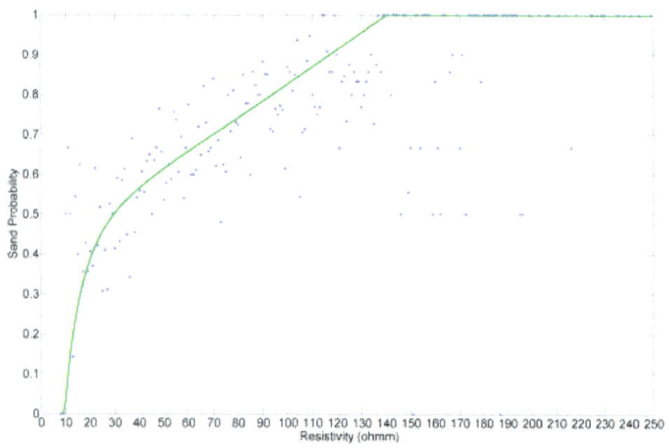

Figure 4. Non-linear regression of sand occurrence against SkyTEM data. Blue points are sand occurrence from borehole data compared to resistivity. The green line is the regression function (Eq. 6), coefficient of determination is 0.80 and RMSE is 0.08.

Due to data noise, this least-squares curve fitting was only performed for resistivities between 10 and 140 $\Omega\cdot$m. The coefficient of determination is 0.80 and the root mean square error (RMSE) is 0.08, which indicates that the function represents the correspondence between resistivity (10–140 $\Omega\cdot$m) and sand probability satisfactorily. Therefore the polynomial regression model was used to convert the 3-D SkyTEM data (10–140 $\Omega\cdot$m) into a 3-D sand probability map. For resistivities lower than 10 $\Omega \cdot$ m there are only two data points both with sand occurrences of 0, thus sand occurrence was set to 0 below this value. For resistivities higher than 140 $\Omega \cdot$ m, 80 % of data points show sand occurrence at 1, thus the corresponding sand occurrence was set to 1. The membership function is therefore expressed as

$$P_s = \begin{cases} 1 & \text{if } R > 140\Omega\cdot m \\ 0.0863 & \times (\ln R)^3 - 0.958 \times (\ln R)^2 + 3.759 \\ & \times (\ln R) - 4.596 \\ & \text{if } 10\Omega\cdot m \ll R \ll 140\Omega\cdot m \\ 0 & \text{if } R < 10\Omega\cdot m \end{cases} \quad (6)$$

where R is the resistivity. Figure 5 shows Quaternary sand probability distributions as derived from Eq. (6) for lateral and vertical cross sections.

Training image

Construction of 3-D TI is challenging since geological patterns are usually described in 1-D or 2-D. In this study two kinds of 3-D training images were generated, denoted as TI1 and TI2, respectively. TI1 was directly converted from SkyTEM data. Based on the resistivity the subsurface was divided into Quaternary sand and Quaternary clay by using a critical resistivity value. According to the 525 borehole logs, the proportion of Quaternary clay is 0.33. This proportion is assumed to be representative of the study area, although it could be slightly biased due to the uneven distribution of boreholes. Figure 6 shows the cumulative distribution function (CDF) of resistivity data. The Quaternary clay proportion of 0.33 corresponds to 41.6 Ω m on the CDF curve. Thus, for resistivities below this value the sediment is categorized as Quaternary clay, while for resistivities above this value the sediment is categorized as Quaternary sand. Figure 7 (left) illustrates cross sections of TI1. TI2 was generated by the TiGenerator (Maharaja, 2008), which is a utility package in SGeMS. The TiGenerator (Maharaja, 2008) provides a method for generating 3-D training images with parametric shapes using non-iterative, unconditional Boolean simulation. The user-defined geometry and orientation of simulated objects can be deterministically or statistically described. With the TiGenerator in SGeMS version 2.1, geobodies with sinusoid, ellipsoid, half-ellipsoid and cuboid shapes can be defined by given geometric parameters such as maximum, median, and minimum radius. The

Table 1. The mean length distribution in X, Y , and Z directions and proportion of Quaternary clay in TI1 and TI2.

	X (m)			*Y* (m)			*Z* (m)			Clay proportion
	Min.	Mean	Max.	Min.	Mean	Max.	Min.	Mean	Max.	
TI1	50	400	2950	50	400	3200	2.5	23	60	0.33
TI2	50	525	2850	50	454	3300	2.5	21	65	0.33

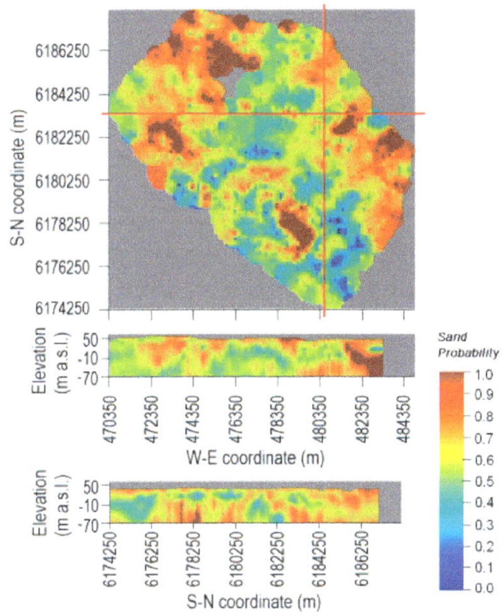

Figure 5: Cross sections of the 3-D map showing the probability of Quaternary sand converted 3 from SkyTEM data. The horizontal plane is taken at 10 m a.s.l..

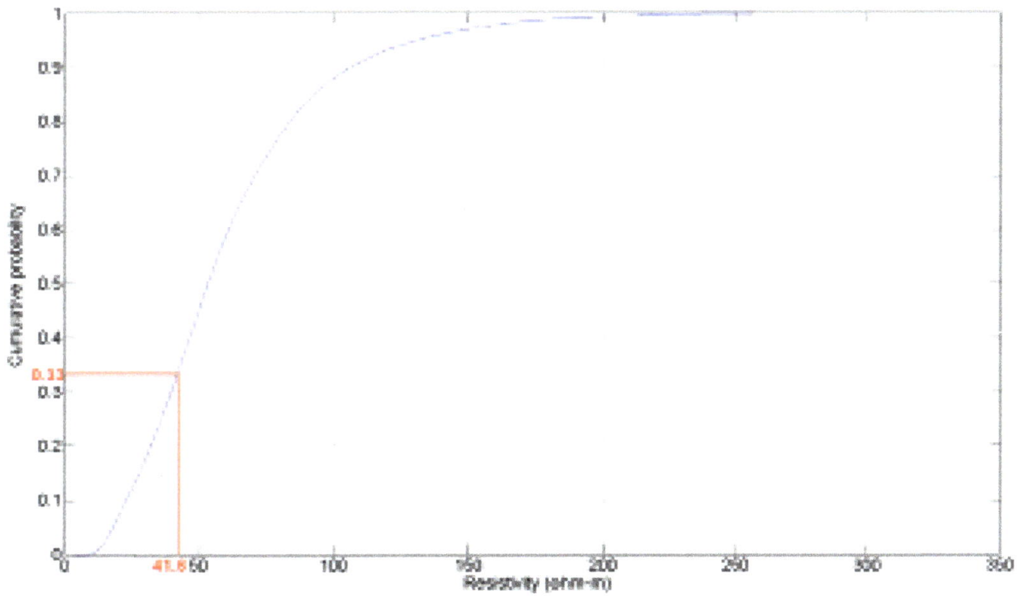

Figure 6. The CDF of SkyTEM data. Resistivity of 41.6 Ω m corresponds to a proportion of 0.33, which is the proportion of Quaternary clay according to borehole data.

geobody is drawn by following a random path until the facies proportion is fulfilled in the simulated grid. The sinusoid shape corresponds to features such as meandering or curvilinear channels, while the cuboid is rectangular. According to the patterns in TI1, the clay lenses are neither channelized nor strongly rectangular, therefore the ellipsoid shape was chosen for the simulation. In TiGenerator the Quaternary sand was used as background facies and the Quaternary clay bodies were simulated with ellipsoid lenses. The target proportion of clay was set to 0.33 as in TI1, and the geometric parameters were determined through the inversion aiming at minimizing the difference between simulated structure and the structure in TI1. The target could also be the information retrieved from borehole data, but in this study the 3-D TI1 depicts the structure in better detail. Therefore, the information on mean length from TI1 was used in the objective function:

$$R = \sqrt{\frac{\sum_{i=1}^{j} (E - \tilde{E})^2}{j}} \tag{7}$$

where E represents the mean length (minimum, mean and maximum) of clay bodies in X, Y and Z directions from the realizations, and E represents the corresponding item from ˜ TI1. j is the total number of items being included and equals 9 in this case. The optimization was based on 20 realizations and the best realization was selected as TI2. The distribution of clay body mean length in TI1 and TI2 is shown in

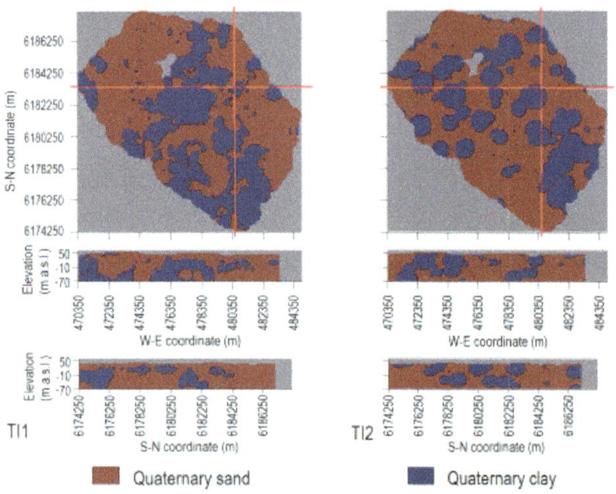

Figure 7. Cross sections of TI1 and TI2. TI1 is converted from SkyTEM data while TI2 is generated by TiGenerator. The horizontal plane is taken at 10 m a.s.l

Table 1. It shows that both TIs have the same clay proportion and minimum mean length of clay bodies in all directions. While it is difficult and time consuming to match the mean and the maximum length in all three directions, the optimized TI2 has a mean length of 525, 454 and 21 m in X, Y and Z directions, and the corresponding values from TI1 are 400, 400 and 23 m. For the maximum mean length, the difference between TI2 and TI1 is 3 % in X and Y directions, and 8 % in the Z direction, which indicates that the simulated clay body is in the same size range as that from TI1. Figure 7 (right) shows the cross sections of TI2. Compared to TI1 (Fig. 7, left), the shapes of the clay bodies are more homogeneous and ellipsoidal than the TI1 clay lenses where the size and shape of clay bodies are more heterogeneous. However, the distribution of clay lenses given by TI2 capture the general trends displayed by TI1 fairly well. Both in the horizontal and vertical direction TI2 shows a tendency of clustering, which is also found in TI1.

Groundwater model

The Groundwater Modelling System (GMS) with the groundwater modelling code MODFLOW-2000 (Harbaugh et al., 2000) was used to assess the effect of the geological structures on the groundwater flow regime, explicitly on simulated hydraulic head. The model area is characterized by high elevations in the central part with several streams flowing towards the model border, and thus no natural hydraulic boundaries could be identified. Instead, a fixed hydraulic head boundary interpolated from observations was employed. The model bottom is Palaeogene clay which serves as no flow boundary. In order to resemble the finely discretized geological model (100 m × 100 m × 5 m), the Hydrogeologic-Unit Flow (HUF) package (Evan and

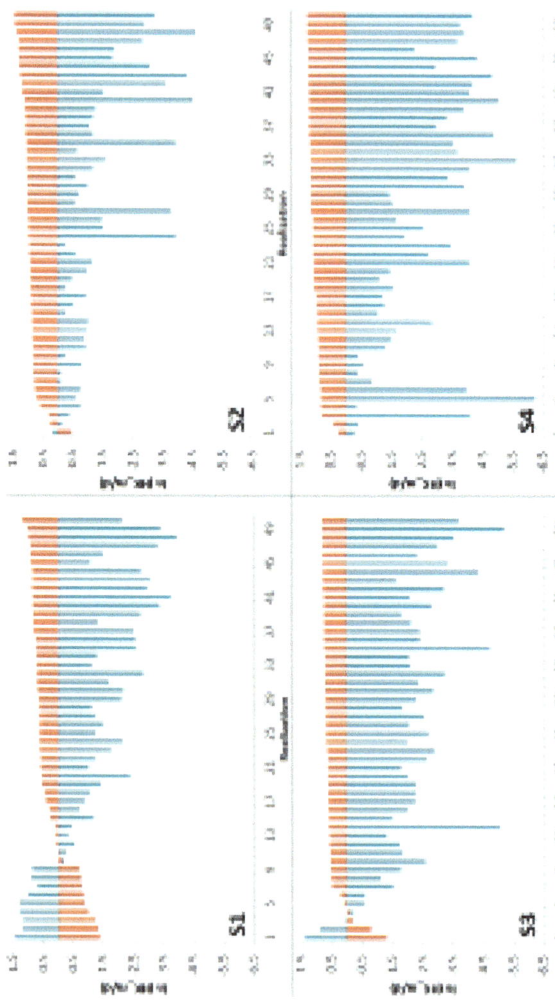

Figure 8. Estimated hydraulic conductivity of Quaternary sand and Quaternary clay using PEST. The realizations are ranked according to the hydraulic conductivity of the Quaternary sand.

Mary, 2000) was used. The groundwater model was discretized into 63 layers with cells of 100 m × 100 m horizontally, resulting in a total of 792 603 active cells. A constant layer thickness of 5 m is used from layer 6 to 63, and to avoid dry cells the top layer is 13 m thick on average (including the unsaturated zone), while the thickness of layer 2 to 5 is 4 m on average. Groundwater recharge from the DK-model (Henriksen et al., 2003) was applied, resulting in an average recharge rate of 611 mm year^{-1} in the model area. Groundwater discharges to river, drain and abstraction wells, accordingly the RIV (River), DRT (Drain Return) and WEL (Well) packages were enabled in GMS. The nearest stream discharge station is Q25.32, located to the northeast of the model area (Fig. 1). Based on 16-year data (1990– 2005), the average discharge which is scaled to the model area is 59 900 m3 d^{-1}. A drain was specified for the entire model area with drain level set to 1 m below terrain, and the drain time constant (1.7×10^{-2} s^{-1}) was taken from the DK-model (Henriksen et al., 2003). There are 165 abstraction wells in the model area with a total pumping rate of 8900 m^3 d^{-1} (JUPITER). The groundwater model was calibrated against 219 observations of hydraulic head and the river discharge to station Q25.32 using the inversion code PEST (Doherty, 2005). Based on the calibrated flow model, the particle tracking post-processing code MODPATH (Pollock, 1994) was used to simulate the probabilistic groundwater capture zones. Three wells with screens in Quaternary sediments are included in this simulation (Figs. 9–11). Well 1 is located to the south east and filtered at a depth of −2.2 m a.s.l., Well 2 and Well 3 are both located in the central part where higher land surface elevations are found, Well 2 is filtered at −47.5 m a.s.l. and Well 3 is filtered at −52.5 m a.s.l. A total of 100 particles were assigned to each well and the backward particle tracking was employed.

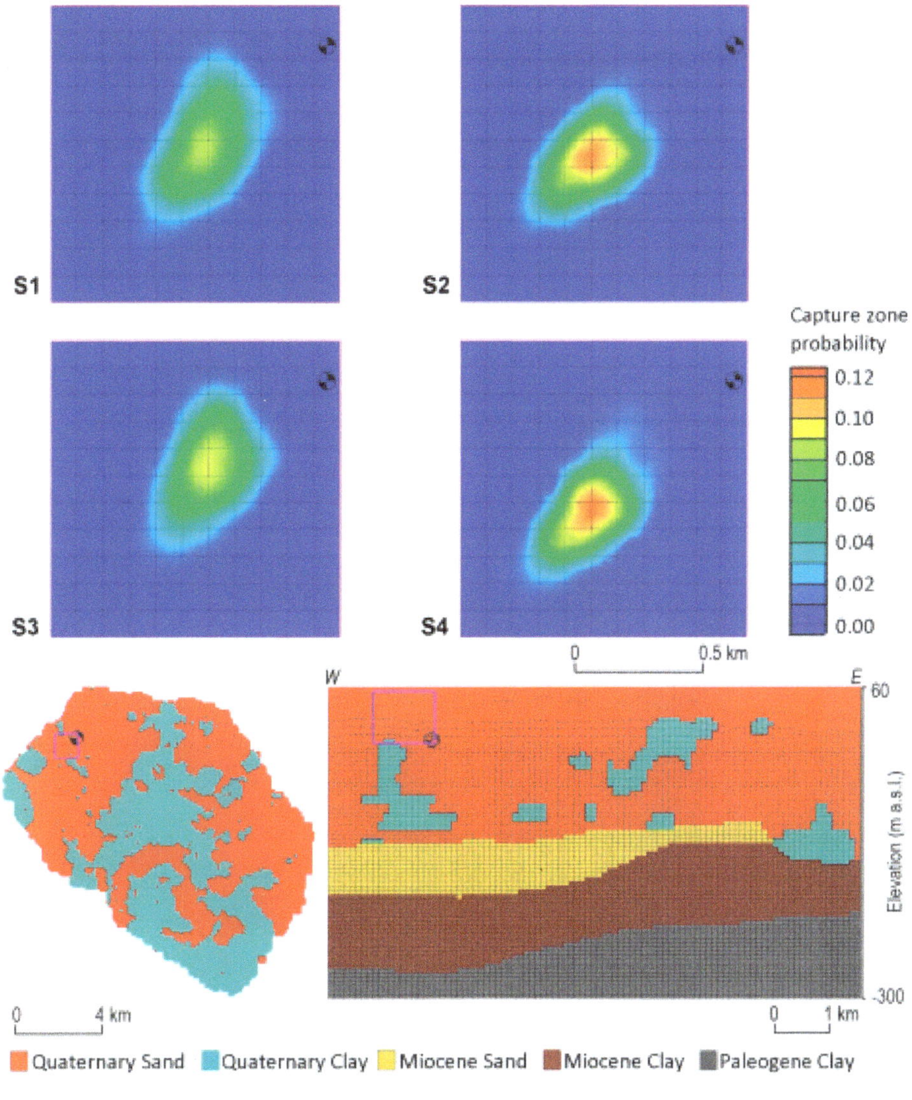

Figure 9. The probabilistic capture zone of Well 1 from each scenario. The well location is illustrated in relation to the geological structure of the TI1.

Ensemble analysis

To test the effect of the training image and soft conditioning on geostatistical realizations, multiple-point geostatistical simulations were carried out by applying four different combinations of TIs and soft conditioning (Table 2). The first two scenarios (S1 and S2) were simulated with TI1. Soft conditioning was applied in S2 but not in S1. The other two scenarios (S3 and S4) were simulated with TI2. Soft conditioning was applied in S4 but not in S3. According to He et al. (2013) the standard deviation on simulated hydraulic head, based on multiple realizations of the geology, converges to a fixed value as more models are accumulated. A stable value is approached after 30 realizations and in this study 50 realizations were therefore generated in each scenario and subsequently anchored to the steady state groundwater model.

RESULT AND DISCUSSION

Model calibration and groundwater head simulations

With 50 realizations in each scenario, the four scenarios amount to 200 groundwater models. Each model was calibrated against hydraulic head measurements using the inversion code PEST (Doherty, 2005). Initially, a sensitivity analy

Table 2. Four scenarios of geological realizations used for the multiple-point geostatistical simulations.

Scenario	Training image	Soft conditioning
S1	TI1	No
S2	TI1	Yes
S3	TI2	No
S4	TI2	Yes

Table 3. RMSE of simulated hydraulic head against observations. The table lists mean, μRMSE, and standard deviation, σRMSE, over 50 realizations (where each RMSE value is based on 219 observation wells).

Scenario	μ_{RMSE} (m)	σ_{RMSE} (m)
S1	2.6	2.2
S2	2.6	2.1
S3	2.6	2.2
S4	2.6	2.2

sis was carried out to identify the parameters to include in the calibration. Based on this analysis the hydraulic conductivities of Quaternary sand, Quaternary clay and Miocene sand were selected as calibration parameters. The RMSE of simulated hydraulic heads of the calibrated models were used as indicators of model performance. In Table 3 the mean, μRMSE, and the standard deviation, σRMSE, of the RMSE of the 50 models in each scenario are listed. The μRMSE is 2.6 m in all four scenarios while the σRMSE is 2.2 m in S1, S3 and S4 and is 2.1 in S2. This implies that models of each scenario are equally well calibrated. The estimated hydraulic conductivity values of the Quaternary units are shown for each scenario in Fig. 8, while the mean and standard deviation of the estimated values are listed in Table 4. The estimated hydraulic conductivities for Quaternary sand fall in the interval 0.2–4.5 m d^{-1}. These are relatively small values for a sandy aquifer material, especially when the lower range of this interval is considered. However, values in the high range of the interval may be within a realistic range for this area (Harrar et al., 2003). A more critical issue is the relative magnitude of Quaternary sand and clay. The hydraulic conductivity of sandy materials should be higher than the corresponding clay material and if this is not the case, it could be an indication of errors in the model structure. In S1 there are nine cases (Fig. 8) where the estimated hydraulic conductivity of clay is higher than the hydraulic conductivity of sand, and there are three such cases in S3.

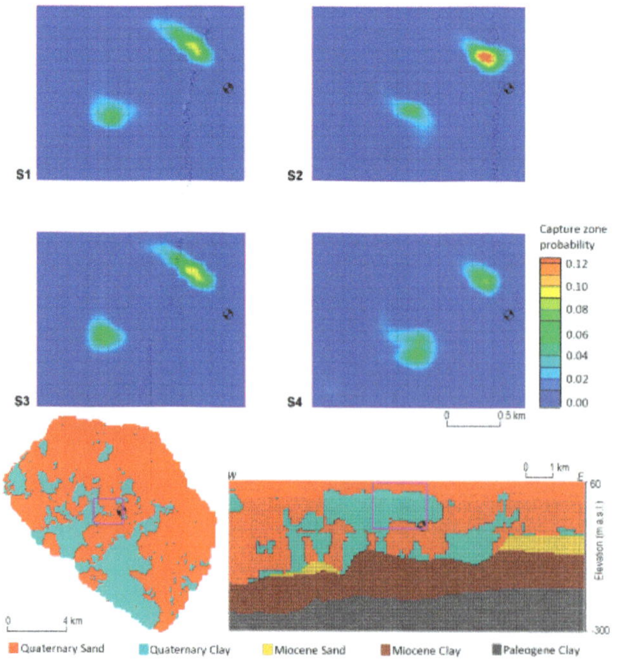

Figure 10. The probabilistic capture zone of Well 2 from each scenario. The well location is illustrated in relation to the geological structure of the TI1.

However, this problem appears only once in S2 and is not found in S4. Hence, around 18 and 6 % of the realizations of S1 and S3, respectively, show indications of errors in the geological model, while this is only observed in 2 and 0 % of the cases in S2 and S4.

Table 4. Estimated hydraulic conductivity of Quaternary sand and Quaternary clay using PEST. The values listed are mean and standard deviations σ of estimated values over 50 realizations of each scenario (unit: m d^{-1}).

Scenario	Quaternary sand		Quaternary clay	
	Mean	σ	Mean	σ
S1	1.7	0.8	0.8	1.1
S2	2.7	0.7	0.4	0.3
S3	1.8	0.4	0.3	0.6
S4	2.9	0.5	0.2	0.2

In addition, the ratio of hydraulic conductivity between sand and clay material is also an indication of the quality of the estimated geological structure where a high contrast would be expected to be the most reasonable result. The mean horizontal hydraulic conductivities of Quaternary sand and clay are listed in Table 4, and it shows that the largest ratio is found for S4 (ratio of 14.5; 2.9–0.2 m d^{-1}), and the second largest ratio is estimated for S2 (ratio of 6.8; 2.7– 0.4 m d^{-1}), while the lowest ratio is found for S1 (ratio of 2.1; 1.7–0.8 m d^{-1}). The difference between the hydraulic conductivity of the two units is also visualized in Fig. 8. These results also clearly show the effect of soft data in the simulation of geological architecture as the most sound results are obtained for scenarios S2 and S4. The impact of the training images is seen by comparing the first row (S1 and S2) to the second row (S3 and S4) in

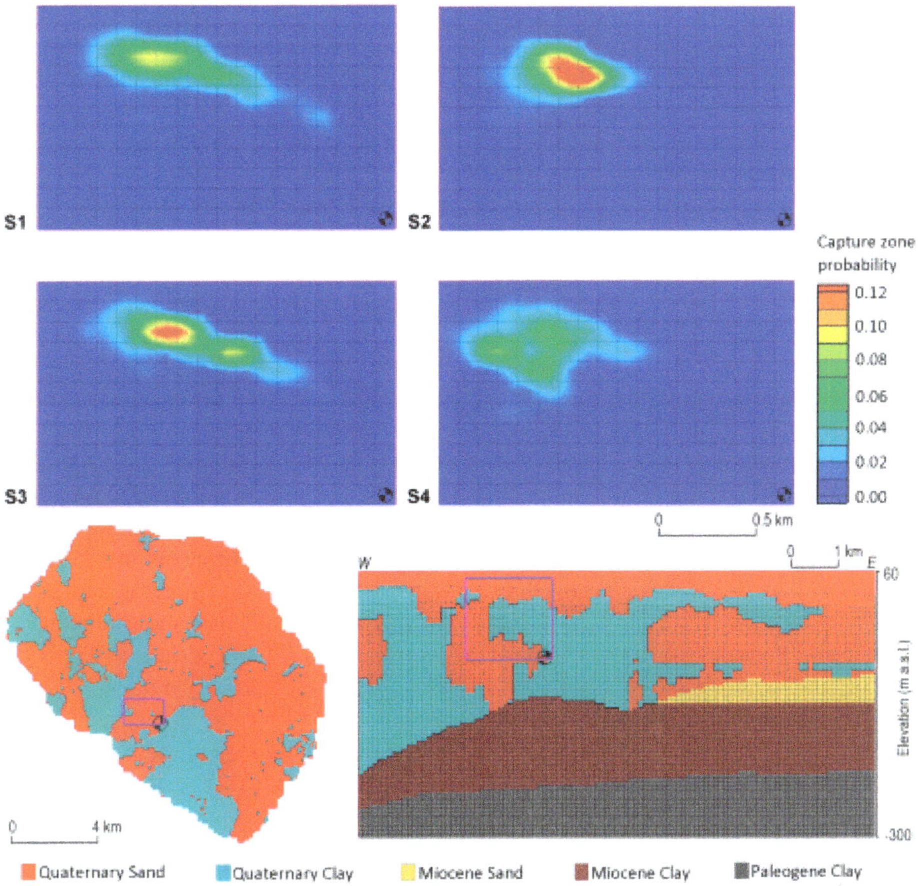

Figure 11. The probabilistic capture zone of Well 3 from each scenario. The well location is illustrated in relation to the geological structure of the TI1.

Fig. 8. Scenarios on the first row use TI1 while the scenarios on the second row use the TiGenerator-based TI2. No apparent statistical discrepancy can be spotted between these two rows, which means that even though TI2 is not directly derived from field data but from a statistical simulation, the TI2-based realizations yield acceptable results. Both with respect to the calibrated groundwater model and the estimated hydraulic parameter values, the choice of training image has a relatively small effect. However, it should be emphasized that TI2 was selected to maximize the similarity between the simulated training image and the field-based training image (TI1), see Table 1.

CAPTURE ZONES

Probabilistic capture zones were simulated for three abstraction wells screened in both shallow and deep layers. Figures 9–11 show the probabilistic capture zone for these three wells. In each figure the first row illustrates the simulated capture zones based on TI1 (S1 and S2), while the second row represents the corresponding simulations based on TI2 (S3 and S4). As mentioned previously, the thickness of the top numerical layer is 13 m on average, which is incoherent with the rest of the layers (5 m thickness on average). The thicker top layer was designed to avoid dry cells, but this could also diminish the variability of simulated capture zones, as the geological heterogeneity has been averaged out. Well 1 (located in layer 4) is more sensitive to this effect than the other two wells, since Well 2 and Well 3 are located in layer 13 and layer 14, respectively, and the geological heterogeneity in all the other layers still has a significant contribution to the variability of simulated capture zones. Generally, there is no distinct trend between the first and second row (Figs. 9–11), which again implies that the simulation is not sensitive to the differences between TI1 and TI2. For all three wells, the area of the capture zone is smaller for S2 and higher probabilities are found compared to S1. However, this is not the case when comparing S4 to S3. In fact, S3 has a more concentrated probabilistic capture zone than that of S4 for Well 2 and Well 3. In the study of He et al. (2013), the simulation of particle travel time was applied on 100 groundwater models with different geological architecture, in combination with a parameter uncertainty analysis of each model. One discovery was that the geological architecture is the most critical factor regarding particle travel time, because it affects the particle travel path. Therefore, the more constrained the probabilistic capture zone is, the less dispersed the path lines are, which is a result of less variation between the structures of the geological realizations. The results illustrated in Figs. 9–11 suggest that there is much less variation among the realizations of S2 than those of S1. This is also indicated by the standard deviation of estimated hydraulic conductivity (Table 4), where the uncertainty of S2 is smaller than that of S1, especially for the hydraulic conductivity of sand. However, such tendency is not seen in simulations using TI2 (S3 and S4). Apparently the soft conditioning results in less variation between the realizations, and it decreases more in the TI1-based simulations than in TI2-based simulations. This is also supported by Fig. 12, where the standard deviation of the 50 realizations of each scenario is illustrated. The standard deviation is computed from the binary geological units (0 for Quaternary sand and 1 for Quaternary clay). Red colour indicates regions with higher standard deviation, corresponding to higher variability between the 50 realizations. It is obvious that by conditioning on soft data (S2 and S4), the area with higher variability decreases dramatically. S2 is the scenario with the smallest area of high standard deviations (red colours) among the four scenarios. The explanation for this is data redundancy. Realizations from S2 have been over conditioned and therefore the uncertainty on the probabilistic capture zone has been constrained much more than any of the other scenarios. The effect of data redundancy is illustrated in Figs. 13 and 14. Figure 13 shows the E-type map (cell-wise arithmetic average) of each scenario. The "E" stands for "expected value" or precisely "conditional expectation" (Remy et al., 2009, p. 37). On the E-type map the cells are more likely to be sand as the colour turns red, and more likely to become clay as the colour turns blue. The cells with dark red or dark blue colours are cells where hard data (borehole data) are located. The effect of soft data can be seen by comparing Fig. 13 to the soft data (sand probability) in Fig. 5. The E-type of S1 and S3 shows almost no similarity to the soft data in Fig. 5, while the E-type of S2 and S4 show patterns that are very similar to the soft data. Figure 14 illustrates the CDF of soft data together with a Q–Q plot of the E-type map from each scenario. In the Q–Q plot the x axis shows the sand probability from the soft data, while the y axis shows the corresponding probability from the E-type map. The black line is the CDF of sand probability based on soft data. The closer the Q–Q plot is to the black line, the more similar the sand probability distribution of the realizations is to the soft data. While the plot for S1 (yellow symbols) is far from the black line, the plot for S2 (blue symbols) is much closer. The red plot (S4) is also closer to the black line than the purple plot (S3). Both scenarios with soft conditioning (blue and red symbols) are much lower than the black line on the low sand probability part, while higher than the black line on the high sand probability part. But for the other scenarios, the plots are always lower than the black line. This shows the effect of soft conditioning. In addition, the curve representing S2 (blue symbols) is higher than for S4 (red symbols), and this can be explained by data dependency. Both TI1 and soft data are derived from SkyTEM data, and although they were processed in different ways, they are not totally independent. In Eq. (2), the parameter τ is used to adjust the dependence between the two data sets B and C (representing the training image and the soft data in the current case). To make S2 and S4 comparable, the parameter τ is specified to 1.0 in both cases. Liu (2006) pointed out that using a value of 1.0 generates the most robust results. In Fig. 14 the plot for S2 (blue) is lower than the black line in the low probability part, while it is located above the black line in the high probability part, indicating that the S2 realizations are over conditioned. In contrast, as TI2 is remotely related to the soft data, the soft conditioning on S4 improves the simulations, but the effect of data redundancy is weaker than that in S2. Therefore, integration of the training image constructed using the TiGenerator (TI2) with the SkyTEM-based soft data is more sound than combining TI1 and soft data that both are based on SkyTEM data. Journel (2002) found that when $\tau < 1$ the influence of additional data is decreased. Therefore, another scenario (S5) was added where TI1 and soft data were used, but the parameter τ is set to 0.5. S5 corresponds to the green plot in Fig. 14, and it is located right between S1 and S2, which illustrates the effect of decreasing τ.

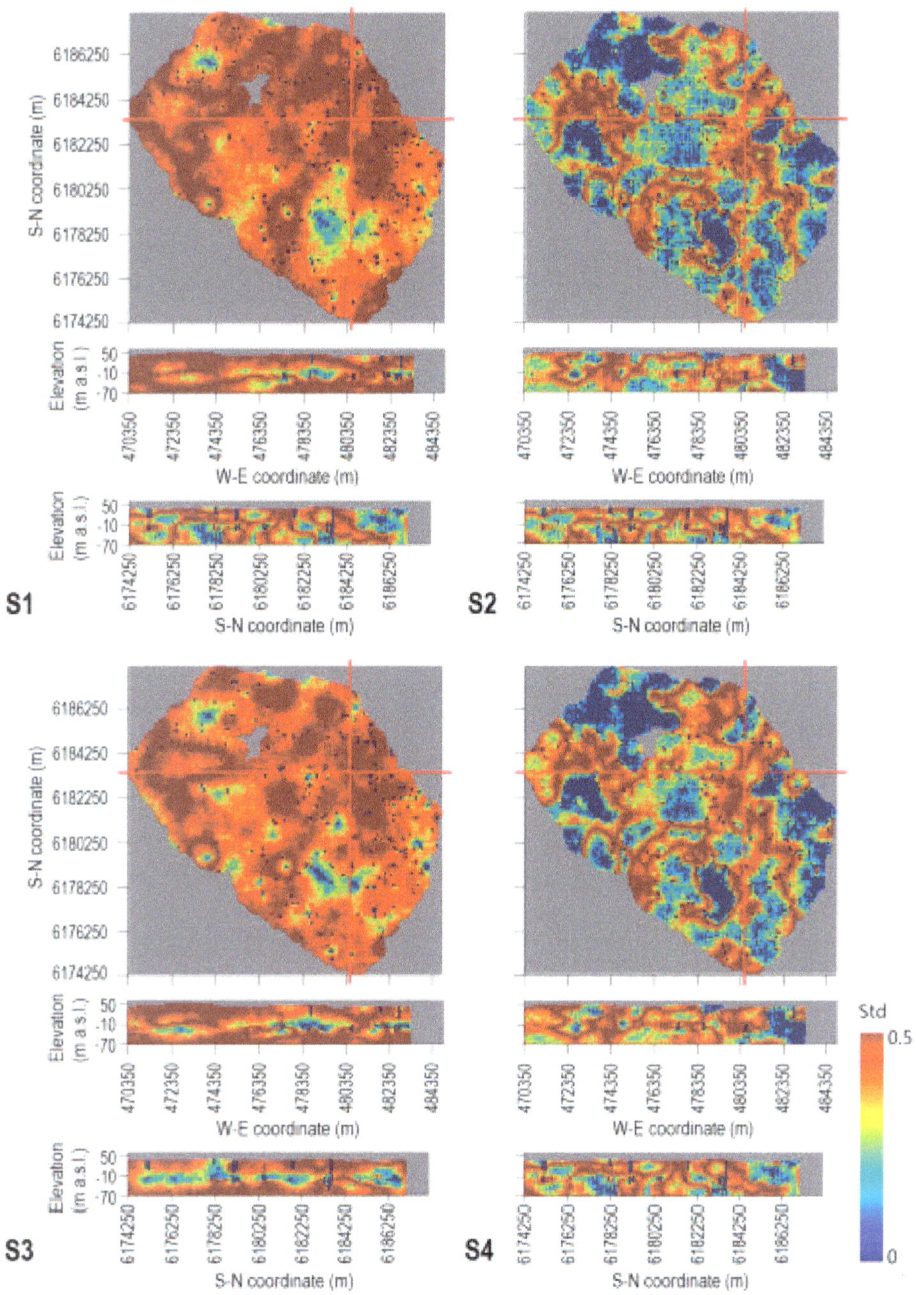

Figure 12: Map of standard deviation over the 50 realizations of each scenario. The computation of standard deviation is based on binary geological units (0 for Quaternary sand and 1 for Quaternary clay). The standard deviations are 0 at cells where the boreholes are located. The horizontal plane is taken at 10 m a.s.l.

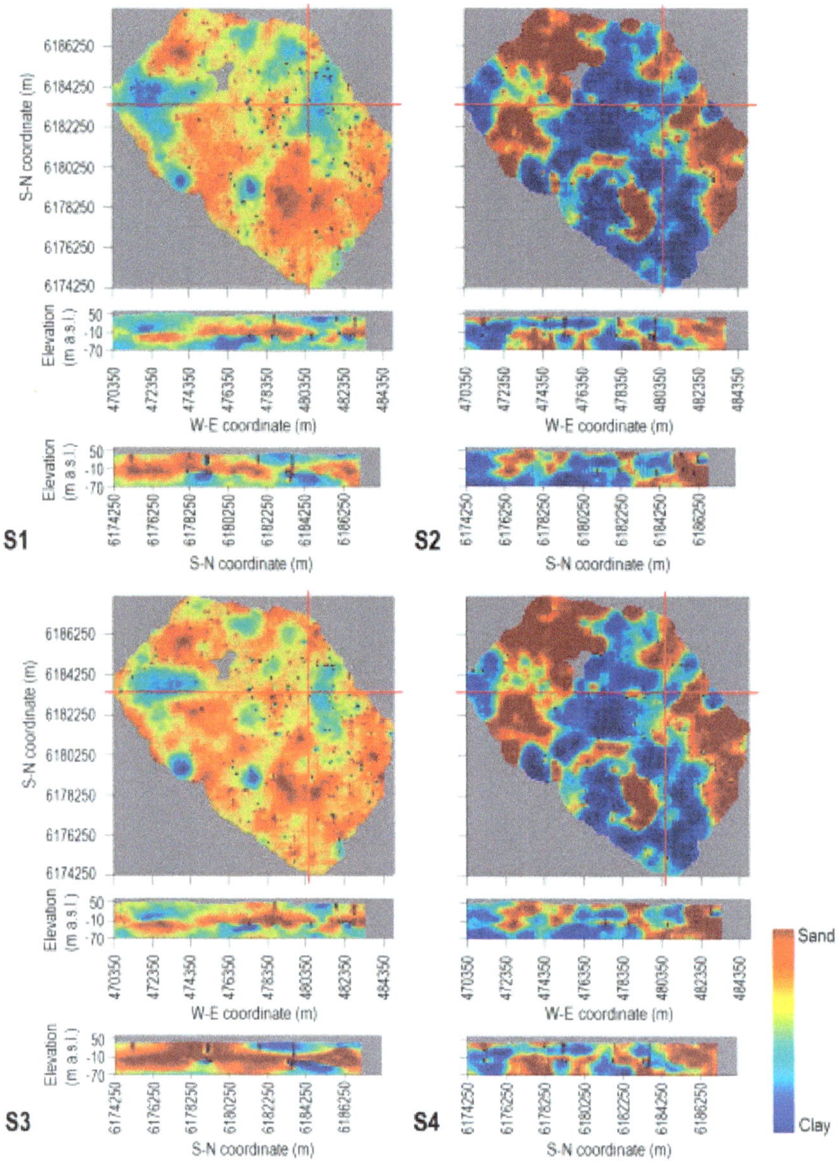

Figure 13: E-type estimation maps generated from 50 realizations of the geology of each scenario. The calls where the boreholes are located have the deepest colours. The horizontal plane is taken at 10 m a s l.

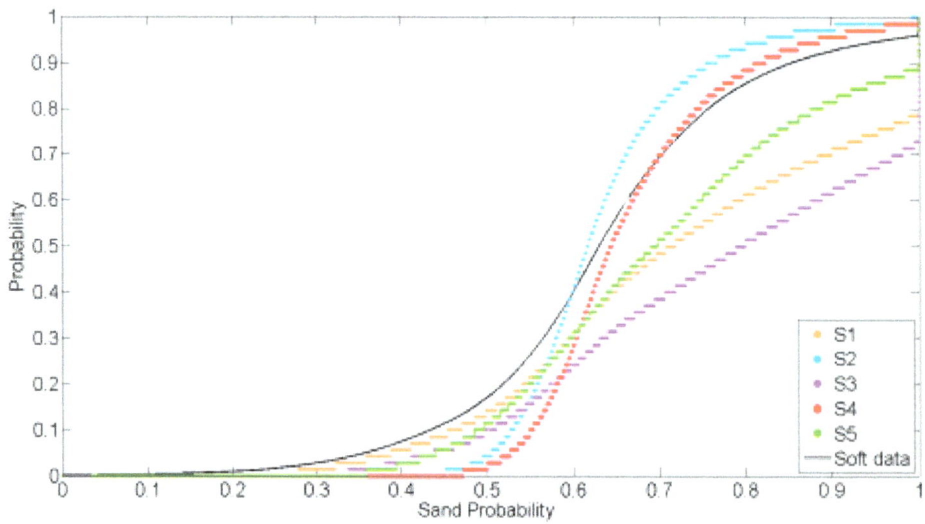

Figure 14. Black line indicates the CDF of the sand probability from soft data. Point plots are Q–Q plots of the average of realizations from each scenario against soft data

CONCLUSIONS

This study is one of the first to evaluate the effect of highresolution airborne electromagnetic data in multiple-point geostatistical simulations. It demonstrates how the 3-D highresolution airborne electromagnetic data can be used as the TI as well as secondary data for soft conditioning. The sensitivity of model predictions to the TI and soft conditioning has also been analysed. The SkyTEM-derived training image which resembles the actual geological structure of the site is evaluated against the parametric training image generated by the object-based TiGenerator. The TI from the TiGenerator is an abstract depiction of the geological structure with parameterized geometry. Although it is not derived directly from field data, it is found to be a reasonable input to MPS. The comparison of simulations of groundwater head and capture zone indicates that in cases where knowledge of the overall geological structure is available, an object-based training image may be just as good as a training image based on geological interpretations of the actual site. Although this conclusion was based on groundwater heads and capture zones, it is expected that similar conclusions would be obtained from simulations of groundwater age or solute transport as these variables are, similar to capture zones, very much dependent on path lines and velocities, which are sensitive to geological heterogeneity (He et al., 2013). This study was based on a highly heterogeneous two-facie geological system where the exact shape and location of the clay lenses may not affect the results from the groundwater model significantly. However, if systems with curvilinear structures such as buried valleys are considered, the geological architecture may have a larger impact on groundwater flow and the choice of training image could therefore be more important. However, more testing is required to assess if the conclusions from the current study can be transferred to other types of geological environments. By further conditioning on secondary data, the simulated geological structure is improved. This is revealed through the results from the parameter calibration where the likelihood of obtaining a realistic parameter set is much higher if soft data is used for conditioning. However, when applying soft conditioning, the dependence of different data sets has to be taken into account. If the training image and the soft conditioning data are based on the same source of information, the resulting geological realizations may be over-conditioned. Over conditioning may not decrease the accuracy of the individual geological model, and this is proved by the calibration where each model from S2 yields reasonable parameter estimates. However, over-conditioning decreases the uncertainty among simulated geological models, which results in under prediction of the uncertainties in transport simulations. This is demonstrated for the examined case where high-resolution SkyTEM data is used as both training image and soft conditioning which results in a higher reduction in capture zone uncertainty than if SkyTEM was only used for soft conditioning. Therefore, it is recommended to use independent data sources for generating the training image and the soft data. Based on the present study, the best result is obtained if the TiGenerator is used for defining the training image and the geophysical data for soft conditioning. The information required by the TiGenerator may be derived from expert knowledge of the geological structure of the site or derived from analysis of mapped geology at a site with comparable geology. Alternatively, a Q–Q plot analysis could be carried out and the contribution from soft data on the realization should be decreased. It is, however, (1) problematic to estimate the optimal level of information which should be obtained from the soft data and (2) time consuming to carry out the analysis.

REFERENCES

1. Bourges, M., Mari, J.-L., and Jeannée, N.: A practical review of geostatistical processing applied to geophysical data: methods and applications, Geophys. Prospect., 60, 400–412, 2012.
2. Carle, S. F. and Fogg, G. E.: Transition probability-based indicator geostatistics, Math. Geol., 28, 453–76, 1996.
3. Clement, W. P. and Ward, A.: GPR surveys across a prototype surface barrier to determine temporal and spatial variations in soil moisture content, Chap. 23, in: The Handbook of Agricultural Geophysics, edited by: Allred, B. J., Ehsani, M. R., and Daniels, J. J., American Society of Agricultural Engineers, CRC Press, 305–315, 2008.
4. Comunian, A., Renard, P., Straubhaar, J., and Bayer, P.: Threedimensional high resolution fluvio-glacial aquifer analog – Part 2: geostatistical modeling, J. Hydrol., 405, 10–23, 2011.
5. Comunian, A., Renard, P., and Straubhaar, J.: 3D multiple-point statistics simulation using 2D training images, Comput. Geosci. 40, 49–65, 2012.
6. Coz, M. L., Genthon, P., and Adler, P. M.: Multiple-point statistics for modeling facies heterogeneities in a porous medium: the Komadugu-Yobe Alluvium, Lake Chad Basin, Math. Geosci., 43, 861–878, 2011.
7. Delhomme, J. P.: Spatial variability and uncertainty in groundwater flow parameters: a geostatistical approach, Water Resour. Res., 15, 269–280, 1979.
8. Deutsch, C. V. and Journel, A. G.: GSLIB Geostatistical Software Library and User's Guide, Oxford University Press, New York, 1992.
9. Deutsch, C. V. and Wang, L.: Hierarchical object-based stochastic modeling of fluvial reservoirs, Math. Geol., 28, 857–880, 1996.
10. Doherty, J.: PEST: Model Independent Parameter Estimation, 5th Edn. of user manual, Watermark Numerical Computing, Brisbane, Australia, 2005.
11. Evan, R. A. and Mary, C. H.: Documentation of the hydrogeologicunit flow (HUF) package, US Geological Survey Open File Rep 00-342, US Geological Survey, Denver, CO, USA, 2000.
12. Goovaerts, P.: Geostatistics for Natural Resources Evaluation, Oxford University Press, New York, 483 pp., 1997.
13. Guardiano, F. B. and Srivastava, R. M.: Multivariate Geostatistics: Beyond Bivariate Moments, in: Geostatistics Tróia '92, Quantitative Geology and Geostatistics, edited by: Soares, A., Kluwer Academic Publishers, Dordrecht, Netherlands, 133–144, 1993.
14. Harbaugh, A. W., Banta, E. R., Hill, M. C., and McDonald, M. G.: MODFLOW-2000, the US Geological

Survey modular groundwater model – user guide to modularization concepts and the Ground-Water Flow Process: US Geological Survey Open-File Report 00-92, US Department of Interior, Reston, VA, USA, 121 pp., 2000.

15. Harrar, W. G., Sonnenborg, T. O., and Henriksen, H. J.: Capture zone, travel time, and solute-transport predictions using inverse modeling and different geological models, Hydrogeol. J., 11, 536–548, 2003.

16. He, X., Sonnenborg, T. O., Jørgensen, F., Høyer, A.-S., Møller, R. R., and Jensen, K. H.: Analyzing the effects of geological and parameter uncertainty on prediction of groundwater head and travel time, Hydrol. Earth Syst. Sci., 17, 3245–3260, doi:10.5194/hess- 17-3245-2013, 2013.

17. Henriksen, H. J., Troldborg, L., Nyegaard, P., Sonnenborg, T. O., Refsgaard, J. C., and Madsen, B.: Methodology for construction, calibration and validation of a national hydrological model for Denmark, J. Hydrol., 280, 52–71, 2003.

18. Hoffmann, J.: The future of satellite remote sensing in hydrogeology, Hydrogeol. J., 13, 247–250, 2005.

19. Hu, L. Y. and Chugunova, T.: Multiple-point geostatistics for modeling subsurface heterogeneity: a comprehensive review, Water Resour. Res., 44, W11413, doi:10.1029/2008WR006993, 2008.

20. Huysmans, M. and Dassargues, A.: Application of multiple-point geostatistics on modeling groundwater flow and transport in a cross-bedded aquifer (Belgium), Hydrogeol. J., 17, 1901–1911, 2009.

21. Huysmans, M. and Dassargues, A.: Modeling the effect of clay drapes on pumping test response in a cross-bedded aquifer using multiple-point geostatistics, J. Hydrol., 450–451, 159–167, 2012.

22. Høyer, A.-S., Lykke-Andersen, H., Jørgensen, F., and Auken, E.: Combined interpretation of SkyTEM and high-resolution seismic data, Phys. Chem. Earth A/B/C, 36, 1386–1397, 2011

23. Johnson, N. M.: Characterization of alluvial hydrostratigraphy with indicator semivariograms, Water Resour. Res., 31, 3217–3227, 1995.

24. Journel, A. G.: Geostatistics: roadblocks and challenges, in: Geostatistics Tróia '92, edited by: Soares, A., Springer Netherlands, Dordrecht, 213–224, 1993.

25. Journel, A. G.: Combining knowledge from diverse sources: an alternative to traditional data independence hypotheses, Math. Geol., 34, 573–596, 2002.

26. Journel, A. G.: Beyond covariance: the advent of multiple-point geostatistics, in: Geostatistics Banff 2004, Quantitative Geology and Geostatistics, edited by: Leuangthong, O. and Deutsch, C. V., Springer Netherlands, Dordrecht, Netherlands, 225–233, 2005.

27. Klise, K. A., Weissmann, G. S., McKenna, S. A., Nichols, E. M., Frechette, J. D., Wawrzyniec, T. F., and Tidwell, V. C.: Exploring solute transport and streamline connectivity using lidar-based outcrop images and geostatistical representations of heterogeneity, Water Resour. Res., 45, W05413, doi:10.1029/2008WR007500, 2009.

28. Liu, Y.: Using the Snesim program for multiple-point statistical simulation, Comput. Geosci., 32, 1544–1563, 2006.

29. Liu, Y., Harding, A., Abriel, W., and Strebelle, S.: Multiple-point simulation integrating wells, three-dimensional seismic data, and geology, AAPG Bull., 88, 905–921, 2004.

30. Liu, Y., Harding, A., Gilbert, R., and Journel, A. G.: A work- flow for multiple-point geostatistical simulation, in: Geostatistics Banff 2004, Quantitative Geology and Geostatistics, edited by: Leuangthong, O. and Deutsch, C. V., Springer Netherlands, Dordrecht, Netherlands, 245–254, 2005.

31. Lubczynski, M. and Roy, J.: Hydrogeological interpretation and potential of the new magnetic resonance sounding (MRS) method, J. Hydrol., 283, 19–40, 2003.

32. Maharaja, A.: TiGenerator: object-based training image generator, Comput. Geosci., 34, 1753–1761, 2008.

33. Okabe, H. and Blunt, M. J.: Pore space reconstruction using multiple-point statistics, J. Pet. Sci. Eng., 46, 121–137, 2005.

34. Okabe, H. and Blunt, M. J.: Pore space reconstruction of vuggy carbonates using microtomography and multiple-point statistics, Water Resour. Res., 43, W12S02, doi:10.1029/2006WR005680, 2007.

35. Pollock, D. W.: User's guide for MODPATH/MODPATH-PLOT a particle tracking post-processing package for MODFLOW, the US Geological Survey finite-difference ground-water flow model (SuDoc I 19.76:94-464). The Survey Earth Science Information Center, Open-File Reports Section, distributor, 1994.

36. Rasmussen, E. S., Dybkjær, K., and Piasecki, S.: Lithostratigraphy of the Upper Oligocene – Miocene succession of Denmark, Geological

37. Survey of Denmark and Greenland Bulletin, Geological Survey of Denmark, Copenhagen, Denmark, 22, 92 pp., 2010.

38. Remy, N., Boucher, A., and Wu, J.: Applied Geostatistics with SGeMS: A User's Guide, Cambridge University Press, 2009.

39. Scharling, P. B., Rasmussen, E. S., Sonnenborg, T. O., Engesgaard, P., and Hinsby, K.: Three-dimensional regional-scale hydrostratigraphic modeling based on sequence stratigraphic methods: a case study of the Miocene succession in Denmark, Hydrogeol. J., 17, 1913–1933, 2009.

40. Stisen, S., Sonnenborg, T. O., Højberg, A. L., Troldborg, L., and Refsgaard, J. C.: Evaluation of climate input biases and water balance issues using a coupled surface–subsurface model, Vadose Zone J., 10, 37–53, 2011.

41. Strebelle, S.: Conditional simulation of complex geological structures using multiple-point statistics, Math. Geol., 34, 1–21, 2002.

42. Strebelle, S. B.: Sequential simulation for modeling geological structures from training images, in: Stochastic Modeling and Geostatistics: Principles, Methods, and

Case Studies, Volume II: AAPG Computer Applications in Geology 5, edited by: Coburn,

43. T. C., Yarus, J. M., and Chambers, R. L., AAPG, Tulsa, OK, USA, 139–149, 2006.

44. Strebelle, S. B., Payrazyan, K., and Caers, J.: Modeling of a deepwater turbidite reservoir conditional to seismic data using multiple point geostatistics. Society of Petroleum Engineers (SPE) Annual Conference and Technical Meeting, SPE 77429, San Antonio, TX, 2002.

45. Sørensen, K. I. and Auken, E.: SkyTEM – a new high-resolution helicopter transient electromagnetic system, Explor. Geophys., 35, 191–199, 2004.

46. Wingle, W. L. and Poeter, E. P.: Uncertainty associated with semivariograms used for site simulation, Ground Water, 31, 725–734, 1993.

INDEX

A

Adaptive Neurofuzzy Inference System (ANFIS) 201
Afforestation 87
Aforementioned 45
Alfeios Hydrosystem 121
Animal husbandry 142
artificial neural network (ANN) 198
artificial neural networks (ANN) 197
autocorrelated model (AR) 197

B

Bayesian Model Averaging (BMA) 153, 154, 160
Bioinfiltration cell 175
Bioretention system 175, 176, 177, 180, 181, 182, 183
Bootstrap method 99, 100

C

Calibration process 2, 6, 13
Chance-constrained programming (CCP) 114
Channelization 69
Cluster analysis (CA) 140
coefficient of efficiency (CE) 200
Containing Ratio 157
Continual degradation 185
Continuous simulation 1, 2, 14
Correlation comparison 94
cumulative distribution function (CDF) 214

D

Daily Root Mean Square Error 156
Drainage network 67
Drought severity index (DSI) 17, 18

E

European Union (EU) 115
Evapotranspiration 2
Extended Transport (EXTRAN) 80

F

Flocculation process 49, 50, 51, 53, 55
Flood disaster 67
Fuzzy-boundary interval-stochastic programming (FBISP) 115
fuzzy C-means (FCMs) 198
Fuzzy (FP) 113

G

Generalized Likelihood Uncertainty Estimation (GLUE) 153
Geomorphologic Instantaneous Unit Hydrograph (GIUH) 77, 84
Ghyben-Herzberg assumption 187
Gibbsian model 78, 79, 80, 81, 82, 83, 84
Gibbsian network 79, 83
ground penetrating radar (GPR) 211
Groundwater management 139
Groundwater Modelling System (GMS) 216
Groundwater quality 185

H

Hydraulic conductivity 186, 187
Hydraulic infrastructure design 1, 12
Hydrodynamic 99, 101, 106, 110
Hydrogeochemical processes 145, 148, 151
Hydrogeologic-Unit Flow (HUF) 216
Hydrological model 175
Hydrologic analysis 77
Hydrologic Manual 180, 184
Hydrology mostly focused 153
Hydropower station (HPS) 119

I

Inexact two-stage stochastic programming (ITSP) 115
Infiltration rate 178, 179, 181, 182
Instantaneous Unit Hydrograph (IUH) 77
Intergovernmental Panel on Climate Change (IPCC) 163, 173, 174
Internal model bias 169
Irrigation canal 141

J

Joint probability distribution 99, 100, 102, 106, 110

L

Levenberg Marquardt (LM) 198

M

Marginal distribution 99, 100, 102, 104, 105, 106
Markov chain Monte Carlo (MCMC) 153
Mean sea level (MSL) 187
Mesoproterozoic Era 140
Meteorological data 163, 164, 165, 166, 169, 172
Meteorological variable 177
Minimum inter-event period (MIP) 3
Model simulations 170
Monte Carlo integration method 100
Monte Carlo simulation 2, 6
multilayer perceptron (MLP) 197, 201
multilinear regression (MLR) 197
Multistage stochastic programming (MSP) 113
Multivariate statistical techniques 143

N

Nash-Sutcliffe efficiency 71
National Climatic Data Center (NCDC) 165
Network configuration 68, 71, 73, 74

O

Objective function (OF) 155
Optimal pipe design 62
Optimal water 113

P

Peak ground acceleration (PGA) 58
Prechange period 90, 94, 95
Precipitation elasticity 88, 90, 95, 96, 97
Precipitation elasticity method 90, 96
Precipitation variability 96
Pressure-driven analysis (PDA) 60
Pressure redundancy 64
Principal component analysis (PCA) 140
Probabilistic structure 17
Probability density function (PDF) 154
Probability distribution 2, 6, 7, 9

Q

quantitative estimation 88, 96

R

Rainfall generator 44, 45, 46, 47, 48
Rainfall realisation 46, 47
Rapid urbanization 175
Runoff generation 1, 5, 12, 15
Runoff propagation 5

S

Saltwater intrusion (SI) 185
Seismic reliability estimation model (SREM) 58
single normal equation simulation algorithm (SNESIM) 211
Soil Moisture Accounting and Routing Model (SMAR) 155
Spatial distribution 144, 163, 169
Standardized precipitation index (SPI) 17, 18
Stochastic analysis 46
Stochastic (SP) 113
Storm Water Management Model (SWMM) 77, 78, 79, 84
Synthetic rectangular catchment 79

T

Temporal variable storm 47
Total dissolved solids (TDS) 142
total maximum daily load (TMDL) 33, 36
training image (TI) 211
Tributary 88

U

United Nations Environment Programme (UNEP) 164
Urban catchment 43, 67, 68, 73, 75
Urban drainage network 67, 68, 74, 75, 78, 84, 85
Urban drainage problem 77
Urban pluvial flooding 43

W

Wallingford model 45
wastewater treatment plants (WWTPs) 37
water distribution system (WDS) 57, 58

X

Xinanjiang model 87
Xinanjiang Rainfall-Runoff Model (XAJ)155